Design and Construction: Building in Value

To Irene and Wally Zagoridis

Design and Construction: Building in Value

edited by Rick Best and Gerard de Valence
University of Technology Sydney

OXFORD AMSTERDAM BOSTON LONDON NEW YORK PARIS
SAN DIEGO SAN FRANCISCO SINGAPORE SYDNEY TOKYO

Butterworth-Heinemann
An imprint of Elsevier Science Limited
Linacre House, Jordan Hill, Oxford OX2 8DP
225 Wildwood Avenue, Woburn MA 01801-2041

First published 2002

British Library Cataloguing in Publication Data
A catalogue record for this book is available from the British Library

Library of Congress Cataloging in Publication Data
Design and construction: building and value/edited by Rick Best and Gerard de Valence.
p. cm.
Includes bibliographical references and index
1. Building–Cost effectiveness. 2. Buildings. 3. Building–Planning.
I. Best, Rick. II. De Valence, Gerard.
TH153.D4723 2002
690–dc21 2002026022

ISBN 0 7506 51490

For information on all Butterworth-Heinemann publications visit our website at www.bh.com

Composition by Genesis Typesetting, Rochester, Kent
Printed and bound in Great Britain

Contents

Acknowledgements

We would like to thank some people who have helped in various ways in the production of this book:

Eliane Wigzell, who helped so much with the first volume and gave us the opportunity to extend the series, and Alex Hollingsworth, who carried on the good work.

Associate Professor Steve Harfield at the Faculty Design Architecture and Building at University of Technology, Sydney (UTS) for his assistance with the whole project.

Craig Langston, formerly at the University of Technology Sydney, now at Deakin University, for all his support.

Sally Beech and Melissa Price for their hard work in preparing the many diagrams and tables.

The contributors who accepted our criticism and suggestions with good humour and stayed with us through the long process.

List of Contributors

Editors

Rick Best – Senior Lecturer at the University of Technology Sydney (UTS). He has degrees in architecture and quantity surveying and his research interests are related to low energy design, information technology in the AEC industry and energy supply systems. He recently completed a Masters degree for which he investigated the potential of cogeneration and district energy systems in Australia. He has begun research for a PhD that looks at comparative costs of construction in an international context.

Gerard de Valence – Senior Lecturer at UTS. He has an honours degree in economics from the University of Sydney. He has worked in industry as an analyst and economist, and as an economist and task leader in the Policy and Research Division of the Royal Commission into Productivity in the Building Industry in NSW. His principal areas of research activity and interest include the measurement of project performance, the study of economic factors relevant to the construction industry, the analysis of the construction industry's role in the national and international economy, the study of interrelationships between construction project participants, and the impact of emerging technologies.

Contributors

Glenn Ballard – Research Director for the Lean Construction Institute and an Adjunct Associate Professor in Construction Management at the University of California, Berkeley. He is a founder of both the Lean Construction Institute and the International Group for Lean Construction. His experience includes stints with Brown & Root and Bechtel as an internal management consultant, and 11 years as an independent management consultant. His degrees include an MA, an MBA and a PhD, the latter in engineering management from the University of Birmingham, United Kingdom.

Chen Swee Eng – Professor of Building at the University of Newcastle, Australia. Before becoming an academic, he practised as an architect and project manager for several years in Singapore and Australia. His research interests include building process performance, complexity perspectives of project performance and buildability/constructability.

Eric Collier – licensed architect and graduate of the Construction Engineering and Management program at Stanford University. He studied architecture at UNC Charlotte and went to Indonesia on a Fulbright scholarship. He has worked for various architectural firms, including Daniel, Mann, Johnson and Mendenhall on wind tunnel remodelling projects at NASA Ames in Mountain View, California. He currently works in design and construction management for the Salt Lake City International Airport.

Grace Ding – Lecturer in Construction Economics at UTS. She has a Diploma from Hong Kong Polytechnic, a Bachelors degree in quantity surveying from the University of Ulster and a Masters degree by thesis from the University of Salford. She has practised as a quantity surveyor in Hong Kong, England and Australia. Grace is currently completing a PhD studentship at UTS and is involved in research and teaching in the area of environmental economics and general practice.

Martin Fischer – Associate Professor of Civil and Environmental Engineering, and, by courtesy, Computer Science at Stanford University, United States. He is the Director of the Center for Integrated Facility Engineering and teaches Project Management and Related Information Technologies. His research goals are to improve the communication of design and schedule information for engineered facilities and to make construction schedules more realistic. His research focuses on 4D (3D plus time) computer building models and computer-based analysis tools for evaluation of 4D models with respect to cost, interference, safety and other important project objectives.

Jon Hand – architect and a Senior Research Fellow in the Energy Systems Research Unit at the University of Strathclyde. He has been involved in solar architecture, energy conservation and building thermal assessments in the United States, Africa, Europe and Australia for almost two decades. In the area of building thermal assessments he has worked with and been part of the development efforts on several simulation tools, the latest being ESP-r. Particular areas of interest are the training of simulation users and the evolution of design decision support tools.

Steve Harfield – Associate Professor in the Faculty of Design, Architecture and Building UTS. His expertise and research interests lie in the areas of architectural and design theory and philosophy, in particular the theory of design processes, the ontology of design, the analysis of normative positions in thinking systems and issues dealing with creativity and problem solving.

Edward Harkness – an architect by profession, he has both masters and doctoral degrees in building science. Apart from running his own practice/consultancy in Sydney, he has worked extensively in South-east Asia and Saudi Arabia. He is the author of several books and teaches in various programs at three of Sydney's major universities.

Patrick Healy – Senior Lecturer at UTS. Originally from an engineering background, he is currently the Director of the Construction Management Program at UTS. He is the author of *Project Management: Getting the Job Done on Time and in Budget.*

Greg Howell – co-founder and Managing Director of the Lean Construction Institute, a non-profit organization devoted to production management, research in design and construction founded in 1997. He worked as a project engineer on heavy construction and general building projects and headed his own construction consulting firm for 10 years prior to his appointment as a Visiting Professor in Construction Management at the University of New Mexico in 1987. He was the Associated General Contractors of

America's Outstanding Educator in 1994 and Eminent Scholar at the Del E. Webb School of Construction in 1996.

George Hurley – Project Executive with DPR Construction, the largest general contractor in California. He holds a BS in civil engineering and is a Member of the American Society of Civil Engineers. He has managed projects worth many hundreds of millions of dollars in the USA and Canada, including airports, data centres, and education, correctional and health care facilities. His activities have included site acquisition, programming, construction management and general project management.

Khalid Karim – is the Deputy Director of the Australian Centre for Construction Innovation. As a civil engineer he gained wide ranging industry experience before developing a broad-based interest in research in construction management. During the last few years his research interests have focused on performance measurement and benchmarking, waste minimization, management of quality, and information technology.

Lauri Koskela – works at VTT Building and Transport, a unit in VTT Technical Research Centre of Finland. He is a founding member of the International Group for Lean Construction. He has written over 130 papers and reports on construction management, construction robotics and lean construction. His present research thrust is in the formulation of a new, theoretically sound project management methodology for construction.

Craig Langston – Professor of Construction Management at Deakin University, Australia. Before commencing life as an academic, he worked for 9 years in a professional quantity surveying office in Sydney. His PhD thesis was concerned with life-cost studies. He developed two cost planning software packages (PROPHET and LIFECOST) that are sold throughout Australia and internationally and is the author of two text books concerning sustainable practices in the construction industry and facility management.

Rima Lauge-Kristensen – has a scientific research background in electrochemistry, solar energy, physiological perception of human comfort and textile physics. She obtained her PhD at the School of Physics at the University of New South Wales, Sydney, in the field of fabrication of efficient photovoltaic solar cells and also obtained a BA in Architecture from UTS. She is currently carrying out research in the Faculty of Design Architecture and Building at UTS and administering energy efficiency/renewable energy strategies at a local government establishment as part of a greenhouse gas abatement programme. Her special interests are in environmental issues, renewable energy and energy efficient architecture.

Helen Lingard – currently at the University of Melbourne, her research interests include behaviour-based safety management in construction, work stress and burnout, and the interaction between work and family life in the construction industry. Having been involved in several major infrastructure projects in Hong Kong, such as the Tsing Ma Bridge and the new Hong Kong International Airport, she moved to Melbourne, where she has taught occupational health and safety, management and law, to property and construction students at RMIT University and the University of Melbourne, as well as working as a consultant.

Martin Loosemore – is an Associate Dean (Postgraduate Studies) and Professor of Building and Construction Management at the University of New South Wales, Sydney. He received his PhD from the University of Reading in the UK and his research interests are in the area of risk management.

Ray Loveridge – a former chairman of the technical committee responsible for the development of the Building Code of Australia, chartered builder, accredited building surveyor, fire engineer and university lecturer, he has been involved in the development of building regulations and performance-based building designs for approximately 15 years. He is currently working for Arup Fire in Sydney.

Low Sui Pheng – Associate Professor and Vice-Dean (Academic) with the School of Design and Environment, National University of Singapore. A chartered builder by profession, he received his BSc (Building)(Hons), MSc (Eng) and PhD degrees from the National University of Singapore, University of Birmingham and University of London, respectively. He has conducted extensive research into quality management systems for the construction industry and is particularly interested in how quality management may be integrated with other management systems to help achieve greater productivity.

Caroline Mackley – joined Bovis Lend Lease in 2000 as the first Sustainable Development Life Cycle Cost-planner. Her research interests are in the systematic quantitative analysis and reporting of the implications of sustainable development initiatives and practices. She has developed comprehensive environmental performance benchmarks for a range of built facilities that are being used to inform the design process as to what constitutes improved environmental performance. She has published widely in various journals and books and her research has been presented at a number of international conferences.

Marton Marosszeky – is an Associate Professor in the Faculty of the Built Environment at the University of New South Wales, Sydney and Director of the Australian Centre for Construction Innovation. He is also Chair of the NSW Construction Industry Consultative Committee, the industry's peak body for industrial relations. He has a background in civil engineering in road and building construction and early in his career he specialized in the structural design of tall buildings. His primary research interest is into innovation, especially in relation to construction project and process management. He has a particular interest in performance measurement and benchmarking as well as safety risk management, quality management and waste management.

Rabee M. Reffat – is a Lecturer at the Faculty of Architecture, University of Sydney, Australia. He is specializing in teaching and researching design computing, virtual architecture and computer aided architectural design. His research work has been published in international journals and international conferences held in Australia, USA, UK, Germany, Japan, Malaysia and the Middle East. His research and teaching interests include architectural design, artificial intelligence and machine learning in design, computer aided architectural design, virtual architecture, intelligent agents, and creativity in design and architecture.

Peter Smith – Senior Lecturer in Construction Economics at UTS. Prior to his current appointment Peter worked for 7 years in a large professional quantity surveying practice and 3 years in an international construction and property development company. He has

Bachelors and Masters Degrees in quantity surveying and is currently completing a PhD. He is involved in a wide range of research and consulting activities including consumer investment advice in the residential property industry.

Stuart Smith – holds a BSc, MSc (Electronics) and MMgt. His expertise is in the area of Workplace Integration – the integration of business strategy and organizational objectives with technology, organizational culture and strategic facility planning. He has over ten years experience in the analysis of intelligent buildings and systems coordinating the development and analysis of technical performance specifications. He is involved with a number of universities contributing to a broad range of disciplines. These include Management, Innovation, Technology, Electronics, Sustainability, Workplace Integration and Facility Performance. He has written extensively on the concept of workplace integration.

Paul Strachan – Lecturer and Deputy Director of the Energy Systems Research Unit (ESRU) at the University of Strathclyde, Glasgow, Scotland. His primary research interests are in the development, validation and application of simulation tools for use in the performance evaluation of building energy systems. He has participated in a number of international research projects involving building-integrated photovoltaics, advanced glazing, passive cooling systems, model validation and design tool integration.

Melissa Teo – Senior Development Officer in the Building Construction Authority in Singapore. She received her Masters of Building from the University of New South Wales, Sydney and her research interests are in issues of sustainability and waste in the construction industry.

Iris Tommelein – Professor of Construction Engineering and Management at the University of California, Berkeley. She holds a PhD in civil engineering and an MS in computer science from Stanford University. Her research focuses on developing principles of project-based production management for the architecture-engineering-construction industry. She is an expert on construction site layout and logistics, supply-chain management and e-commerce. She is an active member of the International Group for Lean Construction and serves on the Board of Directors of the Lean Construction Institute.

John Twyford – Senior Lecturer in Construction Economics at the UTS. He was admitted as a solicitor in 1965 after completing a Diploma of Law. He holds a doctorate in Juridicial Science. After 5 years in private practice he became the Legal Executive Officer of the Master Builders Association of New South Wales. Whilst at the MBA John helped develop the standard building contracts presently in use in Australia. He is the author of *The Layman and the Law in Australia* and is recognized as a leading construction contracts expert. His special legal interests are dispute resolution and the theoretical underpinning of contractual relations.

Foreword

As US architect Frank Lloyd Wright wrote in the *New York Times* Magazine (4 October 1953):

> The physician can bury his mistakes, but the architect can only advise his clients to plant vines.

In the last 50 years, buildings have become more complicated, technology has continued at a relentless pace, work practices have changed, litigation abounds, globalization is a fact of life and we are all swamped with information.

So why read another book about building best value buildings?

The simple answer is that this book is vital to enhancing your knowledge. Today's professional has to keep abreast of all the new developments in design, technology, building codes, and new materials and processes. Rick Best and Gerard de Valence have hit on the winning combination of knowing a lot about the design and construction industry and bringing together experts to write about the important issues in building today. Their last book, *Building in value: pre-design issues*, used the same format and has proved very successful. The editors have moulded a diverse range of topics into a coherent and interesting text.

Everybody likes to think they are operating at the leading edge of their subject area, whether it be design, cost management, materials technology, or construction. If you want to stay at the leading edge, then you must read this book – it has taught me a lot about a wide range of topics. Remember – you don't know what you don't know!

But building in value for whom? For the investor, developer, owner, designer, contractor, specialist contractor, or supply chain, who all played a part in the design and construction phases? Or for the customers, and the customers' customers who will be involved in the whole life of the building? The editors have clearly considered these questions and have compiled a collection of expert views that cover the whole spectrum of construction, from inception to construction to operation.

Today's speed of change in new materials and communication technologies means we can no longer use the events of the past to predict the future. However, whatever the speed at which new technology is created, the changes are incremental, they do not all happen at once. Take, for example, the construction of tall buildings in the world: Petronas Towers in Kuala Lumpur relied upon advances in lift manufacture, e-commerce relies upon faster and faster processor speeds and communication advances, while nanotechnology provides an opportunity for scientists to make single molecules act like transistors – which makes computers the size of a sugar cube a possibility!

The book recognizes the speed of change and includes sections on managing change and innovation, and future technologies, as well as dealing with today's issues of constructability, occupational health and safety, lean construction, cost management, waste management, and quality assurance.

The book represents value for money for both students and professionals alike.

Roger Flanagan
The University of Reading

Preface

This is the second book in a series devoted to the concept of value in buildings, and how those who involved in building procurement may ultimately produce buildings that represent the best 'value-for-money' outcomes. Like its precursor, *Building in Value: Pre-design Issues*, it is intended both for students in construction- and property-related courses at the tertiary level and as a useful resource for industry professionals: property developers, project managers and cost engineers among others.

Once again a wide range of topics have been included, however, the design and construction of buildings is a broad and complex area and no single book could cover all the relevant areas. Once again, though, it brings together in one volume an introduction to many of the parts of the building procurement process that make up the design and construction phase. The reference and bibliography lists at the end of each chapter point readers to a wealth of related material and will thus facilitate in-depth self-directed study of selected topics, while the editorial comments that precede most of the chapters tie the content of individual chapters to the central theme of 'building in value'.

The book is broken into three parts: the first is a review of recent trends and changes in building procurement, the second examines specific functional and procedural issues, while the third looks to the future and examines some of the innovations that are emerging in areas such as the production of 'space age' materials, automation of the construction process and new methods of space conditioning.

Contributions have come from many countries including Australia, New Zealand, Singapore, the USA, England, Scotland and Finland, and the authors once again include academics and practitioners.

The first book in the series has been a success as it gives readers a convenient one-volume reference that provides a basic but solid grounding in the topic areas covered by the various authors. We believe that this book will serve a similar purpose and hope that students and professionals will find it equally useful.

Rick Best
Gerard de Valence
November 2001

1

Issues in design and construction

Gerard de Valence* and Rick Best*

1.1 Introduction

Getting 'value for money' is the basic goal attached to more or less all the transactions that we undertake. The value embodied in the outputs of the various sectors of the construction industry, and how that value can be maximized, is the common theme in this series of books entitled 'Building in Value'. In this, the second in the series, the focus is on the design and construction phase of building procurement.

It is worth noting again, at the outset, that maximizing value in buildings is not merely about minimizing cost but about satisfying clients' needs in the best way possible. The process will invariably be governed by a cost limit, usually a budget or maximum price, that must be observed and that places obvious constraints on those involved in the design and construction of buildings. The challenge is to use the funds available to the best advantage and by so doing achieve the best possible 'value for money'.

1.2 Managing design and construction

It is fairly obvious that avoiding wastage of resources, including those that are purely financial, will be a fundamental part of any attempt to get maximum value, and that good management of the procurement process will play a large part in the success of any such exercise. This book, however, is not a 'how to' book on the management of building procurement; instead it takes a more general, perhaps philosophical, approach to many aspects of the design and construction process.

Building procurement, from inception to commissioning, is a complex undertaking, bringing together the set of skills and knowledge that are required for successful completion of building and construction projects. There has been significant development

* University of Technology Sydney

in this field over several decades, as projects have become more demanding and clients' expectations have increased. Further, the industry has seen a continual flow of new products and materials introduced by the manufacturing suppliers, many of which have delivered significant improvements in building performance or site productivity. There has also been development of new and more capable plant and equipment for use on site as well as great progress in computer-aided design and an explosion of powered hand tools available for use by site workers.

These developments are reflected in the material covered in books dealing with the management of design and construction. The core subjects found include aspects of site management, programming and scheduling, operation of plant and equipment, contract administration and so on. There are significant differences in the extent and focus of topics covered in these works: some construction management books emphasize construction methods (e.g., Nunnally, 1998) while others focus on the human side of construction management (e.g., Fryer, 1997). A new area that has emerged since the Latham (1994) and Egan (Construction Task Force, 1998) reports in the UK is managing teams (e.g., Blockley and Godfrey, 2000).

This book takes a somewhat wider approach to the various tasks involved in construction management and differs from the majority of books in the field by not concentrating on the management of design or on-site activities, or the range of activities that building design and construction involves. Instead, the concern is largely with tasks that are emerging as important to the effectiveness of management of building procurement during design and construction. There are more of these types of tasks as projects become more complex and new tools and techniques appear. As these are applied to building and construction processes, many of those processes change. This is creating a series of cycles of change in an industry that in many respects did not change greatly over the course of the last century.

A comparison of the contents of some representative construction management books shows the typical focus very clearly. The emphasis is rightly on techniques for planning and controlling the construction process. These are often followed by discussions of work study and/or activity sampling methods, plant management and cost control. These days, chapters on quality control, environmental management, marketing and financial management are also common.

Oxley and Poskitt's (1996) text, now in its fifth edition, presents a broad range of management techniques as applied to construction projects. The book stresses the importance of an integrated information system and illustrates, by case study, integration of estimating and production processes. It covers topics such as construction planning and control, project network techniques, work study, bidding strategies, computer applications and cost control.

Harris and McCaffer (2000), one of the leading construction management books, is representative of the field. Also now in a fifth edition, this covers management techniques for production planning and cost control of projects, including work study and plant management. The book also has management techniques for company organization and control, including planning techniques, production process improvement, estimating and tendering, workforce motivation and cost control. Section 2 covers procurement, bidding, budgets and cash flow, economic assessment and plant management. Section 3 is concerned with head office activities such as business development, global construction, the role of information and finance. The main changes between the fourth and fifth editions were new chapters on management of information and knowledge, and the globalization of construction activity. The book also discusses recent developments in regulation, procurement, performance improvement, public–private partnerships, benchmarking and lean construction.

The purpose of McGeorge and Palmer's 1997 book is stated in the Preface:

> . . . we felt there was a need to bring together, for the first time in a single volume, most, if not all of the management concepts currently being advocated for use in the construction industry. The concepts which we have selected for detailed scrutiny are value management; total quality management; constructability; benchmarking; partnering and reengineering.

This is a somewhat different approach to the previous books; each of the six concepts they cover is defined and explained in detail.

The changing nature of clients was classified as a strategic issue for the industry in a 1988 report (CSSC, 1988). That report found that the role of large organizations that are regular and experienced clients of the industry will become increasingly important. This was exactly right, as through the 1990s major clients began using a wider range of procurement systems, many of which were based on collaborations or partnering with contractors (de Valence and Huon, 1999). Both McGeorge and Palmer (1997) and Blockley and Godfrey (2000) focused on these changed forms of relationships between clients and the industry, and within the industry between major contractors, specialist contractors and suppliers.

Cornick (1996) examined the fundamental changes in design and design management that have been occasioned by the development of personal computers. The opportunity for electronic sharing of design information, including the ability for all members of the design team to work on a shared project model, is altering the way in which building design is undertaken at a very basic level. There is some way to go, however, before designers are routinely designing using 3D modelling software rather than treating the computer as an electronic drawing board, and there are still interoperability problems to be resolved before truly seamless exchange of information is possible across all platforms.

Gray and Hughes (2001) concentrate on the management of the design process in a general sense and provide checklists of activities to help managers maximize efficiency during the design phase. They do not, however, have much to say about the changing nature of design, in particular the move from two dimensions to three and the use of shared electronic project models.

1.3 The design and construction phase

Under the traditional lump sum style of construction contract, the design and construction phases of procurement were largely separate. Once the design documentation was complete, tenders were called and evaluated, the contract awarded and construction commenced. Designs were amended and details finalized after the start of construction but the majority of the design generally remained unchanged.

In recent years, on many projects, the design and construction phases have been pushed together, with work often commencing on site well before the completion of design work. This has been driven by client demands for shortened delivery times and has given rise to new forms of procurement such as fast-tracking (usually under a construction management arrangement) and design and construct (D&C) or design and build (D&B). In many instances, the whole design and construction phase is controlled on behalf of the client by a project manager. This entity, whether an individual or a firm, manages most aspects of the entire operation, including selection and appointment of consultants, specialist sub-contractors and suppliers, design development, approvals from statutory authorities and so on. The topics covered in this book include many that will be of particular interest to those who act as project managers and are therefore concerned with the whole project.

It is divided into three parts but, as is the case with any complex system, any subdivision of the whole is somewhat artificial as there is inevitably a good deal of overlapping and interdependence. The first part deals broadly, in a conceptual way, with issues related to the nature of buildings and their design. Part 2 moves more into the mechanics of design and construction and looks at issues that underpin activities such as the administration of building contracts, the philosophy of lean construction, investigations of how risk is allocated among the various parties in building contracts, and attitudes to waste management and site safety among industry personnel.

The final part looks ahead and describes areas of innovation in construction including the development of new materials and techniques, automation of construction processes and advances in building services.

1.4 The nature of buildings

There are many forces that help to determine why individual buildings come to be the way they are and they affect a variety of aspects of any one building: materials, shape, colour, size, form, style (or lack of it), choice of engineering systems, structural system, and more. Function or purpose explains much about the nature of many buildings and availability of resources (particularly money and materials) is often the primary determining factor.

Many specific parts of a building's nature are, however, determined by less obvious influences: a desire to build in an environmentally responsible manner, or a client's wish to say something about image or status, perhaps by constructing a building larger or taller than that of a competitor. Much emphasis is now placed on building 'intelligence' and on streamlining construction to allow for faster completion. Integrated design, with multidisciplinary teams working closely together, is producing buildings quite unlike any before, with building fabric and services functioning together to minimize energy use and improve the health and productivity of the occupants.

1.5 Sustainable construction

There is increasing pressure on all who are concerned with construction, including developers, designers (including engineers of various disciplines) and contractors, to work towards achieving a sustainable construction industry. Major areas of concern are energy use in buildings, selection of materials, resource depletion and waste management. These concerns are being addressed by researchers in many countries with particular emphasis on reducing operational energy in buildings (e.g., energy used for heating, cooling and lighting) and reducing the amount of waste generated during construction and demolition that is disposed of to landfill.

Advances in computer modelling of building performance have made possible substantial reductions in the energy needed to run buildings, with an increasing number of non-residential buildings being constructed with smaller mechanical plant, or none at all. Daylight is utilized more effectively, reducing the need for artificial lighting, while natural ventilation systems based on sophisticated modelling of airflows in and around buildings are becoming more common.

The effect of these developments on building value will become apparent as environmental controls tighten and the environmental costs associated with energy production and use are progressively internalized through the introduction of measures such as carbon taxes (which are already in place in some countries).

1.6 Statutory controls

Government regulations affect construction, not only through the imposition of taxes and charges and control of waste disposal, but in a number of less obvious ways. These include the granting or denial of development approvals, land use control through zoning, setting maximum building heights and floor space to site area ratios, and safety and quality controls.

There has been a worldwide trend in recent years in the nature of building regulations that cover concerns such as structural integrity, fire safety provisions and health and amenity issues, away from prescriptive controls to performance-based controls. There have also been developments in the regulation of occupational health and safety (OHS) on construction sites and an increasing incidence of remediation of contaminated land, necessary before sites can be safely redeveloped.

1.7 Management techniques

As mentioned earlier, the focus here is not directly on the planning and management of projects but rather on the more abstract underpinnings of those activities. Consequently there is discussion of the basic nature of project management as an activity or discipline but little about the precise mechanisms that make it work. Similarly, with the emerging concept of lean construction and the introduction of 3D modelling of projects, there is more background on why these techniques may be of benefit than instruction on how to implement them.

One area of special interest is the integration of 3D CAD models with scheduling software to produce 'four dimensional' models of projects, which allow management to view 3D 'snapshots' of their projects at any future point during the construction period. Tools such as this will give designers, contractors and managers the opportunity to improve their performance enormously as clashes are identified and resolved, disputes are avoided and the whole process streamlined.

1.8 Innovation

The search for new and hopefully better ways of doing things pervades most aspects of life – whether it is the proverbial 'better mousetrap', or more effective pharmaceuticals, or a more responsive golf club or tennis racquet – it is innovation that fuels most industries and gives businesses a competitive edge. While the construction industry has a reputation for being slow to change and there are processes (such as bricklaying) and materials (such as clay tiles) that are quite similar to their ancient counterparts, there has been enormous progress in many aspects of the industry since the Industrial Revolution. In the so-called Information Age, which we are living in now, there is ongoing change across the industry. Two major influences in the recent past have been the dramatic developments in computer hardware and software, and the consequent establishment and expansion of the Internet.

As computers have become more powerful, yet less costly, they have found application in most areas of daily life and the design and construction of buildings has changed dramatically as a result. With the advent of the Internet many businesses that were little more than cottage industries have become transnationals and many contractors and consultants are now competing for work in parts of the world where they would not previously have imagined that they could operate.

It is also through advances in electronics that many of the innovations in robotics and building services have been made possible: building management systems, security and environmental monitoring systems, 3D and 4D project models, industry-specific hubs and portals that provide web-based procurement and project management, and cheap and virtually instantaneous international information and document exchange through e-mail are just some of the outcomes that have changed, and continue to change, the basic functioning of the industry. Anyone in the industry who has hopes of doing anything more than designing and building single houses will need to move with these changes if they are to succeed.

Even greater changes in the way buildings are constructed and operated are possible in the future, as nanocomputers and nanomaterials are developed – there is even speculation that at some point buildings will erect themselves as smart materials and nanotechnology are combined.

1.9 Conclusion

The central concern of the first 'Building in Value' book (Best and de Valence, 1999, p. 9) was

> . . . the value that is embodied in buildings, and how that value can be increased through the application of appropriate measures before any design work actually begins.

In the following chapters the central concern remains the same but now the authors introduce a broad range of activities and issues that should be considered during the design and construction phase.

Value still depends on balancing the three factors of time, cost and quality against the client's requirements and the basic ideal remains, i.e., to complete the project for minimum cost, in the shortest possible time, with the best possible quality, while satisfying the client's needs, including those related to function, aesthetics, business goals and image.

Minimizing cost and time through good design and effective management involves making appropriate design choices, monitoring design development, applying suitable management practices to activities on and off site, building relationships with subcontractors and suppliers, and more. Quality is unfortunately often sacrificed due to the pressures of time and cost but its importance as a measure of value should not be lost – too often the 'short-termism' that characterizes many building projects sees the traditional values associated with craftsmanship abandoned, as there is not sufficient time available for the job to be done well.

The topics addressed in the remainder of this book are just some of those that could be included – the complexity of building procurement means that some pertinent issues are not mentioned and those that are could each be the subject of a separate book. The aim is to provide an introduction to the basic concepts and, hopefully, to prompt readers to explore areas of interest in more detail through the references given at the end of each chapter.

References and bibliography

Best, R. and de Valence, G. (eds) (1999) *Building in Value: Pre-design Issues* (London: Arnold).

Blockley, D. and Godfrey, P. (2000) *Doing it Differently: Systems for Rethinking Construction* (London: Thomas Telford).

Construction Task Force (1998) *Rethinking Construction (The Egan Report)* (London: Department of the Environment, Transport and the Regions).

Cornick, T. (1996) *Computer-integrated Building Design* (London: E. and F.N. Spon).

CSSC (1988) *Building Britain 2001* (Reading: Centre for Strategic Studies in Construction).

de Valence, G. and Huon, N. (1999) Procurement Strategies. In: Best, R. and de Valence, G. (eds) *Building in Value: Pre-design Issues* (London: Arnold).

Fryer, B.G. (1997) *The Practice of Construction Management*. Third edition (Oxford, Cambridge, MA: Blackwell Science).

Gray, C. and Hughes, W. (2001) *Building Design Management* (Oxford: Butterworth-Heinemann).

Harris, F. and McCaffer, R. (1995) *Modern Construction Management*. Fourth edition (London: Blackwell Science).

Harris, F. and McCaffer, R. (2000) *Modern Construction Management*. Fifth edition (Malden, MA: Blackwell Science).

Latham M. (1994) *Constructing the Team, Final Report of the Government/Industry Review of Procurement and Contractual Arrangements in the UK Construction Industry* (London: HMSO).

McGeorge, D. and Palmer, A. (1997) *Construction Management: New Directions* (Oxford: Blackwell Science).

Nunnally, S.W. (1998) *Construction Methods and Management*. Fourth edition (Upper Saddle River, NJ: Prentice Hall).

Oxley, R. and Poskitt, J. (1996) *Management Techniques Applied to the Construction Industry*. Fifth edition (Oxford, Cambridge, MA: Blackwell Science).

PART 1

Design and construction

2

Beyond Modernism

Rick Best*

Editorial comment – Chapters 2 and 3

It is widely believed that the distinctive roof shells of the Sydney Opera House were designed to echo the shape of the sails of the many yachts that are a feature of Sydney Harbour – this may or may not be true. Saarinen's much-photographed TWA terminal in New York is thought to represent the shape of a bird in flight, and it seems that at least in this case that the architect may well have been looking for such clear symbolism, and so we have a comfortable explanation for why the building looks as it does. But what of the thousands upon thousands of other buildings that have been built, why do they look the way they do, and what forces are at work that make architects decide on one 'look' rather than any other?

There are two approaches to this question, and in the following chapters both are explored. In the first, the emphasis is on the influence of individuals and their philosophies, particularly during the past century, and where that has led architectural design in recent years. In the second, the author considers, in a general sense, the basic influences of knowledge (what we know about how to build), materials (what we have available to build with) and so on, and how the many and complex influences that impinge on the designer lead to a particular design.

The two approaches are complementary; without the advances in building technology that began during the Industrial Revolution, neither the philosophy of Mies van der Rohe nor the metal and glass buildings that it produced would have been possible, i.e., without the materials and knowledge of how to use them, the curtain-walled skyscrapers of the 1950s and 1960s would have remained as fanciful as Mies' 1923 design for a multi-storey glass-clad office building with transparent walls and daring curvilinear façade.

The effect that architectural style has on the value of a building varies enormously – in the case of an industrial building it may not be a factor at all, while for a house it may be

* University of Technology Sydney

the most important factor, at least in the eyes of some owners or potential buyers. For the majority of buildings, however, it will be only one of many factors. It is often said that the three most important factors governing the value of property generally are location, location and location, and for many buildings this is true – a retail outlet in a back street, regardless of any supposed architectural merit, is unlikely to be as valuable as one of similar size but nondescript appearance located on a busy High or Main Street, close to the railway and bus interchange.

This is not to suggest that buildings in back streets or on industrial estates should not be designed with the same care as a new corporate headquarters or retail complex situated in the heart of downtown – in the long run buildings are for people to use, to live and work and play in, not simply to be admired (or abhorred) from the outside. In fact, many very successful buildings look unremarkable on the outside but work wonderfully well for their occupants, providing comfort, security, a productive workplace, social space, private space, activity space and so on. For their occupants these buildings are of greater value than a meticulously detailed building created with some largely artificial set of 'rules' in mind, some 'ism' or another that may look good on paper and look well in glossy architectural magazines (typically full of pictures of buildings notable for the complete absence of the most important element – people), that leads to cold, sterile and basically uncomfortable buildings.

With so many conflicting influences and pressures, architecture is undoubtedly the most difficult art to practise. Architects are in the unenviable position of having their creations on public view at all times – no-one has to listen to Beethoven if they prefer Chuck Berry, nor be confronted by a Jackson Pollock if they would rather admire a Constable, yet buildings are around us all the time. We cannot ignore them and it is very easy to criticize them and their designers, yet it is no less true of buildings than of any other item or artefact to say that 'beauty is in the eye of the beholder'. Seldom when the merit of a building is being assessed will we hear anyone talk about how well the building functions apart, perhaps, from discussing the shortcomings of the air conditioning or how often the escalators are out of service. Indeed the better a building works the less likelihood there is that anyone will notice, but if there is a problem then everyone will have an opinion about the cause and most likely they will blame the architect.

Architecture has many parts: technology, common sense, art, ergonomics, visual perception, the grand vision, attention to detail, waterproofing, persuasion. All these and many more are concerns of the architect as a design evolves from a formless notion to build in the mind of the client through the long period of briefing, initial design, making contractual arrangements, getting approvals and so on until the building is complete and the occupants have moved in. The value of the finished product exists at several levels: use, exchange and symbolic value are all important to varying degrees in respect of individual buildings. However, a building that is perceived as a 'good building' will most likely be seen as a valuable building, even if its value is more closely related to its location than its physical appearance.

2.1 Introduction

The question is sometimes asked: 'What is the difference between architecture and building?' One facetious answer is: 'If the roof leaks, it's architecture.' The basis of this response lies in the architect's quest for new and different ways of satisfying humanity's basic need for shelter and security. Innovation always carries some risks and the risks attached to the pursuit of innovative approaches in the design and construction of buildings are exacerbated by the very nature both of the products (i.e., buildings) and of the industry that produces them.

The leaky roof symbolizes the problems that architects and builders face as they try to explore new designs and utilize new materials and construction methods – neither the designers nor the contractors are completely familiar with the new material or system and there are no previously completed installations to look at for guidance. The result is that new methods are only truly tested when they are used in 'live' projects and it may not be until a number of projects have been completed that any new method, material, component or system is completely satisfactory in use.

One famous, or perhaps infamous, case occurred in Boston in the early 1970s. The John Hancock Tower, a 60 storey office building with a mirror glass façade comprising 10 334 individual double-glazed units, had many of its windows fracture while the building was still under construction, with some falling to the ground. Eventually more than 4000 m^2 of its façade was replaced with plywood while a permanent solution to the problem was sought. While it took many years to resolve the dispute, engineers were able to establish that the failure of the double-glazed units resulted from an inherent design fault in the units themselves (SGH, 2000). The designers, in their search for a new solution, caused a major problem for many people, and one which led to 17 years of litigation before it was finally resolved.

2.2 Buildings as prototypes

Apart from some domestic scale buildings and a few repetitive industrial buildings, the majority of construction projects are unique. In many cases it can be said that the finished product is in fact a prototype, but a prototype from which no production run will follow, so the designers and builders have no opportunity to identify and rectify their mistakes, or even any chance to refine or improve aspects of their work, at least not in the first instance. In subsequent projects they can apply the lessons learnt on previous jobs, but that is of little help to those who own or occupy the original building. The leaky roof may be repaired but that may only be done at great cost, and all too often the cause of the problem remains unknown, with the result that the same problems arise again and again.

The nature of the construction industry is a problem in itself as it is generally fragmented, with many players involved in design and construction and much of the detailed work carried out by independent design consultants (including engineers of many disciplines) and on-site sub-contractors. A large construction project may involve several hundred individual contracts for the execution of the works, from the clearing of the site to the installation and maintenance of mechanical plant. The problems associated with the co-ordination of so many activities and the efficient flow of timely and accurate information to a large group of people add to the difficulties which are inherent in the adoption of new and often untried approaches.

Contractors typically operate on narrow profit margins, particularly during the cyclical downturns in activity which characterize the industry, while design consultants are often working under very tight fee structures that allow minimal time for serious research and testing of options. The constraints of time and cost have a marked impact on the quality of the product and hence the search for innovation when constrained by these factors often leads to problems.

The construction industry is generally seen as being very conservative and slow to adopt new ideas. The cyclical nature of the industry and the generally low profit margins militate against spending on research and development (R&D) by construction companies. When the industry is buoyant and work is plentiful companies often find that they do not have time to engage in R&D, and during downturns there is often no money available to fund R&D. Further to those constraints, the 'leaky roof syndrome' encourages designers and contractors to stick to tried and true methods that have much lower risk of failure, as such failures as the John Hancock Tower façade are costly and can cause irreparable damage to a firm's reputation.

In spite of the potential problems, designers still seek new ways of doing things and as technology has advanced, particularly since the Industrial Revolution, the design and construction of buildings has changed enormously. Designers have been quick to utilize new technologies and in many cases have initiated the development of new products and pushed the boundaries of existing manufacturing and construction practice through their innovative designs. Some of the most famous architects of the twentieth century are recognized not only for the way their buildings look, but for their daring use of new materials and technologies; yet in many instances the buildings in which they pioneered these innovations were not very successful in terms of fulfilling the clients' functional requirements.

The owner of Mies van der Rohe's Farnsworth House at Plano, Illinois found it so expensive to live in that she tried (unsuccessfully) to sue the architect (Watkin, 1992), yet the building appears in every collection of landmark buildings of its era. While Le Corbusier was a leader in the use of reinforced concrete, many of his most famous projects (such as the Unité de Habitation in Marseilles and the government complex at Chandigarh) in which he used his famous *béton brut* concrete façades are now in need of extensive repair as the concrete has deteriorated badly over time. Once again, however, these projects are considered to be of great significance, partly because of the innovative use of off-form concrete. The Glass House designed by Philip Johnson in the late 1940s was reportedly too hot in summer and too cold in winter, yet, in spite of its functional shortcomings, it is praised as a perfect expression of the minimalist approach that was predicated on functionalism.

2.3 The modern movement and the international style

So what was it that gave rise to the look of twentieth century buildings? Mies and Le Corbusier are two of the towering figures of twentieth century architecture, giants of the so-called 'modern' movement that shaped so many of the buildings in which we live and work today. So pervasive was the influence of the modernists that the term 'International Style' was coined in the 1930s to describe the plain, 'bare bones' type of design that was becoming increasingly visible in cities all round the world. The simple form and plain appearance of these buildings was no accident but was driven by strongly held beliefs and philosophies espoused by Mies, Le Corbusier and others, notably Walter Gropius.

By no means were all the buildings of the last 100 years designed with the modern movement in mind: the 1930s gave us Art Deco, the 1950s and 1960s the New Brutalism, while some architects still looked to earlier styles such as Gothic and Renaissance and produced their own somewhat updated versions of them. It is fair to say, however, that in a very general sense there emerged a common language of architecture that became evident particularly in commercial buildings: offices, large-scale residential buildings, hotels and so on, that was a direct derivative of the philosophies of the modernists.

In 1908 Adolf Loos, an Austrian architect, wrote (Opel, 1997) that decoration represented 'crime, savagery and depraved sexuality', and that the 'evolution of human culture implies the disappearance of ornament from the object of use'. Le Corbusier announced that the 'house is a machine for living', while Gropius was inspired by the production line approach pioneered by Henry Ford and believed that similar industrial processes could be applied to the mass production of housing. It was from these stated positions that the modern movement of the twentieth century sprang, a movement or style or approach to building design that is evident in practically every major city in the world and which influenced the work of many thousands of architects.

The characteristics of the buildings of the modernists are smooth walls, plain colours (white, grey), simple forms, flat roofs, large areas of glass; famous examples include the Bauhaus (Dessau, Germany, 1925), the Villa Savoye (near Paris, 1929), Illinois Institute of Technology (Chicago, 1939 onwards), Lake Shore Drive Apartments (Chicago, 1951), Lever House (New York, 1951) and the Seagram Building (New York, 1954). There are any number of unremarkable buildings worldwide that are derivatives of these revolutionary buildings but many people find the geometric bareness of many modern buildings, with their plain façades and lack of colour or visual interest, particularly at a micro level, dull and oppressive. Tom Wolfe (1983, p. 4) captioned a picture of a number of faceless office towers on the Avenue of the Americas in New York as 'row after Mies van der row of glass boxes'. A visitor to any city, from Singapore to Cape Town to Moscow, will see endless copies of those same 'glass boxes', some with at least a little refinement of detail or proportion but with, sadly, the majority depressing in their sameness.

2.4 Post-Modernism and high-tech

In 1955 the City of St Louis completed a large public housing project – the Pruitt Igoe flats. Designed by Minoru Yamasaki (who also designed the World Trade Centre towers in New York) and based firmly on the principles of the great modernists such as Le Corbusier, the development won an American Institute of Architects prize. By 1972, just 17 years later, they had been 'vandalized, defaced, mutilated and had witnessed a higher crime rate than any other development of their type' (Nuttgens, 1997, p. 284) and were demolished. The well-known author and critic, Charles Jencks, suggests that the modern era ended in 1972 with the demolition of the Pruitt Igoe buildings. Since then many other tower blocks have been demolished after a similarly short life.

The fundamental problem with the buildings of the modern movement was that the strict functionalism of the designs was at odds with what people feel comfortable with – there is great irony in the observation that the interiors of many offices encased in a sternly functional steel-and-glass curtain wall are filled with timber panelling, decorative set plaster ceilings and Greco-Roman columns and pediments. In residential buildings people want to be able to personalize their space by adding gardens, and pergolas, and fountains, and window boxes and so on; consequently the grand visions of the architects who seek to control the very lives of those who occupy their buildings produced environments that made people uncomfortable and angry. Mies van der Rohe even expected tenants in his Lake Shore Drive apartment buildings to keep their blinds (all of the same standard colour) at the same height so that the elevations of the buildings appeared properly ordered (Nuttgens, 1997).

Not surprisingly Modernism and the International Style, and their rigid supporting philosophies, began to be questioned by some in architectural circles. The publication in 1966 of Venturi's book, *Complexity and Contradiction in Architecture* (Venturi, 1966), and in 1972, coincidentally the year of the demolition in St Louis, of *Learning from Las Vegas* (Venturi *et al.*, 1972), marked the beginning of a move away from the strict functionalism that had driven so many to design the faceless boxes that still dominate many of our cities today.

What has emerged in the last 30 years can hardly be called a 'style' but there are, perhaps, two streams that may be identified at least in broad terms. Jencks (1981) suggested the title 'post-modern' to cover the amalgam of stylistic fragments that appeared as a reaction against the austerity of the modern school. Today we see buildings that are colourful, displaying many textures and forms, using elements from all historical periods side by side; a classical column here, a stylized broken pediment there, even caryatids in the form of cartoon characters, and these often made from the most modern of materials such as glass reinforced plastic or epoxy. As well as this seemingly *ad hoc* mix, and sometimes as part of it, there is the so-called 'hi-tech' approach that may see designers do away with the skin of the building and instead display the bones and the circulatory and nervous systems for all to see. Look no further that the Lloyds building in central London where the services pipes and ductwork dominate the exterior of the building, gleaming metal and glass lifts run up the outside of the building and even the cleaning gantries become 'decoration' at roof level.

Of course, when we talk of an architectural style it is easy to lose sight of a most important fact: the vast majority of buildings around the world are not designed in any architectural sense, or style, at all. Influential Dutch architect Rem Koolhaas claims that each year in the Pearl River Delta region of China around 500 sq km of 'urban substance' are created, much of it in nondescript tower blocks perhaps reminiscent of the grand designs of Le Corbusier but 'designed' on personal computers in a week or two and then built as quickly as possible to meet the heavy demand for space (Wolf, 2000). New houses, whether cottages in Surrey, tract houses in the suburbs of Houston, or bungalows on the outskirts of Sydney or Melbourne are most often based on standard designs, often with detailing to appeal to the common view of what a family home should look like: a front garden with a path to a cheerily painted door flanked by small paned windows with lace curtains. There is precious little of the Villa Savoye, the Farnsworth House or the Vanna Venturi house in any of these but they dominate vast areas of our modern cities.

The same is true of the majority of buildings being constructed in many places – from houses to hotels, fast food outlets to high rise offices, from warehouses to whorehouses. While there may be the name of an architect on the documentation, the design will be driven more by commercial imperatives, such as corporate identity (a McDonald's outlet in Moscow looks remarkably similar to one in Tokyo), least cost, or ease of construction, than by any architectural philosophy or vision. In Australia, in recent years, perceived savings in construction time and cost have, for instance, led to the widespread use of a method, which originated in the USA, called 'tilt-up' construction for low rise industrial/commercial buildings. A concrete floor slab is poured on the ground and wall slabs are poured flat then hoisted into an upright position, braced and bolted into position – the erection of a steel-framed roof, fabricated off-site, is a quick operation and the building is well on its way to completion. Visually, the result is uniformly awful: the construction method dictates that walls are flat, the overall form is strictly rectilinear, decoration is restricted to banal points and scallops along the top of the wall and there are no shadows or rhythm in the façades. Yet they flourish because they provide a cost-effective solution to the basic problem of providing large enclosed spaces with flat floors suitable for lift trucks; no architect is needed save, perhaps, for the preparation of documents for approval by the relevant controlling bodies (local council, municipal authority or whatever) and a signature to give the project legitimacy.

Of course there are visionary buildings being designed and constructed today, such as Frank Gehry's Guggenheim Museum in Bilbao, but they represent only a tiny fraction of the building work that is going on at any one time. These buildings do, however, exert some influence on the rest, just as the glass boxes of Mies spawned thousands of similar buildings, so elements of the landmark buildings of the late twentieth century are appearing everywhere: on blocks of flats, supermarkets, institutional buildings and hotels. The general rule now is that there are no rules, and architects, when given the opportunity to actually design a building, rather than simply plan some spaces to suit the client's operation, have a freedom to 'mix and match' or perhaps 'plug 'n' play', as they tackle the eternal problem of enclosing space for their clients' needs in a way that is functional yet satisfying for the people who see and use their buildings every day.

2.5 Pluralism

If there is a label that can be applied to contemporary architecture then perhaps it is pluralism. Post-modernism has been used to describe the move away from the geometric approach of the Modernists and so included the assembly of a range of historical elements mixed, in many cases, with a playfulness that saw such buildings as the Portland Public Services building (architect: Michael Graves, completed 1983) challenge the ideal of the rational steel-and-glass tower. The Pompidou Centre in Paris (architects: Richard Rogers and Renzo Piano, completed 1977) represents a parallel stream, that of the so-called high-tech style, where structure and services dominate the visual experience of the building. These buildings often use the structure as the main visual interest, with exposed trusses, frames, masts and tension stays acting as virtual exoskeletons; in some cases with ducts, pipes, cooling towers and elevators all visible.

Naturally what we see in the everyday buildings around us ends up being an amalgam of these buildings and the styles and philosophies that they represent – some exposed ductwork and an external elevator may well be juxtaposed with some Modern white walls framed by an applique Greek temple front – there are no rules, hence the adoption by some of the label 'pluralism'.

2.6 A general view

It has been said we get the government we deserve, and it may also be said that we get the buildings we deserve. Tom Wolfe (1983) observed that for much of the twentieth century those who commissioned and approved buildings were persuaded to accept designs put to them by architects in spite of their unsuitability – designs driven by the opinions of a few influential thinkers, opinions that somehow became treated as immutable laws. Auberon Waugh, reviewing *From Bauhaus to Our House*, described the products of these laws as 'monstrosities of ugliness and ineptitude', and in doing so voiced the feelings of many who have no love for the offspring of Le Corbusier and Mies that inhabit cities all over the world.

Building design at the beginning of the new millennium is, then, less constrained stylistically than at any time in human history. Architects have an unimagined range of materials and systems available and are free to draw on the full spectrum of previous styles and elements and use them in any combinations that they may choose. The historical styles available of course include the Modern, as the transition from Modern to Post-Modern was not as drastic as it may sound – as Klotz (1988, p. 128) put it (paraphrasing Venturi and Brown): 'The protest against modernism is not a determinate and rigid "No"; rather it is a "Yes, but".'

Increasingly we can expect to see building design shaped by the need to curb energy use and slow the use of resources. These constraints need not necessarily be reflected in the visible design, i.e., how the building looks, and in fact there is often a conscious effort made by designers, as they try to meet the demands for more environmentally sensitive buildings, to avoid producing buildings that 'look green'. In practice, however, the addition of photovoltaic arrays and extensive sunshading, which are characteristic of many of the new low energy buildings, do make some mark on their overall look. Similarly an increasing awareness of the energy 'bound up' in building materials and components (embodied energy) and attempts by designers to look for and use less energy-intensive products may exert some influence on the appearance of buildings. This is likely to be minimal, however, as commonly used materials such as steel, concrete and glass will not be displaced but rather be recycled or re-used.

Visually most buildings will probably remain largely undistinguished and all but undistinguishable from the bulk of those around them, based on a simple vocabulary of materials, forms and finishes. A shopping mall in Woking may not only look very similar to one in Sydney or Seattle but will even contain a significant number of the same shops as the multinational character of franchises and chain stores becomes ever more apparent.

The public's demand for space for homes, businesses and recreation will be met mostly by construction that is driven by market forces and the need to build quickly in order to satisfy demand. How many of these buildings will, in any sense, stand the test of time remains to be seen; some will go the way of Pruitt Igoe, others will be regularly refitted internally and perhaps be given a new façade at intervals, some will fail due to poor construction or specification of unsuitable materials or bad design detailing. Certainly there must be some concern about the durability of many new buildings – for whatever reason many new buildings begin to look shabby very shortly after occupation, and the increased use of colour on building exteriors must raise questions about the long-term effects of weather and pollution and the maintenance that will be required if they are not to look tawdry in just a few years.

2.7 Conclusion

The value that those who build will ultimately gain from their buildings will vary but the initial design will only be one of a number of the factors that contribute to eventual value outcomes – changing tastes, scarcity or oversupply of alternative space, changing demographics, competition from other parts of town, traffic patterns – the list goes on. The majority of factors on the list are essentially beyond the control of the building owner, which emphasizes the basically risky nature of construction. However, if designers look back at the successful buildings of the past there are basic attributes that appear to have contributed to their endurance: simple planning and the use of durable materials combine to produce buildings that follow the principle of 'long life, loose fit'. If 'low energy' is added to that formula then there is good reason to suspect that the buildings that result, regardless of the 'ism' that may be tacked on to them and therefore regardless of their superficial outward appearance, will be valuable assets for their owners in the longer term.

References and bibliography

Jencks, C. (1981) *The Language of Post-Modern Architecture.* Third edition (New York: Rizzoli).

Klotz, H. (1988) *The History of Postmodern Architecture* (Cambridge, MA: MIT Press).

Nuttgens, P. (1997) *The Story of Architecture.* Second edition (London: Phaidon Press).

Opel, A. (ed.) (1997) *Ornament and Crime: Selected Essays (Adolf Loos)* (Riverside, CA: Ariadne Press).

SGH (2000) *Glass Failure on the John Hancock Tower*, Simpson Gumpertz and Heger, Inc., www.sgh.com/practiceareas/buildingtechnology/hancock/hancock.htm

Venturi, R. (1966) *Complexity and Contradiction in Architecture* (NewYork: Museum of Modern Art).

Venturi, R., Brown, D. and Izenour, S. (1972) *Learning from Las Vegas* (Cambridge, MA: MIT Press).

Watkin, D. (1992) *A History of Western Architecture* (London: Laurence King Publishing).

Wolf, G. (2000) Exploring the unmaterial world. *Wired.* www.wired.com/wired/archive/8.06/koolhaas_p

Wolfe, T. (1983) *From Bauhaus to Our House* (London: Sphere Books).

3

How buildings come to be the way they are

Steve Harfield*

3.1 From simple beginnings

Let us start with some simple observations, observations that may appear so self-evident that they are not usually made at all. When we look at any building – whether it is one of those that surround us in our daily lives, those that we see while travelling in different places, that have been produced by different cultures, or those that we might claim to know something about from history and are thus from a different time – we can make a number of assertions; three spring to mind immediately. First, the building has an appearance, an outward form, i.e., it looks a certain way; second, it is made of certain materials; and third, it has or had a structural system that makes it stand up. Two further observations can be made that are perhaps even more basic: each building is an artifact, i.e., rather than being a naturally-occurring object, it was made by, though not necessarily for, human beings; and we infer that it had a function, i.e., it served a purpose, it was constructed knowingly and intentionally to house specific activities, even if we can no longer be certain what those purposes or activities were.

Finally, one further observation, obvious but fundamental as a springing point for asking a series of important questions: each building has a specific spatial and temporal location – it was built on a site somewhere at some point in time. While self-evident, this last observation is important in at least two respects. On the one hand it alerts us to the relationships between the building and the society and cultures that produced it. Cultures and societies differ with time and place: cultural values, customs and traditions differ; religious, political, ethical, economic and other societal beliefs and requirements differ; specific knowledge and technological availabilities differ; and, of course, artistic or aesthetic or formal preferences differ. All of these, we might infer, will have an overwhelming impact on the forms of the buildings produced and the functions to which these buildings respond.

* University of Technology Sydney

On the other hand, the specificities of time and place lead us to ask questions about, and thus begin to appreciate the significance of, such basic issues as the nature of the particular climatic conditions and thus the environmental determinants of human shelter in a particular location, the nature of the specific site conditions, such as its geological and topographical details, and the availability of materials from which to construct buildings. While each of the above – climate, site conditions and material availability – is essentially physical, and thus seemingly independent of culture, cultural factors inevitably play a significant role in determining how such physical factors are addressed and/or how they may be utilized.

3.2 The lessons of the castaway: climate and technology

Were one to ask so broad a question as 'What has shaped built form over the last 5000 years?' one might – given the necessary and almost unreasonable simplicities that any such answer must engender – be offered a reply something like this:

If one were to be washed up, Robinson Crusoe-like, on one of those mythical 'desert islands' so beloved of story tellers and reality TV, then the first determinant of shelter – leaving aside for the moment any notion of 'architecture' – would undoubtedly be climate. If conditions proved to be hot during the day, one might feel the need for shade; if the island proved to be cold at night or suffered from torrential rainfall, one might seek shelter to ameliorate such conditions, and so on. Such conditions do not, of course, force one to consider the making of a shelter; the shade of a tree or the warmth and shelter of a cave may be sufficient. If, however, one does wish to make a shelter, or if there are no suitable trees or caves, then two further determinants are immediately brought into play.

The first is materials – on our desert island we can only make our shelter out of those materials that are available to us; it is of little use for the Inuit to think of timber huts if the only material available is snow. Any discussion of materials must lead us to a second determinant and its key lies in the word 'available'. What does it mean to describe a material as being 'available'? In terms of our desire to make a shelter, material availability must be conjoined with technology, as even if our island has abundant trees, if we have no means of cutting them down, or of moving or lifting any trees that have fallen naturally, or of processing the wood from these trees into sizes and shapes suitable for making our shelter, then we can hardly describe wood as being an available material. Material availability is thus also determined by our level of technology.

One further point should be made here immediately. It might seem that here the term 'technology' corresponds to the more prosaic word 'tools': we can only use timber if we have tools to work it, or, perhaps better expressed, the extent to which we can use wood and the sophistication with which we can use it, depends on the availability of appropriate tools and their sophistication. Technology, however, isn't simply tools, at least in a physical sense – technology must also embrace knowledge. We can only usefully exploit material availability within the limits of our constructional and structural knowledge: while we may have the tools to chop down a tree, we may have neither the knowledge nor the skill necessary for us to use this material to make a shelter.

It is at this point that we might replace the (deliberate) phrase 'make a shelter' with the single term 'build', implying here not simply a desire to but a knowledge of how to. In this context 'how to' means a *societal* knowledge of how to, and it is worth noting the distinction between the knowledge held in a given society at any particular time, and individual or personal knowledge. On the desert island the vast technological and intellectual resources of the early twenty-first century count as nothing if no-one on the island has such knowledge personally. While we may know of extremely sophisticated timber technologies, our own knowledge and skill levels might approximate those of the most so-called 'primitive' societies when it comes to actually building a shelter.

Of course, climate, materials and technology are not the only determinants of built form, but before mentioning others it is as well to consider how changes over time might affect each of these three. Climate we might safely ignore as even though climatic change over time is a recognized phenomenon – made manifest most clearly to the society of today by the increasing evidence of global warming – the processes and effects of such change, significant as they may be, are too gradual. The other determinants, however, are highly susceptible to rapid short-term change. In terms of materials we might imagine at first glance that these have been ever-present and unchanging, yet resource availability changes in two dramatic ways, one related to sustainability, the other to technology. It is easy to imagine the situation in which certain materials, previously readily available, have ceased to be so; and therefore timber, once a ready source of building material on the island, ceases to be so if all the trees are cut down and we have no means of replacing them.

Resource availability should also be considered with relation to changes in technology; over the course of time other materials might become available precisely because of advancements in technology. Again, these improvements might be seen as increasingly sophisticated or powerful *means* (i.e., improved tools) or *developments in information and theory* (i.e., improved knowledge). Thus, for example, the island may always have been a ready source of iron ore, but we could hardly have made use of iron (let alone steel) as a building material before we

- knew of its existence
- had the means of extracting it
- had the means of processing it
- knew what to do with it.

While each of these four points is of significance, it is perhaps the last point, crudely stated here, that is crucial. Knowledge, just as much as physical material, is a major determinant of architecture, i.e., the development of iron as a building material is not just an issue of the extraction and refining of a previously unavailable natural resource, but an intellectual development in its own right, while the development of plastics, for example, is not an issue of recognizing and utilizing a natural material at all, but the invention of a completely new one, with all the intellectual and theoretical foreknowledge that this implies. Similarly, and as just one example, developments in the theories of structures, whether based on empirical evidence (trial and error?) or rational intellectual speculation, have both preceded and precipitated changes in built form. In addition – and in some ways to pre-empt our argument – it should be noted that the very idea of technology has often been its own imperative. One might easily claim that architects have long embraced the motto 'what can be done *will* be done', while the suggestion that the Modernists pursued the new as an end in itself, with material and technological advancement being a direct representation of the 'spirit of the times', may be overly-simplistic but nevertheless not without merit.

3.3 Investigating function

The above, of course, are far from the only determinants of built form. As a useful comparison it is worth noting that a building's performance as a climate modifier and environmental filter represents but one of four categories developed by Hillier *et al.* (1972) to describe the variety of functions that may be fulfilled by a building. Beginning with the question 'What, in theoretical terms, is a building?', and on the grounds that 'buildings are not gratuitous but entirely purposeful objects', the authors propose four generic ontological categories: the building as climate modifier (as described above); the building as a container of activities; the building as 'an addition of value to raw materials'; and finally the building as 'a symbolic and cultural object'. While such categories are neither mutually exclusive nor exhaustive they do present a useful point of departure for further argument regarding the determinants of built form.

First, a matter of terminology: although Hillier *et al.* quite appropriately present each of the above categories as being functions performed by buildings, it is perhaps the second of these, the building as a container of activities, that corresponds most closely to the meaning intended when the word 'function' is used in everyday or conversational speech, particularly by builders, architects, clients and the general public. The question 'what is the function of the building?' is most usually answered in terms of the activities it is intended to house, or the uses to which it is to be put.

The varieties of functions or uses ascribed to buildings throughout history are legion. Early suggestions might include: to offer defence, to store food, to house animals, or to provide facilities for public gatherings or meetings. Others, depending on the specific culture and time from which we select our examples, might include the satisfaction of spiritual needs (ritual, religion, worship), for specialization in work or trade-related needs (e.g., tanneries, smithies, mills, bakeries) or for health, education, entertainment or sporting purposes.

A number of points are worth making here. Firstly, while there are examples of functions or activities that are intensely culturally-specific and are thus entirely non-generic, many others, with perhaps the most obvious being the need for some form of housing, might conveniently be regarded as universal insofar as we imagine that they always have been a fundamental requirement of any society and may be expected to remain so in the future. However, the identification of certain generic functions, like 'housing for people', does not imply that the specific needs and requirements, nor their specific means of attainment, are in any sense shared or common. Secondly, therefore, we might note that universal generic needs nevertheless produce very different specific requirements that vary greatly with society and culture.

Thirdly, and as a parallel to this, it is worth noting that functional needs or activity types or use requirements respond to cultural and technological changes and thus are intensely time-specific. Before the invention of the steam engine there was no need for, and thus no conception of, railway stations; before the invention of powered flight, no need for airports, and so on. Conversely, technological changes may bring about the end of previous building types, e.g., once horse or horse-drawn transportation was replaced by the automobile there was no longer a need for extensive stabling facilities or a rich network of coaching inns.

Fourthly, and most importantly, while the specific functional or activity or use requirements of a building undoubtedly contribute to the final form of that building, and to the nature and the arrangement or disposition of its constituent parts, they play a far less significant role than might be expected. Certainly functional requirements must be met; but how they should be met in terms of the formal, spatial and technological (i.e., material, constructional and structural) characteristics of the outcome does not of necessity flow from an understanding of the functional or use requirements themselves. The same set of requirements may engender a vast array of formal outcomes, each satisfactorily meeting the requirements but via different formal means. So, while functional requirements are perhaps the most obvious of a range of factors that influence designers in their formal and spatial decision-making, there is clearly something more at play here, that influences designers in the development of their schemes.

3.4 The building as an economic commodity

The third category suggested by Hillier *et al.* (1972) considers the building as an addition of value to raw materials. The term 'value' might be approached in two ways: as an abstract quality denoting 'improving amenity', howsoever that is defined, and as an economic characteristic measured in strictly financial terms. The first of these naturally overlaps with both earlier categories because if the design and making of a building is specifically purposeful, then the successful response to a variety of climatic and functional needs must be perceived not merely as assembling raw materials but as adding amenity through such an assemblage. In terms of the determinants of form, however, the understanding of value in strictly financial terms is more significant. This in turn may be examined under two distinct headings, corresponding to the more prosaic terms 'budget' and 'profit'.

The first of these is to a large degree self-evident: all projects cost money, whether measured in terms of human time and energy resources, or material and technological resources, and the allocated budget for any given project will, at least in part, influence both the form, or at least the choices informing the form, and the content of the final outcome. While 'more money' does not necessarily equate with more expensive materials, more sophisticated technological solutions, more generous spatial allocations, or more indulgent briefs on the part of the client, 'less money' certainly rules out possibilities that would otherwise be available to both the client and the designer. This is no bad thing in design terms, and, contrary to some prejudices, it would be facile to assert that all projects could have been better if only more money had been available. Nevertheless, it is indisputable that money available is inextricably linked to outcome possibilities and thus must be seen as yet another factor contributing to the final formal outcome of the building.

In the above, of course, the term 'budget' has been used to indicate what might be termed a unidirectional flow of money – from the client or owner to that range of providers (designers, consultants, builders, statutory bodies, and so on) charged with taking the project from inception through to final completion. Such a financial flow is inevitable in any project – buildings cost money. Yet what should not be neglected in the analysis of how buildings get to be the way they are is that the exact opposite may also obtain. This is not just to make the banal observation that the financial flow is not inescapably one-way, i.e., that as well as costing money, buildings sometimes make money, but rather to remind ourselves that there are many situations where profit is the prime, indeed the only, motivation for the erection of a building. In the case of many commercial buildings, high-rise residential developments, speculative housing estates and shopping centres of whatever hyper-size, it can be said that without this motivation there would be no building and in fact that the very purpose of the building is as a vehicle for the creation of wealth; and that those 'other' functions, the actual activities or uses the building provides for, are secondary to and conditioned by this profit motive. This is not meant to imply any comment about good and bad buildings or good and bad motives, but rather to establish that in such cases the design of the building itself is conditioned by, and firmly tied to, prospective returns. For designers this is not simply a matter of minimizing budget and maximizing profits, rather it is part of a complex economic equation concerning asset and investment management in which the decision to produce a building is weighed against other possibilities offering better rates of return and so on. Developers are not, after all, philanthropists – they are financial investors for whom a building is a commodity of financial exchange. For them, 'what to build?' and 'how to build?' and 'what will be the appropriate form to build?' are questions whose answers are not informed solely by 'designerly' factors but by stringent financial and market ones as well.

3.5 Introducing clients and designers

From the preceding discussion it is evident that a number of determinants are at play. Climatic factors, functional requirements, and technological and material means all contribute to the final form of the building. Economic issues and the requirements of statutory bodies provide restrictions and/or opportunities to which the building must respond. Each of these is influenced by, and must be seen within the context of, both time- and location-specific social and cultural parameters that must play a significant role in structuring the final result.

Yet it is apparent that in order to understand the determinants of built form more clearly it is necessary to make explicit a number of contributing factors that have so far been only implicit. These factors arise as we move from the general to the particular, i.e., from the rather abstract question of 'what causes buildings to be the way they are?', to the more personal and concrete one of 'what causes any particular building to be the way it is?'. This can be done without compromising any of the general factors discussed earlier, all of which continue to play a major role in the determination of each individual building. This personalization or move towards specificity, however, allows us to invoke a series of simple everyday terms that are commonly used when the design of buildings is discussed, and which have hardly appeared in the discussion thus far: client and designer, needs and wants, the design brief and the design problem.

It is usual to assume that buildings proceed from clients; they have need of a building and at a general level are usually able to specify quite clearly the building type they want, its functional requirements, site availability, and, potentially, the economic circumstances affecting the building. Clients then, on the assumption that they lack the skills themselves, employ designers. These designers are not necessarily architects; while it may reasonably be argued that all buildings are designed, only a proportion of the world's buildings are actually designed by architects, and then most often in collaboration with other experts.

Putting aside the significant contributions to the process made by other experts, it is the role of the designer to design the building. Given the earlier observation that, in general, the client is able to specify quite clearly the building type, its functional requirements and certain financial or budgetary information relevant to the project, then we may say that the client approaches the designer with a set of needs that must be fulfilled. These needs may be simple or complex depending on the nature of the project, but they are inevitably both incomplete in a number of regards, and plastic, i.e., not yet fixed and rigid, but still capable of being changed or moulded as circumstances may dictate. As the project develops – from the general requirement for a shopping centre to the design of *this particular* shopping centre – then some of the needs may be modified. Indeed, and as might reasonably be expected, many requirements will emerge within the project that were previously unknown to the client and would not have formed part of the general list of needs. It is presumably one of the key roles of designers, and a condition of their expertise, that they will be able to identify, advise on and deal with such contingent needs specific to the individual project. Two simple examples will illustrate this point. Firstly, each specific project is built on a specific site, and specific physical site conditions, including soil type, slope, orientation, access and surroundings, will contribute to the requirements to be met by the designer in a variety of ways. Secondly, that the very notion of a specific physical location will introduce a series of statutory regulations with which the future building will be expected to comply, from specified height limits and percentage allowable site coverage, to requirements related to fire safety and egress provisions, or to the number of on-site car parking spaces that must be provided, and/or a range of factors addressing heritage, environmental, aesthetic and social concerns. All fall within the purview of the designer and all may be understood by the client as being additional to the generic specification of needs determined at the inception of the project.

Of course, not all factors bearing on the project fall neatly under the heading of needs and without wishing to draw a firm distinction between words, all designers know that clients arrive well-armed with a collection of wants – desires for the project that go beyond the (simple or complex) meeting of physical requirements. This is neither surprising nor inappropriate and, as will be seen, it is precisely this notion of 'what one wants to do with the project' that drives this discussion. It is a major task of the designer to interrogate and scrutinize the perceived needs and wants to be addressed in the project and to discuss them with the client, so that the latter may be better informed and some agreement may be reached as to how to proceed with the project. Simply put, somewhere between the initial needs, the additional needs and the specified wants lies the project brief.

3.6 Constructing the brief

While commonplace within discussions of design and designing, the term 'brief' has been introduced here quite deliberately to signal two further and closely intertwined issues of importance in understanding how buildings get to be the way they are. The first of these relates to the generally held idea that design is a problem-solving activity and thus that the brief represents an unmediated specification of the problem. The second relates to the view that the brief somehow lies 'outside' the control of the designer and simply informs him/her in proceeding to a solution.

Whether design activity is merely a problem-solving activity or not is a moot point. Recent research indicates that the 'design = problem solving' equation is as likely to be endorsed, whether explicitly or tacitly, by a large range of design theorists in the disciplines of architecture, industrial design and engineering, who are taken to be experts in the field, as by new first year architecture students who are taken to be novices, as yet untrained and thus largely uninfluenced by the assumptions inherent in the literature (Harfield, 1997).

What is significant is not whether design outcomes are somehow solutions to problems, but rather the belief that design problems, albeit falling under the category of wicked or ill-structured problems rather than tame or well-structured ones, are nevertheless specified within the brief.[1]

To a significant extent this is true, at least insofar as the brief might be expected to enumerate that extensive and complex list of requirements that the final design solution is meant to address. This may be identified as *the problem as given* – that set of data presented to, and often developed in conjunction with, the designer, and upon which the design solution is predicated. But to accept that the data provided in the brief informs, and to some extent concretizes, the problem is a far cry from asserting either that the data requisite for the solution inheres within the brief and therefore that a skilled designer can simply move from problem to solution, or, and perhaps more importantly, that the problem 'solved' by the designer corresponds to the problem as given.

The second point, that the perception that the brief somehow lies outside the control of the designer and simply informs him/her in proceeding to a solution, suggests that the designer is somehow neutral in respect of the problem, that the problem is simply given to the designer, who proceeds by skill and training, to 'solve' it. This does not, however, suggest identical solutions since such a position tacitly accepts the fact that all but the most simple or most determinate of design problems will likely fall under the category of ill-structured problems and will thus produce what are often referred to as 'satisficing' solutions. Unlike the solutions to those problems which can be assessed as being unequivocally right or wrong, satisficing solutions – sometimes called 'better or worse' solutions – can only be judged on the extent to which they satisfy some set of criteria, based presumably on those requirements and complexities contained within the brief.[2]

The obvious result is that different designers will produce different outcomes based on the same brief and any architectural studio or architectural competition provides ample evidence of this. This is entirely consistent with the idea of satisficing solutions.

Recast in the language of problem solving, this means that different solutions may be derived from the same problem. While this appears obvious and barely worthy of comment, it is based, in the case of design, on the tacit assumption that it *is* the same problem, i.e., that the same brief inevitably yields the same problem, irrespective of designer; that each designer, whether expert or novice student, has taken the problem as given and simply 'solved' it; that the designer is effectively neutral in respect of the problem and introduces nothing into it but skill and knowledge; that the varieties of outcome are precisely the result of differences in such skill and knowledge; and thus, pre-eminently in the student situation, that the variety of schemes, judged one against the other, is a measure of the different ways in which a control group has solved the same problem! Yet this is not necessarily how design and designers work, or how they could work, even if they were so inclined. This suggests a significant distinction between *the problem as given* (the brief, its requirements, etc.) and *the problem as 'taken'* or, perhaps better, *the problem as design goal*, i.e., the individual designer's overlay on the brief that determines what he or she wants to do. This personal decision as to 'what to do' is often referred to as 'problematizing' since it relates not to issues of how to solve the problem established by the requirements in the brief, but rather, in addition to those requirements, what is the architectural or design problem that I as designer *choose to adopt and solve*? Such problematization – different for each designer and for each project – lies at the very heart of design and will inform and constrain both the design activity and the final outcome in ways that are not dictated by the brief itself.[3]

This idea of the problem as design goal is essential to an understanding of how buildings get to be the way they are – while not marginalizing the brief and the host of specific requirements that undeniably inform the solution, it does begin to reinstate the designer as the central formalizing agent in the design process and may be used to explain not only why different designers produce different designs to what is ostensibly the same problem, but how, within this range of differences, common languages and approaches are clearly evident. This leads to the question, 'Does the designer know in advance of its production what design outcome he or she will produce?'

At one level, and as already discussed, the answer must be 'yes'; the final outcome or goal state is necessarily pre-defined in that the designer, whether the young architectural student or the seasoned professional, knows in advance that they are to design a shopping centre, an opera house, or a private dwelling. Yet it is clear that such goal identification is almost always generic rather than detailed and specific, and irrespective of such information, the designer does not know, at the outset of the process, what the final design will be, i.e., precisely what it will look like, its ultimate formal disposition, and so on. To know such detail in advance would be to obviate the need for design; the final outcome would already have been designed before the process had even started. This is central to what is often referred to in discussions of creativity as the problem of foreknowledge.[4]

The problem is simply stated but difficult to resolve. If the designer knows the outcome at the start then the creation/design is already complete and no further designing or design process is required. If, however, they do not, then this raises a further critical question: 'How is the creative process controlled without some foreknowledge?'.

It is evident that these questions do influence how design outcomes come to be the way they are. We cannot readily accept the affirmative to the first question: as Abel (1976) notes, 'Does the artist *know in advance* what his work will be? If he did, he would already have it completely in hand!'. At the same time neither can a simple negative suffice: 'The artist is never sure . . . he can [never] predict how his work will turn out.' This is not to deny the accuracy of Abel's conclusions, rather to indicate that more sophisticated analysis is required to account for processes that undeniably occur, hence also the appropriateness of Hausman's splendid summary (Hausman, 1969, p. 2) of the same issue:

> The paradox of creation can also be seen in the activity of the creative artist. The artist begins a creative process without a preconceived plan or concept of the form he will produce. If he were to start with such a plan, then creativity already would be complete in his mind. But the creator does begin with a certain talent and set of established habits of work. At first he senses that certain elements are required in the future product, but he does not yet know what these are. And as he creates, he somehow discovers what he wants to create. He makes his plan at the same time that he comes to see what that plan is – at the same time that he sees what is required to complete the process he started. Paradoxically, then, the creator must at once create and discover. He must generate a necessitated product which is not

> necessitated by anything given to the creator with which he can generate it. In short, the creator must act as an agent which is a cause without a prior cause, a cause which causes itself.

Yet this does not go far enough. In architectural terms, the designer does not start only with 'a certain talent and set of established habits of work', although this is certainly the case, nor only with the generic instruction or intention to 'design a house', nor with only that highly determinate set of requirements, i.e., the specific functions and uses, budgetary constraints, site conditions, statutory and regulatory requirements, and so on which constitute the client brief and the problem as given, and all of which are expected to be known by the designer and which both affect and effect the design outcome. In addition, the designer brings to bear on the problem a viewpoint or a position – a set of formal and aesthetic and technical sensibilities based on prior experiences and preferences and prejudices which determine not only how the problem at hand will be solved, as if it is somehow neutrally presented for the most efficacious solution, but just what problem the designer will *choose* to solve. As Bryan Lawson (1979) notes: 'Clearly throughout the history of architecture designers have sought to explore ideas which transcend the particular problem in hand as well as attempting to meet the immediate brief.' Thus we return not only to the centrality of problem solving in the design process, but also to the simultaneous centrality of what might be neatly encompassed by the heading 'ideology'. As Lawson (1979) quite rightly observes: 'The role of ideology in design has been sadly neglected by methodologists but it would be absurd to maintain that the designer can or should entirely divorce his thinking on a specific problem from his general more philosophical beliefs.'

Deriving from its now archaic meaning of the study of the origin and nature of ideas, ideology may be used as a generic descriptor for any system of ideas concerning specified phenomena and thus it represents 'the manner of thinking characteristic of a class or an individual' (OED, 1972). Class is used here not in its social sense, as a division of society according to status, but in what might be described as its mathematical sense, to represent a set, a group of persons who may be so identified by virtue of their shared belief in, commitment to and utilization of a common body of ideas on a particular subject or issue.

3.7 Ideology and framing

In its usage here then, and in terms of design, ideology refers to that collection of ideas and theories, knowledges and assumptions, preferences and prejudices that are present to or deeply embedded within the mind and actions of the designer by virtue of past education, cultural milieu, location in space and time, personal disposition and professional self-evidence; those ideas and positions – ethical, social, aesthetic, formal and technological – that each designer brings to bear on her/his design, and which informs and structures their design response, often without their full cognizance.

Undeniably valuable and valued – it is a truism that architects and designers, cries for freedom and creativity notwithstanding, embrace ideologies in the manner of life-rafts in a turbulent sea of aesthetic indecision and theoretical pluralism – the decision-making potential of agreed-upon means or norms or positions is not to be overlooked. This is not to suggest cynicism: the moral imperatives of architectural theory from the nineteenth century green architecture, the formal dictates of aesthetic puritanism from modernism to minimalism, the *laissez-faire* but ironically conformist freedoms offered by postmodernism, or the supposedly deconstructive 'slash-in-the-plan' schemes claiming derivation from certain persuasions of contemporary philosophy, are, no doubt, genuinely believed by their adherents as right, proper, true and responsible for their time. Like all ideologies, however, systems of design ideas are normalized and naturalized by familiarity, expertise, pedigree and the dictates of an often unreflective 'common sense'. In the taken-for-granted, uncritically-self-evident world of the professional designer, established knowledge, the attractiveness of contemporary views and – let us not deny it – the seductiveness of the fashion system, of which we remain so familiar in terms of its appearance and yet so ignorant in terms of its workings and strategies, are highly influential in conditioning design thinking, making personal and group ideologies central to understanding how buildings get to be the way they are. This centrality of beliefs, not only in informing design decisions but in actively effecting and structuring them, is paralleled in Rowe's discussions of 'enabling prejudices' and 'normative positions that guide design thinking' (Rowe, 1987).[5]

It is this notion of adhering to or constraining within norms that sets ideology apart from that collection of skills and technical knowledge that the designer draws upon to 'solve the problem'. Thus, while the differences between two or more designs based on the same brief *may* reflect differences in skill levels between their designers, they are more likely to reflect the fact that the designers are not, in fact, solving the same problem at all. Instead, they may have problematized the brief quite differently, one from the other, each applying to her/his program – aesthetic, formal, moral, technological – different normative positions. They are working within different, and often conflicting, ideological frames, and their results are thus correspondingly different. Each is still a satisficing solution, yet each a solution to a different problem, one that is set not by the client alone and the determinate requirements of the brief, as convention might dictate, but set by the designer as a reflection of her/his creativity and her/his professional and personal beliefs.

This notion of problematization – differentiating the 'problem as design goal' from the more prosaic 'problem as given' – raises the awareness of control and choice. To what extent, we might ask, is the designer aware of the ideological frame within which he or she works? How cognizant is the designer of the extent to which her or his belief system and choices are always already conditioned by past education and experiences, by the influences of parents and friends, by the mores and customs of the society or sub-culture or profession to which they belong, and by the assumptions, preferences and prejudices that such groups engender? To this extent it may be said that we all, inescapably, work within a frame.

The term 'frame' has gained great currency over the last two decades in relation to contemporary philosophy and discourse studies. Its usage in the current context melds the intellectual, emotional and physical meanings attributed to the term by most dictionaries such that it supplements the intellectual identification of 'an established order, plan, scheme, [or] system' with both the human qualities of a 'mental or emotional disposition or state' and the more physical meaning of 'a skeleton structure or support'. Significantly the term also has the meanings of 'to give shape, expression, or direction to', 'to adapt, adjust, fit *to* or *into*', and thus 'to conform'. In this sense the frame is literally that which surrounds, borders or contains – that which sets boundaries. As such it must be realized that the effects of such framing range from positive direction and focus in our thinking on the one hand to restriction and blind conformity on the other. Of the latter, the key questions are how does one recognize that one is constrained within a frame, how does one see outside it, how does one reveal on what foundations the framing propositions – self-evident, sensible, natural and apparently unquestionable in themselves – are based and how they are grounded, and why should one do this at all? In this sense, then, frames act like spectacles: we see the world through their lenses, without being aware of them, and with the firm belief that we see the world better for their assistance – undoubtedly with greater clarity but otherwise unchanged. Were that this were so! As many scholars from a variety of disciplines have shown us over the last 40 years – Thomas Kuhn on paradigms, Michel Foucault on discourse and regimes on truth, or Stanley Fish on communities of interpretation, to cite but three – while spectacles undoubtedly allow us to see more clearly, the view through them is inescapably mediated and precisely *what* we see clearly and *why* and *how*, are questions that should not be overlooked.

3.8 Style, language and concept

Given the simple starting point of 'making a building' how do these rather esoteric notions of ideology and framing manifest themselves in the day-to-day world of the architect, the designer and the client? The answer is: ubiquitously, inevitably and with beguiling simplicity. Three common design terms might be introduced to explain this: style, language and concept.

The first of these, style, is widely understood. While it may be used with the implication of 'stylishness', it is more widely employed within architectural and design circles to indicate that range of characteristics, evident at a macro level and usually associated with formal or aesthetic properties, by which buildings, despite a host of individual differences, including differences in function and scale, may be recognized as similar to each other or, in Wittgenstein's sense, as sharing 'family resemblances' and thus 'belonging together'. In this sense architects may be said to work within a style – though this style may change from project to project. Putting aside the question of whether such large-scale formal and aesthetic systems as classicism and gothic can be so easily subsumed under the heading of style, or, much closer to home, whether such portmanteau terms as modernism and postmodernism are too broad and catholic to allow of such a designation, and represent cultures within which a host of sub-cultures flourish and compete, the importance of style, or rather, of an adherence to a style, is quite clear. Whether adopted consciously, as a deliberate decision of choice, or because the style is so transparent to its users that it appears not as a style at all but as the language of the everyday – the normal, the natural, the self-evidently 'only' mode in which to work – working within a style implies the imposition of a particular set of formal and aesthetic conditions upon one's problem solving and thus upon the act of design. In this sense, then, a style offers to the designer a set of quasi rules; a set of criteria and preferences and requirements on which decisions may be made; a focus and a pathway by which a generic problem may be framed as a design goal. It does not solve the problem but it does provide a frame within which it is to be solved, and thus a set of aesthetic and formal building blocks from which the solution can be constructed. Like ideology, then, it is both valuable and restrictive – or perhaps for some, valuable precisely because of its restrictiveness. As one of the many dictionary definitions indicates, style constitutes 'a method or custom of performing actions or functions, esp. one sanctioned by usage or law' (OED, 1972).

To this extent also the idea of a style is consistent with the notion of a language. A style may provide the words, the grammar, the syntax within which one composes; and just like any normal language, what one says and how one says it, is dependent to a large extent on who is doing the saying; on their skill and refinement, wit and imagination, as well as on what they have to say – if anything! It is perhaps in the area of meaning that a distinction may be drawn between language and style. If style may be defined as 'those features of literary composition which belong to form and expression rather than to the substance of the thought or matter expressed', then perhaps language pre-eminently should be concerned with substance and with the communication of that substance. This putative distinction notwithstanding, the choice of a language necessarily means choice of and adherence to a set of rules; the provision of direction, focus, reduction, and precedent at the cost of exclusion, restriction, containment and conformity. This is, of course, neither to suggest that the adherence to a particular language is a bad thing, nor that its opposite, choosing not to subscribe to any language in the vain hope that one can invent one's own is even a remote possibility. We are all bound by language, as much in our abilities to think as to speak and write. The issue here is only the realization that just as language is inescapable, so language choices have inescapable implications for design outcomes. If minimalism is our preferred style or language, chosen at the outset of the problem-solving process, then, regardless of the qualities of the design that eventually emerges and regardless of the brief on which it was based, certain outcomes are precluded and others predetermined.[6]

If we have an ethical commitment to sustainable design – and regardless of the desires of some to see this not as an issue of ideology but rather as one of self-evident responsibility – then again there are, from the outset, certain inclusions and certain exclusions, and certain strategies, preferences and prejudices that will be imposed on the supposed neutrality of the problem as given. Similarly, if your predilection is for high-tech, then certain themes, certain technologies and certain appearances will be evident: they will have come out of the problem precisely by being introduced into it.

The final term, concept, is similarly common within design discourse and design education, but is rarely made explicit. In its usage here it is taken to refer to that central structuring or organizing principle, usually determined – or 'discovered' – by the designer at an early stage of the project, and which may be said to drive the actual design. Akin to, but not identical to, the notion of a 'primary generator', a concept is that spark or idea around which the design coalesces, that proposition that literally forms its central core, and that leads the designer towards her/his design. A concept is thus that decision that allows other decisions, that choice and commitment against which other decisions and proposals can be checked (Darke, 1979). In this sense then, and like both language and style, the identification and active choice of the concept which will condition the growth, development and resolution of the design is a species of foreknowledge, a powerful but plastic *a priori* of the design process. Both informing and refusing, allowing and constraining, the concept assists the designer not only in the movement from initial conditions to final outcome, but in identifying and actively determining what the design goal state should be, i.e., not what is the problem as given, but what will be the problem as taken.

3.9 Conclusion

The journey from the inception of a building project to its final manifestation is thus a complex one, rich in influences and views. Many and diverse routes are possible, and if at the end of the journey a single final outcome emerges, then that outcome is both inescapably determined by the actual path chosen and yet alive with the myriad possibilities of roads not taken. As we have seen, the forms and appearances of buildings are determined by much more than simple function, by the supposed basics of shelter and climate and activities – what might be defined as *use value*. Ironically, buildings are the result of both choice and lack of choice, of freedom and confinement, of what can be done, what must be done and what one doesn't know one is doing! Buildings come to be the way they are as a result both of defining ideas conceived in the mind of the designer in response to a brief of opportunities and of the prescriptions and restrictions, prejudices and transparent self-evidences of style, taste, language and concept.

Buildings are, of course, representations of those individuals, institutions or states that cause them to be built, inescapable cultural and symbolic indicators – and hence have what might be called *sign value*. Whether demanded by or commissioned by the church, the king, the state, the multinational corporation, or the small-scale client struggling for their own slice of the good life, buildings from all eras – the Pyramids, the Parthenon, Chartres Cathedral, Versailles, the Eiffel Tower, Auschwitz, the Hong Kong Shanghai Bank, the World Trade Towers or the dream home in the suburbs – are the physical embodiment of ideologies. Even when we believe that architecture is driven by the creative energies of the architects themselves, of their freedom to choose, to conjure from the air and to surprise us – which they regularly do – dominant and time-specific ideologies prevail and influence: if the white planes and stripped aesthetic of the Modernists was really a natural and inevitable part of the world, a transparently universal truth to be recognized, then how could Bramante and Borromini have got it so wrong?

Endnotes

1 For more information on wicked or ill-structured problems, see Reitman (1964), Churchman (1967), Buchanan (1992) and Simon (1977).

2 For a general background to satisficing solutions and connected topics see Simon (1977) and Simon (1979).

3 As a sub-set of the latter we should also note what might best be termed the problem of expectations. While no less applicable to seasoned professionals, this is especially pertinent to design education where the qualities of the expected solution are neither simply defined nor determined by the brief, nor are they explicitly stated by the tutors, but which are implicit in the whole idea of 'being a designer'. Hence there is a tacit expectation – sometimes unappreciated by early year students – that it is not just 'meeting the brief' that is significant, but the manner in which that 'meeting' has been

couched. 'Designerly' solutions are often praised, while prosaic ones, while undoubtedly meeting the brief in a physical and pragmatic sense, are criticized as 'not demonstrating a designerly sense' or 'not showing any indication of an architect at work'! It is perhaps not unreasonable to suggest that the imperative that accompanies this view drives many, if not most, experienced architects.

4 Compare with Abel (1976), Hausman (1969) and Briskman (1980).
5 See also Alexander's use of the term 'program': 'Finding the right design program for a given problem is the first phase of the design process' in Alexander (1966).
6 See also Cerver (1997) and Meyer (2000).

References and bibliography

Abel, R. (1976) Creativity. In: *Man Is The Measure* (New York: The Free Press), pp. 266–7.

Alexander C. (1966) *Notes on the Synthesis of Form* (New Haven: Harvard University Press), p. 84.

Baker, G.P. and Hacker, P. (1980) *Wittgenstein: Meaning and Understanding* (Oxford: Basil Blackwell).

Briskman, L. (1980) Creative Product and Creative Process, *Inquiry*, **23** (1), 83–106.

Buchanan, R. (1992) Wicked problems in design thinking. *Design Issues*, **8** (2), Spring, 5–21.

Cerver, F.A. (1997) *The Architecture of Minimalism* (New York: Hearst Books).

Churchman, C.W. (1967) Wicked problems. *Management Science*, **14** (4), December, B141–2.

Darke, J. (1979) The primary generator and the design process. *Design Studies*, **1** (1), 36–44. Reprinted in Cross, N. (ed.) (1984) *Developments in Design Methodology* (Chichester: John Wiley), pp. 175–88.

Davis, C. (1988) *High Tech Architecture* (London: Thames & Hudson).

Fish, S. (1980) *Is There a Text in this Class?: The Authority if Interpretive Communities* (Cambridge, MA: Harvard University Press)

Foucalt, M. (1972) *The Archaeology of Knowledge* (New York: Pantheon Books).

Foucalt, M. (1980) Truth and power. In: Gordon C. (ed.) *Power/Knowledge* (New York: Pantheon Books), pp. 109–33.

Harfield, S. (1997) Off the top of their heads: the nature and processes of architectural design as perceived by new intake students. *Form/Work*, **1**, October, 34–49.

Hausman, C.R. (1969) Mystery, paradox, and the creative act. *Southern Journal of Philosophy*, **7** (3), 2.

Hillier, B., Musgrove, J. and O'Sullivan, P. (1972) Knowledge and design. In: Mitchell W.J. (ed.) *Environmental Design: Research and Practice*, Vol. 2 (Irvine: University of California).

Kuhn, T. (1962) *The Structure of Scientific Revolutions* (Chicago: University of Chicago Press).

Lawson, B. (1979) The act of designing. *Design Methods and Theories*, **13** (1) (Jan–April), 29–30.

Lomborg, B. (2001) *The Skeptical Environmentalist* (Cambridge: University Press).

MacLachlan, G. and Reid, I. (1994) *Framing and Interpretation* (Melbourne: Melbourne University Press).

Meyer, J. (2000) *Minimalism: Themes and Movements* (London: Phaidon).

OED (1972) *Shorter Oxford English Dictionary.* Third edition (Oxford: Clarendon Press).

Papanek, V. (1995) *The Green Imperative: Ecology and Ethics in Design and Architecture* (London: Thames & Hudson).

Reitman, W.R. (1964) Heuristic decision procedures, open constraints, and the structure of ill-defined problems. In: Shelly M.W. and Bryan G.L. (eds) *Human Judgements and Optimality.* (New York: Wiley), pp. 282–315.

Rowe, P. (1987) *Design Thinking* (Cambridge: MIT Press).

Simon, H.A. (1977) The structure of ill-structured problems. In: Simon, H.A. (ed.) *Models of Discovery* (Dordrecht, Holland: D. Reidel), pp. 304–25. First published 1973 in *Artificial Intelligence*, **4**, 181–200.

Simon, H.A. (1979) *Models of Thought* (New Haven: Yale University Press).

Vale, B. and Vale, R. (1991) *Green Architecture: Design for a Sustainable Future* (London: Thames & Hudson).

Wittgenstein, L. (1958a) *Philosophical Investigations*. Second edition (Oxford: Basil Blackwell).

Wittgenstein, L. (1958b) *The Blue and Brown Books* (Oxford: Basil Blackwell).

4

Intelligent buildings

Stuart Smith*

Editorial comment

The term 'intelligent building' has emerged only fairly recently and as yet has no universally accepted definition. To some, building intelligence is closely allied to adaptability, while another view is related to how the building responds to change. The two views are connected but the difference is a matter of emphasis – adaptability or responsiveness may be considered in terms of accommodating climatic variations, for example, automatically compensating for changes in outdoor temperature, or lighting controls that alter artificial lighting levels in response to variations in the level of daylight entering through the windows. Alternatively the emphasis may be on how the building can adapt to changes in layout, corporate structure or new work practices. In fact, as we move on into the twenty-first century, a truly intelligent building should be able to respond and adapt in all these ways and it will only be able to do so if it has been designed to from the outset.

There are a variety of factors that are driving intelligent building design and they include rapidly evolving communications technologies, constraints on energy use, radical changes in how people work and a growing awareness of the connection between the indoor environment, worker productivity and company profits. Consequently it is possible to design a building which is extremely flexible in terms of its spatial layout and is highly energy efficient yet lacks the communications infrastructure necessary to service the needs of a modern commercial organization and so could hardly be described as 'intelligent'; similarly a building with extensive cabling and network capacity that is permanently divided spatially according to an outdated corporate structure and so can not be readily adapted to suit the current structure would not be classed as 'intelligent'.

Today building intelligence is directly related to building value – whether building for sale or lease, or as an owner-occupier, a client will be hoping to have a building that will

* Sydney, Australia

have the attributes that will either make it attractive to prospective buyers or tenants, or provide an environment which will promote maximum profitability for their own business. The attributes of the intelligent building are precisely those that make the building a success from the client's point of view.

In this chapter the author looks at the general concept of building intelligence, how specific systems such as lighting, communications and heating, ventilation and air conditioning (HVAC) determine the level of building intelligence, and describes some tools that can be used to measure building intelligence.

4.1 Introduction

The concept of the intelligent building is not new. In the earliest context, buildings were 'intelligent' in that they met the cultural and environmental conditions of the time. These buildings were simple in form, flexible in function and responsive to needs as they changed. In the late twentieth century, certain buildings were described as 'intelligent', not because of any underlying ability to satisfy cultural or environmental needs, but as a marketing tool to promote buildings that had adopted the latest in computer technology and automation. These buildings were complex in form and provided very little flexibility.

Organizations are discovering that their ability to keep up with a changing market is limited not only by the speed at which they can incorporate innovation in their processes, but by the inability of the buildings they occupy to change with them. Intelligent organizations understand that the relationship between organizational success and the strategic use of space is uppermost in any strategic facility plan.

The intelligent building concept equates performance with market value and functionality through the integration of space, people and technology. The ability of a facility to enhance the performance of an organization through the improved performance of its workforce is a fundamental pre-condition of the design of an intelligent building.

Building designers, owners and occupiers are facing a new challenge: to meet the needs of the 'new' organization in the 'new economy'. Issues currently on the agenda are:

- the virtual organization is a reality and some organizations function in a 24 hour cycle. The intelligent building concept must integrate new organizational forms
- it is easier to source a new building than plan for sustainability or future organizational scenarios. The question in this case, when many organizations view buildings as consumables, is how to reconcile the difference between the lifespan of the organization and the lifespan of the building
- the value gained through efficiency (a focus on cost) does not equate to the value gained by focusing on effectiveness (a focus on function)
- equating the value in operation and value in exchange (sale) is a key attribute of intelligent buildings. The question is: can the notion that a client, or building owner, places greater value on the use of the building than in the exchange of the building be made a reality?
- intelligent buildings enhance the relationship of building systems technology to organizational technology, process and productivity. If the cost of intelligent buildings is comparable with other premium grade office space, what other incentives are needed for organizations to adopt the intelligent building concept?

4.2 The demand for intelligent buildings

There are a number of factors behind the increasing demand for buildings that may be described as 'intelligent' and they reflect not only significant advances in technology but changing attitudes to work and the workplace.

4.2.1 What do people want?

Demand for intelligent buildings and intelligent office space has been driven by clients looking for energy savings, a need for increased worker productivity, and the expectation of a healthier work environment. Tenants want greater utility of floor space. They are factoring in the impact (cost) of restructuring space as the organization restructures to either match market demands or match new organizational forms that aid in establishing a competitive advantage.

The results of post occupancy research by Ahuja (1999) indicate that people want a greater influence on the environmental controls and spatial layout of the workplace. Duffy (1998) indicates that this does not necessarily mean more complexity and greater control but rather that the processes that govern the effectiveness of the workplace be more intuitive. For the intelligent building, this has meant greater distribution of systems and hence a greater need for integration and intelligence, creating or demanding systems that are capable of managing the complexity and range of attributes effectively, without impacting on the functionality of the building or reducing the building's capabilities.

4.2.2 What is an intelligent building?

Can an intelligent building be defined? It is not necessarily a hi-tech, multi-storey building; it may be a 'motivational building' or a 'quality built environment', or perhaps it is better described as a 'sustainable building'. For the purpose of this discussion the term 'intelligent building' will be used and some possible definitions given. Hartkopf (1995) states that, in a general sense, an intelligent building could be described as:

> A building which responds to the requirements of its occupants.

The implications of this statement are many and varied. Certainly a variety of views and explanations are possible, each correct in its own right and all incorporating aspects of other views.

The Intelligent Building Institute (IBI) adopted the following definition (GK Communications, 2001):

> An intelligent building is one that provides a productive and cost-effective environment through the optimization of its four basic elements: systems, structures, services and management and the interrelationship between them. The only characteristic that all intelligent buildings have in common is a structure designed to accommodate change in a convenient, cost effective manner.

By contrast, the European Intelligent Building Group (EIBG, 2001) states:

> An intelligent building creates an environment that allows organizations to achieve their business objectives and maximizes the effectiveness of its occupants while at the same time allowing efficient management of resources with minimum life-time costs.

Both definitions point toward similar conclusions but their approach (the effective means to create an environmentally sensitive and productive centre) is different.

Stubbings (1988) limits his definition to 'a building which totally controls its own environment'. The implication here is that there is an overriding technical mastery of the building.

Lush (1987) takes another perspective focusing on a multi-dimensional view of spatial efficiency, allowing for the inclusion of extrinsic aspects of the building (exterior spaces), and melding them with its intrinsic fabric, functionality, layout and its degree of responsiveness:

> An intelligent building would include a situation where the properties of the fabric vary according to the internal and external climates to provide the most efficient and user friendly operation in both energy and aesthetic terms.

4.2.3 Intelligent architecture

Kroner (1995) describes intelligent architecture as architecture that is responsive. That is, the architectural components of the building can be replaced and/or modified as the building's use changes.

Buildings have become products of well-established practices and principles. Most do not challenge the status quo, being designed for the ebb and flow of marketability. Investors, developers, architects, engineers, occupiers and the community all influence the supply and demand to varying degrees, influencing the design and specification of the building. Each group has a competing interest that depends on the form of value it seeks to extract from the development process. The identifiable aspects of these buildings are not limited to external appearance or internal fit-out but also include building environments, both internal and external.

Buildings cost money but only function can add value. Minimizing design costs and time has created an environment that is not conducive to adding value. The design of intelligent buildings has value enhancement at its core. To achieve added value, integrated design that enhances functional and physical effectiveness is needed.

4.2.4 Intelligent technology

The use of intelligent architecture in isolation cannot provide the interaction and integration that allows the building to respond to the requirements of its occupants and meet performance objectives. Intelligent technology is needed to complement the intelligent architecture already present. It is through the technology, now forming part of the building structure, that the building endowed with an actual architectural functionality is able to exploit its capabilities.

Traditionally, the engineer views the building from the perspective of structure and performance, while the architect is more concerned with function and aesthetics. Both professions are fluent in the language of building technology specific to their field, but tend to see themselves as co-existent rather than complementary. In essence, they often do not understand the linkages that exist between the parts they play in the process of building.

Co-existence creates *a cycle of innovative filtering*, whereby ideas are progressively discarded at each stage of the building's development (Smith, 1999a). Innovative filtering tends to result in the construction of a sealed, inert structure, shrouded in an inert façade. Inside this dormant shell is a mechanical juggernaut automating and sterilizing the space with an eye on the delivery of a cheaper, energy efficient environment that may be at odds with the health of its occupants. In this case there is no scope to add value to either the organization's business activities or the building's functionality.

Loftness (1995) defines intelligent technology as technology that encompasses the use and integration of building systems, architectural structures, office automation, and information technology, 'plug 'n' play' furniture systems, management practices and operational processes. Intelligent technology is founded on sophisticated communications systems enabling more adaptive technology or technical functionality. The focus of technology is now knowledge rather than control of the environment and the emphasis is on user interaction and integration (Smith, 1999b). In this way the intelligent technology

Table 4.1 Performance criteria for intelligent buildings

Performance criteria	Technology
Spatial quality	• a dynamic mix of space types and integrated circulation • switchable (clear to opaque) glass panelling, ergonomic and adaptive furniture • column free construction and shallow floor plates • movable partitioning co-ordinatedwith zoning grid modules
Visual quality	• ambient uplighting and adjustable perimeter lighting • natural light, interior shading and exterior shading devices • dynamic shading/light diffusing • fully adjustable desk-based task lighting
Building integrity	• personal environmental monitoring – occupant control • superwiring – integrated relocatable power and IT cabling • floor-based relocatable infrastructures • universal cabling – to integrate voice and data • micro zoning
Air quality	• fresh air architecture • constant volume ventilation, displacement ventilation
Acoustical quality	• acoustic ceiling systems • white noise generators and sound masking
Building management	• integrated building management systems (IBMS) • facility management information systems (FMIS) • computer aided facilities management (CAFM)
Thermal quality	• displacement ventilation systems, mixed mode HVAC • chilled ceilings, in floor HVAC • passive heating and cooling, radiant cooling • floor diffusers and radiant façades

considers the business and social context and the technical performance required in the building. Performance of the building is a critical value-adding measure. Physical performance, functional performance and financial performance are criteria that need to be understood to drive building success.

4.2.5 Total building performance

Hartkopf (1995) asserts that total building performance is composed of six building performance factors: spatial quality, thermal quality, air quality, acoustic quality, visual quality and building integrity. Integrated building systems and energy effectiveness programs underscore this. Table 4.1 outlines the performance criteria for an intelligent building.

4.3 Intelligent building systems

Developments in intelligent buildings have not been limited to advances in technology in the areas of computers, communications and building engineering. Changes in societal attitudes that reflect a higher standard of living have highlighted issues associated with the provision of a healthy working environment. This is being reflected in an increasing demand for high quality office space, spread across all classes of buildings, and a need for advanced information processing and communications systems. Hartkopf (1995) outlines the technologies and services that are part of the intelligent building infrastructure. Some of these are shown in Table 4.2.

Table 4.2 An outline of the technologies and services that form part of the intelligent building infrastructure

Infrastructure	Services
Enclosure	• load balancing • solar control • heat loss control • daylighting • passive and active ventilation • passive and active solar heating
Interior	• spatial quality • thermal and air quality, visual quality • acoustic quality in the individual workstation • new workgroup concepts • shared services and amenities
Telecommunications	• external connectivity and command centres • vertical chases and satellite closets/rooms • horizontal networks and horizontal plenums • service hubs and shared equipment • conference hubs, connectivity
Site	• transport • streetscape, public access and thoroughfares • relationship with the community

Developers of building systems of all types have been challenged to deliver on these demands. The successes to date have created a new opportunity for adding value to the capabilities already inherent in the building. Intelligence has become 'distributed' enabling micro zones to exist independently of the rest of the building. For organizations, this is an important consideration. Organizations modify space to suit business needs. People, by their nature, tend to modify their work environment to suit personal tastes and corporate identities. The ability of building systems to interact using distributed intelligence enables a successful outcome to be achieved.

Site specification is important in terms of the delivery of services. It is not enough to assume that technology and site specification represent separate criteria and can be dealt with individually; rather, it should be assumed that the site and the technology are intimately linked with the allocation of services. Lehto and Karjalainen (1997) focus on the user-orientated approach to site and technology in order to deliver energy management through extensive analysis of the brief, addressing all aspects of the building's function, to ensure the building can be effectively and efficiently structured to provide optimal functionality.

4.3.1 Capabilities of the intelligent building

The overriding function of the intelligent building system is to support the capabilities inherent in it (Figure 4.1). Clearly it is necessary to consider an intelligent building as a single entity unifying objectives of the owner in delivering the building's desired capabilities with the adaptability and functionality desired by the occupants.

As with the systems present in the intelligent building it is possible to list the capabilities. They include:

- sensing human presence and/or occupancy characteristics in any part of the building and controlling the lighting and HVAC systems based on appropriate pre-programmed responses

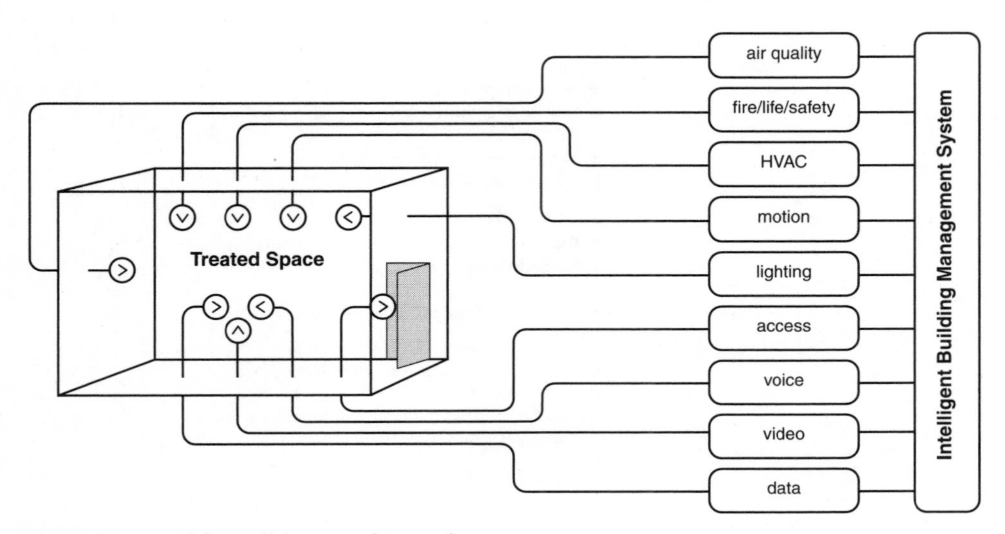

Figure 4.1 Typical functional zone and its attributes.

- performing self-diagnostics on all building system components
- alerting security and fire alarm systems and monitoring the location of occupants in case of emergencies
- sensing the intensity and angle of light and solar radiation, temperature and humidity, and adjusting the building's envelope according to the desired interior performance levels
- monitoring electrical outlets for malfunctioning equipment
- monitoring access to the building and individual building spaces
- detecting odours and pollutants and responding by increasing ventilation rates
- distributing electric power to computers on demand or in accordance with a preset priority schedule and automatically activating reserve batteries or back-up systems
- selecting the least cost carrier for long distance telephone calls
- activating ice-making or heat storage systems when the utility signals that discounted rates are in effect
- providing better acoustic privacy by activating white noise systems to mask background noise.

4.3.2 Heating, ventilation and air conditioning (HVAC)

HVAC systems directly influence productivity through the health of the occupants and so are a major factor in the operation of both the building and business. Research on intelligent control strategies by Zaheer-Uddin (1994) has shown improved performance can be achieved by reducing the HVAC zone size. This has been achieved both by the introduction of improved direct digital control (DDC) technology, distributed processing and more adaptive spatial planning in general. Systems have been developed that deliver HVAC using both the traditional approach, where the ducting is installed in the plenum space and air is directed downward, and in floor-ducted systems that supply air in an upward direction. Personal environmental control systems that integrate technologies to deliver cooling/heating to the individual are also available and are impacting greatly on the development of future HVAC systems.

The distribution of control zones and hence the number of zones that can exist on each floor is aligned with the type of HVAC system used (Table 4.3). This may pose difficulties if a broad distribution of functional zones is created in the process of providing the desired work places, thus limiting the opportunity to align building system control zones with

Table 4.3 Attributes and types of HVAC systems used in intelligent buildings

HVAC attributes
• Horizontal distribution: air, water, none
• Horizontal distribution: ceiling, floor, furniture supply/return
• Environmental load management and load balancing
• Split ambient and task conditioning: individual controls
• All air systems and mixed mode systems
• High performance systems
• Air quality management and energy management
• Chilled ceilings
• Geothermal ground water systems

building use zones. Furthermore, Bell *et al.* (1991) shows that a disparity between building system control zone boundaries, functional boundaries and building use zones may produce areas that can never be balanced effectively.

4.3.3 Lighting

The primary objective of lighting is to provide acceptable levels of illumination for all aspects of occupation. Lighting should reflect the different needs of occupants and the ambient level of natural light to ensure that optimal time-based illumination levels are achieved. Aligned with this is an aim to reduce energy consumption, without compromising energy effectiveness, and to provide flexibility in operations for people in the building. Lighting control, or more specifically, the effective utilization of daylighting, is a critical factor in any energy strategy.

Too often the idea of 'acceptable levels of illumination' has meant monotonous banks of fluorescent lights strategically located in a regular matrix throughout the entire floor. This lighting scheme fails to consider the individual needs of people and is destined to contribute to an unhealthy working environment. Haessig (1993) has identified a number of important attributes to consider when integrating lighting systems (Table 4.4).

Table 4.4 Attributes and types of lighting systems used in intelligent buildings

Lighting attribute
• Fibre bundles • High frequency fluorescent lights • Internal and external shading devices • Integrated control systems • Occupancy sensing • Daylight harvesting • Light tubes and light shelves • Atriums and daylight bridging • Variable refractive index glass • Fenestration

Just as the threshold of cold and hot differs with individuals, so does the perception of changes in illumination. It has been shown that creating a visually dynamic space is no less energy efficient and, more importantly, encourages lower absenteeism and greater productivity. Hence it is valuable to talk about 'the quality of light', where quality encompasses intensity, interaction with natural light and the use of a dynamic visual plane, one that encourages variation in the intensity of the light across the floor. Opposing this is the reality of lighting in most office environments. The first list focuses on the requirements. The second list focuses on reality.

The requirements:

- minimum guaranteed light levels at every workstation dependent on task
- no glare
- no concentrations of light

- user control (different levels) at every workstation
- use of daylighting wherever possible
- end users have the facility to control the microenvironment; central control can change the light levels at the macro level
- better use of vertical lighting
- better use of direct/indirect lighting that is co-ordinated with the building's structure
- 400 lux across the floor, 800 lux (via a task light) to work on paper and 200 lux in corridors.

The reality:

- unacceptable direct source glare on computer screens
- no facility to vary the light levels at an individual workstation
- no minimum guaranteed light levels at each and every workstation
- light distribution is dramatically variable, with 'peaks and troughs', due to partitioning, columns, walls, despite uniform ceiling grid layout
- light distribution is not evenly diffused
- widespread reflections on computer screens
- offices are cave-like with gloomy walls and ceilings.

4.3.4 Access, egress and security (AES)

Francis (1992) states that the security strategy used is driven, and in some cases imposed by, the physical restrictions of the building (i.e., site specification and function). Because of this spread in the focus of responsibility, the functional variation of each space ensures that no one AES system is appropriate over the entire building.

Office spaces are identical in many respects. As with HVAC and lighting, however, the dynamics of a space are very different once it is occupied. Each tenant may require a different security regime to meet particular aspects of the business, whether that is security of equipment, data and information, and/or personnel. This is further complicated by the inclusion of several floors or part floors of the building that may or may not be adjacent in an individual business domain. Such a regime adds additional complexity when integrating AES with HVAC and lighting occupation zones.

To satisfy the security aspirations of the occupants without restricting the movement of people throughout the building, various systems and devices are used that are controlled by both the tenant and the facility manager. These systems, particular to their role in the building, establish the complexity of any subsequent integration of AES systems and the transparency of AES systems to the building's occupants. Those functions that are building driven include:

- intelligent lift control
- automated alarm points
- emergency warning information systems (EWIS)
- CCTV and video surveillance of building forecourts and public spaces.

The desired level of security is dependent on the type of security required and how the devices used to achieve this interact with other building systems. Examples of the devices and techniques used are:

- smart cards, either swipe or proximity
- video surveillance
- occupancy sensors
- keypad-coded doors and door grouping
- user authorization
- remote monitoring
- biometric devices (voice, fingerprint, iris)
- phone/security access
- secure lifts.

Unfortunately, factors such as tenant risk and lack of integration of AES systems across the domain of the building can reduce the effectiveness of the installed AES system components.

4.4 The intelligent building communications network

A sophisticated communications system is fundamental to the success of the intelligent building. In many instances this is the most important aspect of such a building and without the link over which information is gathered and disseminated the building would fail to fulfil its capabilities.

In the new organizational scenario of a flatter organizational structure, information moves laterally, not vertically. Communications infrastructure must be responsive to this. The base building design must be able to absorb the expected increase in technology due to more flexible work scenarios; high performance workplaces and practices will fail unless high performance communications are included.

Some organizations are geographically dispersed, servicing geographically separate markets. These organizations operate from multiple sites but they are virtually integrated through the use of communications technology. The buildings they occupy will be 'interchangeable' in matching operational priorities while still being able to 'reinvent' themselves for new organizational strategies.

The implication for buildings is that they will need to respond at the same rate if not faster. This responsiveness extends to systems integration: the lights will know when the meeting is imminent and the HVAC system will have already conditioned the space.

4.4.1 The application of business resources and telecommunications

To ensure that the objectives of organizations and those of the intelligent building communications network are met, certain characteristics of the communications infrastructure are necessary. Kelly (1990) defines these as:

- universal connectivity: all applications are supported by a limited set of protocols and input adapters with the quality of service (QOS) assured
- network intelligence: for the system to function effectively the network must be capable of supporting a wide variety of communications services without major modification to circuits or switches

- flexible topology: the network must be adaptable to allow reconfiguration if the deletion and addition of services is to be encouraged
- transparent media: the network must accommodate advances in technology and be able to be constructed on a variety of media without loss of generality.

The blurring of boundaries between business resources and telecommunications has led to a steady increase in the complexity of local area network (LAN) services. These services are not just a means of providing a connection between one point and another, but are becoming important elements of the business resource. With this in mind, it is important to specify telecommunications requirements prior to occupation. Questions that can be asked include:

- what types of services are needed?
- what is the functional layout of the building?
- what are the strategic communications needs of the business?
- what are the problems the implemented system intends to address?
- how has the structure of work/the office changed and where is it heading?
- what effect will this have on communications planning decisions?
- what opportunities exist for implementing a ubiquitous communications network?

It should be noted that the difficulty for organizations in determining their communications needs is not understanding the technical specifications but identifying how communications can create competitive advantage and how the implemented communications infrastructure can be made to adapt to change.

4.4.2 Trends and opportunities in communications technology

Communications technology is being driven by the development of digital communication services that provide multimedia (voice, video and data) applications using Internet protocols. A primary feature of this development is the migration from fixed bandwidth allocations to 'bandwidth on demand'. This will produce a fundamental change in the application and management of telecommunications.

The convergence of communications services through Internet technologies has changed the way that applications are implemented and presented. Convergence and interoperability have usurped the concept of data, voice and video as separate modes of information. This is presenting opportunities for the delivery of communications services in new and innovative ways. Convergence of technology and communications systems is a key driver in the value adding associated with intelligent buildings.

Some of the technologies on offer enabling intelligent building functionality are:

- shared tenant networks and serviced application management that use revenue-sharing mechanisms to offset the cost of providing a broadband universal network – this includes building management systems and controls
- wireless PABX systems that allow unrestricted roaming throughout the building enabling connectivity with single-user, single-zone applications – this provides a zero cost scenario inside the building and a seamless connection to a communications provider at the building boundary
- interoperability between business resources and environmental control systems that allows occupants to modify the operation of their space at will.

Key trends and technologies that need to be considered include:

- bandwidth
- the Internet, intranets and extranets
- virtual private networks
- asynchronous transfer mode (ATM) access – (multi/single site) videoconferencing
- asynchronous digital subscriber line (ADSL) and other digital telephony systems
- e-commerce and supply chain management
- LAN-based telephony – computer telephony integration
- call centre technology
- power line communications
- wireless local area networks
- wireless application protocol (WAP) for web-based mobile multimedia transmission
- location-independent working and virtual networking
- infrared and microwave data networks
- shared tenant (network) services
- cross-platform integration.

In addition, there are other factors affecting intelligent building communications technology. Some of these are:

- space use intensification – which encourages and supports flexible working/learning
- the growing use of high-end workstations and the resultant demand for bandwidth
- a reduction in the use of personal storage space and paper-based document storage
- mobile computing and communications
- intelligent building control systems integrated with business systems
- the development of broadband universal digital networks.

4.4.3 Communications technology and network infrastructure

The physical network infrastructure (cables and connectors) in buildings provides the connectivity between system devices and individual systems. It is primarily based on copper wires (unshielded twisted pair [UTP]) at the desktop and optical fibre in the backbone. The installation of fibre optic cables has added increased throughput in the support of the communications networks due to its increased bandwidth. On another level, the widespread implementation of wireless networking, coupled with voice networks using Internet protocol technology (voice over IP; VoIP) will fundamentally change how telecommunications operates in the intelligent building.

For this reason it is now considered impractical to rely on the implementation of a single UTP network philosophy. However, given that lead-time before fibre to the desk is a common occurrence, copper will maintain its position as the dominant medium for network connections of less than 100 m.

Studies by McKinley, as early as 1993, indicated that the overall level of traffic generated in commercial office towers is in excess of 1.3 Gbit/s and hence will be beyond the capabilities of copper as a suitable medium by early in the new century (McKinley, 1993). However, the level of traffic generated on a particular floor will not grow substantially beyond the current UTP limit of 100 Mbit/s in the same time frame. Anecdotal evidence suggests this view has not changed.

Even so, facilities managers are, in instances where demand may warrant it, specifying optical fibre connections to the desk along with UTP to reduce any future impact on the cost of refitting the cabling infrastructure.

4.4.4 Fibre cabling networks

The use of optical fibre cables for in-building networks has increased dramatically in the last 5 years due to the added capacity of fibre optic cables, improved methods of termination and the reduction in cost of amplifiers and transceivers.

Advantages of fibre include:

- security: the nature of optical fibre prevents tapping of the signal without causing serious degradation of the signal/noise ratio and hence easy detection
- very low electromagnetic interference (EMI): fibre optic cables are not affected by external electromagnetic fields, eliminating crosstalk
- insensitivity to environment: the fibre is not susceptible to chemical, radiation or thermal stresses that are inherent in copper conductors
- very high bandwidth capacity.

While providing a means of accommodating the backbone requirements in new buildings, optical fibre possesses disadvantages that limit its application. For example:

- insufficient riser space
- limited bend radius restricts the routing of fibre if the riser space is not designed as a single vertical shaft and the horizontal raceways are tightly configured
- large protective cladding may limit the number of fibres in the riser space or may restrict the movement of the fibre between floors through pre-configured pathways
- poor quality terminations can produce large attenuations.

4.4.5 Wireless networks

The use of wireless networks to augment, if not replace, existing cabling networks will revolutionize the delivery of information throughout the building. It will reduce the complexity of cabling systems and mean that communications between systems is not hampered by the physical location of any device.

Intra-office data networks based on 10 Mbit/s wireless networks are already widely available, however, there are a number of limitations to the growth of wireless in-building networks. The transceivers are still relatively expensive and cannot match either the bandwidth or the market dominance of UTP, and data rates of only 1Mbyte/s to 4Mbit/s can be reliably guaranteed.

Two new standards have recently been recognized by the International Telecommunications Union (ITU). When coupled with digitally enhanced cordless technology (DECT) full wireless connectivity will be a reality. Presuming that these developments flow through to the consumer, in the long term, and there is an uptake of WAPs, wireless networks are expected to become the dominant media for low to medium bit rates (100 Mbit/s) in many applications. It is expected that higher speed wireless protocols such as 'Bluetooth' for device to device communications (up to 2 m) will enable a complete vertical and horizontal 'in-building wireless local loop' to be created.

4.4.6 Distribution of cabling networks in intelligent buildings

In most buildings, the core of the building houses the riser space along with other building systems and services, with all horizontal distribution fanning out from the backbone. This is beginning to change; while the path taken to provide a point-to-point connection to the end-user is arbitrary, the suitability of a core-based regime is becoming increasingly problematic.

In addition, an older building may be burdened with a large and cumbersome cabling infrastructure that is inadequate for current applications. For facilities managers, the tasks of administration and management, refurbishment and development of network services cabling infrastructure have become a major concern in terms of the economic considerations and communications engineering.

It is envisaged that raised (access) floor systems will become a standard requirement in the construction of intelligent buildings as the need for adaptable telecommunications infrastructure increases. They offer greater opportunity to add value by providing the flexibility for relocation of services across occupancy and functional zones, with minimal interference to the function of the space and thus provide a long-term cost advantage. There are, however, still considerable obstacles, mostly financial in nature, that must be removed before systems flexibility is supported by access floors.

Figure 4.2 shows a cut through of the floor of a typical intelligent building. It reveals under-floor HVAC, raised floor access and structured cabling. The overhead lighting is

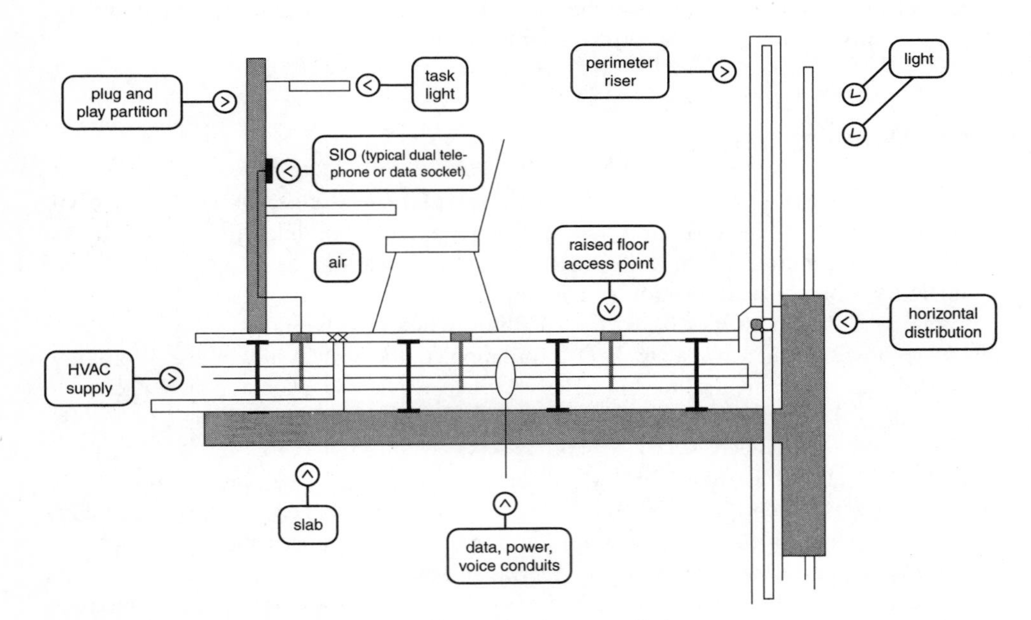

Figure 4.2 A typical intelligent building cabling and services installation.

supplemented by task lighting, good use of natural light and minimal thermal loading. Note that the perimeter is used as the riser space in this particular installation. The workstation system uses 'plug 'n' play' components that allow for easy reconfiguration and hence lower churn cost.

In this distribution layout, each zone is served by a single distribution path and is terminated at the zone boundary. The boundary may be determined by geographical factors, tenant boundaries, intra-tenant occupancy zones, intelligent building zone boundaries or architectural boundaries. The choice of the most efficient and cost-effective distribution regime is determined by the particular use for that zone of the building.

An even more appropriate form of cable management uses the functional zone boundaries to determine the distribution of the cabling infrastructure. This is appropriate in a building where flexible space architecture has been established. It will not be appropriate in buildings that have fixed occupancy zones, where the functional zones cannot change. In these cases saturated cabling is necessary. Combining the features of block zone cabling and saturated cabling has led to the development of structured wiring, or structured cabling, systems that are an integral part of the building infrastructure.

Structured cabling systems provide a complete cabling infrastructure that contains all necessary cable and connectors. These cables and connectors comply with international standards that allow different types of equipment, connected with broad bandwidth requirements, to be installed as part of a multi-product, multi-vendor environment. Steloner (1998) suggests that further benefits of structured cabling systems are obtained if they are used in conjunction with raised access floors and wireless networks. This allows for a fully accessible cabling infrastructure to be installed and provides continual flexibility in the location of desks, partitions and service outlets. The flexibility is made possible by establishing an access grid based on occupancy zones that in most cases provides the least disruption to the function of the workplace, dramatically reducing churn costs.

4.4.7 Intelligent building services systems (IBSS) communications networks

The use of LAN technology to support building services has become an integral part of the operation of intelligent building communications networks. The impetus for this is the growing number of sensor/actuator control nodes necessary to support energy management, HVAC, spatial comfort, lighting, security and the transfer of building diagnostic information.

A consequence of the increased interaction between building systems in the intelligent building communications network is that a simple approach to network implementation is not possible. It was not so long ago when hard wiring was all that was available. Even though developments in communications technology now support many of the functions of the intelligent building, the level of integration of disparate building systems communications networks is limited and as a result provides limited value to the occupant. This is due to a number of factors:

- reliance on low-speed communications systems
- fragmentation of the building communications industry
- use of proprietary distributed control systems and protocols
- minimal use of shared network services and facilities

- limited migration to digital telecommunications services
- failure to address the issues of 'throughput', instead focusing on the use of larger computers and more complex control procedures
- difficulty in defining the structure of a universally acceptable intelligent building communications protocol structure.

If a fully integrated ubiquitous network – one which is suitable for a wide variety of bandwidths – is to be developed, then other types of communicative mechanisms that utilize standard physical layer topologies and technologies are needed.

Given that the fundamental needs of the building systems communications network remain in the low bandwidth category (under 1 Mbit/s), the use of higher bit rate networks is not yet warranted. This is expected to change with the inclusion of voice and video services in a building (one application is security and public address). The application of higher bit rate transmission structures and switched network services will place IBSS communications system on the same level as other telecommunications networks and position the intelligent building network into a more generic telecommunications environment. This will ultimately ensure interoperability of intelligent building communication systems.

4.4.8 An integrated communications platform

The integration of communications networks will provide a mechanism to connect distributed controllers from a wide variety of building systems. To ensure interoperability and connectivity across the network, data from individual building systems is converted to a transmission structure that operates over the physical link. The physical link must be transparent to the functional aspects of the data or the media from which it is derived. This enables the network to access information across media boundaries, different LAN segments and nodes transparently and without loss of generality. The end result of this action is that it gives the network the ability to function independently of the operating environment, while possessing a high level of network intelligence.

In order to integrate additional services under current regimes, functional applicability needs to be re-appraised and the hierarchical model replaced by one that is 'broader', or at least by one which operates in a LAN environment where devices can be connected and disconnected at will. This also applies to site specification, where integration may be constrained by previous installations. Any new installation will be technically hampered by the necessity to either provide a high level interface or install a system that is compatible with other systems present in the building.

An important outcome of an integrated communications network is the reduction in the number of building systems communications networks and the ability to support a wide array of bandwidth requirements.

4.5 Measuring the 'intelligence' of an intelligent building

Measuring performance systems and processes is an important management function. It enables comparisons to be made between leading edge indicators and the 'quality' of work that is being performed. As with other measuring techniques, what you measure and how you measure it affects the final outcome.

All buildings contain common elements such as lighting, HVAC and communications. How these elements are packaged is unique for each building. In intelligent buildings it is expected that there is a high degree of interaction. This ensures that the benchmarking outcomes, and the inherent value of the building, are above an average building norm.

Building intelligence can be 'measured' in a number of different ways. The methods mentioned focus on specific properties of the building; as such they do not present an overall index or singular result. Considering the generic definition proposed earlier, 'a building that responds to the requirements of its occupants', it is appropriate for intelligent buildings that there is no single index. However, there is always the potential to integrate outcomes to produce a holistic numeric figure. The question is, does it provide any further value?

4.5.1 The Intelligent Building Score

Arkin and Pacuik (1995) have developed the Intelligent Building Score as a mechanism that allows a building's performance to be quantified in terms of the building systems installed and the level of integration that exists between them. This score provides a readily understandable comparison of buildings for the purpose of evaluating a building's level of 'intelligence'.

The score is based on a rating scale for systems integration with the lowest rating reserved for buildings with no systems integration (level 5), and the highest rating (level 1) reserved for the comprehensive integration of building systems across the entire building information spectrum.

The score, defined as the Magnitude of Systems Integration (MSI) is a simple index that averages the total rating given to a building against the number of systems in the building. To determine the level of systems integration the systems in the building are evaluated against the criteria listed in Table 4.5.

Table 4.5 Magnitude of systems integration in intelligent buildings

Level	Type of integration	Attributes and description	Rating
5	No integration		1
4	One directional	no feedback; safety codes	2
3	Interface discrete systems (one-to-one)	advanced smoke control; PABX; LAN	3
		other modem;	4
		sophisticated controls; integration with smart structural elements	5
2	Building automation Emergency management Office automation communications management	office automation (OA); shared tenant services (STS)	6
		building automation system; energy management; etc.	7
		as above + smart billing; integrates smart structural elements; personal environment control	8
1	Intelligent building management system	partial intelligent building system (IBMS)	9
		full IBMS includes maintenance and billing and smart structural elements	10

The index provides a score that is applicable for each building type only. A more appropriate score is one that has been normalized for the building type in question. In terms of a multi-storey building the normalized index, or MSI-reference (MSIR) is an index that considers the raw score against a 'reference' score for a multi-storey building with a high degree of integration.

4.5.2 Building Intelligence Assessment

The Building Intelligence Assessment index covers a number of broad characteristics of the building. Each of these characteristics is made up of components that focus on particular aspects of the building. Using this systematic approach it would be expected that a truly intelligent building would score well in each characteristic. These are:

- site specification >50: close to transport, amenities and where possible the workforce
- ergonomics >50: workspace size
- operational cost <30: churn, energy effectiveness, reuse of basic materials, water, heat and air
- occupational health and safety >50: low incidence of sick building syndrome (SBS), quality work spaces and risk assessment procedures in place
- identity >50: leading edge organizations and high degree of marketability
- access and security >50: high degree of seamless personal and organizational security.

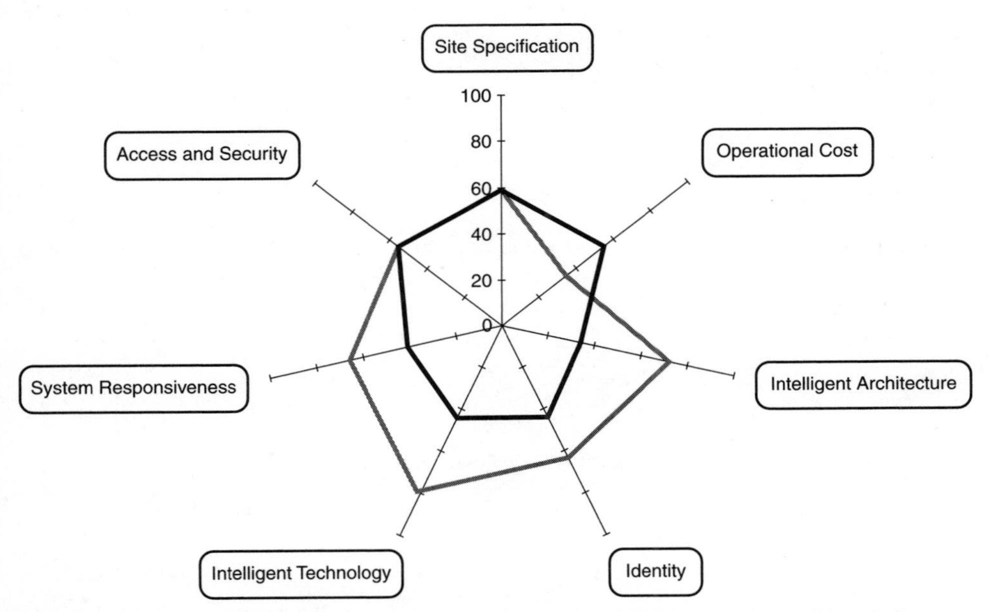

Figure 4.3 A typical building intelligent assessment method (BIAM) index for an intelligent building and a speculative building.

In an ideal intelligent building all of the characteristics are optimized. It is expected that as characteristics that produce positive results improve then the operational costs remain static.

The visual index this produces is shown in Figure 4.3. The inner line is for a speculative building. Note that for nearly all of the measures the speculative building score is lower. Not unexpectedly, the score that is higher is operational cost.

The Intelligent Building Score and the Intelligent Building Assessment method consider only the tangible aspects of buildings. In many instances, however, it is the intangible attributes that contribute more to the overall success of the building.

4.5.3 Reframing

Reframing (Smith, 1999c) is a technique that evaluates the enabling ability of the building to meet organizational objectives through the alignment of 'the way we do things around here' and the workplace environment that supports work practices. Many of the attributes may be intangible or are not readily measurable using simple financial tools. As an example, one methodology for measuring communications in the workplace is the Quinn's Competing Values Questionnaire (Quinn *et al.*, 1996). Other methods include before and after staff surveys to gauge the response to the new/changed workplace.

The process begins with a statement of organizational objectives such as 'to improve communications between functional groups' or 'to capture ideas and knowledge across the organization'. The statement is then analysed from four different frames: organizational structure, politics, human resources and culture, using focus groups to extract information. The information is then considered in terms of the capabilities of the building to support the objectives. Like the building assessment method a radar diagram representation is used to support a written evaluation (Figure 4.4).

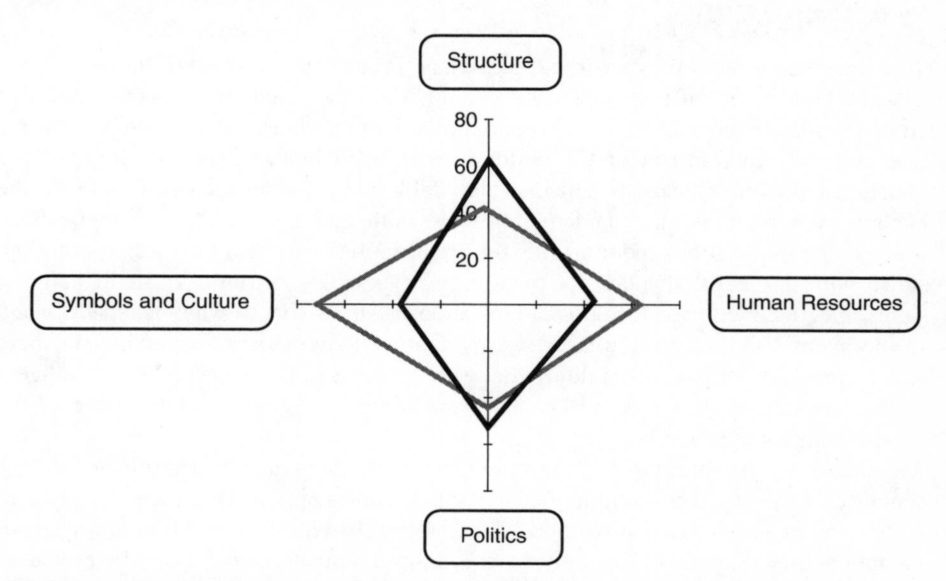

Figure 4.4 A typical 'Reframing' index for both an intelligent building and speculative building.

Given that intelligent buildings are designed to respond to the changing needs of organizations, it would be expected that a skewed 'kite', tending toward the culture and human resource, is the optimal result. This indicates that the structural element of an organization possesses a degree of 'flexibility' in order to re-align in some other form. It also indicates that the development of coalitions that limit openness is minimized.

4.5.4 Quality Facilities Strategic Design

Quality Facilities Strategic Design (QFSD) (Smith, 1999c) is a numerical evaluation. QFSD uses a ranking process to establish an order of priorities for the stakeholder requirements. These requirements have been 'distilled' to (normally) ten core requirements. The process compares the requirement of stakeholders with the ability of the inherent capabilities of the building to satisfy the stakeholder wants and needs. A 'high', 'medium' or 'low' relationship is applied to each item for comparison. Additional information can be included that takes into consideration attributes of other buildings and specific organizational requirements. Figure 4.5 illustrates a typical QFSD evaluation. The matrix indicates that the most demanding requirements are micro zoning on 50% of the building, the use of personal environmental monitors and utilizing performance specifications for the environmental office of the future. These items are the formal targets for the capabilities of integrated office design and selective environmental control, respectively. QFSD provides a very comprehensive analysis of the properties of the building for a large number of parameters. The matrix is scalable up or down, however, a matrix with more than ten stakeholder requirements can become difficult to manipulate.

4.6 Conclusion

Delivering value to both the owner and the tenant through the capabilities inherent in the intelligent building is difficult given the varying and often mutually exclusive goals of each of the participants. There are two parts to delivering value, smart building systems that enable the environment of the building to respond to the changing needs of the occupants, and communications networks that deliver the technology when and how the occupants require it. As the demand for bandwidth and the number of applications increases, communications requirements for organizations will be a key design criterion. Aligned with this is the demand for diverse and flexible work settings. Smart building systems integrated with advanced communications systems will be what occupants want, i.e., intelligent buildings. Value is driven by functionality, and by responding to these requirements the intelligent building delivers value to its stakeholders. Intelligent buildings are designed in such a way that responding to changing needs is one of the capabilities inherent in the building.

Measuring the performance of the intelligent building is a key determinant in any acceptance of the intelligent building as a viable building option. There are a number of performance measures that can be used. The measures listed focus on different aspects of the intelligent building. QFSD analyses the design characteristics. The Magnitude of Systems Intelligent Building Score and the Building Intelligence Assessment Index focus

The Integrated Office

		1 low level churn	2 flexible workspaces	3 integrated office design	4 quality design and finishes	5 selective environmental control	6 enhanced energy management	7 adaptive office systems	8 virtual networking	9 breakout zones	10 efficient thouroughfares	11 central to services	12 value added FM	13 technology transfer	14 max. building utilities space	15 max. available floor space
minimize number of moves	1	●	◐	◐	◐	○	○	◐	○	○			○	◐	○	○
build team spaces quickly	2	◐	●	◐	◐	◐	○	◐	○	◐	◐			◐		
variation in environmental quality	3	○	○	●	○	●	◐	◐					◐		○	○
aesthetics, image, quality	4			○	●	○	○	○		○	◐	○	◐		○	○
healthy environment	5	○	○	○	○	●	◐	○		○	○	○	◐			
lower energy costs	6	○	◐	◐	○	◐	●	◐	◐	◐	◐	◐	○	○		
lower ongoing facilities cost	7	◐	◐	◐	◐	◐	◐	●	◐	○	○		◐		◐	◐
flexible working arrangements	8	○	◐	◐	○	◐	○	◐	●	◐	◐	◐	○	◐		
staff and visitor services	9	○	◐	○	◐	○	○	◐	◐	●	◐	◐	◐	○	◐	◐
quick access to all areas	10		○	○	○	○	○	○	◐	○	●	◐	◐	○	○	○
attractive location	11			◐	◐	○	○	○	○	○	◐	●	◐		◐	◐
responsive tenant services	12	◐	○	◐	◐	○	◐	○	○	○	◐	○	●	○		
de-layer organization	13	○	◐	◐	○		○	○	◐	◐	◐	○		●	○	○
net occupancy space tenant > 85%	14	○	◐	◐	◐		◐	◐	○	◐	◐		◐		●	◐
net lettable area client > 85%	15	○		○	◐	◐	○				◐		○	○	◐	●
technical importance	1	1	2	3	4	5	6	7	8	9	10	11	12	13	14	15
importance to design parameters	2	321	271	300	278	283	209	264	183	153	171	107	207	142	127	125
future target values	3	less than 10%	structured cabling	50% micro zoning	10% micro zoning	50% PEM + EEOF	passive design, BREEAM	50% PnP workstations	30% teleworking, IVPN	5% B and 10% H zones	organizational affinity	near CBD, transport	on site FM CMMS	non-hierarchical, IT	20% of services core	20% perimeter riser

		1 **stakeholder importance**	2 tenant (0.8)	3 knowledge worker (0.4)	4 owner/client	5 community/environment (0.2)	6 weighted average importance	7 **competitive assessment**	8 speculative development	9 current building	10 planned building	11 improvement index	12 market leverage factor	13 **overall importance**
minimize number of moves	1	12.7	5	2	2	1	2.5		2	3	4	1	1.5	62.9
build team spaces quickly	2	10.4	4	4	1	1	2.5		2	3	4	1	1	31.2
variation in environmental quality	3	9.6	4	3	4	1	3		2	2	4	1	1.5	60.5
aesthetics, image, quality	4	8.3	3	4	4	3	3.5		3	3	4	1	1.2	41.8
healthy environment	5	7	3	4	3	2	3		3	2	4	1	1.2	35.3
lower energy costs	6	6.7	4	2	4	4	3.5		2	3	4	1	1.5	42.2
lower ongoing facilities cost	7	6.6	5	2	3	2	3		2	2	4	1	1.5	41.6
flexible working arrangements	8	4.9	4	4	1	1	2.5		2	2	3	1	1	16.2
staff and visitor services	9	4.8	4	5	2	4	3.8		2	3	4	1	1.2	16.2
quick access to all areas	10	4.2	3	3	3	3	3		2	3	3	1	1	13.6
attractive location	11	3.9	3	3	4	3	3.3		4	3	4	1	1.2	19.7
responsive tenant services	12	3.9	5	4	3	2	3.5		2	2	3	1	1.2	18.3
de-layer organization	13	3.9	3	3	2	7	2.3		2	3	3	1	1	9.8
net occupancy space tenant > 85%	14	3.6	5	4	3	1	3.3		4	4	4	1	1.2	10.8
net lettable area client > 85%	15	3.3	3	1	5	1	2.5		4	4	4	1	1.2	9.9
technical importance	1	1	2	3	4	5	6	7	8	9	10	11	12	13

relationship code
high (value 9) - ●
medium (value 3) - ◐
low (value 1) - ○

Figure 4.5 The planning matrix to evaluate the stakeholder requirements against the nominated technologies.

on the structures and systems associated with intelligent buildings, while 'reframing' analyses the intangible aspects of organizations and the relationship with the building they occupy – what is evident is that there is no absolute measure.

References and bibliography

Ahuja, A. (1999) Design-build from the inside out. *Consulting Specifying Engineer*, **25** (7), 36–8.

Arkin, H. and Pacuik, M. (1995) Service systems in intelligent buildings. In: Lustig, A. (ed.), *Proceedings 1st International Congress of Intelligent Buildings*, Israel, pp. 19–29.

Bell, S.V., Murray, T.M. and Duncan, K.T. (1991) Design of direct digital control systems for building control and facilities management. In: *Proceedings of Southeastcon '91*, pp. 674–6.

Bernard, R. (1999) Designing access control. *Buildings*, **93** (8), 26–7.

Burger, D.E. (1999) High performance communications systems in buildings. Presented at the *Joint IEEE/IREE and IEE meeting*, March, Eagle House, North Sydney.

Caffrey, R.J. (1988) The intelligent building – an ASHRAE opportunity. *Intelligent Buildings, ASHRAE Technical Data Bulletin*, pp. 1–9.

Dillon, M.E. (1999) Is integration safe? *Consulting Specifying Engineer*, **26** (3), 52–4.

Duffy, F. (1998) *The New Office* (London: Conran Octopus).

Dunphy, D. and Griffiths, A. (1998) *The Sustainable Organisation* (Sydney: Unwin & Allen).

EIBG (2001) European Intelligent Building Group. www.eibg.net

Francis, J.A. (1992) Designing security into the intelligent building. *Real Estate Finance Journal*, Winter, **7** (3), 77–9.

GK Communications (2001) *What is an intelligent building?* www.tricom.co.uk/concept/ibs/ibs.htm

Guy, S. (1998) *Alternative Developments: The Social Construction of Green Buildings* (London: Royal Institute of Chartered Surveyors).

Haessig, D. (1993) Integrating lighting control. *Energy Engineering*, **90** (6), 59–74.

Harrison, A. (1992) The intelligent building in Europe. *Facilities,* **10** (2), 14–19.

Hartkopf, V. (1995) High-tech, low-tech, healthy intelligent buildings; new methods and technologies in planning and construction of intelligent buildings. In: Lustig, A. (ed.), *Proceedings 1st International Congress on Intelligent Buildings*, Israel, pp. 98–136.

Huston, J. (1999) Building integration: mastering the facility. *Buildings,* **93** (12), 51–4.

Kelly, T. (1990) Defining the indefinable – quality of service in telecommunications. *Telecommunications,* **24** (10), 92–6.

Kroner, W. (1995) Towards an intelligent architecture: the knowledge building. In: Lustig, A. (ed.), *Proceedings 1st International Congress on Intelligent Buildings*, Israel, pp. 60–75.

Lehto, M. and Karjalainen, S. (1997) The intelligent building concept as a user oriented approach to the office – focusing on the indoor air and the energy consumption. In: *Proceedings of Healthy Buildings/IAQ '97 – Global Issues and Regional Solutions*. Washington, DC, USA.

Loftness, V. (1995) Environmental consciousness in the intelligent workplace. In: Lustig, A. (ed.), *Proceedings 1st International Congress on Intelligent Buildings*, Israel, pp. 108–18.

Lush, D.M. (1987) Intelligent buildings: the integration of building services and fabric. Arup R&D internal paper.

McKinley, A.D. (1993) The characteristics of electronic communication traffic in office buildings. Carleton University, Ontario, MEng. thesis.

Quinn, R., Faerman, S., Thompson, M. and McGrath, M. (1996) *Becoming a Master Manager.* Second edition (New York: Wiley).

Smith, S. (1995) The use of ISDN as a platform to integrate building services in the intelligent building. In: Lustig, A. (ed.), *Proceedings 1st International Congress on Intelligent Buildings,* Israel, pp. 35–46.

Smith, S. (1999a) The impact of communications technology on the office of the future. *FM Magazine,* April, p. 42.

Smith, S. (1999b) Integration [agent of change]. *Property Australia*, May, p. 38.

Smith, S. (1999c) The use of quality functional deployment in the design of the office of the future. In: Karim, K. (ed), *Proceedings 2nd International Conference on Construction Management*, Sydney, July, pp. 146–55.

Steelcase (1999) Assessing workplace intangibles: techniques for understanding. Steelcase Knowledge Paper, www.steelcase.com/knowledgebase/intangb.htm

Steloner, M. (1998) Structured cabling: the road to intelligent building. *Buildings,* **92** (9), 90–1.

Stubbings, M. (1988) *Intelligent Buildings; An IFS Executive Briefing* (Bedford: Springer-Verlag, Blenheim Online).

Tarricone, P. (1998) Best practices make perfect. *Facilities Design & Management*, March, 50–4.

Zaheer-Uddin, M. (1994) Intelligent control strategies for HVAC processes. *Building Energy,* **19** (1).

5

A new era in cost planning

Craig Langston*

Editorial comment

> Annual income twenty pounds, annual expenditure nineteen nineteen and six, result happiness. Annual income twenty pounds, annual expenditure twenty pounds ought and six, result misery. The blossom is blighted, the leaf is withered, the god of day goes down upon the dreary scene, and – and in short you are for ever floored.
>
> (Charles Dickens, *David Copperfield*, Chapter 12).

The importance of balancing expenditure against available funds is as important in the context of the construction industry today as it was for Mr Micawber. While Debtors Prison no longer awaits those who cannot meet their financial obligations the consequences of financial mismanagement for clients and contractors is no less devastating than it was for Dickens' famous optimist.

Clients demand a high degree of certainty with regard to how much their projects will ultimately cost and tied closely to this is a requirement for certainty about when their projects will be completed – time and cost are very much dependent on each other and the point is often made that one may get a quick, cheap job done but one should never expect a *good*, quick, cheap job.

For many years clients have been demanding firmer guarantees of building cost before the commencement of construction. Buildings are commercial assets that produce income for their owners either through their direct use, or by generating rental income and just as a client would not buy a new machine without knowing its cost, so they expect to know exactly what a new building will cost before they proceed with construction. Most building projects are paid for with borrowed money and minimizing interest paid on those

* Deakin University, Australia

borrowings is a key concern for clients. When a client cannot pay, or a contractor becomes insolvent before a project is completed, the results are disastrous and far-reaching and generally the only people who benefit in any way are the legal personnel who inherit the task of sorting out the mess.

Many of the characteristics of the construction industry make the formulation of precise predictions of final building cost all but impossible. However, estimates within reasonable tolerances can be produced and these provide clients with a workable basis for cost management during design and construction.

As buildings have become increasingly complex, so cost estimating and planning techniques have become both more complex and more sophisticated. As the author notes in this chapter, cost estimating and cost planning are not the same thing and clients are becoming more and more interested in the latter as they continue their search for better value for their money. Construction cost consultants can now call on a variety of techniques that enable them to advise clients and designers on how they can achieve better value for money not only when a building is first built but throughout its useful life. Similar techniques can be also be applied to the selection of materials, processes and components that are less damaging to the natural environment, and to the design of engineering systems – while the disciplines may have different names, many of the procedures, such as cost benefit analysis, cost engineering, and value management, follow similar algorithms and all have the same goal, i.e., helping clients get better value for their construction dollar.

In this chapter Craig Langston looks at the changing face of cost planning and suggests that there is now, more than ever before, the opportunity for clients, by utilizing the expertise of cost professionals who are supported by increasingly powerful computer technology, to maximize the value that they gain from investing in construction.

5.1 Introduction

Cost planning, as a tool to aid the process of design development, has been around for the best part of half a century, however, it is only in recent times that the technique has undergone a considerable facelift. It forms an important weapon in the fight to manage project costs and to ensure that clients receive value for money. It is now clear that costs are not confined to initial construction, but also relate to operational aspects such as maintenance, cleaning, energy usage and functional occupancy. In this context cost planning, or more appropriately life-cost planning, needs to be applied as a matter of routine from the very earliest stages of the design process. It is only through the proper consideration of cost and its relationship to other criteria such as function, aesthetics, performance and environmental impact that society in general and clients in particular will receive value for money from their investments.

This chapter explores the cost planning technique and its evolution from traditional paper-based methods through to sophisticated modelling techniques that integrate cost with other design criteria. Today cost planning is widely practised, but there is still considerable opportunity for improvement, particularly in regard to the proper investigation of design choices. Computer software is now instrumental in any effective cost management process.

5.2 The aims of cost management

The overall purpose of cost management is to ensure that resources are used to their best advantage. Quantity surveyors (or other cost consultants/engineers) are increasingly employed during the design stage of buildings to advise architects on the probable cost implications of their design decisions.

The quantity surveyor's function is to ensure that the client receives value for money in building work. Advice may be given on the strategic planning of a project which will affect the decision whether or not to build, where to build, how quickly to build and the effect of time on costs or prices and on profitability. During the design stage, advice is needed on the relationship of capital costs to maintenance costs, and on the cost implications of design variables and differing constructional techniques. Value for money is defined as an effective balance between project performance and cost.

The main aims of cost management are:

- to give the client good value for money – this will lead to a building that is soundly constructed, of satisfactory appearance and well suited to perform the functions for which it is required, combined with economical construction and layout
- to achieve a balanced and logical distribution of the available funds between the various parts of the building – thus the sums allocated to cladding, insulation, finishings, services and other elements of the building will be properly related to each other and to the class of building
- to keep total expenditure within the amount agreed by the client – this is frequently based on an approximate estimate of cost prepared in the early stages of the design process. There is need for strict cost discipline throughout all stages of design and construction to ensure that the initial estimate, tender figure and final account sum are all closely related.

5.3 Cost planning

5.3.1 Definition

Cost planning is the means that enables the objectives of project cost management to be achieved. It is a process used during the design stage of a particular scheme to help minimize the cost of construction and subsequent usage, and maximize the functionality that is anticipated by the client. Cost planning is in fact a system of procedures and techniques used by quantity surveyors. Its purpose is to ensure that clients are provided with value for money on their projects; that clients and designers are aware of the cost consequences of their proposals; that if they so choose, clients may establish budgets for their projects; and the designers are given advice that enables them to arrive at practical and balanced designs within budget. It helps to ensure that clients can realize their expected profits and benefits and that their financial requirements stay within anticipated limits. Therefore cost planning monitors and helps direct design and organizational decisions in order to achieve the client's cost objectives.

Cost planning is a process that brings cost information to bear systematically upon the evolution, construction and maintenance of a project to highlight the relationship between capital and operational investment, quality, function and appearance, and to provide a framework for the management of costs in order to deliver value for money to the client.

The cost plan itself is one of the principal documents prepared during the initial stages of the cost management process. It is a statement of the proposed expenditure for each section or element of a building related to a definite standard of quality. Costs, quantities and specification details are itemized by element and sub-element and collectively summarized. Measures of efficiency are calculated and used to assess the success of the developing design. The elemental approach aids the interpretation of performance by comparison of individual building attributes with similar attributes in different buildings, and forms a useful classification system.

Table 5.1 Elemental cost plan summary (excluding siteworks)

Code	Element	% BC	$/m^2	Quantity	Unit rate	Total cost
01SB	***Substructure***	4.93	41.72	1,232 m^2	564.34	695.264
	Superstructure					
02CL	Columns	2.20	18.58	17,311 m^2	17.89	309,635
03UF	Upper floors	19.33	163.61	15,376 m^2	177.32	2,726,495
04SC	Staircases	1.31	11.05	64 m^2	287.28	184,148
05RF	Roof	2.94	24.91	1,294 m^2	320.81	415,125
06EW	External walls	8.80	74.44	5,714 m^2	217.18	1,240,543
07WW	Windows	7.18	60.78	1,070 m^2	946.63	1,012,899
08ED	External doors	0.15	1.30	29 m^2	747.00	21,663
09NW	Internal walls	5.45	46.13	2,221 m^2	346.13	768,756
10NS	Internal screens	0.82	6.92	199 m^2	579.53	115,327
11ND	Internal doors	0.96	8.10	160 m^2	843.64	134,982
	Superstructure	49.14	415.82	–	–	6,929,573
	Finishes					
12WF	Wall finishes	1.90	16.49	10,154 m^2	27.06	274,806
13FF	Floor finishes	3.77	31.87	15,464 m^2	34.35	581,114
14CF	Ceiling finishes	3.47	29.33	15,464 m^2	31.61	488,784
	Finishes	9.18	77.69	–	–	1,294,704
	Fittings					
15FT	Fitments	1.31	11.08	–	–	184,598
16SE	Special equipment	0.81	6.84	–	–	113,969
	Fittings	2.12	17.92	–	–	298,567
	Services					
17SF	Sanitary fixtures	0.50	4.20	244 No	286.86	69,993
18PD	Sanitary plumbing	0.31	2.63	854 Fu	51.32	48,831
19WS	Water supply	0.35	2.95	16,285 m^2	3.02	49,162
20GS	Gas service	–	–	–	–	–
21SH	Space heating	–	–	–	–	–
22VE	Ventilation	0.82	6.93	16,285 m^2	7.09	115,488
23EC	Evaporative cooling	–	–	–	–	–
24AC	Air conditioning	12.43	105.20	16,285 m^2	107.65	1,753,158
25FP	Fire protection	3.37	28.51	16,285 m^2	29.18	475,119
26LP	Electric light and power	7.37	62.34	16,285 m^2	63.79	1,038,866
27CM	Communications	0.22	1.87	16,285 m^2	1.91	31,164
28TS	Transportation systems	8.51	72.04	–	–	1,200,506
29SS	Special services	0.75	6.36	–	–	106,030
	Services	34.63	293.03	–	–	4,883,317

Cost planning depends heavily on a technique known as *cost modelling*. Modelling, in a generic sense, is the symbolic representation of a system; a cost model represents a project expressing the content of that particular system in terms of the factors that influence its cost. Its objective is thus to 'simulate' a situation in order that the problems posed will generate results that may be analysed and used to make informed decisions. Each item of cost is generally regarded as a 'cost target' and is usually expressed in terms of cost per square metre as well as total cost[1]. An example of the sort of information presented in a cost plan is shown in Table 5.1.

5.3.2 Estimating

Estimating is an integral part of effective cost planning. During the early stages of the design process, estimates should be regarded as target figures within which costs should be managed rather than as cost predictions. It is recommended that estimates be produced on an elemental basis. Elemental estimating using cost data based on priced bills of quantities or actual construction information can produce accurate estimates, at least for the general run of building work.

Although other forms of estimating may be applied at the concept stage, as soon as drawn information is available, estimates should be produced on an elemental basis, sub-divided into 'approximate quantities'. Ideally there should be sufficient measured items to ensure that the cost is not highly dependent on a small number of calculations. For example, any particular item that exceeds 1.5% (say) of the total project cost should be further considered using one of the following methods:

- break up the item into two or more parts so that no part exceeds 1.5% of the total project cost
- break up the unit rate into two or more parts so that the composite rate is more reliable – this may require building up the unit rate in terms of labour, material, waste, plant, etc.
- undertake an analysis of three or more similar items in other projects and calculate an average unit rate, adjusted for time, market and locational differences as necessary
- obtain a reliable quote for the work from a specialist sub-contractor.

If these simple rules are followed when producing an estimate, a high degree of accuracy is likely to be forthcoming. Nevertheless, estimating is just one aspect of the cost planning process.

5.4 Cost planning philosophies

The philosophy of cost planning can be divided into two main ideological groups – those that pertain to 'costing a design' and those that pertain to 'designing to a cost'. The difference is significant. Costing a design refers to methods of estimation that can arrive at a cost for a given (predetermined) design. Naturally some formalized drawings need to be in existence for this process to function. Designing to a cost on the other hand means guiding the development of a scheme so that it not only sets a budget amount and ensures that the project does not exceed same but provides the client with the greatest value for money possible.

5.4.1 Costing a design

Costing a design covers the majority of cost planning techniques whose main purpose (in theory) is to develop the most economic form of a given design. The emphasis is intended to be placed on the phrase 'of a given design' since this cost planning philosophy does not investigate alternative design solutions. It does, however, identify cost-significant items contained in the scheme that may be deleted or substituted during the project's design stage.

The principles of costing a design are well known, as this philosophy is easily the most common process now in use in most countries. But its advantages, limited as they may be, are not being fully capitalized upon at the present time.

The initial decisions that significantly involve the developing design are commonly made by the architect. They may be based on aesthetic considerations, previous projects, experience, or even an assumed economic configuration. But no objective analysis is undertaken and no facts are compiled to verify the choices made. The sketch plans are prepared as a result of the initial decisions. This is the point at which the quantity surveyor usually becomes involved. His/her task is largely pre-defined as estimating the cost of a project using one of the many methods available and thence preparing the initial cost plan. The quantity surveyor monitors the costs as the drawings progress and makes suggestions (although these are not necessarily implemented) about possible savings. Very little constructive criticism pertaining to design decisions takes place for reasons that perhaps are more concerned with maintenance of professional relations than anything else. The level of interaction between cost consultant and designer is largely governed by whether or not the developing cost is remaining within the original budget allowance.

During the pre-tender period the cost plan is continually revised, and as the number of unknowns is reduced, since there are obviously less unknowns, the design contingency (an allowance for future detailing – also called design risk) is progressively reduced. The final cost plan should provide an estimate of the total cost equivalent to that produced by the successful tenderer who has priced a bill of quantities based on the same set of drawings, although the cost plan is presented in a completely different format. If the cost plan is founded on the pricing of measured quantities, and there has been due consideration of market conditions, then this will usually be the result.

It was suggested that in theory costing a design refers to the majority of cost planning techniques whose main function is to develop the most economic form of a given design. In practice, however, this is not always the case. Costing a design in the most part is the control of a project's developing construction cost with reference to a pre-determined budget amount. If the derived cost is below the budget then often no further investigation at that particular stage is carried out. If the derived cost is above the budget then activity does occur and items of cost significance are identified and either deleted or substituted with less expensive alternatives.

Government projects frequently have the budget set by a government-employed quantity surveyor who is independent of the consultant quantity surveyor involved in the cost planning process. The budget amount is derived from a standard procedure that, in simplified terms, utilizes information from past projects. Any new design, ignoring special factors and site works which are both separately allowed for, should be capable of being built for the same price as an average of previous projects escalated and adjusted to the new tender date.

Costing a design is widespread. Very little optimization takes place on an objective and formalized level as intuition seems to play an important role in the decision-making process. Generally alternatives are analysed only on specific instruction, not as a matter of course. In crude terms, the philosophy of costing a design when used in practice is merely a process of cost estimation. This situation is considered an unsatisfactory one and fails to achieve the overall cost planning objective of providing value for money.

5.4.2 Designing to a cost

Designing to a cost covers the majority of cost planning techniques whose main purpose, unlike the costing a design approach, is to develop the most economic design solution possible. It is implied that alternative methods of achieving the intended result are examined objectively as an essential aid in the decision-making process. Since neither the architect nor the quantity surveyor nor any other member of the design team can reasonably fulfil this objective in isolation, a co-operative effort is required.

In order that this philosophy can function properly it must be implemented before any formalized attempt at documentation has begun. A feasibility study optimally examining all possible design solutions in respect to cost is ideal, but in practical terms the scope of the investigation should be confined to a much smaller selection.

The architect, in association with the quantity surveyor and other consultants, formulates a number of proposals that all have the potential to achieve the stated design objectives. The quantity surveyor then investigates each such proposal from a cost point of view and ranks them in order of merit. After further discussions with the architect, and perhaps further investigation, an optimum configuration in respect to the original selection, based on cost and performance, is nominated as the concept design. Using this information the sketch plans are prepared.

During the design process, the architect, the quantity surveyor or others may suggest more specific ways in which the project may be further optimized. Such suggestions should take the form of a collection of proposals and each proposal should be objectively investigated using quantitative techniques. As each cost plan is prepared an opportunity is provided to review the status of the project. It is important to note that the level of interaction is largely governed by the co-operation achieved between architectural and economic constraints.

Designing to a cost is generally not common in practice. The philosophy relies fundamentally on co-operation between the architect and the quantity surveyor at the concept stage as well as during the development of a selected design. It is therefore necessary that the quantity surveyor takes a more prominent role in the design of a project, especially before sketch plan stage has been reached. It is only during this period that significant cost savings are possible, for after commencement of documentation the basic concept is fixed and design alterations are thus non-productive and impractical to implement. Until initial decision making is changed from the role of the architect to that of all members of the project team this efficient cost planning process will remain merely of academic interest. It is the client, not the architect, who can change the current situation.

5.4.3 Comparing cost planning philosophies

Design variables is a term applied to the fundamental characteristics of a given project, e.g., plan shape, height, floor area. The advantage of the 'designing to a cost' philosophy is related to the mechanisms that allow the design team to search for and identify the optimum design variables that are capable of guaranteeing value for money. Hence options can be individually appraised prior to commencement of drawings in order to select a building configuration within available funding limits. Attempts at optimization after drawings have been started are of little practical benefit.

5.5 The budgeting process

The rising cost of building in recent years has emphasized the need for careful management of cost. The budget is a fundamental part of the mechanism by which control can be achieved; it provides a standard for comparison, without which there can be no control.

Budgeting is the disciplined pre-determination of cost, and the formulation of a budget is a significant first step of planning and control for any project or business. It shows the likely consequences of design decisions in terms of money and identifies major cost factors in the project.

The budget acts as the framework within which further cost planning activities can function. It is used to establish initial feasibility and is a basis for regulating and reviewing project cost as design development proceeds. Budgets set cost targets, usually in elemental form, based on past experience and client objectives. Accountants readily recognize the importance of budgeting in business concerns; in these cases the budget provides management with the right tool to make effective use of the capital at its command. The basic budget is seen by management as both an early warning system and a frame of reference for evaluating the financial consequences of operational decisions. The degree of sophistication in a company's budgeting system will usually depend on its size, but even small businesses will find that an elementary level of budgeting will assist in the control of the business.

So, too, construction professionals realize that budgeting for building projects provides the same advantages. As in accounting budgeting, a budget prepared for a building project gives the client and design team the means to plan, co-ordinate and manage the project and establishes a system for evaluating the consequences of design decisions. The budget establishes cost limits for key areas or elements of the building as well as for the whole project. While the total project may be within budget, examination of key cost areas may show major discrepancies.

The budget sets the framework within which the client and the designers must work to achieve their objectives. It also provides a considered standard for comparison and judgement of results and so it is useful for those who manage the cost of the project from design through to completion of construction.

In effect a budget is a plan, in money terms, of the future outcome of the project and thus establishes limits for the project. The client uses the budget to plan appropriate cash needs or evaluate more viable investment alternatives. The preparation of a budget forces the client to look at the realities of current and future business conditions and the practical options available.

The cost of a building is determined by every decision made and, just as initial design decisions will affect a continuing series of interrelated developments throughout the design period, decisions on costs will also affect and be affected by the development of design ideas. Through the budget, cost becomes a design limitation. Without a budget the limiting factors of design are function and aesthetics, with cost merely a result of decisions based upon the considerations given to those limiting factors. Without a budget the designers are not aware of the extent of the project in terms of what the client can afford. The budgeting tool relates design development to costs and gives the designer the means to forecast the probable outcome of design decisions. It allows investigation of alternative design solutions which can be compared to money available as set out in the budget thus keeping the design within the existing budget limits.

One of the reasons for the increasing importance of cost planning in recent years is the fact that construction methods are becoming increasingly complex. The use of budgeting within cost planning can ensure that the complexity of the construction procedure is readily measurable in dollar terms. A budget helps to ensure that resources are being used to their best advantage. It clarifies to the client the cost of the building that he/she wants and clarifies to the designer the cost within which he/she must work. Without a budget, the client may find that he/she does not receive what he/she wants from the project. An 'open purse' attitude will inevitably result in an over-design which may fit the client's criteria functionally but be unnecessarily expensive. In this respect the budget can ensure that value for money is a continuing objective.

Budgets should be realistic. The setting of unrealistically low budgets may initiate a large amount of investigation into alternative solutions, but may in the end lead to an overrun of cost or a reduction in quality. Hence the budget should not be excessively low, and generally it is based on the performance of previous designs of similar function. For example, if a hospital was built for $1500 per square metre of floor area in the past, then a proposed hospital should also be capable of being built for an equivalent amount, with due allowance for time-related cost fluctuations.

5.6 The evolution of cost planning

5.6.1 The past

Evolution is the process of gradual change over a long period of time involving survival of the strong and adaptable species, and extinction of those that are inherently weak or unsuited to the demands of their changing environment. Systematic improvement is the means by which the evolutionary progression can be utilized to arrive at the optimum end result.

Cost planning has evolved from crude methods of estimation to highly complex and time-consuming activities that must conform to a wide variety of procedural rules and constraints. Economic demands now make it essential to look at a number of alternative design solutions in order that the overall construction cost and/or life-cost can be minimized. The time spent on effectively examining options must be weighed against the time allowed for the task, which is perhaps a function of the level of fees paid by the client. Education of clients in the range and benefits of cost planning activities is important.

Traditional methods of project cost management refer to cost planning activities undertaken before the advent of computer assistance. The majority of cost planning activities undertaken in the recent past could hence be classified as traditional. They have still managed to continue in popular use because the limited amount of investigation that takes place enables the process to fall within the available fee margin.

The inadequacies of traditional practice have led to the realization that the cost planning process is currently undergoing significant evolutionary change. While the mechanics of cost management are well established, the amount of time associated with their employment precludes thorough investigation. The primary role of cost planning today seems to be costing a design and then reducing or eliminating components of high cost instead of instructing the designer at an early stage in ways that would enable the project to comply with the initial budget.

Traditional methods also suffer from the inherent problem of human error. While any system that interacts with people will possess this disadvantage, the extent of its influence and ways in which it can be reduced are of interest when discussing systematic improvement. In a science that relies heavily on manipulation and sorting of cost data, areas, percentages and the like, such a component must be of significance to any procedural changes that may be made.

Guidelines are required to control the directions for efficient investigative techniques and reduce the time associated with them. Either the number of cost significant areas needs to be reduced or the pace at which these areas can be analysed increased. Computers are an obvious means by which time-minimization can be achieved while maintaining or expanding the range of design alternatives researched. The speed and accuracy of the computer, its accessibility and in particular its versatility, are considered as prime attributes that reflect and further promote this not-so-new technology.

5.6.2 The present

Although change can come slowly to some sectors of the community, the introduction and development of computers has been a relatively rapid and widespread innovation. The present low cost of personal computers has been a major reason for this trend together with the development of computer software capable of performing many specific labour-saving activities.

Cost planning in 'the present' is characterized by the widespread use of computers. Their use to date has been mostly computerized estimating and the storage and presentation of specific cost plan information in electronic databases or computer storage systems.

Although computer storage systems have a definite role and possess numerous advantages that save time and reduce errors, they do not assist the investigation of design alternatives other than by time-intensive methods based on trial and error. They certainly do not analyse or optimize the concept of the design prior to the development of a formalized scheme nor show the effects on cost of plan shape change, building height or number of storeys.

5.6.3 The future

Future evolutionary trends in cost planning must involve computers. There can be little doubt that computers will become more powerful and more flexible and the amount of marketable software will continue to increase. New technology will enable greater amounts of information to be processed and allow complex tasks to be handled with more efficiency. Simultaneously the size and cost of computers will continue to decrease.

From an evolutionary point of view, the frontier for cost planning may be classified as computer guidance systems. Since only about 20% of a project's cost can be affected after documentation has begun (see Figure 5.1), it is important that cost planning activities be concentrated on the 80% that can be manipulated during the concept stage. Computer guidance systems must be directed to this area of investigation – an area that, in the past, has been largely ignored.

Experience to date has highlighted that attempts at design optimization must be initiated at an early stage before valuable time is spent documenting inherently expensive schemes. For practical reasons, as mentioned earlier, more detailed exploration of alternative solutions requires either that fewer areas are investigated or that we reduce the time needed to test each alternative. The second option is clearly preferred and the use of computers to carry out cost management activities means that this can be done without diminishing the depth of an investigation nor over-simplifying the effect on cost of a particular design decision.

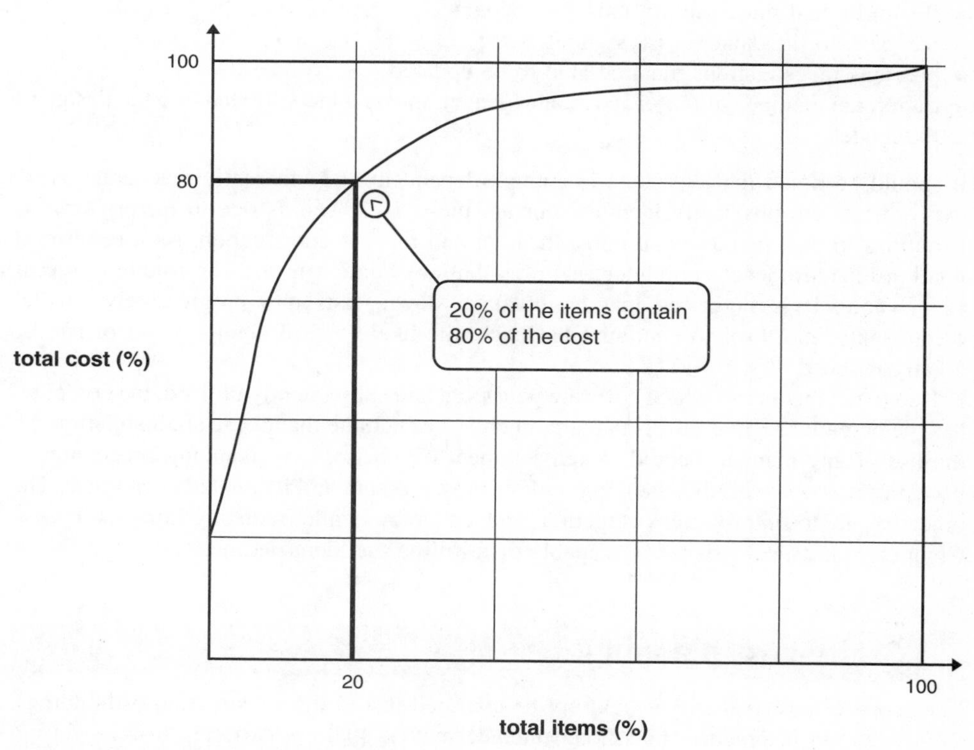

Figure 5.1 Pareto's law of distribution.

It is clear that current practice can be significantly improved through the utilization of computer guidance systems at all stages in the design of a project. Computer guidance systems must be able to simulate a situation in order that the problems posed will generate results that may be analysed and used in the decision-making process. They are hence prime examples of cost models. A range of cost modelling techniques can then be employed to examine the impact of a wide range of design options and their comparative merits quickly assessed.

Some of the other attributes that a computer guidance system must possess include:

- the mechanics of cost generation need to be 'transparent' so that cost consequences can be traced through all inter-relationships
- the user must be allowed to interrogate the results in great detail
- the model should facilitate a dynamic decision-making process
- the costing mechanism must be highly flexible and adaptable to the most innovative design solution
- cost information should not interrupt the 'flow' of design
- the model must link the user to a central database (or store of information)
- the model should relate a proposed design to comparable completed projects and indicate the variability of that relationship
- it should assist in solution generation
- information must be presented in a consistent and familiar form
- the system must be easy to use and must significantly reduce time
- flexibility and practicability must be encouraged
- feedback to the database is essential
- previous investigations must be able to be updated
- traditional methods of project cost management must be incorporated into the design of the model.

It should be noted that cost data is context dependent and inherently inaccurate as no two projects are physically identical nor are their costs. This is due to factors such as variations in site, market conditions, location and time of construction. As a result cost relationships are never completely manifest but are implicit in the construction process as a whole. It is suggested that by breaking down costs into progressively smaller components, and through continual feedback, the likelihood of significant error can be greatly reduced.

Due to the manner in which costs are generated and subsequently utilized, the computer has the potential to give results that are much more accurate than those attainable through the use of any manual method. A reassessment of existing cost planning procedures is necessary to cope with the changing building environment and its related economics. The objective is to improve investigation and accuracy while reducing time and cost. Computer guidance systems are capable of fulfilling the identified needs.

5.7 Life-cost planning

The scope of cost planning should not be confined just to the construction of buildings, as is common in practice, but should include matters that are expected to arise during the life of the project. The process is aimed at improving value for money for the investor and/or future users through comparison of alternatives that meet stated objectives and qualities at reduced expense.

Life-cost planning is similar in concept to capital cost planning except for the types of costs that are taken into account and the need to express all costs in common dollars, i.e., costs that are adjusted to reflect the effects of escalation over time. The aim is to prepare a document that describes the composition of the building in a manner that is of use to the investor or owner. A cost plan in this format can be used to demonstrate the relationship between initial, replacement and running costs and to assist in the choice of specification and design details.

The primary objective of a comprehensive methodology of life-cost planning is to create and maintain an up-to-date picture of options available to the project client and subsequent users. The objective of life-cost planning is not necessarily to reduce running costs, or even total costs, but rather to enable investors and building users to know how to obtain value for money in their own terms by knowing what these costs are likely to be and whether the performance obtained warrants particular levels of expenditure. A project is cost effective if its life-costs are lower than those of alternative courses of action that would achieve the same objectives.

Expenditure commonly associated with commercial type buildings includes acquisition, cleaning, energy, rates and charges, maintenance and replacement. These costs can apply to the building structure, its finishes, fitments, services and external works. Establishing realistic costs for items of plant and equipment, in particular, requires technical data such as performance statistics and energy demands. The establishment of such life-costs may be difficult without historical data or expert knowledge.

The life-cost plan should be prepared on at least an elemental basis showing all quantities and unit rates. It should be set out in a fashion that enables extraction of totals for each type of cost category, including capital cost, along with costs per square metre and percentages of total cost. The method of presentation needs to be designed to bring out the underlying cost relationships, so that the analyst can see how to develop the design and make value for money improvements. Life-cost planning, like any other form of cost planning, is most effective in the early stages of design.

The two main objectives of applying life-cost studies to the construction of buildings and their subsequent usage are:

- to facilitate the effective choice between various design solutions for the purpose of arriving at the best value for money – this is a comparison activity
- to identify the total costs of acquisition and subsequent usage of a given design solution for the purpose of budgeting, planning and controlling actual performance – this is a measurement activity.

As the investigation of cost, and subsequent action, is more effective when undertaken at an early stage in the design process, it is not surprising that life-cost studies have concentrated on the first objective, to the virtual exclusion of the second. Discounting methodology is routinely used for comparison and selection of alternatives, however, if value for money over the life of a project is to be vigorously pursued, then life-cost studies must additionally become concerned with the measurement of costs. This requires the presentation of costs as real values rather than comparative (discounted) values.

The measurement of real value involves collection of the total expenditure arising from a project over the study period. Obviously future costs cannot be merely added without some form of adjustment taking place. Inflation is the appropriate 'exchange rate' for this purpose. Present value (as opposed to discounted present value) translates future cost into real terms, enabling both capital and operating costs to be properly interpreted. Present value (defined as today's value) thus forms a suitable basis for the measurement and control of life-costs.

The life-cost plan is used to record the present value of both initial and recurrent costs for the chosen design. Determination of the present value of goods and services over the life of a project can enable quantification of the potential cost liability to the owner or investor, which is of great use in the management of financial resources.

Comparative life-cost studies should be undertaken systematically on significant areas of expenditure using a discounting approach. The results of these studies lead to selection of materials and systems and are ultimately reflected in the life-cost plan. Although discounting is a suitable means of assessing the impact of timing on expenditure, the results obtained do not represent real values and there appears little need to present information in this form in the life-cost plan.

Investment appraisal often must include the consideration of aspects other than cost. Form, function, aesthetics and environmental interaction are some of the more common issues that will contribute to the selection of optimal designs. For this reason comparative life-cost studies are best undertaken as part of a value management process.

5.7.1 Value management

Value management involves the identification of function and the selection of solutions that can maximize this function at minimum life-cost. Ultimately an effective balance is struck between function and cost and it is this balance that is known as value for money. Commonly life-cost is considered, along with other design criteria, as a percentage of the final decision. Each design criterion is weighted as to its importance and the various alternatives being appraised are rated against each criterion using a numeric score. The multiplication of design criterion weight and performance score when totalled for each alternative provides the basis for identification of optimal value. An example of the manner in which subjective and objective issues are collectively analysed and judged is illustrated in Figure 5.2.

Value for money can be determined through division of the value score by the calculated comparative life-cost. The value score represents functional performance issues and is exclusive of matters that can be measured in monetary terms. Life-cost is judged as representing 40% of the decision in the presented example, but this can be altered to reflect various client motives. The higher the value for money index (or benefit ratio) the better is the balance between function and cost. Some form of risk analysis would be undertaken to indicate the probability of the identified value for money being realized.

5.8 Conclusion

Cost planning is operating in a new era. The focus has changed from the analysis of construction costs to embrace not only future commitments over the life of the project, but also non-monetary considerations such as function, aesthetics, performance and

Subject	external paving for uncovered paths			
Primary Function	to define and facilitate pedestrian travel over site			
				Weighting (0-10)
Design	A	=	safety	10
	B	=	stability	8
	C	=	appearance	5
	D	=	water shedding ability	4
	E	=		
	F	=		
	G	=		

Alternatives scored (0-5)	**Design Criteria**							**Design Evaluation**		
	A *10*	**B** *8*	**C** *5*	**D** *4*	**E**	**F**	**G**	**Value Score 60%**	**Life-Cost ($/m²) 40%**	**Benefit Ratio**
brick pavers on sand base	3 / 30	4 / 32	5 / 25	5 / 20				107	$66.87	2.40
brick pavers on concrete slab	4 / 40	5 / 40	5 / 25	5 / 20				125	$75.55	2.48
concrete pavers on sand base	3 / 30	4 / 32	4 / 20	5 / 20				102	$61.74	2.48
concrete pavers on concrete slab	4 / 40	5 / 40	4 / 20	5 / 20				120	$70.43	2.56
concrete pavers with wood float finish	5 / 50	5 / 40	3 / 15	5 / 20				125	$49.14	3.82
concrete slab with broom finish	5 / 50	5 / 40	4 / 20	5 / 20				130	$49.64	3.93
concrete slab with aggregate finish	5 / 50	5 / 40	5 / 25	5 / 20				135	$67.02	3.02
concrete slab with quarry tile finish	5 / 50	4 / 32	5 / 25	5 / 20				127	$113.03	1.69
concrete slab with cement topping	5 / 50	4 / 32	5 / 25	5 / 20				127	$65.53	2.91
natural stone paving	3 / 30	4 / 32	5 / 25	5 / 20				107	$169.41	0.95
precast concrete blocks	3 / 30	4 / 32	4 / 20	5 / 20				102	$221.40	0.69
pine bark mulching	2 / 20	1 / 8	5 / 25	3 / 12				65	$28.69	3.40
loose aggregate	1 / 10	1 / 8	5 / 25	3 / 12				55	$22.68	3.64
asphaltic concrete	5 / 50	4 / 32	2 / 10	4 / 16				108	$33.44	4.84

Asphaltic concrete is the selected alternative. Loose aggregate is the value standard.

Figure 5.2 Value management study evaluation technique.

environmental impact. The increased role of computer simulation to assist in the comparison and evaluation of options is evident. Cost management is now a broader discipline that recognizes the need to balance often opposing objectives. Value for money is the key criterion and is inclusive of a wide range of issues that ultimately combine to represent a successful project.

Endnote

1 The example shown in Table 5.1 is based on a standard format used in Australia; other countries may use systems that are different in their detail but perform a similar function in general. In the UK, for instance, the SFCA (standard form of cost analysis) provides a uniform basis for apportioning building costs in a standard elemental format.

References and bibliography

Ashworth, A. (1999) *Cost Studies of Buildings.* Third edition (Longman).

Best, R. and de Valence, G. (1999) *Building in Value: Pre-design Issues* (London: Arnold).

Bull, J.W. (1992) *Life Cycle Costing for Construction* (Thomson Science and Professional).

Dell'Isola, A.J. and Kirk, S.J. (1995a) *Life Cycle Costing for Design Professionals.* Second edition (McGraw-Hill).

Dell'Isola, A.J. and Kirk, S.J. (1995b) *Life Cycle Cost Data.* Second edition (McGraw-Hill).

Ferry, D.J., Brandon, P.S. and Ferry, J.D. (1999) *Cost Planning of Buildings.* Seventh edition (Blackwell Science).

Flanagan, R. and Norman, G. (1983) *Life Cycle Costing for Construction* (Surveyors Publications).

Flanagan, R. and Tate, B. (1997) *Cost Control in Building Design: An Interactive Learning Tool* (Blackwell Science).

Flanagan, R., Norman, G., Meadows, J. and Robinson, G. (1989) *Life Cycle Costing: Theory and Practice* (BSP Professional Books).

Langston, C. (1991a) *The Measurement of Life-Costs* (Sydney: NSW Department of Public Works).

Langston, C. (1991b) *Guidelines for Life-Cost Planning and Analysis of Buildings* (Sydney: NSW Department of Public Works).

Langston, C. (1994) The determination of equivalent value in life-cost studies: an intergenerational approach, University of Technology, Sydney, PhD dissertation.

Mooney, J. (1983) *Cost Effective Building Design* (Sydney: New South Wales University Press).

Seeley, I.H. (1997) *Building Economics* (Macmillan Press).

Smith, J. (1998) *Building Cost Planning for the Design Team* (Deakin University Press).

6

The energy of materials

Caroline Mackley*

Editorial comment

Buildings are major energy users. It is only recently, however, that researchers have begun to explore the concept of embodied energy, that is the energy which is used to extract raw materials, to transport and process those materials into building materials and components, to power the on-site processes of construction and even to demolish and dispose of buildings at the end of their life. Where it was once thought that this energy would be far outweighed by that used to operate and maintain buildings, contemporary research is proving that this is not the case. There are opportunities for significant reductions in the energy embodied in buildings if designers can make informed choices when selecting from alternative available materials, and such choices can contribute to a general reduction in energy use and resultant environmental damage.

As we look for ways to pursue the goal of sustainable development the energy of building materials cannot be overlooked. As the author of this chapter shows, however, making choices between alternative materials and components is not as straightforward as it may appear at first. There are a number of major problems to be addressed, not the least of which is just how should we go about measuring embodied energy, and how can we account for vast differences in impacts associated with energy use that are the result of manufacturers in different areas using different energy sources.

A classic Australian example involves aluminium production at two major smelters, one in Queensland, the other in Tasmania. Aluminium smelting is extremely energy intensive, with large amounts of electricity required to drive the process – such are the energy needs that aluminium has sometimes been referred to as 'congealed electricity'. In Queensland, where there are extensive coal reserves, the bulk of the electricity is produced in coal-fired thermal power stations, with low conversion efficiencies and high levels of emissions of pollutants including waste heat and greenhouse gases. By contrast, the bulk of Tasmania's

* Bovis Lend Lease, Sydney, Australia

electricity is produced by hydro-electric plants with minimal pollution problems but a range of other environmental impacts that are extremely difficult to quantify, such as loss of habitat for native flora and fauna when rivers are dammed to provide water to drive the turbines. This presents analysts with a dilemma: how are they to provide a true basis for comparison of aluminium products with products made from other materials, such as steel or timber, when the impacts of the energy used in the centres of production are so different. It is very hard to know where the aluminium used to manufacture window frames in Sydney, say, is sourced and so while it may be feasible to arrive at an average figure in MJ/kg for aluminium production it is basically impossible to determine a standard embodied energy rating that is truly meaningful for what is a very common building component.

Just because it is not easy does not, however, mean that the problem should not be addressed. Researchers in various countries are grappling with the problems of this analysis and some designers are already using published data as a basis for some of the choices that they make. The value-for-money implications of these choices for clients are by no means clear cut at this time but it is likely that in the future, as externalities associated with energy production are progressively incorporated into energy costs (by mechanisms such as carbon taxes), that the selection of materials and components with less embodied energy will produce capital cost savings. In the short term there is the potential for authorities, via building codes and regulations, to require the use of less energy intensive alternatives in design and construction. In any case the work described in the following chapter is important as we establish a more sustainable industry and look for better value-for-money outcomes in our buildings.

6.1 Introduction

The topic of discussion in this chapter is embodied energy and its relationship to building materials. The reason that this relationship is worthy of exploration is that it enables us to understand the associated environmental impacts that flow from the production of both individual materials and the complex products known as buildings. The aim of this analysis is to determine whether more energy efficient materials, products and buildings are feasible and in doing so, to contribute to reducing the ecological impact of the built environment. Our contemporary concern for the embodied energy intensity of building materials stems from the association between greenhouse gas emissions (GGE) and energy. If the energy intensity of a particular process is unknown, GGE cannot be quantified.

It has always been held that the energy associated with the production of materials as they are embodied into buildings is irrelevant in comparison to their operational energy consumption. This fallacious proposition has endured mostly because of inadequate measurement methodologies and ignorance within the building industry. As a global average, it is estimated that 50–60% of the world's annual energy production is used by the industries of the built environment (Janda and Busch, 1994; Berkebile and McClennan, 1999) of which 36% is consumed directly in the operation of buildings (IEA, 1997). A simple calculation then tells us that between 14 and 24% is embodied in the buildings and materials themselves. It is likely that these figures understate the full impact as transportation energy is measured within a different boundary.

Advances in quantification techniques have significantly increased our awareness of the proportional representation of embodied energy for a given building's life cycle. The rate of increase in the (operational) energy efficiency of buildings also means that the relative importance of embodied energy will continue to increase with time.

Historically, it was held that embodied energy typically comprised perhaps 10% of the life cycle energy requirements of buildings. That is, the building would need to use energy in operations for 3–5 years before the operational energy equalled the energy embodied in the building. Contemporary research has repeatedly challenged this view, demonstrating that in some cases the embodied energy is ten to twenty times the annual operational energy consumption (Pullen, 1995; Treloar, 1997; Mackley, 1998). In the case of office building construction, recent work appears to suggest that a number as high as thirty-five times the annual operational energy consumption may be closer to the truth (Mackley, 2000). These advances are significant for a number of reasons:

- they provide clear proof that the embodied energy and associated GGE are a significant part of the global environmental impact of building construction
- the opportunity to facilitate significant GGE mitigation is a reality with advanced calculation methods
- under a carbon trading regime the cost of building construction is likely to increase by a minimum of 6–12% (Mackley, 2000) with the core building materials, that have historically enjoyed the highest levels of government subsidy (cement, aluminium and steel), increasing in cost by the greatest proportion.

6.2 The energy context

In order for us to understand the concept of embodied energy, we first need to consider the broader concept of energy and its relationship to our existence.

Bullard and Herendeen (1975, p. 268) observed that 'when you consume anything, you are consuming energy'. The physical laws of thermodynamics dictate the theoretical energy efficiency of all processes. Energy analysis (EA) is the term used to describe the methodology associated with the measurement and assessment of a process in relation to its energy requirements. It provides a means by which the relative energy efficiency of a process can be assessed, thereby enabling the implementation of strategies to reduce energy use and associated emissions.

The concept of the measurement of energy that flows through goods and services is not new; energy analysis can be traced back as far as the early nineteenth century. Its original function was associated with the process economics of commodity production. Its reappearance in the 1950s and 1960s was driven by an awareness of the finite nature of our energy resources and led to the development of methods for assessing stocks and the potential for their conservation. Since this time, EA of resource stocks has been largely used by governments to provide policy direction in relation to conservation and issues of national security.

Its application to the assessment of energy that flows through the built environment for the purpose of developing mitigation strategies is more recent and is founded on the relationship that energy has with GGE. Each stage of the production, distribution and use of energy that is based on carbon (i.e., fossil fuels such as coal and natural gas) including the extraction of the fuel itself, results in the liberation of greenhouse gases, including CO and CO_2. It is now widely accepted that environmental damage and the enhanced greenhouse effect and its associated climate change are largely the result of mankind's accelerating consumption of fossil-based resources.

It has been observed (Chapman, 1974, p. 91) that 'all that man's activities accomplish is a temporary change from a stock of raw material or flow of solar energy, into products such as automobiles and food which, in time become discarded materials and dissipated energy'. Our historically cheap access to abundant fossil energy resources has been linked to our continued economic growth and development. The dual problem of resource depletion and emissions pollution means that there is a pressing need to drive efficiency increases in processes dependent upon non-renewable resources and create incentives to foster growth of alternative activities that are less energy intensive.

Energy sources may be classed as either primary or secondary, and renewable or non-renewable. Primary energy sources are those that can be used in their natural state in order to provide useful energy, e.g., coal, oil and natural gas. Secondary energy sources are those that result from the processing of primary sources; electricity, petrol (gasoline) and liquid natural gas are examples.

Non-renewable resources are those for which it is believed there are finite reserves, and include natural gas, oil (including synthetic derivatives), coal and methane (from coal seams), all of which are fossil- or carbon-based and which liberate carbon dioxide when burnt. Renewable fuel sources on the other hand are those without practical limits, such as water, solar, wind and wave energy. The energy directly produced from these sources is carbon free and without direct GGE impacts, however, they still have an indirect impact as they require built infrastructure for processing and distribution. Deforestation that is a result of dam construction also contributes to GGE as carbon sink capacity is reduced and therefore the associated GGE must be ascribed to the energy produced. There is also evidence that suggests that the dams used by hydro-electric schemes have a finite life due to problems of sedimentation.

Biomass fuels, although from renewable sources, are also carbon-based and thus liberate carbon dioxide and water vapour through combustion, contributing to emissions. They are, therefore, not as 'green' as they may appear.

6.3 Embodied energy and materials

Embodied energy may be defined as 'the energy consumed in all activities necessary to support a process, including upstream processes' (Treloar, 1997, p. 375). It may be divided into two components: the direct energy requirement and the indirect energy requirement.

> Direct energy includes the inputs of energy purchased from producers used directly in a process (including, in the case of a building, the energy to construct it). Indirect energy includes the energy embodied in inputs of goods and services to a process, as well as the energy embodied in upstream inputs to those processes (Treloar, 1997, p. 376).

In simple terms, this means all of the energy needed in order to produce any good or service, including all of the energy necessary to make all of the machines that make the

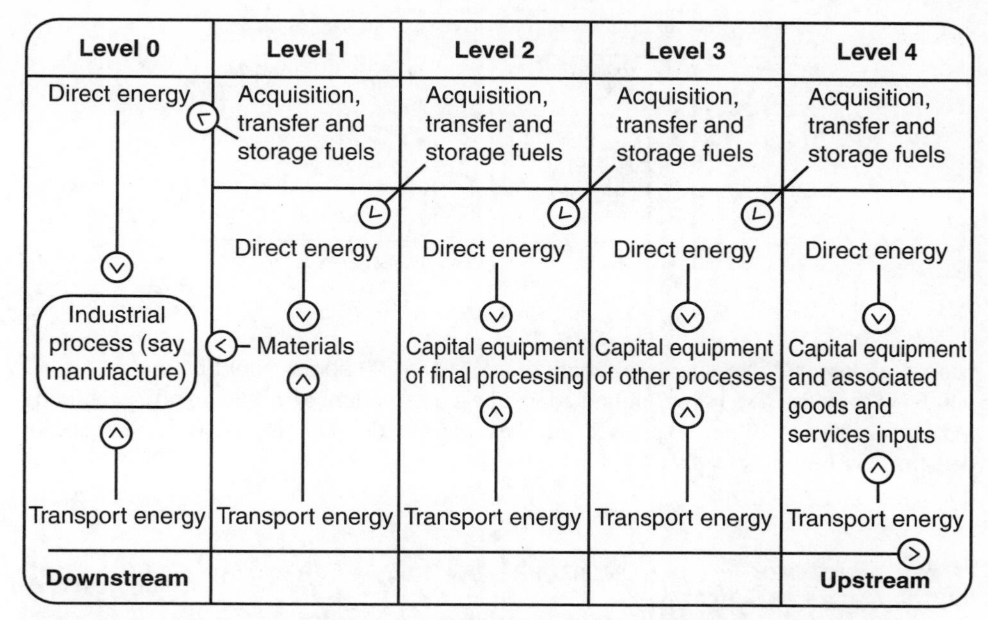

Figure 6.1 General schematic of the embodied energy notion (adapted from Treloar, 1996a).

goods. If the process is followed right to its origin, which in the case of metals is the ore-bearing ground, then it includes the energy used to extract the ore. Figure 6.1 provides a schematic of the embodied energy notion.

It is critical to the validity of any energy analysis that it is systematic and has a clearly defined boundary set, within which the measurements are completed. The International Federation of Institutes for Advanced Study (IFIAS, 1974) provided the original international methodological framework for defining the stages and system boundaries for the analysis to ensure the validity of the study in question. This framework in broad terms remains but has been developed to incorporate life cycle considerations and is now also bound by the internationally accepted methodology established by ISO 14040:1997 and ISO 14042:2000.

In the context of the built environment, analysis of embodied energy is concerned with the flows of energy through buildings (including that related to materials, components and systems) over the course of their entire life cycle. Conventional thinking suggested that the energy embodied in a building was relatively small in comparison to that used in its operation.

Historically, a ratio of life cycle embodied energy to operational energy has been used to demonstrate the relative significance of these two factors. As the operational energy efficiency of buildings has generally increased over time, this ratio has continued to tip towards the embodied end of the scale. Table 6.1 demonstrates this case and provides an indication of the significance of embodied energy in relation to energy use over the life cycle of buildings.

Whilst these figures may appear inconsequential when expressed in absolute terms for a single building, when aggregated the total numbers can be daunting. For example, it has

Table 6.1 Ratio of life cycle operational energy to embodied energy*

	Range	Non-efficient	Efficient
House	Upper	5.8:1	3.6:1
	Lower	10.8:1	5.1:1
Office	Upper	10:1	1.5:1
	Lower	8:1	2:1

*Ratio calculated on the basis of MJ/m^2 of GFA (Mackley, 1997; 2000).

been demonstrated that a town of approximately 185 000 homes could be operated for a whole year with the energy embodied in the construction of a single office building (Mackley, 2000). When expressed in these terms the relevance of the impact of construction becomes clear.

6.4 Embodied energy intensity extraction methodologies

The development of embodied energy intensity extraction methods has significantly advanced the understanding of the importance of embodied energy to the life cycle energy balance of a building and the potential for energy savings through materials selection. Questions remain, however, as to the reliability of embodied energy intensities due to significant variance in published data. These differences are largely the result of the various extraction methodologies and reporting conventions. This has contributed to the construction industry's hesitancy in adopting and implementing wide-ranging energy use-reduction measures (Mackley, 1999).

There are three main methods of embodied energy intensity extraction: process-based, input–output-based and hybrid analysis. They vary significantly in terms of their system boundaries and reporting conventions, and this has led to significant confusion among researchers and within the industry. These methods and their relative advantages and disadvantages are briefly discussed below.

6.4.1 Process analysis

This can be described as embodied energy analysis that uses 'any other source of information other than input–output tables' (Treloar, 1997). It is generally accepted (Pullen, 1995; Alcorn, 1995; Treloar, 1996a) that its major advantage is the degree of accuracy possible for the precisely defined system to which it relates (e.g., the production of kiln-dried timber from a particular mill). Published figures derived by this method are, however, incomplete for a number of reasons:

- it is difficult to account for direct energy inputs that occur more than two stages upstream of the process being analysed, and these are therefore not included in the calculations

- indirect energy associated with the provision of capital goods required for the process both directly and upstream from the process are usually excluded
- the energy lost in energy conversion processes (e.g., the energy lost as waste heat during electricity generation) is excluded from the calculation – if primary energy were to be reported, rather than delivered or end-use energy, then for many common building materials the energy intensity may be 30 to 40% greater (Pears, 1995).

6.4.2 Input–output analysis

This method uses a mathematical approach in an attempt to sum all of the upstream energy requirements, and so is a more complete method than process analysis. There are, however, problems related to the validity of the results as a range of assumptions must be made in regard to the homogeneity of some or all of the sectors of the economy relevant to the analysis, the energy tariffs paid by the various sectors, and the prices of the commodities that each sector produces. Most sectors of the economy are not homogeneous and consequently significant variations in embodied energy intensities may appear when the results obtained using this analysis are compared to those derived by applying process analysis to individual producers/manufacturers.

6.4.3 Hybrid analysis

As the name suggests this method is a combination of the previous two. It utilizes 'the completeness of input–output analysis and the reliability of process analysis' (Treloar, 1997). The significant energy pathways are identified and analysed using input–output analysis, then data for a specific process is used to calculate the energy intensity of a particular material. This avoids many of the problems that arise when either method is used in isolation. Importantly, the results are reported in terms of primary energy use and potential double counting of direct energy inputs is avoided. This method can produce results that are up to 90% complete (Treloar, 1998), although there are still unresolved problems related to energy tariffs, primary energy factors and sector homogeneity.

In spite of these remaining questions the major advantage of this method is the consistent quality of the data, and the energy intensities that are produced provide the most complete and constant indicators of the energy required to produce individual materials on a national basis.

Unfortunately these methods generally lack credibility in the construction industry, partly because few people have a clear understanding of the methodologies involved. Treloar (1996b) suggests that it is the 'black box' nature of the processes that makes the uninitiated suspicious of them. They are also questioned because they provide average material intensities but make no allowance for the relative efficiency of individual manufacturers. Ironically this is the greatest advantage of hybrid analysis.

It is evident that the validity of any reported embodied energy figures will depend on the method of analysis used, and on the system boundaries being clearly defined. This is particularly important if comparisons are to be made between embodied energy figures from a variety of sources. Table 6.2 demonstrates this point clearly: embodied energy intensities for rough processed timber, calculated by a number of different analysts are listed.

Table 6.2 Embodied energy intensity for rough sawn timber (After Salomonsson, 1996)

Source	GJ/m³	GJ/t	Method	Country of origin
Alcorn (1995)	1.4	2.8	Process	New Zealand
Beca *et al.* (1976)	2.7	5.3	Process (DEO)	New Zealand
Forintek Canada Corporation(1993)	2.7	5.3	Process (DEO)	Canada
Partridge and Lawson (1995)	1.8	3.6	Process (GER)	Australia
Treloar (1999)	7.0	14.0	Hybrid input–output (D&IE)	Australia
Salomonsson (1996)	6.1	12.2	Process (DEO)	Australia

DEO, direct energy only accounted for; GER, gross energy requirement but excluding direct energy; D&IE, direct and indirect energy accounted for.

The differences in these figures highlight the problem. They vary by a factor of 5 from a low of $1.4\,MJ/m^3$ to $7.0\,MJ/m^3$. An average brick veneer home of $120\,m^2$, with timber frames to the inner half of external walls, all internal walls and roof/ceiling but excluding timber for trims and doors, contains an average of around $8\,m^3$ of timber framing. This would produce figures for total capital embodied energy for timber ranging from 11.2 GJ to 56 GJ. Pullen (1995) suggests that the average total capital embodied energy for typical residential buildings is $4–6\,GJ/m^2$ – based on this average the variance shown above for timber alone would produce an error of $\pm 0.37\,GJ/m^2$. In the case of a $4\,MJ/m^2$ house this would result in an error of ±9.3%; for a $6\,MJ/m^2$ house the error would be ±6.2%.

The potential for further compounding differences is enormous given that there are a multitude of individual parts in each building, and similar variances could be expected for each part.

6.5 The relevance of embodied energy analysis

Thus far the discussion has focused on energy and the methodologies that are available for accounting for the embodied energy of materials. It has also touched on the association between operational and embodied energy in buildings over their life cycle. However, we must now make the link between embodied energy analyses and the opportunities that arise once a systematic assessment framework is established.

The methodological difficulties discussed above are particularly relevant to our ability to account for reductions in carbon dioxide emissions that result from energy use. The Kyoto Protocol sets targets for nations to reduce their total carbon dioxide emissions to 1990 levels, plus or minus a specific country margin. However, the discussion above shows that there is a range of methods available to account for energy and emissions. Actions taken by individual companies and industry groups can qualify for what is known as 'early action credit'. This means that emissions reductions will have some financial benefit, and therefore there is some concern that the commercial interests of those who do reduce their emissions could be skewed depending on the method of analysis that is used.

Industry argues that where an Australian average (hybrid input/output) energy and carbon intensity is used, then specific high efficiency plants or firms will be disadvantaged because the hybrid method cannot differentiate at the individual plant level. For example, take the case of a builder who wants to get financial credits for a building using aluminium window frames from a 'sustainable' source.

Table 6.3. Comparative analysis of economic impact of energy and carbon mitigation for aluminium windows

		Embodied energy (GJ)	Carbon emissions (t)	Value of carbon emissions	
				Lower US$20/t	Upper US$120/t
National average (HIO method)	1 tonne	264.2	15.85	US$317	US$1902
A	Whole building 90.5 t	23 910	1434	US$8/m^2	US$48/m^2
Efficient factory	1 tonne	105.68	6.34	US$127	US$760
B	Whole building 90.5 t	9564	574	US$3/m^2	US$20/m^2
Variance		(14 346)	(860)	(US$5)	(US$28)

In this example, the building has 3530 m^2 of aluminium windows. Assume that the efficient factory recycles used aluminium into new aluminium, uses efficient plant and uses some renewables in energy production, thus achieving a 60% reduction in both energy use and carbon emissions. International consensus currently suggests a carbon emission value of between US$20 and $120 per tonne (Table 6.3). The task is to estimate the likely economic impact in $/m^2 of windows for using aluminium from this efficient plant rather than from an 'average' producer.

In this scenario, the building has approximately 90.5 tonnes of extruded aluminium in its glazed windows. A total of 1434 tonnes of carbon is emitted on the basis of the Australian industry average figure (Treloar, 1999). For scenario A, the potential increase in the square metre cost of windows in this case is estimated to be between US$8 and US$48.

Scenario B identifies the energy, carbon and capital cost characteristics for the windows where the efficient plant is assumed. The 'variance' demonstrates the potential mitigation and economic variance.

6.5.1 Reducing embodied energy and carbon emissions in building elements

There is a danger in selecting materials on the basis of their embodied energy and carbon dioxide emissions without considering their function relative to the building element and building system that they are part of over their life cycle. Discrete mitigation initiatives should always be balanced by an analysis of the implications of each initiative on the total building.

It has been demonstrated that design decisions taken specifically to control operational energy consumption have an upstream impact on the energy and carbon intensity of the whole building over its life cycle (Treloar, 1996b; Mackley, 1997, 2000). It is therefore essential that we account for energy use in buildings and resultant carbon emissions on a holistic basis over the entire life cycle. Similarly, a review of mitigation options or the optimization process should be completed with the objective of balancing the functional objectives of the building as a whole. The following example illustrates this.

An option for reducing energy use and carbon emissions is proposed for a high rise building that involves substituting aluminium for timber in the window systems. To compare the two options, the energy and carbon intensity for the proposed systems needs to be calculated on a functional unit basis. This enables a life cycle comparison to be effected.

The following assumptions are made: the standard window unit is 1200 × 900 mm overall, with a typical frame section of 100 × 50 mm; a service life of 50 years is selected with re-painting every 10 years the only maintenance required.

The physical mass of materials for each scenario is first calculated. Carbon and energy intensities are then applied to the material components (e.g., glass, aluminium, timber, silicon, paint, etc.). Similar quantities are calculated for materials necessary for maintenance. Table 6.4 presents the results of the findings.

Table 6.4 Comparison of energy and carbon life cycle analysis for a window

	Timber		Aluminium	
	EE (GJ)	CO_2 (t)	EE (GJ)	CO_2 (t)
Capital				
Frame only	0.93	0.100	2.50	0.40
Paint to frame	0.10	0.001	0.20	0.02
Sub-total capital	1.03	0.101	2.70	0.42
Operational				
Paint	0.50	0.005	1.00	0.10
Gaskets/seals	0.96	0.040	0.96	0.04
Sub-total operational	1.46	0.045	1.96	0.14
Life cycle total	**2.49**	**0.146**	**4.66**	**0.56**

Note that any theoretical 'credit' that may be achieved through recycling part or all of the frame at the end of its life cycle is excluded from this analysis.

As can be seen from the 'Life cycle total' line above, the aluminium window has an embodied energy almost twice that of the timber windows. Carbon emissions are almost four times greater. However, it is clear that there is more energy and carbon arising from the care and maintenance of the timber window than it takes to construct it initially. In this case, it *could* be concluded that timber would be a much better option for a high rise building but if the functionality is considered it is likely that the conclusion would change; timber windows for commercial buildings would not necessarily be either appropriate or physically feasible.

However, in the case of an average 120 m^2 three-bedroom residence with 37 m^2 of windows, the potential savings, if timber windows are chosen over aluminium windows, is 74.34 GJ of energy and 15.3 t of CO_2. This example demonstrates the effect of changing the measurement 'system boundary' and the potential downstream effects that are not identified if the life cycle view is ignored.

As previously noted, a life cycle approach facilitates an overall view of the relative proportions of the embodied and operational energy consumed by a building, thus enabling the quantum of potential savings to be assessed. The main benefit in adopting an energy life cycle approach is that it facilitates comparative analysis between alternative designs and the identification of the means of making a particular option more efficient (in terms of both cost and energy). From a sustainability standpoint it enables the achievement of a balance in the distribution of energy over the various parts of the project. In the absence of a life cycle approach a clear understanding of the impact that a particular design feature may have on the upstream or downstream phase of a building's life cycle cannot be fully comprehended.

6.5.2 Case study – energy life cycle analysis

The discussion above deals with individual elements but mentions the importance of the holistic approach. The following case study demonstrates a holistic comparative analysis of the capital embodied energy of two dwellings analysed on an elemental basis. The base case (BC) is a full timber house, the other is of cavity brick construction (CB). Figure 6.2 presents the data graphically, grouped by elements identified in the Australian National Public Works code. These results are then tabulated (Table 6.5) and presented as a percentage of the total and compared to the cost by element in the same manner.

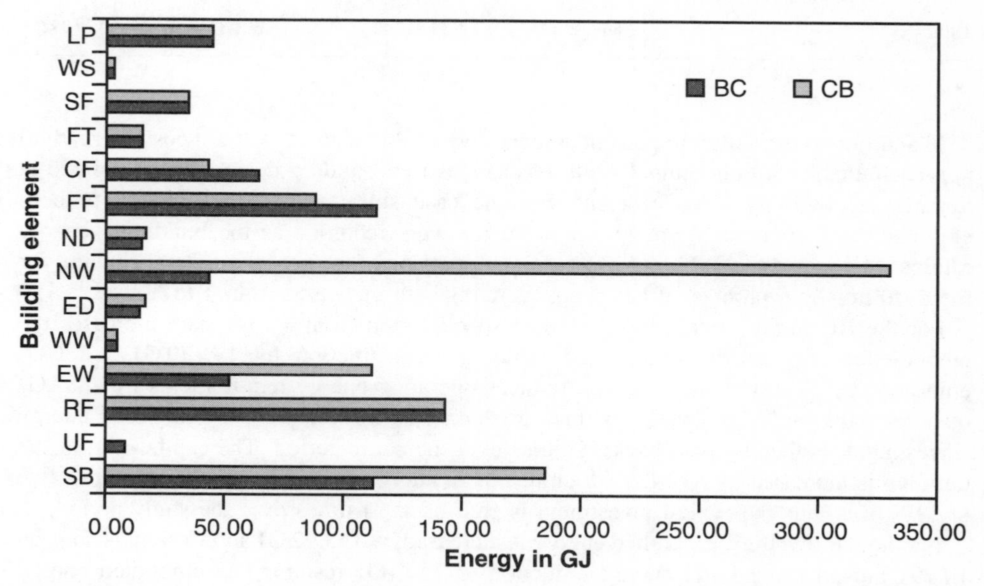

Figure 6.2 Comparative analysis of capital embodied energy in BC and CB dwellings. See Table 6.5 for key to initials.

What is clear from these two analyses is that for the CB dwelling, the internal wall (NW) element requires twice as much energy and cost than any other single element. In reviewing this report it becomes clear that there is a significant opportunity to save both energy and money by changing the specification of the internal walls. In this case, if the specification were changed to match the BC dwelling a total of 287.28 GJ of energy, 27.81 tonnes of CO_2 and $29 000 would be saved. The energy savings would equal approximately 2 years of operating energy and capital cost savings would represent around 12 years of energy costs.

Table 6.5 Comparative analysis of elements as proportion of totals

		BC		CB	
		EE by element as % of total capital EE	**$ by element as % of total $**	**EE by element as % of total capital EE**	**$ by element as % of total $**
Sub-structure	SB	17.00	14.11	17.96	17.52
Upper floors	UF	1.15	4.76	–	–
Roof	RF	21.40	13.47	13.80	12.01
External walls	EW	7.83	15.67	10.92	9.41
Windows	WW	0.69	6.13	0.44	5.50
External doors	ED	1.96	1.09	1.42	1.04
Internal walls	NW	6.59	10.60	31.90	26.35
Internal doors	ND	2.27	4.65	1.65	4.19
Floor finishes	FF	17.09	5.38	8.47	4.84
Ceiling finishes	CF	9.65	6.27	4.16	5.38
Fitments	FT	2.28	6.38	1.47	5.73
Sanitary fixtures	SF	5.13	6.50	3.31	5.83
Water services	WS	0.36	0.51	0.23	0.53
Electric light and power	LP	6.62	4.47	4.27	4.69
Total		**669.99 GJ**	**$152 542**	**1039.1 GJ**	**$179 092**

In addition to the ability to present comparative analysis, an LCA (least cost alternative) approach enables a longitudinal study of changes to a building design that are aimed at making the building more efficient. For the case study above, the changes to both embodied and operational energy requirements were recorded as the buildings' energy ratings were increased. Table 6.6 presents a comparison of the net longitudinal changes for the dwellings in a number of key areas as ratings are increased from 1 to 5 stars.

For the BC house, increasing the 'star' specification from 1 to 5 stars increases the embodied energy attributable to the building's construction by 191.79 GJ and CO_2 emissions by 17.02 t. This, however, reduces operating energy requirements by 2882 GJ over the building's life cycle (in this case taken as 50 years). The embodied energy 'investment' is thereby paid back 15 times over the study period. The capital cost of the building is increased by $12 422 which results in energy cost savings with a future value of $177 663, thus the capital investment is paid back 14 times over the study period.

For the CB dwelling, embodied energy is increased by 153.92 GJ and CO_2 emissions by 13.19 t but an operational energy reduction of 2148 GJ results. The embodied energy

Table 6.6 Comparative life cycle analysis of BC and CB dwellings with increased energy rating from 1 to 5 stars

	Increase in capital EE (GJ)	**Increase in CO_2 emissions (t)**	**LC Op E savings (GJ)**	**Increase in capital ($)**	**DV of reduced LC Op E ($)**
BC	191.79	17.02	2 882	12 422	177 663
CB	153.92	13.19	2 148	13 793	132 415

EE embodied energy DV discounted value Op E operational energy investment is paid back fourteen times. The capital cost increase is $13 793, but this produces energy cost savings with a future value of $132 415, and the capital investment is paid back nine times over the study period.

In the case of the BC dwelling, the total of the future value of operational energy savings is actually greater than the initial capital value of the dwelling.

6.6 Conclusion

Whilst energy is not the only environmental impact that must be considered if we are to attain a sustainable building standard, it has been demonstrated that it has great potential to create a real economic incentive for the stakeholder.

The methods of assessing the energy flows over a building's life cycle vary, and vastly different results have been reported to date. Although this has added to industry frustration and slowed support for further research and development in this area, there is some work underway which attempts to deal with these issues. The International Energy Agency's (IEA) Annex 31 program (IEA, 1998) is an example of the support required in order to progress the field of life cycle energy analysis.

The discussion here has shown the magnitude of potential risk and opportunity in relation to energy use and carbon dioxide emissions for buildings. Significant opportunities exist to effect significant and sustained reductions, however, the development of reliable second generation integrated modelling software, which will enable energy LCA to be completed quickly, and concurrently with the building design process, is essential if uptake by industry is to increase. This is the key to ensuring the incorporation of LCA within industry's sustainable management paradigm.

References and bibliography

Alcorn, A. (1995) *Embodied Energy Coefficients of Building Materials* (Wellington: Centre for Building Performance Research, Victoria University).

Baird, G. and Chan, S.A. (1983) *Energy Cost of Houses and Light Construction Buildings and Remodelling of Existing Houses*. Report No. 76 (Auckland: NZ Energy Research and Development Committee).

Bartlett, E. and Howard, N. (2000) Informing the decision makers on the cost and value of green buildings. *Building Research & Information,* **28** (5/6), 315–24.

Beca Carter Hollings & Ferner Ltd. and Scott, G.C. (1976) *Forest Industry Energy Research Summary.* Report No. 12. New Zealand Energy Research and Development Committee, Auckland.

Bender, S. and Fish, A. (2000) The transfer of knowledge and the retention of expertise: the continuing need for global assignments. *Journal of Knowledge Management,* **4**, 125–37.

Berkebile, R. and McClennan, F. (1999) The living building. *The World & I,* **14** (10), October, 160.

Bon, R. and Hutchinson, K. (2000) Sustainable construction: some economic challenges. *Building Research & Information,* **28** (5/6), 310–14.

Bordass, B. (2000) Cost and value: fact and fiction. *Building Research & Information,* **28** (5/6), 338–52.

Bradbrook, A.J. (1991) *The Development of Energy Conservation Legislation in Australia.* ERDC 91/94, August (Canberra: Energy Research and Development Corporation).

Bramslev, K. (2000) *Eco-efficiency in the Building and Real Estate Sector* (Oslo: GRIP/ØkoBygg).

Bullard, C.W. and Herendeen, R.A. (1975) Energy impact of consumption decisions. In: *Proceedings of Institute of Electrical and Electronics Engineers*, March, 63:3.

Chapman, P.F. (1974) Energy costs – a review of methods, *Energy Policy*, Vol. 2, No. 2, June, p. 93.

Cole, R.J. (1993) Embodied energy and residential building construction. In: *Proceedings of Innovative Housing*, Volume 1, Technology, Vancouver, 21–25 June, pp. 49–59.

Cole, R.J. (1999) *GBC 2000 – Changes to the GBC Framework and GBTool*. Report submitted to the Buildings Group/CETC, Natural Resources Canada, March 31.

Cole, R.J. (2000) Cost and value in building green – editorial. *Building Research & Information,* **28** (5/6), 304–9.

Cole, R.J. and Larsson, N. (1997) Green Building Challenge '98. In: *Proceedings of CIB 2nd International Conference on Buildings and the Environment*, Paris, June, 19–29.

Cole, R.J. and Larsson, N. (2000) Green Building Challenge – lessons learned from GBC '98 and GBC 2000. In: *Proceedings of Sustainable Buildings 2000*, Maastrict, The Netherlands, October, pp. 213–15.

Cole, R.J. and Sterner, E. (2000) Reconciling theory and practice of life-cycle costing. *Building Research & Information,* **28** (5/6), 368–75.

Department of Environment, Sport & Territories (1992) *National Greenhouse Response Strategy (NGRS)* (Canberra: Australia Government Printing Office).

Dinesin, J. and Traberg-Borup, S. (1994) An energy life cycle assessment model for building design. In: *Proceedings of CIB TG8, Buildings and the Environment, 1st International Conference*, Session: Materials, Paper 13, BRE Watford, April, pp. 1–8.

Environment Australia (2000) *Tender Specification Brief: Comprehensive Environmental Rating System for Australian Buildings* (Environment Australia). www.ea.gov.au

Forintek Canada Corporation (1993) *Building materials in the context of sustainable development. Raw material balances, energy profiles and environmental init factor estimates for structural wood products*. March 1993, Forintek Canada Corporation, Canada.

Garvin (1993) Building a learning organization. *Harvard Business Review*, July–August, 78–91.

Gick, M.L. and Holyoak, K.J. (1987) The cognitive basis of knowledge transfer. In: Cormier, S.M. and Hagman, J.D. (eds) *Transfer of Learning: Contemporary Research and Applications* (San Diego: Academic Press).

Harris, D.B. (1996) *Creating a Knowledge Centric Information Technology Environment* (Technology in Education Institute). www.dbharris.com/ckc.htm

Hawkins, P. and Lovins, A. (2000) *Natural Capitalism – Creating the Next Industrial Revolution* (USA: Bay Books).

IEA (1997) *Energy Use and Human Activity: Indicators of Energy Use and Efficiency* (Paris: International Energy Agency).

IEA (1998) Energy related environmental impact of buildings, IEA Annex-31. In: *Proceedings of CIB 3rd World Building Congress*, Sweden, Symposium A, June, 983–90.

IFIAS (1974) *Energy Analysis Workshop on Methodology and Convention*, Workshop Report No. 6 (Stockholm: International Federation of Institutes for Advanced Study).

International Framework Committee (1999) *Notes on 6th IFC Meeting*, Toledo, Spain, March 4–5.

Janda, K. and Busch, J. (1994) Worldwide status of energy standards for buildings. *Energy* **19** (1), 27–44.

Langston, C. (1996) Life-cost studies. *Royal Australian Institute of Architects Environmental Design Guide* (Canberra: RAIA).

Larsson, N. (1999a) National Team Roles – Green Building Challenge 2000. Documentation from the International Framework Committee (Green Building Information Council). http://greenbuilding.ca/

Larsson, N. (1999b) An overview of green building challenge. *Greenbuilding Canada*, May.

Larsson, N. (2001) *Improving the Performance of Buildings* (Natural Resources Canada).

Larsson, N. and Clark, J. (2000) Incremental costs within the design process for energy efficient buildings. *Building Research & Information,* **28** (5/6), 413–18.

Lawson, B. (1996) *Building Materials Energy and the Environment – Towards Ecologically Sustainable Development* (Canberra: Royal Australian Institute of Architects).

Leveratto, M.J. (2000) Assessing methodologies for environmental performance of office buildings in Buenos Aries. In: *Proceedings of Sustainable Buildings 2000*, Maastrict, The Netherlands, October 22–25), pp. 728–30.

Levine, M., Martin, N. and Worrell, E. (1995) *Efficient Use of Energy Utilising High Technology: An Assessment of Energy in Industry and Buildings* (London: World Energy Council).

Mackley, C.J. (1997*) Life Cycle Energy Analysis of Residential Construction – A Case Study*. Master of Building in Construction Economics thesis, University of Technology, Sydney (unpublished).

Mackley, C.J. (1998) Environmental economics – an opportunity for us all. In: *Proceedings of Pacific Association of Quantity Surveyors*, Queenstown, New Zealand, June, pp. 214–20.

Mackley, C.J. (1999) Embodied energy and re-cycling. In: Langston, C. and Ding, G. (eds) *Sustainable Practices in the Built Environment*, Second edition (Oxford: Butterworth-Heinemann).

Mackley, C.J. (2000) *Marginal Cost of Implementation of Sustainable Development Strategies for Office Buildings*. Internal publication (Sydney: Bovis Lend Lease).

Mackley, C.J. (2001) *Life Cycle Energy Economics of Office Construction*. Doctoral Thesis (in progress), Faculty of Built Environment, University of New South Wales, Australia.

Mackley, C.J. and MacLennan, P. (2000) Environmental education for the building industry. In: *Proceedings of Sustainable Buildings 2000*, Maastrict, The Netherlands, October 22–25, pp. 174–6.

Oppenheim, D. and Treloar, G.J. (1994) Mass, insulation and infiltration. In: *Proceedings of Solar '94: Secrets of the Sun*, Sydney, November, pp. 214–19.

Partridge, H. and Lawson, B. (1995) *The Building Material Ecological Sustainability Index – a numerical method of assessing the environmental impact of building materials*. Partridge Partners, Sydney.

Pears, A. (1995) Australian residential energy use: what's been happening and where are we going? In: *Transactions of the 1st International Symposium on Energy, Environment and Economy*, Melbourne University, November, Paper 2C3.

Peet, J. and Baines, J. (1992) *Energy Analysis*. Report No. 126 (Auckland: NZ Energy Research and Development Committee).

Prior, J.J. (1993) *BREEAM 1/93, An Environmental Assessment for New Office Buildings*. Report No. BR234 (Watford: Building Research Establishment).

Pullen, S. (1995) *Embodied Energy of Building Materials in Houses*. Master of Building Science thesis, University of Adelaide, Adelaide (unpublished).

Salomonsson, G. (1996) Product Comparison Methods, in *Proceedings of Embodied Energy State of Play Seminar*, Deakin University, Geelong, 28–29 November, pp. 23–31.

Smith, M., Whitelegg, J. and Williams, N. (1997) Life cycle analysis of housing. *Housing Studies* **12** (2), 215.

Treloar, G.J. (1994) *Embodied Energy Analysis of the Construction of Office Buildings*. Master of Architecture thesis, Deakin University, Geelong (unpublished).

Treloar, G.J. (1995) Assessing the Embodied Energy Saving from Recycling and Alternate Materials in Buildings, in *Proceedings of SOLAR '95 Conference*, Hobart.

Treloar, G.J. (1996a) Indirect Embodied Energy Pathways of the Australian 'Residential Building' Sector, in *Proceedings of CIB Conference*, Melbourne.

Treloar, G.J. (1996b) Completeness and Accuracy of Embodied Energy Data: A national model of residential buildings, in *Proceedings of Embodied Energy State of Play Seminar*, Deakin University, Geelong, 28–29 November, pp. 51–60.

Treloar, G.J. (1997) Extracting embodied energy paths from input–output tables: towards and input–output based hybrid energy analysis method. *Economic Systems Research*, **9** (4), 375–91.

Treloar, G.J. (1998) A Comprehensive Embodied Energy Analysis Framework, Ph.D. thesis, Deakin University.

Treolar, G.J. (1999) Embodied Energy Intensities of Materials from the Australian Economy for 1992–93 Input–Output Data, unpublished database for CSIRO, provided to author by Treloar.

Tucker, S.N., Salomonsson, G., Treloar, G., McSporran, C. and Flood, J. (1993) *Environmental Impact of Energy Embodied in Construction*. Report prepared for the Research Institute for Innovative Technology for the Earth (RITE), Kyoto. DBCE DOC 93/39M. (Highett, Vic.:CSIRO Australia Division of Building, Construction and Engineering).

United Nations (1992) *Report of the United Nations Conference on Environment and Development*, Rio de Janeiro, June.

Wells, J. (1998) The US Green Building Council – promoting whole system design. *Building Australia*, **1** (5), September.

Wilkenfeld, G. (1991) *Greenhouse Gas Emissions from the Australian Energy System – The Impact of Energy Substitution* (Canberra: Energy Research and Development Corporation).

World Commission on the Environment and Development (1987) *Our Common Future (The Brundtland Report)* (Oxford: Oxford University Press).

Yates, A. and Baldwin, R. (1994) Assessing the environmental impact of buildings in the UK. In: *Proceedings of Buildings and the Environment, 1st International Conference of CIB TG8* (Watford: Building Research Establishment).

7

Innovation in energy modelling

Jon Hand* and Paul Strachan*

Editorial comment

Around 30 BC the Roman architect Vitruvius wrote his ten books of architecture, the oldest surviving textbooks of building practice. In his famous work Vitruvius wrote of many aspects of building design including instructions to designers as to how they might place a building on a site so that it would catch the cooling breezes in summer and benefit from the warmth of the winter sun. For much of the twentieth century building designers ignored the simple lessons of Vitruvius and built energy-hungry buildings, air conditioned and artificially lit, with scant regard for the costs that their designs imposed not only on building owners and tenants, but on the natural environment.

Recent changes in community attitudes to the burgeoning problems of pollution and environmental degradation are placing designers and developers under increasing pressure to find better ways to heat, cool and light their buildings. Consulting engineers are regularly asked to evaluate alternative design solutions in terms of relative energy usage and to make recommendations based on the perceived benefits of reduced energy requirements. With the advent of affordable yet ever more powerful computer hardware and the development of a wide range of modelling and analytical software, engineers have at their disposal a variety of tools for simulating building operation and performance. As computing power has grown so the power of these simulation tools has expanded, such that today virtual models of building systems of unimagined complexity can be generated, tested, altered, re-tested and interrogated, quickly and cheaply.

Once a commitment is made to use these tools, and to allow the designers the time needed to utilize them efficiently, the opportunity is there for the production of

* University of Strathclyde

buildings that will perform better and use less precious natural resources such as oil and gas. The advantages are obvious with lower operating costs, less environmental impact and so on. The case studies presented in this chapter illustrate a variety of specific yet typical situations where energy modelling tools have been used to provide engineers with data that has allowed them to make much better informed decisions and recommendations regarding selection of appropriate solutions.

7.1 Introduction

The design process thrives on the exchange of ideas and information between members of the design team. Designs evolve by testing ideas against a range of design objectives such as the economical use of space, the visual interplay between the form and façade of the building and its surroundings, ease of access, fitness for purpose, delivery of thermal and visual comfort, as well as capital and life cycle costs. The first volume in this series (Best and de Valence, 1999) emphasized the importance of early decisions and integrated design in the success of the building design. This is certainly true of the energy and environmental issues that are the subject of this chapter.

To understand how practitioners test design ideas there is no better place to start than to look at the design sketches that are a ubiquitous element of the schematic design phase. As a communication mechanism for the design team, such sketches convey the essential structure and understanding of the evolving design. Figure 7.1 shows a typical design sketch where ideas on solar penetration and ventilation are shown.

The problem is in knowing if the ideas encapsulated in such sketches will work. Of course, if the design team has, or knows of, past experience of the particular configuration this should give them confidence. It is believed, however, that for successful innovation

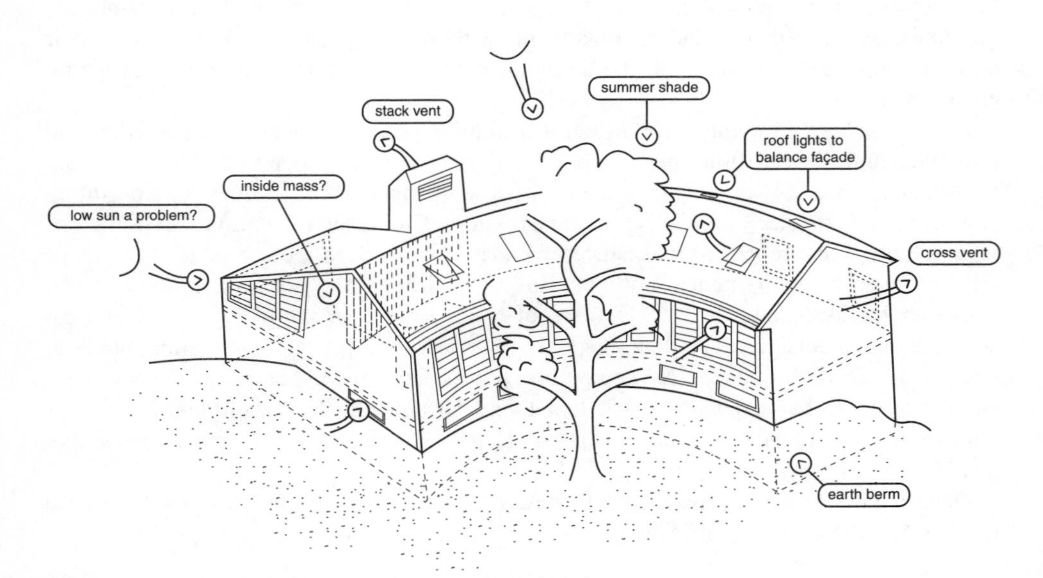

Figure 7.1 Sketch design for solar and ventilation performance.

in design, there is a need for energy and environmental modelling which allows the ideas to be tested.

For many aspects of design, practitioners have been trained to recognize design conflicts for a range of issues such as disabled access, fire protection, structural spans and spatial planning. Design sketches that fall foul of these rules are subject to discussion and 'what if' explorations to resolve problems. It has proved rather more difficult for design teams to recognize when sketch representations of environmental performance (such as that shown in Figure 7.1) are at variance with the underlying physics and in need of further discussion and analysis. There are several reasons for this:

- the energy and environmental performance of a design (in terms of such things as thermal and visual comfort, light distribution, air flow, and the effect of variations in outdoor climate) cannot readily be described by rules of thumb. There is often uncertainty as to whether general guidelines from published literature are applicable to the specific project. The behaviour indicated by arrows of air flow or rays of light in a sketch may result from assumptions that are optimistic and in some cases simply wrong
- the temporal aspects of a design which may be critical to the productivity and comfort of occupants are difficult to express in design sketches. For example, the angle of the sun is constantly changing, as is the speed and direction of the wind; in such cases animated sequences would be more likely to support understanding of design implications
- the curricula of architecture and design degrees have sometimes been deficient in giving members of the design team an understanding of energy and environmental performance issues that arise in current design practice
- energy and environment assessment tools that can test the performance of a design are widely perceived to be inapplicable at the early design stage. Those who have the skills to use such tools may be isolated from the design team and thus poorly equipped to translate design sketches into an appropriate numerical model or interpret the resulting predictions in ways that make sense to the design team.

The life cycle costs of buildings are starting to receive widespread attention, leading to increased emphasis on integrated design. It should not be the case that the mechanical engineer is presented with a flawed design which must, somehow, be coerced into providing the thermal comfort requirements specified by the client. Early testing can not only limit last minute design modifications (which are costly), but also offer the potential for better integration between the building and its environmental systems.

Although this discussion has focused on the need to consider energy and environmental performance in the early design stage, there is also a role for modelling in answering 'what-if' questions in the development of the detailed design. Such questions are often focused on the performance of just one facet of the design: the use of a novel component, or the search for the best choice within a set of design options. It may then be important to revise the model to focus on particular aspects of the design, perhaps with enhanced resolution and analysis techniques.

What is required is for design teams to have access to modellers, either in-house or in outside consulting engineering practice, who can use various design assessment tools to assess the energy and environmental performance of the evolving design at various levels of detail. Without modelling, the designers would need to rely on previously tried and tested designs (or take a large risk!), thereby discouraging successful innovation.

The following case studies illustrate the use of modelling software in informing the design process. Each case study includes a synopsis of the project, issues investigated, the approach taken, the metrics used to judge performance, the model created and results of the analysis.

Before moving to the case studies a brief review of assessment tools is appropriate with some explanation of the requirements and use of such tools.

7.2 Review of assessment tools

There are a large number of assessment tool types: envelope thermal analysis, plant systems and controls analysis, combined building and plant simulation, lighting analysis, ventilation and air movement analysis, and integrated tools which combine one or more of these assessment domains. Developers are actively evolving existing tools and introducing new ones. A current list is held on the US Department of Energy's *Tools Directory* web page (DoE, 2001).

Innovative designs are often more complex in their interactions, perhaps involving elements of daylighting, natural ventilation, passive solar components and thermal mass. Such designs require an integrated approach and detailed simulation models. Simplified tools are often based on 'standard' designs with in-built assumptions and are therefore not so suitable for novel designs.

Integrated detailed assessment tools are distributed as single applications which provide a full range of facilities, e.g., DOE-2 (LBNL, 2001a), or as a suite of co-operative modules, each of which support particular tasks such as model definition, database management, simulation and results presentation, e.g., ESP-r (ESRU, 2001).

In some cases the innovative element of a project may be a new component to be fitted into a building. In such cases there may be some experimental work undertaken to determine the characteristics of the component. The experimental data can be fed into the simulation program to explore how the innovative component would work in the context of the whole building.

7.3 Modelling requirements, procedures and outputs

As for most disciplines, practitioners and developers of assessment tools have evolved terminology to help describe specific tasks. Although each assessment tool has a different 'look and feel' and offers variations in functionality, there are many aspects of project planning, quality assurance, review of assessment data and reporting which are applicable to all uses of modelling.

Each assessment tool has a set of basic entities from which models can be defined. Although entity details and the user interactions required to define them vary from tool to tool, models almost always include the context of the site (site details), computer-aided design (CAD) data such as the geometry of rooms, and attributes specifying the composition and relationship between surfaces. Depending on the analysis being undertaken, additional specifications may include the definition of electrical and mechanical systems, details of openings (deliberate or otherwise) that may affect air flow in the building, and schedules which describe occupancy, small power and environmental control regimes.

Abstraction is the task of extracting the essential characteristics and attributes of the design to include in the model that are sufficient for answering the specific design questions posed by the design team. To accomplish this the practitioner will be asking questions such as:

- which part of the building needs to be described?
- how much detail is required for this assessment in terms of geometry and occupancy?
- can standard guidebook values of air infiltration rate be used or do they need to be calculated?

After the model has been defined, boundary conditions need to be specified. These depend on the particular analysis. For thermal studies they typically include hourly values for climatic factors such as external temperature, solar radiation, wind speed and direction, and relative humidity. For detailed air flow simulation using computational fluid dynamics (CFD) programs, boundary conditions are typically room surface temperatures and characteristics of air flow into the room. For lighting simulations, they would include sky luminance distribution and the characteristics of artificial light sources. The user must select representative boundary conditions to address the specific design question, e.g., hot sunny days for overheating studies, or average climate sequences for annual energy consumption calculation.

In many cases several models are tested. First a base case model is generated which represents the existing building in the case of a refurbishment project, or a reference design in the case of a new building. The model is then adapted to form a series of models that reflect the various design options to be tested.

After the models and boundary conditions have been defined simulation can proceed. The time for this may vary from a few seconds for a relatively simple analysis to several days for detailed air flow or lighting simulations.

A large range of performance metrics can be evaluated to help assess performance of the modelled building. These include:

- annual energy consumption
- heating and cooling plant capacity
- seasonal and daily variations of heating and cooling requirements
- thermal comfort
- ventilation effectiveness
- risk of condensation
- air temperature and velocity distribution in a space
- pollutant distribution
- atmospheric emissions due to use of fuels
- daylight factors
- visual comfort.

As an example, Fanger's predicted percentage dissatisfied (PPD) can be used to assess the risk of thermal discomfort (ISO, 1993). A design may be considered acceptable if the PPD at particular offices does not exceed, say, 20% for more than 40 occupied summer hours.

It is important to recognize that the particular modelling objective influences the model created. For example, to calculate the thermal comfort metric PPD would require a high level of specification of the building, its construction and operation as well as details of occupant activity and clothing levels. In contrast, a study that was concerned only with assessing peak room temperatures and heating demands might be adequately supported by an abstract representation that sought only to preserve the overall volume, mass and surface areas of the design.

Due to the large effort involved in model specification and results interpretation, quality assurance procedures are essential. This typically involves a second person checking the model for consistency with the design brief and for syntax errors. Results are also reviewed against past experience of similar designs, where applicable, and with expected performance.

CIBSE (1998) covers the topics discussed above in more detail, together with other important modelling issues, e.g., software selection, planning the modelling work and report presentation.

7.4 Case studies

The following examples demonstrate some of the situations in which modelling can be used to determine whether a design proposal will work as anticipated, or to help designers make an informed choice between possible solutions.

7.4.1 Solar ventilation pre-heat study

The addition of a sunspace or conservatory to a house has benefits of increased amenity, but the benefits in energy terms are less clear. A conservatory acts as a buffer between the house and the outside environment and can have elevated temperatures, even in winter. If the warm air could be ducted into the house as solar ventilation pre-heat (SVPH) air there should be energy savings. As with many passive solar components however, the warm air is often available when it is not required, for example, on hot summer afternoons. This case study describes a modelling study to quantify conservatory performance. The primary objective was to assess annual energy savings, but risk of overheating and the buffering effect were also investigated.

A model of a detached house was used as a focus for the study. The south façade included an aperture that could accept either the conservatory or standard glazing units. Particular attention was paid to modelling the flow of air through the house (Figure 7.2). The ability of a thermal simulation program to predict conservatory performance had already been demonstrated (Strachan and Baker, 1992). The house plus conservatory was modelled in two operational modes:

- buffer mode, in which there was only a low level of air movement between the conservatory and the house
- SVPH mode, in which air was ducted from the conservatory into the living room; two flow rates were studied, based on 0.5 ac/h for the living room and 0.5 ac/h for the whole house.

Simulations were conducted using the UK example year climate data set.

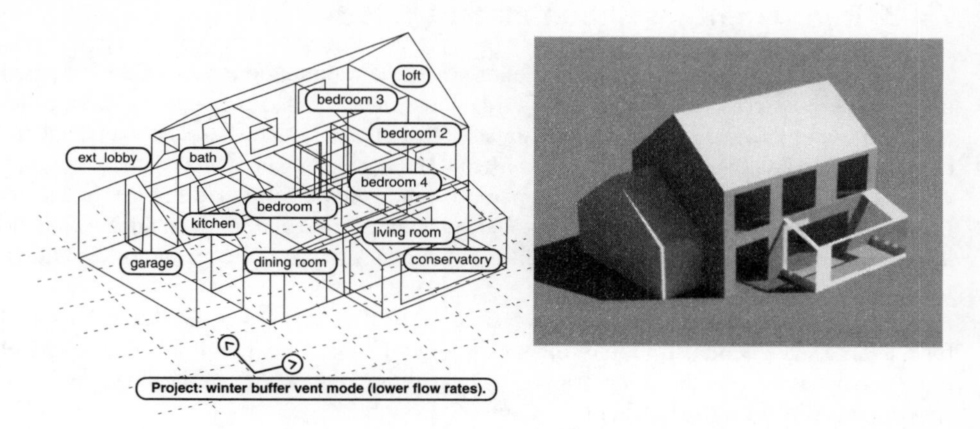

Figure 7.2 Model of detached house with attached conservatory.

Evaluation of the conservatory required the consideration of short time-step performance patterns within the dwellings under specific climatic and operational conditions as well as long-term assessments. The approach taken was to begin with a series of short-term assessments. The results were used to optimize the design. This process of conjecture and testing was similar to that undertaken by a practitioner, but with quantified measures of energy consumption and internal temperatures to aid the evaluation. The improved designs were then subjected to annual simulations to predict energy savings, with results as shown in Table 7.1.

The analysis discussed in this section was also repeated for different locations throughout the UK. The following conclusions were drawn:

- energy savings are small in the middle of winter but increase significantly towards the start and end of the heating season
- to a large extent, the performance is governed by the impact of the conservatory on air flow rather than solar gain
- simulations for more northerly locations had greater wind speeds, resulting in greater levels of air infiltration, and correspondingly greater energy savings with the addition of a conservatory.

Table 7.1 Whole house space heating requirements in kWh

	Oct	Nov	Dec	Jan	Feb	Mar	Apr	Total
Base	530.6	1151.2	1680.9	1789.1	1421.6	1040.4	914.9	8528.8
Buffer	494.7	1139.1	1680.8	1780.9	1392.9	984.9	869.2	8342.4
SVPH (0.5 ac/h living room)	482.1	1090.5	1614.4	1706.5	1335.1	925.3	814.9	7968.8
SVPH (0.5 ac/h whole house)	485.1	1151.2	1695.7	1786.8	1388.2	946.4	817.9	8271.2

7.4.2 Transparent insulation material study

This case study describes the application of modelling and simulation to the study of a novel façade material – transparent insulation (TIM). The material has a honeycomb structure that has the merit of allowing solar radiation to pass through, but which also acts as an insulator by reducing convective and radiative losses from the façade in which it is used.

The objective of the study was to compare the predicted performance against the results from monitoring and then alter certain design details to see if performance could be improved, and to repeat simulations in different climates to see if the technology could be successfully replicated.

The University of Strathclyde passive solar student residences in Glasgow, Scotland form a large scale demonstration of the technology of TIM. They comprise four blocks of flats on a steep south-facing slope, arranged in two east–west lines, offering accommodation for 376 students (see Figure 7.3).

One flat had been monitored in detail and a model of this flat was created, with attention focused on the four bedrooms forming a unit as shown in Figure 7.4. These rooms were considered to be representative of the building as a whole. Bedrooms 1, 2 and 3 have the majority of their south facing external wall covered with TIM, but each room has an openable window.

All geometrical and constructional data were available for the building. Of particular importance were the heavyweight structure of internal walls and floors, and the high levels of insulation (150 mm cavity insulation) on the north façade. All windows were either

Figure 7.3 Transparent insulation façade.

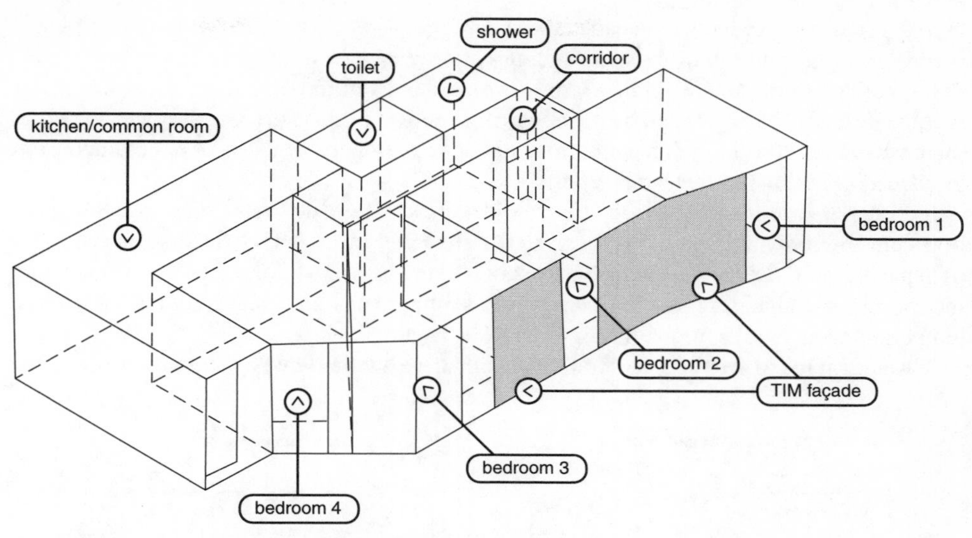

Figure 7.4 Model of part of the student residences.

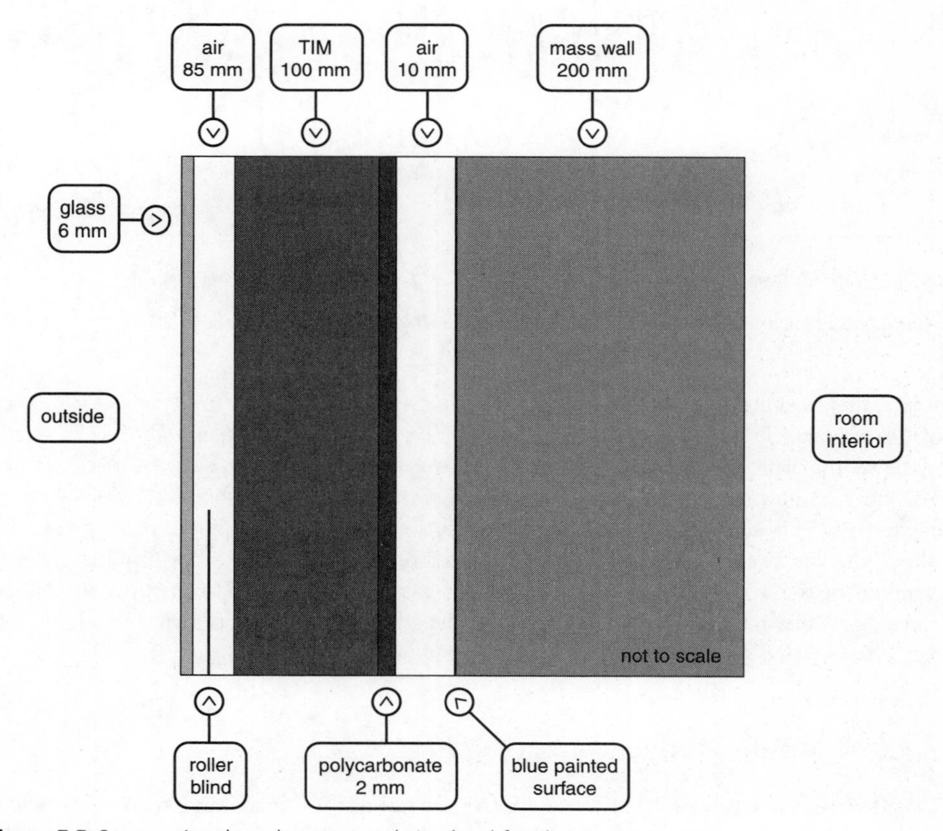

Figure 7.5 Cross-section through transparently insulated façade.

triple-glazed (for the north façade and bedroom 4) or low-emissivity double-glazed (bedrooms 1, 2 and 3 and the south facing window in the common room).

Of particular importance in this study were the details of the TIM façade itself. A cross-section through the construction is shown in Figure 7.5. As TIM was a novel material, there was little information on its thermal and optical properties; these were obtained from published experimental research reports.

In order to model the façade in more detail, separate thermal zones were created between the polycarbonate backing sheet and the mass wall, thus separating the transparent and opaque elements. Measured data was used for external climate and internal heat gains. Figure 7.6 shows the comparisons of measured and predicted temperatures at two locations through the TIM façade.

It was concluded from these comparisons that the model gave acceptable predictions.

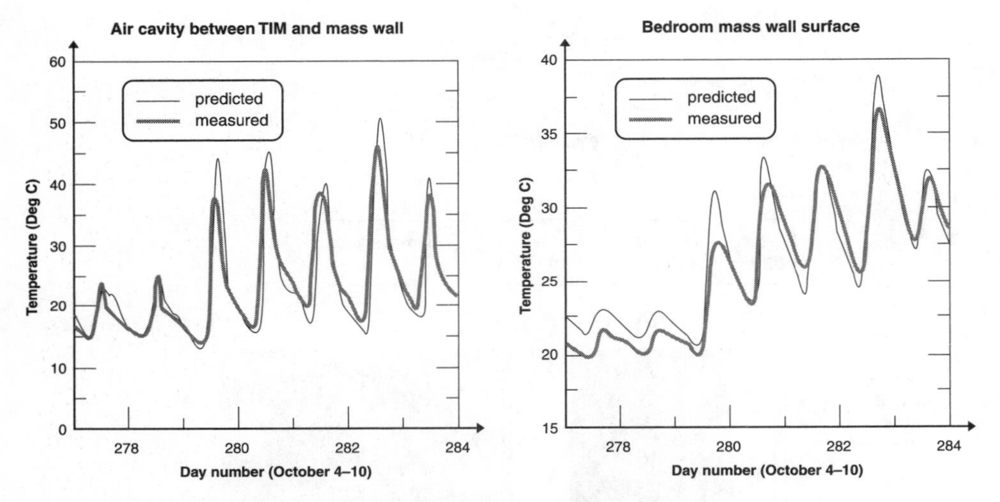

Figure 7.6 Predicted and measured temperatures within the TIM façade.

It was then used to investigate design alternatives. Firstly, the possibility was considered of putting the TIM material directly against the wall without an air gap. This would increase the heat losses at night but eliminate the ventilation losses from the cavity. Secondly, simulations were run with the mass wall being less thick and with a lower capacity, which may have proven a cheaper option. It was found that attaching the TIM directly to the external surface of the mass wall results in improved performance, due to removal of the ventilation along the external mass wall surface. However, reducing the capacity of the mass wall leads to larger, perhaps unacceptable, temperature swings. The study is reported in more detail in Strachan and Johnstone (1994).

7.4.3 Lighting study

Visual simulation can be used to depict the distribution of shade and shadow around a design and the distribution of light within a building. Design questions can also take the form of 'what is my environment like at a particular position within a room?'. In this case study the design team was presented with an office building that the client wished to be visually open to the historical urban context of the surrounding buildings. Although the adjacent buildings obstructed much of the sky, the team was concerned that visual glare might be a problem.

The approach taken was to represent the overall building façade and the adjacent streetscape at moderate resolution and provide a higher resolution at the façade adjacent to the office and for the immediate surroundings of the occupant. Composition of the model required a morning's work and involved importing CAD data from the architect and enhancement of objects near the occupant. Initial renderings were available for a meeting the next day. Figure 7.7 shows the occupant's view at mid-day in January with a clear sky, generated by the RADIANCE visual simulation tool (LBNL, 2001b). The design team was also presented with false-colour renderings that identified glare sources and an

Figure 7.7 Visual environment at a workstation: midday in winter with clear sky conditions.

animated sequence of views throughout the day at several times of year. Processing these sequences required several days, however, the patterns of changing light and glare were critical to the design team's understanding of the design.

7.4.4 Heating systems study

During an early phase of the design of a children's hospital ward, the design team wished to compare a proposed underfloor heating system (which would free up wall space and lessen the risk from hot surfaces) with a typical radiator system. The design team was concerned about control, system capacity and stability of comfort in day rooms, shared occupancy and five-occupant wards. Before proceeding with space planning, the design team also wished to know if performance was sensitive to orientation and exposure of the rooms.

Metrics for the study were temperature extremes and rate of change of average thermal comfort and environmental system response. As detailed spatial planning was not yet complete typical room dimensions, occupancy patterns and room use were used to form a composite representation. Because neither comfort nor system response metrics demanded geometric detail each room was simply represented and furniture and fittings were omitted.

To support direct comparison of the base case and design variant performance, the lower floor of the model (see Figure 7.8) used the underfloor system and the upper floor the standard system (all other details being the same). Most design tools support model copying so that a number of design variants can be composed rapidly. Indeed, once a base

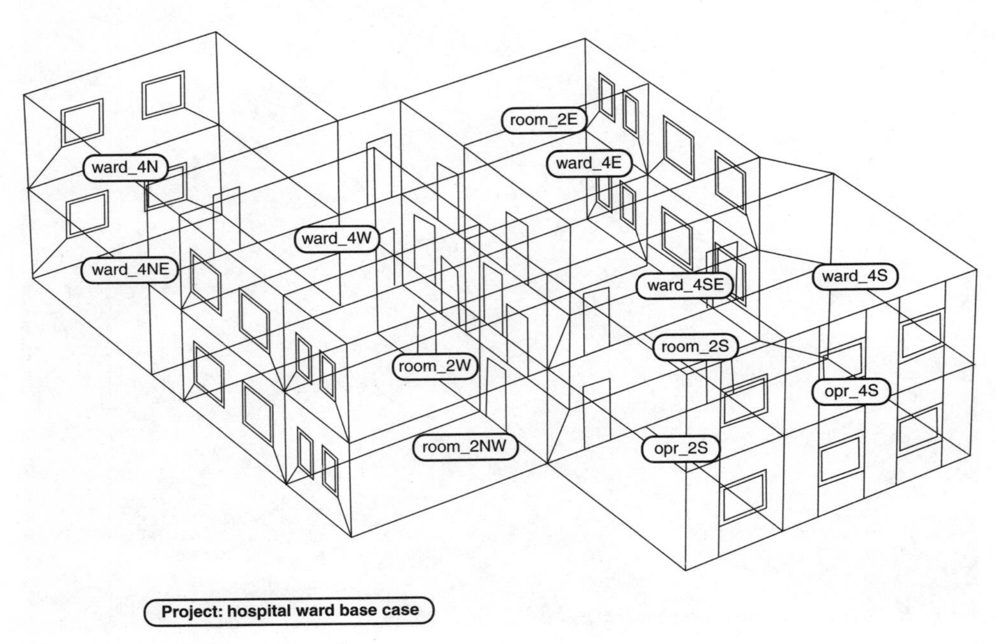

Figure 7.8 Model of hospital ward.

case model is created, the marginal cost of incremental changes and additional assessments is minimal.

Only broad categories of environmental systems were being studied, so only the essential characteristics of the systems were included. For the floor heating system, a heating flux was injected into a layer of the floor rather than into an explicitly defined piping and boiler network. The radiator system was defined as a mixed radiant/convective heat injection rather than as individually specified components. In both cases control was based on average zone dry bulb sensors. To test whether system response and thermal comfort were sensitive to changes in climate, several typical and extreme sequences of climate data were identified.

Having defined the model, the next task was to undertake sufficient performance assessments to understand the benefits and constraints of each design option. This was accomplished by undertaking about six assessments. By checking each performance metric in turn it was possible to build up an impression of the performance of the alternative designs. Figure 7.9 shows example graphs used to judge performance.

On the left is a plot of temperature and system demands for one of the wards with floor heating and on the right is the same data for the standard design. Although peak demands for each case are similar, the timing of the demands are substantially different. The floor heating design is active overnight (potentially using off-peak energy supply) while standard convector demands are distributed throughout the day. This led the design team to consider the use of a mean radiant temperature sensor and PI (proportional integral) controller rather than an on/off controller for the floor heating system. This resulted in finer temperature control and slightly less overall energy demand for the floor-based system than the convector system.

7.4.5 Ventilation study

This case study focuses on a detailed phase design study in an office building undertaken to determine whether a displacement ventilation scheme would be more advantageous than a standard constant volume (CV) diffuser system. The building included a raised floor for services as well as a suspended ceiling system. The design team wondered whether a night purge of cool air using the displacement system would result in sufficient structural cooling to counteract loads during office hours. The project would benefit from lower tariffs overnight if peak temperatures could be reduced in the afternoon.

The performance metrics of this study were system capacity and time of peak demands, the distribution of demands over the day, degree of structural cooling and the number of hours when cooling demands could be met with mechanical ventilation. These metrics required an explicit treatment of the floor and ceiling voids and air flows within the building. As only the demand profiles were of interest the plant system was represented abstractly. Although the building was substantially open plan, each floor was sub-divided into east and west zones in order to determine if early morning and late afternoon solar loads might unbalance demands.

The model of the building is shown in Figure 7.10. The base case model was replicated and the air flow network and control logic modified to represent the design with the CV system.

Lib: childrens_win.res: Results for childrens
Period: Sat 9 Jan @00h45 to Mon 11 Jan 2 @23h45 Year: 1999 : sim @ 30m, output @ 30m
Zones: room 2_NW

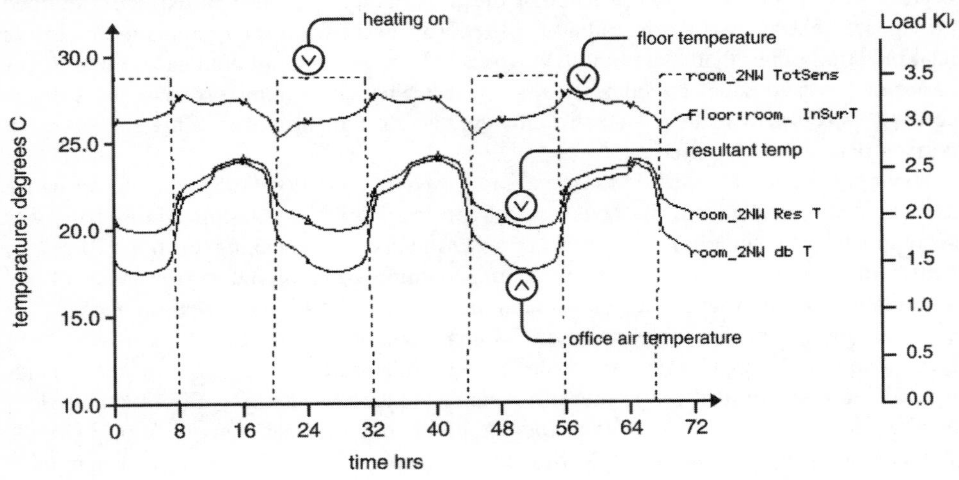

Performance with floor heating

Lib: childrens_win.res: Results for childrens
Period: Sat 9 Jan @00h45 to Mon 11 Jan 2 @23h45 Year: 1999 : sim @ 30m, output @ 30m
Zones: room 2_W

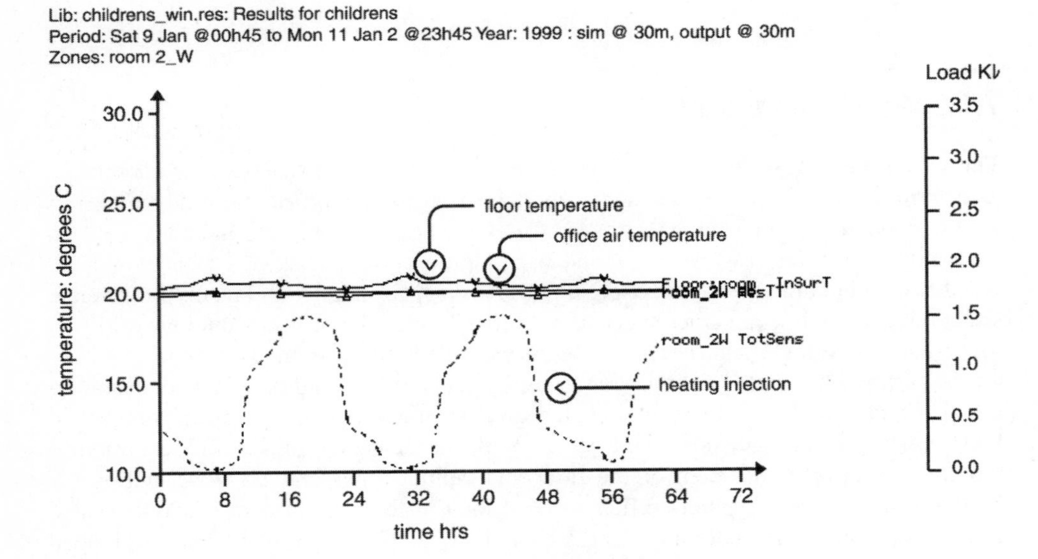

Performance with conventional convectors

Figure 7.9 Performance of floor-heated ward and standard ward.

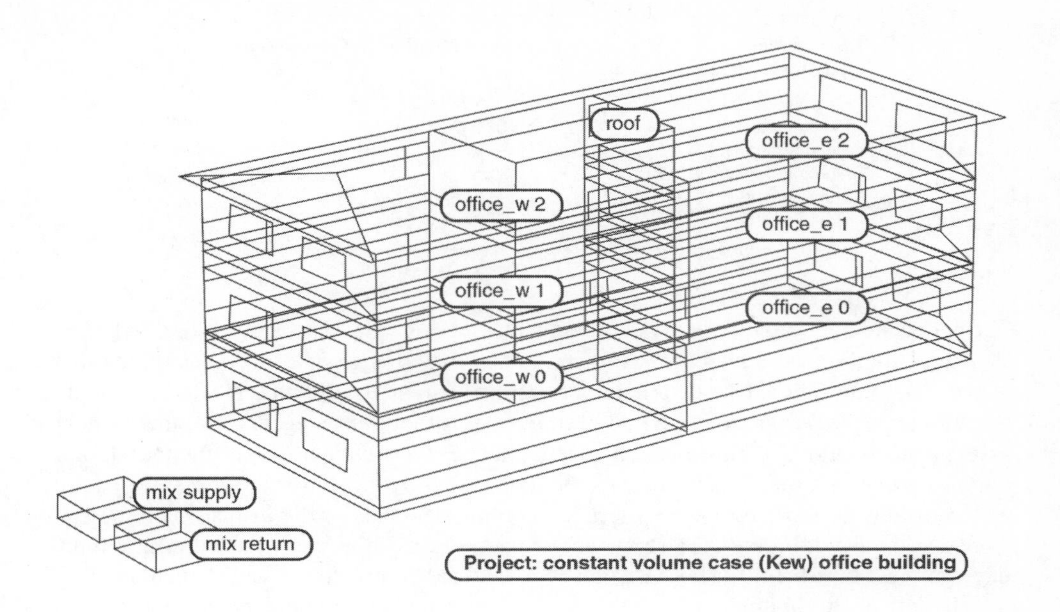

Figure 7.10 Model of office building.

Table 7.2 Performance during a typical four day summer period

Description	Displacement ventilation	CV ventilation
Heating capacity (kW)	9.7	17.1
Cooling capacity (kW)	19.9	22.3
Heating demand (kWh)	39.9	69.1
Cooling demand (kWh)	285.3	298.4
Hours heating	2.5	3.8
Hours cooling	22.9	21.4

Understanding patterns of system demands during the day and testing whether structural cooling was advantageous required only that a range of climate conditions was tested. The team found data covering a fortnight which included both typical and extreme conditions. If the design question had been in terms of long-term energy use, then annual assessments would also have been included. Table 7.2 is a comparison of the two design variants. Note that the displacement ventilation design requires a smaller plant capacity and is somewhat more efficient. Afternoon cooling demands in the displacement design were also reduced. Figure 7.11 shows the temperatures within the offices during a four-day summer assessment for both designs. Notice that the structural displacement cooling (on the left of the figure) limits the overnight temperatures in comparison with the CV variant (on the right of the figure).

The design team went on to study the control logic during the transition between night purge ventilation and the usual occupied hours ventilation rate to reduce the heating required if over-cooling occurred. In the process, they discovered an effective control logic for transition season operation.

7.4.6 External air flow study

This case study relates to the extraction of gases from fume cupboards in two hospital buildings in Edinburgh, Scotland. The purpose of the modelling investigation was to determine whether the proposed venting arrangements were satisfactory in keeping gases clear of window openings and air intakes on the surrounding buildings. It demonstrates how modelling of air flow using detailed CFD can be used to determine air flow patterns around buildings. Such techniques are useful in many ventilation studies. In projects where this is of particular importance, wind tunnel tests of scale models may also be undertaken to confirm model predictions for selected cases.

Previous experience with gas venting had shown that a difficult case might arise in strong winds, where powerful eddies can overcome the upward velocity and buoyancy of the flue gases, and transport pollutants to regions where they might become trapped. A more general problem arises in light winds, where lateral movement of gases is relatively sluggish and pollutants are slow to disperse from the site.

As inputs to the model, the geometry of the buildings and land in the area of the flues was gathered. All relevant buildings were modelled as rectangular boxes, or combinations thereof, to reduce computing times to manageable levels. The gas volume flow rates and exit velocities of the fume cupboard exhaust stacks were obtained.

Lib: impact_bc_sum.res: Results for impact_bc
Period: Mon 2 Aug @00h50 to: Thu 5 Aug @23h50 Year: 1999 : sim@ 20m, output@ 20m
Zones: office_e_0 office_w_0 office_e_1 office_w_1 office_e_2 office_w_2

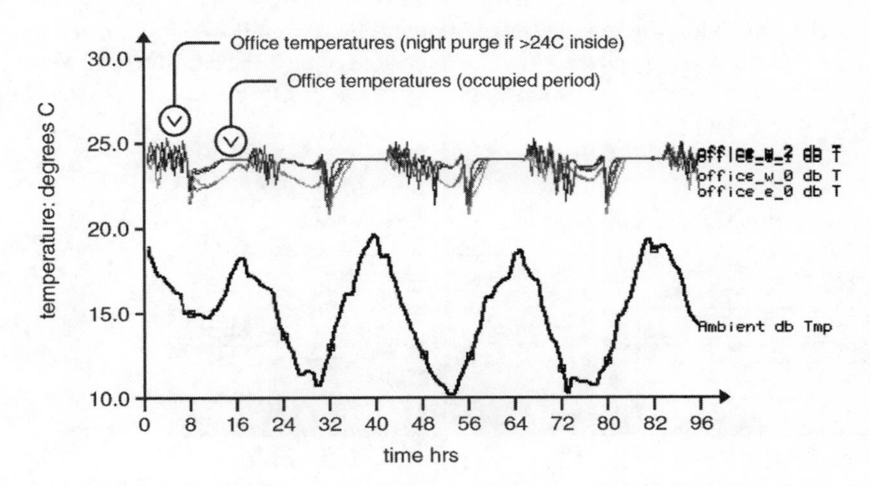

Offices with displacement ventilation and night purge

Lib: impact_cv_sum.res: Results for impact_cv
Period: Mon 2 Aug @00h50 to: Thu 5 Aug @23h50 Year: 1999 : sim@ 20m, output@ 20m
Zones: office_e_0 office_w_0 office_e_1 office_w_1 office_e_2 office_w_2

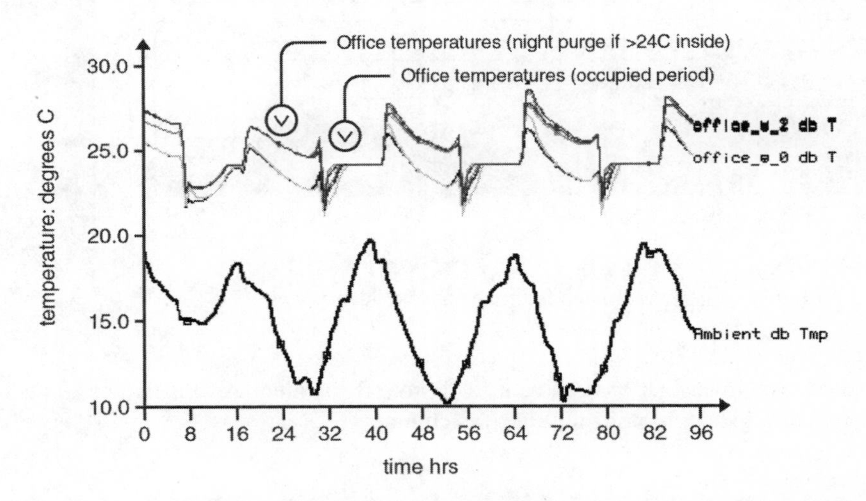

Offices with CV system and no night purge

Figure 7.11 Summer performance assessment (displacement vs CV system).

The CFD analysis built up a 3-dimensional picture of the wind flow around and through the site. Several wind directions were investigated, after consideration of the prevailing winds in the region and the local topography of the site. Two wind speeds, corresponding to light and strong wind conditions, were studied (nominal wind speeds of 2 m/s and 10 m/s, respectively, at a height of 10 m above the ground).

Once the simulation was complete, the 3-dimensional wind flow or pollutant dispersion pattern produced could be viewed in two dimensions at any plane of interest; perspective views can also be generated. As an example of outputs, Figures 7.12 and 7.13 depict wind velocity vectors and gas concentrations.

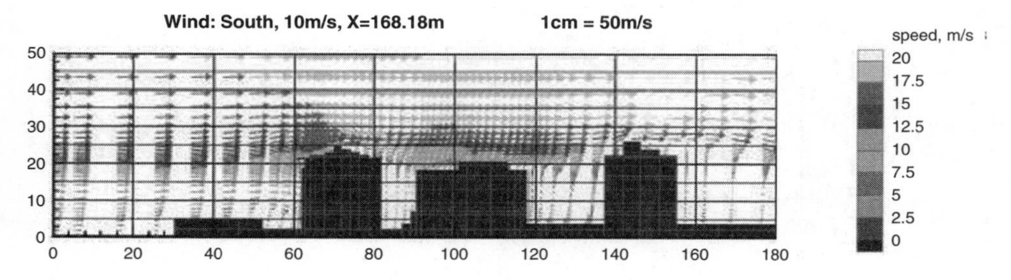

Figure 7.12 Cross-section of wind velocity distribution around the buildings.

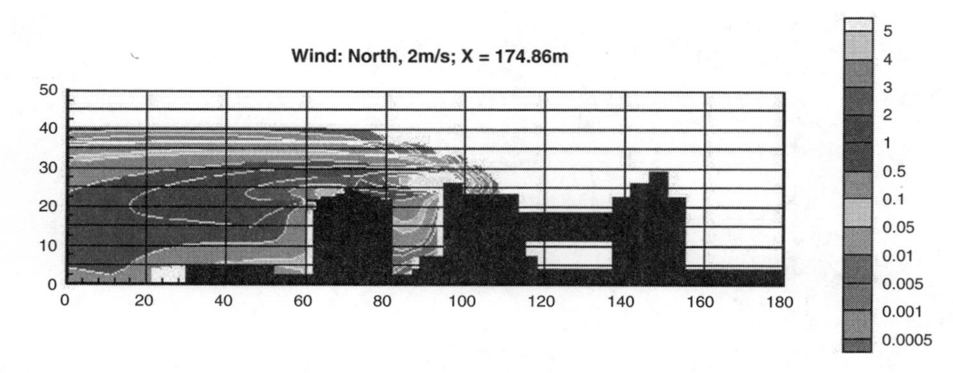

Figure 7.13 Cross-section of pollutant distribution around the buildings.

Clearly, such investigations can be used to investigate alternative locations and heights for optimizing the location of the flue gas chimneys.

7.5 Conclusion

Energy and environmental modelling can be applied to a huge range of situations, a few of which have been mentioned in the foregoing sections. Because every building has unique characteristics it is very difficult to predict how energy efficient and comfortable a building will be without recourse to modelling. In the past, modelling has been something of a speciality, due to the complexity of the software, the long computation times and skill shortage. Such barriers are being overcome, and the obvious benefits of modelling are being realized on a greater number of projects, giving the design team more confidence in the use of innovative designs.

The case studies have provided a glimpse of the integration of energy and environmental modelling within the design process. They illustrate a few of the many uses of such tools and some of the successful approaches that design teams have used. Indeed, the list of possible uses of such tools and their application is limited only by the variety of innovative designs that are being attempted in design practice. Access to performance information inevitably leads to additional explorations, better understanding and confidence in innovative design approaches.

There is every reason to conclude that the use of energy and environmental modelling will become more common within the design process as these applications and the skills of the design and engineering professions mature.

References and bibliography

Best, R. and de Valence, G. (1999) *Building in Value: Pre-Design Issues* (London: Arnold).

CIBSE (1998) Building energy and environmental modelling. In: *Applications Manual AM11* (London: Chartered Institution of Building Services Engineers).

DoE (2001) US Department of Energy. www.eren.doe.gov/buildings/tools_directory

ESRU (2001) Energy Systems Research Unit. www.esru.strath.ac.uk

ISO (1993) *Moderate Thermal Environments – Determination of the PMV and PPD Indices and Specification of the Conditions for Thermal Comfort.* ISO Standard 7730 (Geneva: International Organization for Standardization).

LBNL (2001a) Lawrence Berkeley National Laboratory. http://simulationresearch.lbl.gov

LBNL (2001b) Lawrence Berkeley National Laboratory. http://radsite.lbl.gov/radiance

Strachan, P. and Baker, P. (1992) Comparison of measured and predicted performance of a conservatory. In *Proceedings: North Sun 1992*, Trondheim, pp. 345–50.

Strachan, P. and Johnstone, C.M. (1994) Solar residences with transparent insulation: predictions from a calibrated model. In: *Proceedings – North Sun 1994,* Glasgow, pp. 347–53.

8

Buildability/constructability

Chen Swee Eng*

Editorial comment

Building designers are often criticized for producing fanciful or elaborate schemes with scant regard for the problems that will be faced by the contractors who have to build them. Fortunately the determination of architects, often with the support of visionary engineers, has allowed building design to evolve and made possible the realization of such landmark projects as the Guggenheim Museum in Bilbao, Sydney Opera House and the Hong Kong and Shanghai Bank.

In the past traditional procurement methods have generally not allowed contractors to be involved at any stage of the design process – instead design and documentation would be completed then tenders called and only at that stage would the contractor have any opportunity for input, and often at the risk of submitting a non-conforming tender. This situation produced a number of less than satisfactory outcomes including a lack of 'ownership' of the project in the contractor's mind, an adversarial relationship between builder and designer characterized by conflict rather than co-operation, and a perception of architects as being impractical dreamers and builders as conservative drudges, only interested in doing things as quickly and simply as possible.

With the advent of a variety of new procurement methods, notably design and build, and a move towards more inclusive approaches such as partnering and alliancing, contractors can now become involved in projects much earlier, often from the very start of the procurement process. There is a considerable body of research supporting these changes which has been conducted in various parts of the world aimed at building up a formal structure for assessing and improving the 'buildability' of projects. The underlying aim has been to add an extra dimension to building design by introducing the practical concerns of how a building will actually be constructed into the design process. The benefit of this for clients is clear: improved buildability can not only directly reduce

* University of Newcastle, Australia

construction cost by reducing construction time but can also produce some important flow-on effects such as fewer disputes, fewer variations and improved product quality, all of which translate into better value for money.

This chapter looks at a variety of approaches to buildability and presents some of the theory underpinning a structured approach to assessing and improving buildability.

8.1 Introduction

Buildability and constructability are synonyms for a concept that has evolved over a number of years. McGeorge and Palmer (1997) identified it as the only management concept to have been designed and developed by the construction industry for the construction industry. They suggested that this is because the separation of design and construction processes is unique to the construction industry.

In the early 1960s, the division between the process of design and construction was recognized as contributing to inefficiencies throughout the construction industry. The problem was seen to be that of communication and co-ordination between contractual parties, and the focus was on the need for greater co-operation and co-ordination of the people and processes involved in construction. Emmerson (1962) identified a number of specific factors contributing to potential inefficiency in the construction industry including:

- inadequate documentation of projects before they are put out for tender
- complex and inefficient pre-contract design procedures
- lack of communication between architects and contractors, sub-contractors and other consultants.

Banwell (1964) supported this with his view that in the traditional contracting situation, the contractor is too far from the design stage for his specialized knowledge to be put to use. He suggested that the complexities of modern construction and its requirement for specialized techniques demand that the design process and the construction stage should not be regarded as separate fields of activity.

The Tavistock Report (Higgin and Jessop, 1963) identified numerous problems of miscommunication between contractual parties attributable to the pattern of relationships and the division of responsibility within the building team. It stated that 'effective achievement of the common design task requires full and continuous interchange of information . . . there is a need for more "carry-over" in the co-ordination with respect to design and construction phases'.

8.2 Evolution of the buildability/constructability concept

A number of stages in the evolution of the buildability/constructability concept can be identified and related to research efforts in different parts of the world and by different groups of researchers. These include research by:

- the Construction Industry Research and Information Association (CIRIA) in the UK
- the Construction Industry Institute (CII) in the USA
- the Construction Industry Institute of Australia (CIIA)
- the Building Performance Research Group (BPRG) at the University of Newcastle in Australia.

'Buildability' emerged as an identified issue in construction management in the late 1970s in the UK. Although it was researched and reviewed intensively for about a decade in the UK construction industry, its potential was not fully exploited.

The early approach taken by CIRIA and researchers in the UK, which regarded buildability as a problem that arose simply from the division between designers and builders, led researchers to focus on technical issues such as design detailing, and site and construction planning (Gray, 1983; Adams, 1989; Ferguson, 1989).

The prevailing debate at the time centred very much on looking for causes of buildability problems and blame allocation. Most industry commentators and researchers tended to see buildability as a function that is within the influence or control of the designer. Illingworth (1984) went as far as to suggest that the problem of buildability was exacerbated because designers and others in the professional team resented the involvement of contractors in the design process.

This narrow focus ultimately limited the ability of researchers to identify satisfactory responses to a complex problem. It was interesting to note that at the Conference on Buildability in the Barbican in 1983, the only view advocating a broad systemic approach to buildability was that of the American designer Robert Feitl (Coombs, 1983).

The American CII used a different approach to investigate constructability problems. They used industry case studies that allowed researchers to obtain a holistic understanding of the issues. This led to the understanding that different stages of the project life cycle would relate to different issues in constructability. The research approach was informed by task force members who were, in the main, experienced industry people with intuitive appreciation of the practical issues.

The CIIA adopted the operational model of the CII, so their research approach was also shaped by inputs from experienced industry practitioners. They also used the case study approach, which enabled a holistic perspective to be maintained. In contrast with the CII constructability strategy, which advocated three different sets of concepts for the three main stages of activities investigated, the CIIA proposed twelve principles that would be relevant with different emphases over the five different life cycle stages of a project.

The BPRG at the University of Newcastle in Australia took a totally different research approach that was strongly influenced by the inclination of the research team towards systems methodology. They started with a conceptualization of the buildability problem as one that derived from a complex system. The complex relations that contributed to buildability were then modelled and the principles underlying this model explicated.

The research approaches and results obtained by the CII, CIIA and the BPRG showed remarkable convergence although the initial work of the BPRG was conducted quite independently of the two construction industry institutes.

Several researchers in this field have discussed the difficulty of defining the appropriate boundaries for the buildability model (Gray, 1983; Bishop, 1985; Griffith, 1986). On the one hand, there is the inherent problem of attempting to formulate a simplistic model for universal application, an approach that tends to equate buildability with a set of motherhood statements and usually produces a model that has very little prospe practical implementation. Conversely, very narrowly focused definitions may result models that fail to realize the full potential of the buildability approach (Chen and McGeorge, 1994).

8.2.1 Buildability research by CIRIA

A widely accepted definition of buildability in early research was coined by CIRIA (1983) as:

> the extent to which the design of a building facilitates ease of construction, subject to the overall requirements for the completed building.

This view was taken by CIRIA because of the perception that buildability problems existed '. . . probably because of the comparative isolation of many designers from the practical construction process. The shortcomings as seen by the builders were not the personal shortcomings of particular people, but of the separation of the design and construction functions which has characterized the UK building industry over the last century or so.'

The CIRIA definition focused on the link between design and construction and implied that factors which are solely within the influence or control of the design team are those which have a significant impact on the ease of construction of a project. The concept was recognized as an issue within an integrated design-management context.

Good buildability required the design of a building, structure or other construction project to inherently consider the construction phase with emphasis on the method of construction, the sequence of work, the overlap and interrelation of activities and the way that these are incorporated into the overall design.

A number of other British studies (NEDO, 1983; 1987; Griffith, 1986) in the same era addressed the gap between design and construction and recommended the need to bridge these two functions.

Adams (1989), in further work commissioned by CIRIA, developed sixteen guiding principles for achieving buildability. These are summarized as:

- investigate thoroughly
- consider access at the design stage
- consider storage at the design stage
- design for minimum time below ground
- design for early enclosure
- use suitable materials
- design for the skills available
- design for simple assembly
- plan for maximum repetition and/or standardization
- maximize the use of plant
- allow for sensible tolerances
- allow for a practical sequence of operations
- avoid return visits by trades
- plan to avoid change to work by subsequent operations

onstruction
arly.

he view that buildability is only a design-oriented activity has been
1986) and is reflected in the loss of some momentum in the research
the UK after the end of the 1980s.

8.2.2 Constructability development by the CII

In the same period, the Business Roundtable in the United States, prompted by the declining cost-effectiveness and quality of the American construction industry, addressed the issue of 'constructability' which was later defined by CII (1986) as:

> . . . the optimum integration of construction knowledge and experience in planning, engineering, procurement and field operations in the building process and balancing the various project objectives and environmental constraints to achieve overall project objectives.

The CII task forces that researched constructability used a case study approach and identified different requirements for the conceptual planning stage (Tatum *et al.*, 1986), the engineering and procurement phases of a project (O'Connor *et al.*, 1986), and the constructability improvements that can be made during field operations (O'Connor and Davis, 1988).

Russell *et al.* (1993) later reviewed the implementation of constructability principles, developing a project-level model of procedures to assist organizations in implementing constructability.

The implementation strategy of the CII was based on the promotion of a 'constructability system' incorporating a number of components which are described in a 'constructability concepts file' for program implementation (CII, 1986, 1987a, 1987b, 1993). Six concepts were identified for the conceptual planning stage, seven concepts for the design and procurement phases and one for the field operations phase as follows:

Conceptual planning

- site layouts promote efficient construction
- constructability programs are an integral part of project implementation plans
- project planning requires construction knowledge and experience
- early construction involvement in the development of contracting strategy
- project schedules are construction sensitive
- basic design approaches consider major construction methods.

Design and procurement

- design for accessibility of personnel, materials and equipment
- design for construction in adverse weather and remote locations
- design and procurement schedules are construction sensitive
- design to enable efficient construction
- design elements are standardized
- specifications developed for construction and procurement efficiency
- design for modularization/pre-assembly to facilitate fabrication/transportation.

Field operations

- contractor use of innovative methods.

The CII promoted the concept of constructability as a total system concept, placing emphasis on the commitment and adoption of a total program (CII, 1987b). The essence of the system (Griffith and Sidwell, 1995) was seen to be an understanding of the cost influence curve applied to a systematic program consisting of:

- self assessment of the extent to which the organization achieved constructability
- development of a written policy towards constructability
- senior executives sponsoring and having a commitment to the implementation of constructability
- organizational culture embracing constructability
- knowledge of implementation procedures based on the concepts file
- reviews of projects and lessons learnt through experience
- maintenance of a database of constructability savings for later reference.

The constructability implementation strategy focused on:

- high level corporate commitment to establish functional support, organization and procedures
- identification of constructability barriers and tactical use of 'barrier breakers'
- the development of a 'lessons-learned' database to inform on-going constructability implementation
- evaluation of corporate program effectiveness and modification of organization and procedures.

The cost influence curve, illustrated in Figure 8.1 is a fundamental notion in the CII's approach to constructability. The early consideration of constructability when the ability

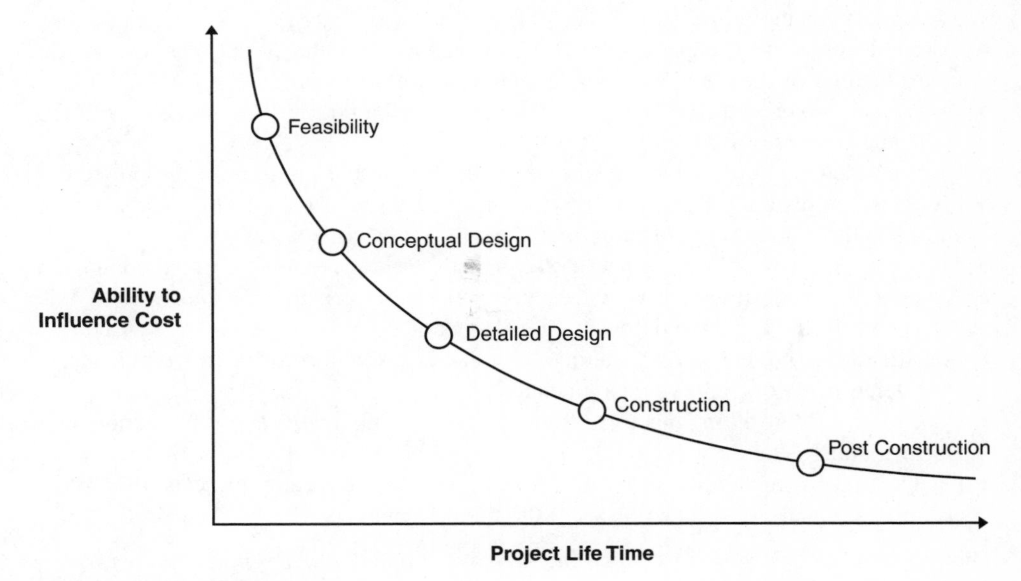

Figure 8.1 Cost influence curve.

to influence project costs is higher was seen to be the key to achieving better performance and better value for money from the client's point of view.

8.2.3 Development of constructability principles by the CIIA

There was no significant research into buildability in Australia prior to 1988 (Hon *et al.*, 1988) and some of the early published work looked to the developments in the UK (Miller, 1990) and the USA for leads.

In the early 1990s, the CIIA followed the CII approach and adapted their constructability process to Australian conditions. In doing so, the CIIA (CIIA, 1992) adopted a slightly modified definition of constructability as:

> . . . a system for achieving optimum integration of construction knowledge in the building process and balancing the various project and environmental constraints to achieve maximization of project goals and building performance.

The CIIA used an industry-based task force to review the work of the CII and developed construction information within the Australian context. This initially led to seventeen principles being identified and tested with feedback from twenty-one experienced construction personnel. This resulted in the production of the Constructability Principles File (CIIA, 1992) comprising twelve overriding concepts of constructability representing current best practice applicable over five project life cycle stages. The twelve principles espoused (Francis and Sidwell, 1996) were:

- Integration: constructability must be made an integral part of the project plan.
- Construction knowledge: project planning must actively involve construction knowledge and experience.
- Team skills: the experience, skills and composition of the project team must be appropriate for the project.
- Corporate objectives: constructability is enhanced when the project team gains an understanding of the client's corporate and project objectives.
- Available resources: the technology of the design solution must be matched with the skills and resources available.
- External factors: external factors can affect the cost and/or program of the project.
- Program: the overall program for the project must be realistic, construction sensitive and have the commitment of the project team.
- Construction methodology: project design must consider construction methodology.
- Accessibility: constructability will be enhanced if construction accessibility is considered in the design and construction stages of the project.
- Specifications: project constructability is enhanced when construction efficiency is considered in specification development.
- Construction innovation: the use of innovative techniques during construction will enhance constructability.
- Feedback: constructability can be enhanced on similar future projects if a post-construction analysis is undertaken by the project team.

The project life cycle stages to which these twelve principles were applied with varying emphases were:

- feasibility
- conceptual design
- detailed design
- construction
- post-construction.

The approach of the CIIA is to encourage project teams to apply practical measures, developed through consideration of key issues rather than to prescribe specific procedures to improve the construction process. It also promoted the cost-influence curve as a fundamental concept of the constructability strategy.

8.2.4 Buildability research at the University of Newcastle

At about the same time, research was being carried out by the BPRG at the University of Newcastle in Australia, quite independently of the work being done by the CIIA.

The BPRG reviewed the earlier work done by CIRIA and re-assessed the concept of buildability as a strategic rather than operational concern (Chen and McGeorge, 1994). They proposed that a workable concept of buildability needed to recognize that there are many factors in a project environment which impact on the design process, the construction process, and the link between design and construction. Indeed these would be factors which are relevant to, and are shaped by, the whole building procurement process.

Decisions which are made upstream of the design stage can impose constraints on the design decision process. These decisions may also identify project goals that influence the decisions of the designer. These are therefore relevant factors if the ultimate ease of construction of the project is affected. At the same time, decisions which may not have been made by the designer concerning intermediate functions between design and construction such as documentation, contractor selection, choice of contract form and so on may have a significant impact on the construction process. Figure 8.2 illustrates this systems view of the design-construction process.

Chen and McGeorge (1994) suggested that a conceptual approach which takes into account the influence of all relevant factors and other project goals is complex but necessary in the development of a buildability model that can be applied to specific projects. Only when the complex interactions of these factors are recognized can the potential of buildability be achieved. This challenged the traditional view of buildability that was primarily concerned with the design and construction phases of a project and proposed that buildability must encompass the whole life cycle of a building. The impact of decisions to reduce construction time may have implications downstream that affect the use or maintenance of the building. They proposed a working definition of buildability as:

> the extent to which decisions, made during the whole building procurement process, ultimately facilitate the ease of construction and the quality of the completed project.

This definition differed from the earlier British concept of buildability (CIRIA, 1983) by:

- not equating buildability simply with the ease of construction but also the appropriateness of the completed product

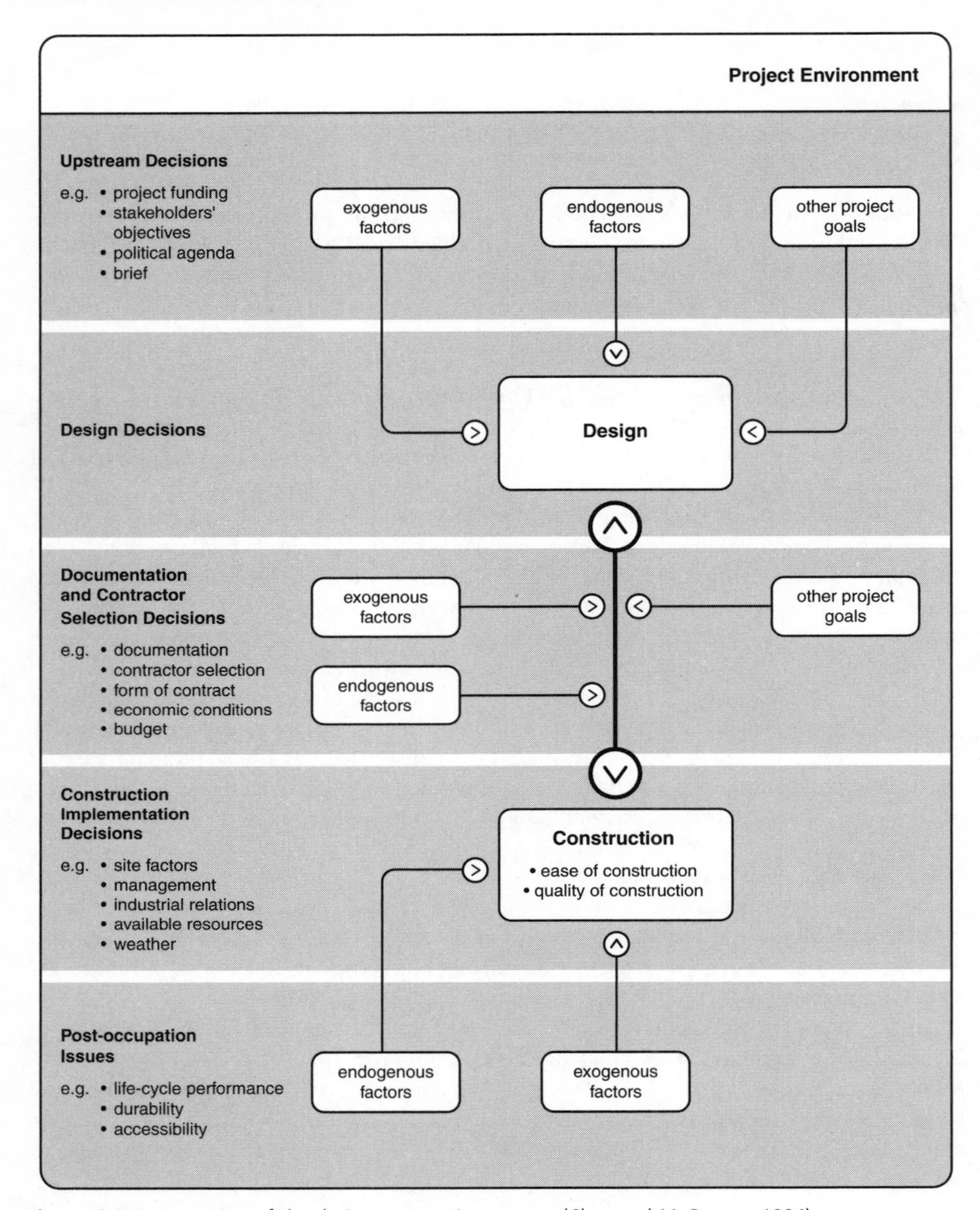

Figure 8.2 Systems view of the design-construction process (Chen and McGeorge, 1994).

- recognizing the life cycle implications of buildability
- identifying decisions rather than technical details as the determinant of buildability.

In addressing the life cycle implications of buildability, the BPRG research also linked the effectiveness of the timing of relevant decisions to the principle that earlier decisions have more potential to influence outcomes than later decisions (the Pareto Principle).

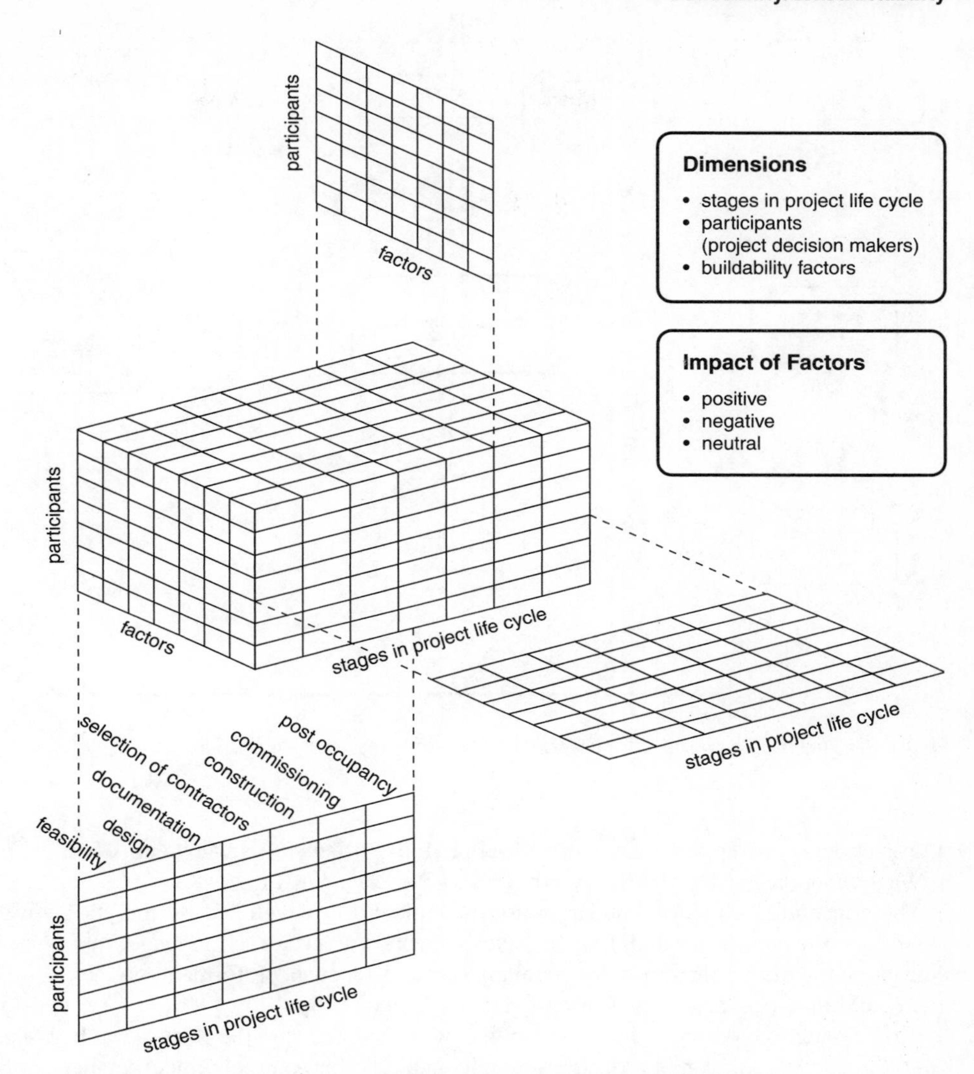

Figure 8.3 Three-dimensional model for buildability decisions (Chen and McGeorge, 1994).

Chen and McGeorge (1994) proposed a three-dimensional conceptual model that mapped out the decision space for buildability decisions. The three dimensions proposed were the project participants (stakeholders and decision makers), the buildability factors (exogenous factors, endogenous factors and project goals) and the stages of the building life cycle (from inception to post-occupation). This model is illustrated in Figure 8.3.

The focus of their buildability research was on the management of the procurement process rather than on construction technology. They advocated that much of the buildability problem was not due to the lack of information but rather the lack of the management of information. In this respect, their research extended to the proposal for

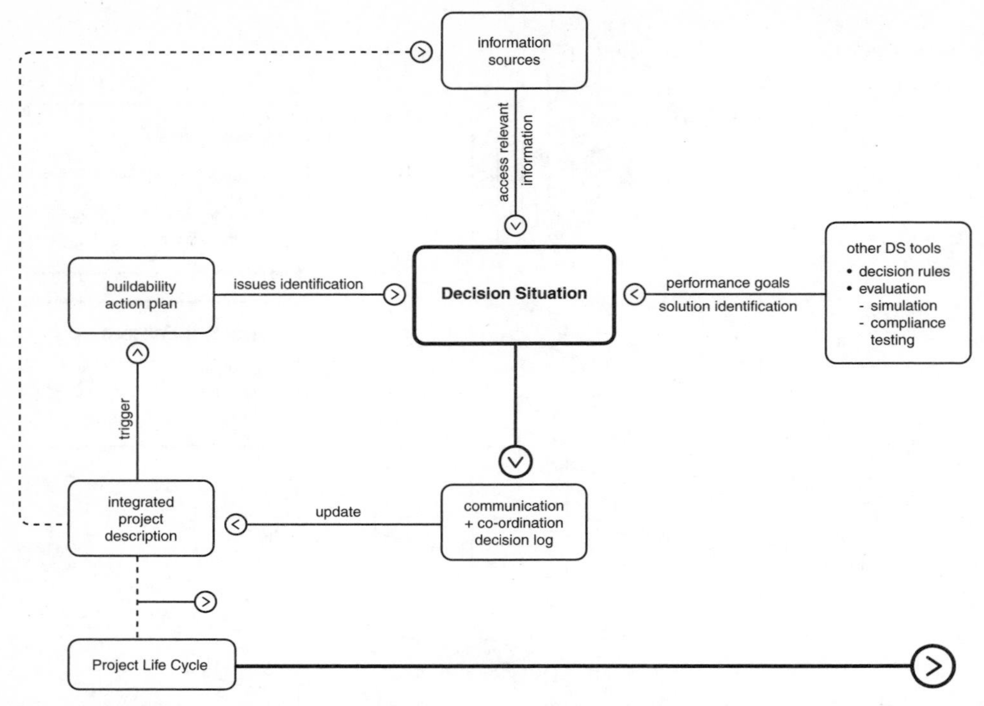

Figure 8.4 Project decision support framework for buildability management (McGeorge *et al.*, 1995).

a project decision support framework for buildability-oriented management (Chen *et al.*, 1993; McGeorge *et al.*, 1994b, 1995). This is illustrated in Figure 8.4.

The conceptual direction and definition of buildability developed by the BPRG was extended to stages of the building life cycle beyond the construction stage, taking into consideration the implications for maintenance and renewal performance (Chen *et al.*, 1993) and post-occupancy evaluation (McGeorge *et al.*, 1994b).

This strategic approach for buildability was adopted by the New South Wales Government as an element of its strategic capital works procurement management program (CPSC, 1993).

8.3 CIIA and BPRG – constructability implementation

The CIIA and BPRG, after developing their approaches to constructability and buildability separately, collaborated in developing implementation strategies that combined their concepts. This was a natural move given that although the two groups approached the issue of buildability/constructability from two very different perspectives, the conclusions of their work had much in common. These include the identification of buildability/constructability management as a life cycle performance concept (Chen *et al.*, 1996a), the adoption of an integrative management approach and the importance of the Pareto principle or cost-influence curve as a fundamental principle for improving performance.

Both the groups identified buildability/constructability as an issue that required the systematic integration of construction knowledge through all stages of the project's life cycle and across functional divisions separating the roles of project participants (Chen *et al.*, 1996b). Decisions which impact on constructability performance are made at different levels, from the strategic to the operational, and on different issues, reflecting the particular combination of factors which are significant for each specific project. The various participants in the project process will have different roles and responsibilities with respect to these factors, and at different stages of a project's life cycle.

The collaborative work done by these two groups on constructability implementation resulted in the production of a *Constructability Manual* (Francis *et al.*, 1996) and a *Client Guide to Implementing Constructability* (Francis *et al.*, 1997) based on the development of a constructability implementation planning framework which identifies and co-ordinates the decision roles of individual project participants through a project's life cycle. Using the three-dimensional strategic management framework (Chen and McGeorge, 1994), the twelve constructability principles are plotted in a co-ordinated fashion across the roles and responsibilities of project participants and the stages of a project life cycle. The structure of this implementation planning framework is illustrated in Figure 8.5.

The twelve principles have been expanded into implementation guidelines which will enable constructability action plans to be developed for individual projects, taking into account the unique demands, objectives and environmental conditions of the project (Chen *et al.*, 1996b).

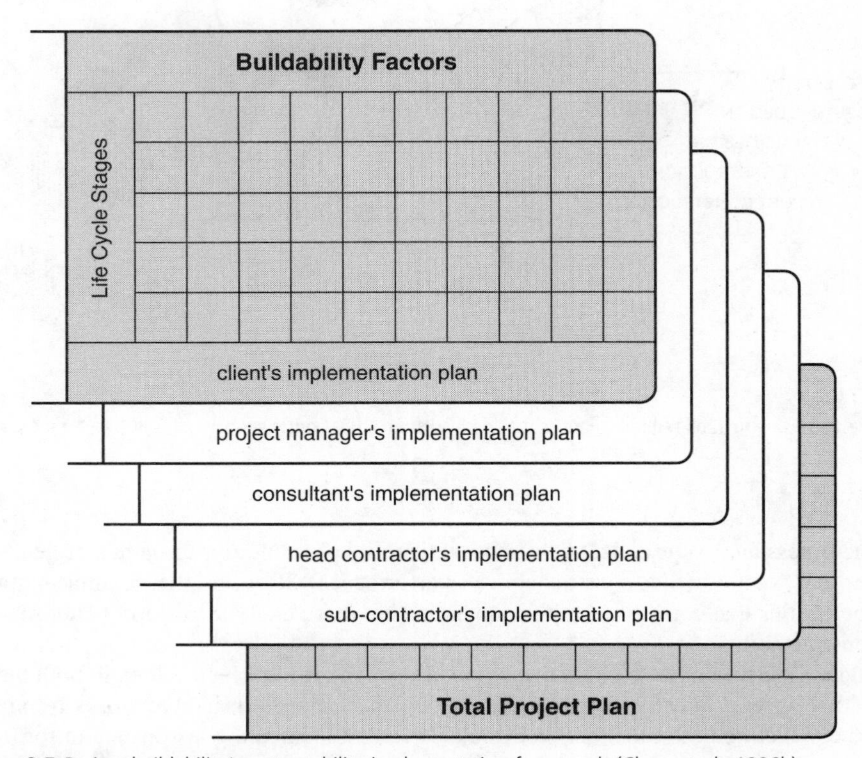

Figure 8.5 Project buildability/constructability implementation framework (Chen *et al.*, 1996b).

8.4 Project fitness landscapes

Both the CII and CIIA approached constructability with systematic strategies advocating concepts and principles that should be addressed appropriately for each project situation rather than provide prescriptive measures. The CII recognized that construction projects go through phase transitions during the project life cycle and therefore researched constructability concepts for three distinct phases of the project life. The phase transitions reflect different activities, objectives and participant roles for the project. Each phase transition is connected to and dependent on the directions and outcomes of the preceding phase.

The systematic strategies advocated by CII and the CIIA essentially embodied the systemic principles for improving constructability. These are not causal rules that would determine whether constructability would be achieved or not. Rather they are interrelated conditions that promote environments in which constructability is supported. The plotting of these principles with appropriate weightings (Francis *et al.*, 1996) against a typical project life cycle produces a fitness landscape (Figure 8.6).

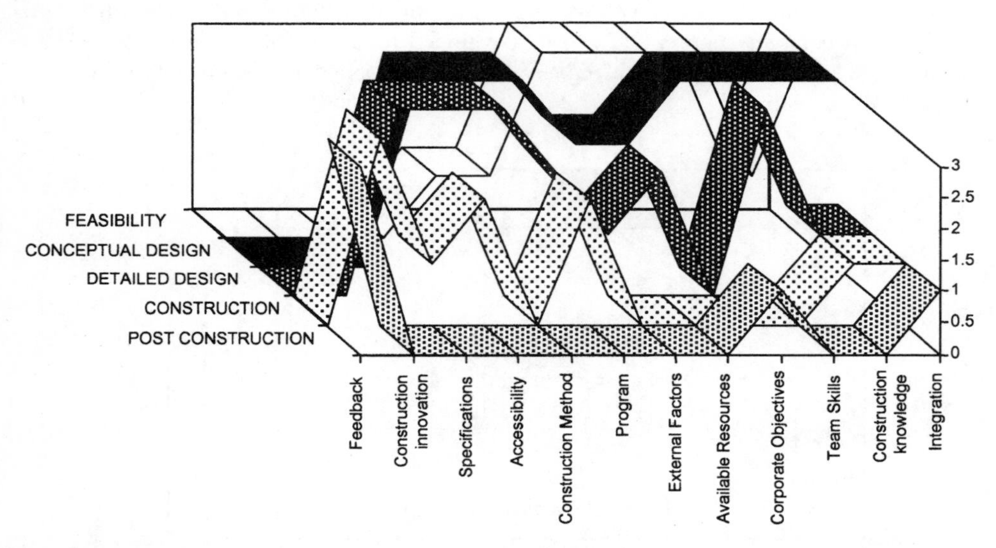

Figure 8.6 Constructability fitness landscape (derived from CIIA Constructability Flow Chart, Francis *et al.*, 1996).

The fitness landscape concept describes the profile of conditions through a project's life cycle that support the constructability performance. Projects that exhibit 'fitness' characteristics that correspond to this landscape are more likely to perform better in terms of constructability.

An important element of this fitness landscape is the feedback principle. In both the CII and CIIA approaches (CIIA, 1992; CII, 1993), the feedback principle works by project teams conducting post-construction evaluations and documenting lessons learnt for future project improvements. This principle ensures the on-going evolution of project constructability performance by learning from experience. Thus an organization can start with very poor 'fitness' levels with respect to the constructability fitness landscape but with repeated iterations of experience and learning we can expect the evolution process to lead to increasingly better performance as more of the landscape peaks are climbed.

8.5 Innovating design and construction with buildability/constructability

The development of the buildability or constructability concept over the past decades has seen it shift from an operational concern at project level, focusing on construction details, to strategic concerns that drive higher level decisions related to such concerns as project procurement methods and project value management.

Project procurement methods such as design and build (or design and construct), and their various hybrid forms, seek to improve performance by integrating the expertise of designers and constructors within an organizational framework. The separation of design and construction is removed by the sharing of overall risks and responsibilities within the same organization. This facilitates the application of relevant construction knowledge to the design process and improves the potential for innovative design solutions that may otherwise attract additional cost premiums in traditional procurement methods.

The strategic view of buildability/constructability as being the emergent result of decision making by all the participants in a project is consistent with the principles of project value management. These principles advocate a shared process by project stakeholders to examine function within a system-wide context and optimize design solutions to meet project objectives (Neasbey *et al.*, 1999).

The current leading edge of the concept focuses on supporting project organizations in integrating design and construction to achieve productivity and quality improvements. The concept recognizes the complexity that is generated through the interactions of different players bringing different contributions to the project process through the whole project life cycle.

Implementing buildability or constructability embraces the following :

- taking the broad view of design as being the sum total of all decisions made during the project process
- establishing a systematic buildability/constructability implementation framework as early in the project process as possible that will carry through the full project life cycle
- ensuring that all project participants address buildability/constructability issues in their decision making, information production and transmission
- ensuring the maximum fitness landscape of the project organization for buildability/constructability through the project life cycle.

Successful and innovative buildability/constructability is about integrating the decisions of all project participants that have been evaluated to enhance buildability objectives.

Current developments in the construction industry that support these principles include partnering strategies, and the development of computer-based information sharing and decision support systems.

A number of countries such as Singapore and Hong Kong have developed buildability rating systems to facilitate the measurement and benchmarking of buildability performance. One limitation of this approach is the assumption that standard elements of measurement can adequately capture the qualities of unique design solutions.

8.6 Conclusion

It seems only logical that the practicality of building design from a construction viewpoint should be considered as part of the pre-construction phase of procurement. In earlier times, when architect, engineer and contractor (or supervisor at least) were the same person, design and construction were integrated, and it was only as the design and construction functions were separated that problems emerged. Following research carried out over the past couple of decades, frameworks have been developed that enable project teams to assess the buildability of their schemes. This does not preclude the adoption of innovative procedures or limit designers to producing schemes based solely on the ease with which a contractor may build them, but it does provide an opportunity for the consideration and incorporation of another set of views of any project that can improve the overall efficiency of the procurement process and save time and money for the client.

References and bibliography

Adams, S. (1989) *Practical Buildability* (London: CIRIA/Butterworths).

Banwell, H. (1964) *The Placing and Management of Contracts for Building and Civil Engineering Work* (London: HMSO).

Bishop, D. (1985) *Buildability: The Criteria for Assessment* (Ascot, Berkshire: CIOB).

Chen, S.E. and McGeorge, D. (1994) A systems approach to managing buildability. *The Australian Institute of Building Papers*, **5**, 75–86.

Chen, S.E., McGeorge, D. and Ostwald, M. (1993) The role of information management in the development of a strategic model of buildability. *Management of Information Technology for Construction Conference*, Singapore.

Chen, S.E., McGeorge, D., Sidwell, A.C. and Francis, V.E. (1996a) A performance framework for managing buildability. *CIB-ASTM-ISO-RILEM 3rd International Symposium: Applications of the Performance Concept in Building*, Tel Aviv, Israel.

Chen, S.E., Francis, V.E., McGeorge, D. and Sidwell, A.C. (1996b) Decision support for constructability. In: Carmichael, D.G., Hovey, B., Rijsdijk, H. and Baccarini, D. *Project Management Directions* (Sydney: Australian Institute of Project Management) pp. 133–40.

CII (1986) *Constructability – A Primer* (Austin, Texas: Construction Industry Institute).

CII (1987a) *Constructability Concepts File* (Austin, Texas: Construction Industry Institute).

CII (1987b) *Guidelines for Implementing a Constructability Program* (Austin, Texas: Construction Industry Institute).

CII (1993) *Constructability Implementation Guide* (Austin, Texas: Construction Industry Institute).

CIIA (1992) *Constructability Principles File* (Adelaide: Construction Industry Institute).

CIRIA (1983) *Buildability: An Assessment, Special Publication 26* (London: CIRIA Publications).

Coombs, M. (1983) Buildability: a delegate's view. *Architects Journal*, **177** (15), 47.

CPSC (1993) *Capital Project Procurement Manual* (Sydney: NSW Government).

Emmerson, S.H. (1962) *Survey of Problems Before the Construction Industries* (London: HMSO).

Ferguson, I. (1989) *Buildability in Practice* (London: Mitchell).

Francis, V.E. and Sidwell, A.C. (1996) *The Development of Constructability Principles for the Australian Construction Industry* (Australia: Construction Industry Institute).

Francis, V.E., Sidwell, A.C. and Chen, S.E. (1996) *Constructability Manual* (Adelaide: Construction Industry Institute).

Francis, V.E., Sidwell, A.C. and Chen, S.E. (1997) *Client Guide to Implementing Constructability* (Adelaide: Construction Industry Institute).

Gray, C. (1983) *Buildability – The Construction Contribution* (Ascot: CIOB).

Griffith, A. (1986) Concept of buildability. Paper presented at the *IABSE Workshop: Organisation of the Design Process*, Zurich, Switzerland.

Griffith, A. and Sidwell, A.C. (1995) *Constructability in Building and Engineering Projects* (Houndmills, Basingstoke: Macmillan Press).

Higgin, G. and Jessop, N. (1963) *Communications in the Building Industry: The Report of a Pilot Study* (London: Tavistock).

Hon, S.L., Gairns, D.A. and Wilson, O.D. (1988) Buildability: a review of research and practice. *Australian Institute of Building Papers*, **3**, 101–18.

Illingworth, J.R. (1984) Buildability – Tomorrow's need. *Building Technology and Management*, February, 16–19.

McGeorge, D., Chen, S.E. and London, K. (1994a) The use of post occupancy evaluation in developing a model of buildability. *Strategies and Technologies for Maintenance and Modernisation of Building*, Tokyo, CIB W70.

McGeorge, D., Chen, S.E. and Ostwald, M. (1994b) A dynamic framework for project decision support – developing soft interfaces for the management of construction projects. *CIB W78 Workshop on Computer Integrated Construction*, Espoo, Finland.

McGeorge, D., Chen, S.E. and Ostwald, M. (1995) The application of computer generated graphics in organizational decision support systems operating in a real time mode. *COBRA 1995 RICS Construction and Building Research Conference*, Edinburgh (Royal Institution of Chartered Surveyors).

McGeorge, D. and Palmer, A. (1997) *Construction Management – New Directions* (Oxford: Blackwell Science).

Miller, G. (1990) Buildability – A design problem. *Exedra*, **2** (2), 34–8.

Neasbey, M., Barton, R. and Knott, J. (1999) Value management. In: Best, R. and de Valence, G. (eds) *Building In Value – Pre-Design Issues* (London: Arnold), pp. 232–47.

NEDO (1983) *Faster Building for Industry* (London: HMSO).

NEDO (1987) *Achieving Quality on Building Sites* (London: NEDO).

O'Connor, J.T. and Davis, V.S. (1988) Constructability during field operations. *Construction Engineering and Management*, **114** (4), 548–64.

O'Connor, J.T., Rusch, S.E. and Schulz, M.J. (1986) *Constructability Improvement During Engineering and Procurement* (Austin, Texas: Construction Industry Institute).

Russell, J.S., Gugel, J.G. and Ratke, M.W. (1993) Documented constructability savings for petrochemical-facility expansion. *Journal of Performance of Constructed Facilities*, **7** (1), 27–45.

Tatum, C.B., Vanegas, A. and Williams, J.M. (1986) *Constructability Improvement during Conceptual Planning* (Austin, Texas: Construction Industry Institute).

9

Building regulation: from prescription to performance

Ray Loveridge*

Editorial comment

Most countries have some form of building codes or regulations in place to ensure that buildings meet acceptable minimum standards with respect to factors such as structural integrity, health and amenity, and energy use. The stringency of these regulations varies and, perhaps more importantly, the stringency of their enforcement varies. In some countries endemic corruption within bureaucracies means that standards can be flouted with impunity, sometimes with catastrophic results when structures fail.

In many countries the codes are formulated and updated by panels of experienced professionals: engineers, architects, planners and so on. In other places, however, codes have developed in different ways. The designers of a cultural centre in the Cook Islands, completed in the early 1990s, were initially constrained by an informal local 'standard' that specified that no building could be taller than the nearest palm tree (Muir, 2001, personal communication – David Muir is a Sydney-based architect). Later in the design process they were alarmed to find that an entrepreneurial engineer had 'adapted' an Australian building code and sold it to the Cook Islands government. The original code called for provision of one handicapped toilet for every 200 standard toilets, under the adapted code they were required to include one handicapped toilet for every 200 people – in a building designed to hold well over 2000 people this meant there would have been more handicapped toilets that standard toilets!

The Cook Islands examples are extreme and, today, somewhat atypical in their approach in that they *prescribe* certain characteristics of the building. The trend in building codes now is very much towards *performance* codes in which required outcomes or performance targets are specified. Such codes tell the designers what the result must be but leave the method of achieving that result to their discretion. This approach provides designers with the flexibility to search for new solutions to design problems but does have its own attendant problems as the author explains in this chapter.

* Arup Fire, Sydney

Clients can benefit from this flexibility as designers can propose more cost-effective solutions, either by reducing initial cost or improving the ultimate performance of buildings, and so improve the value for money outcomes for their clients while satisfying the appropriate regulatory requirements.

9.1 Introduction

During the past decade or so there has been an international trend in the reform of regulations governing the design and construction of buildings. Historically, such regulations have been presented as legislation in a prescriptive format, i.e., legislation prescribing definitive mandatory standards and processes that were applied to the design and construction of buildings.

In recent years many countries have introduced performance-based standards in an endeavour to provide greater flexibility in the design process. In essence, performance-based regulations establish the outcomes or results that must be achieved, with the methods used to achieve those outcomes being discretionary, subject to the approval of the appropriate authorities. By providing this flexibility they provide a platform for the development of alternative and innovative 'tailor-made' building designs and building material technologies. For example, under prescriptive regulations the scope design load criteria and relevant design standards for the structural design of a building would be mandatory; under performance-based regulations the structural design of a building could be covered within a single performance requirement with wording such as: 'a building must be designed to sustain the most adverse combination of loads and other actions to which it may reasonably be subjected'.

As a result of changes to legislative building control systems, industry practitioners have found it increasingly difficult to maintain a current knowledge of relevant legislation, technical documentation and administrative support systems. Additionally, rapid advances in computer technology have made it beneficial, if not essential, for all practitioners to understand the application of this technology to the design and construction of buildings, particularly in the field of fire-engineered building designs.

In this context industry practitioners need to understand the basic content and application of current development control systems, and design and construction requirements. Similarly they need to understand the roles that they may be required to play in these systems, including the development, assessment, acceptance or approval of new concepts, and the obvious need to develop appropriate procedures to responsibly fulfil these roles.

As different countries, and divisions of countries, may well have their own regulatory systems, and while many of these may be reasonably consistent in certain aspects, there will be some that differ markedly and readers should seek information specific to their location.

9.2 An introduction to building control systems

The development, maintenance and application of building control systems is generally a function of government. The level of government that undertakes these roles can vary between countries and may depend on the division of roles established by national constitutions.

In Australia, for example, the Federal constitution establishes the powers, responsibilities and regulatory functions of the Federal Government and the individual State and Territory Governments. Under the constitution the regulation of development and building is a responsibility of individual State and Territory governments. Accordingly, as Australia has six States and two Territories, as a nation it has eight varying building control systems, and while these systems may be reasonably similar in many aspects, the eight individual governments are able to do as they see fit for the benefit of their respective constituencies.

While the Federal government can endeavour to promote and nurture a nationally consistent approach to building control matters, such as the overall objectives or aims of building control legislation, it is necessary for a State or Territory to determine whether or not to actively participate in the implementation of national uniformity.

In Australia there is a third tier of government, referred to as local government, and this tier is generally responsible for administering State government legislation, i.e., the States enact legislation and local government authorities, generally known as local councils, apply the legislation on behalf of the State. In regard to building control matters, it is these local councils that undertake most of the day-to-day operation of the various State building control systems.

In some countries governments establish independent organizations to undertake the development and maintenance of many of their building control responsibilities, other than the passage of relevant legislation. These organizations may be established as a Board, or a Council, whose activities are generally underwritten by government and whose membership may include representatives from private industry and the public sector.

Additionally many countries establish specific standards writing bodies, such as Standards Australia, or become collaborative members of standards writing bodies, such as the International Organization for Standardization (ISO), that are responsible for the development of technical standards that may be incorporated within, or adopted by, building control legislation.

The structure of government, and consequently the distribution of regulatory responsibilities, varies internationally, so it is necessary for practitioners to inquire and discover 'who controls what' in their respective areas of professional practice; if their respective regulatory system does not produce effective outcomes then it may be up to industry practitioners to initiate appropriate reforms. Practitioners must understand the distribution of regulatory responsibilities and the processes for implementation of building control mechanisms in order to initiate regulatory reform.

Until recently most building control systems required a government official to police the regulations and ensure compliance. In recent years, however, there has been an international trend towards enhancing the role of private practitioners in the regulatory process.

9.3 Administrative provisions versus technical provisions

Building regulation systems generally contain regulations that can be described as being either *administrative* or *technical* in nature. The administrative content of each of the systems will vary to some degree depending on matters such as the structure, or existence, of local government and the primary role of the government organization(s) responsible for building control matters.

The type of regulatory provisions generally considered to be *administrative* may include:

- processes for lodging and assessing building applications
- integrated approvals systems
- concurrent approval requirements (where more than the regulatory body must approve)
- application fee structures
- the roles of public and private practitioners
- the role of the fire brigades
- licensing requirements for particular buildings, e.g., hospitals
- processes for dealing with alterations and additions to existing buildings
- mandatory maintenance provisions
- regulatory objection or appeal systems.

Administrative provisions may also exist for matters such as those directly related to support systems associated with the assessment and approval of applications lodged under the performance-based technical codes, such as:

- accreditation of building designs, processes or components
- professional/private certification
- accreditation/registration systems for products and practitioners.

All of these matters are generally addressed through administrative provisions of legislation.

On the other hand, *technical* provisions of building control systems are generally considered to be those contained within specific codes and standards. These performance-based provisions establish the technical design and construction requirements for buildings.

In Australia each of the States and Territories has its own suite of administrative provisions contained within legislation. However, the technical requirements are contained within a nationally adopted document called the Building Code of Australia. This document is discussed in other sections of this chapter.

The effective application of performance-based standards within building control systems relies upon the application of an entire suite of interrelated reforms. In this context, when all components of the suite of regulatory reforms are efficiently operating in unison, the building regulation system will facilitate the development and approval of solutions that are alternatives to traditional design and construction techniques.

9.4 The basic components of a performance-based regulatory system

Recent trends in building control systems have reflected a general move toward national uniformity, the introduction of more flexibility into design and construction processes and enhanced roles for the private sector. However, in order to maximize the potential acceptance and operation of performance-based design and construction standards, there is a need for interrelated administrative reforms.

In essence the effective operation of performance-based building regulations requires that there are also reforms to regulatory provisions that address the following:

- appropriate tertiary qualifications for practitioners
- accreditation/registration of building practitioners
- mandatory insurance cover
- time-related limitations to practitioner liability
- proportionate distribution of liability
- determination of the propriety of alternative solutions.

9.4.1 Appropriate tertiary qualifications for practitioners

Historically there has not been the need for industry practitioners to hold tertiary qualifications, generally taken to be a minimum of a Bachelors degree, within many facets of the building control industry. However, as countries move toward, or adopt, performance-based building regulations, there is an accompanying need for practitioners to be appropriately qualified. This arises from the need for industry practitioners to possess an in-depth understanding of performance-based building regulations, rather than simply possessing knowledge of what is prescriptively required by legislation. Under a prescriptive regulatory regime, practitioners generally only need to know what is prescribed, then in the event that something special is designed or needs to be approved, the approval authority has a specialist procedure for approval matters that are out of the ordinary.

Under a performance-based regime, however, both designers and approval authorities need to understand both *what* is required and *why* it is required in order to develop or approve appropriate alternative solutions. Consequently they need to have appropriate qualifications. For example, most prescriptive regulatory systems specify a maximum distance for a path of travel to an exit of, say, 40 m. It is relatively easy to measure a proposed design and determine if it complies. However, under a performance-based regime there would be no maximum, and therefore a designer, and subsequently an approval authority, would need to know the origins and reasons for the previous prescriptive maximum distance of 40 m in order to approve a greater distance. Consequently, the science behind the previous prescriptive needs to be understood in order to produce appropriate alternative solutions.

9.4.2 Accreditation/registration of building practitioners

As the application of performance-based building regulations generally requires specialist knowledge to be applied to the design or approval of alternative solutions there is a need for practitioners to demonstrate that they have the qualifications and experience to operate effectively within a performance-based regime. Many regulatory authorities have established systems of accreditation or registration of practitioners, through either government agencies or professional practitioner institutions, to monitor and control the performance of practitioners. It would appear from local experiences that the accreditation/registration of all industry practitioners, not just private practitioners, is the most appropriate system.

9.4.3 Mandatory insurance cover

Most systems of accreditation or registration require practitioners to maintain a prescribed minimum level of insurance cover to protect consumers. Under a performance-based regime, a practitioner cannot be expected to be a 'jack of all trades' and have sufficient knowledge to undertake the design or approval of all technical matters addressed within building regulations. Consequently there is a need for most practitioners to rely on other specialists at some time. If a specialist practitioner is accredited in a particular field, then others should be able to rely upon documentation produced by that practitioner without fear of inheriting liability in the event of a consequential occurrence. Most accreditation bodies will, therefore, generally require that a mandatory minimum insurance cover be held by all accredited practitioners in order to protect clients who use their services.

9.4.4 Proportionate distribution of liability

Proportionate liability is a form of liability that is proportionally distributed based on the assessment of a determining authority, such as a court of law. A major benefit of this form of liability is that defendants are generally only held accountable for the portion of the total liability for which they are judged to be responsible. Due to the perceived need for practitioners, including approval authorities, to be reliant upon documentation provided by other accredited practitioners in order to fulfil their regulatory roles, proportionate liability is considered to be an integral component of an effective performance-based building control system.

In some systems there are liability waiver provisions that provide that accredited practitioners are not to be held liable if they rely, in good faith, on documentation produced by other accredited practitioners.

9.4.5 Time-related limitations to practitioner liability

Associated with the application of proportionate liability provisions is the limitation of liability with respect to time. As part of their performance-based building control regime, many regulatory authorities have introduced administrative provisions that set a maximum period, say, of 10 years, within which any claims for latent damages must be lodged. This eliminates the potential for 'life of the building' responsibility to be applied and the subsequent liability problems for designers, constructors and approval authorities that such a system can impose.

9.4.6 Determination of the propriety of alternative solutions

Obviously the introduction of performance-based building regulations can impose significant pressures on certain practitioners in the design and construction process, particularly the approval authorities. Many approval authorities may ask, 'What is in this for me – why should I take a potential risk and approve something new and innovative?' This is often a legitimate question for approval authorities that do not have appropriately trained staff to assess alternative solutions and who may not fully understand the basis upon which the traditional recipes for design and construction have been developed. In this context it would be difficult for a practitioner who does not have the necessary skills to assess the propriety of an innovative alternative solution and unfair that they should be placed in a position where they are required to do so. In such circumstances in is inevitable that the practitioner will refuse to approve an alternative solution and so the re-education of some practitioners is fundamental to the long-term success of the implementation of performance-based building codes.

In order to reduce the implications that such an approach may have on the introduction of performance-based regimes, it is often necessary to establish administrative systems that can be used by both designers and approval authorities to independently determine the propriety of alternative solutions. These systems place the decision-making process in the hands of practitioners who are adequately trained to make such decisions. One such system is a process of peer review through which industry practitioners may have alternative solutions assessed.

These systems may only need to be implemented during the introductory years of performance-based regimes in order to provide sufficient time for approval authorities to gain confidence and experience in making decisions on alternative solutions. Various forms of peer review may be implemented such as formal court systems, or simpler systems supported by voluntary industry experts in their respective fields of practice. Many other forms of administrative systems can be implemented that assist the approval of alternative solutions and these are discussed later.

In summary, performance-based building control systems require the establishment of an integrated suite of administrative provisions that are complementary and essential to the effective operation of technical performance-based building codes. The isolated development and introduction of technical performance-based building codes will not effectively produce reform of the design and construction industry.

9.5 Technical provisions – prescriptive-based versus performance-based

As discussed previously the objectives of building regulations are established by consensus between the respective regulatory authorities and are implemented through legislation. Required outcomes to achieve these objectives are reflected in the content of the specific regulatory provisions that are consequently imposed upon the building industry as mandatory standards.

Historically, building regulations were expressed in a manner that dictated the technical processes by which the required outcomes were to be achieved. In simple terms the regulations prescribed design criteria, or 'recipes', which were to be strictly complied with in order to produce the required outcomes. This form of regulation was referred to as 'prescriptive' regulation.

Obviously prescriptive regulations inhibited innovation. In most instances, if a designer wished to incorporate an innovative concept into a proposal, the approval authority for any subsequent application did not have the power to approve the alternative design. Consequently designers had to seek special dispensations in order to circumvent the limitations or restrictions of the system.

The operation of a prescriptive building regulation system has a number of disadvantages as:

- it contributes to the workload of the various objection or appeal authorities established to assess the propriety of alternative designs
- designers may become frustrated with the system and simply relegate new or innovative solutions to the 'too hard basket' and choose to simply accommodate the requirements of the regulations
- they do not readily accommodate the ability of a number of progressive and supportive approval authorities to approve suitable alternative designs.

As a result the prescriptive approach is not popular with designers, developers or progressive approval authorities for it does not readily allow flexibility or the ready acceptance of suitable alternative designs.

Performance-based regulations provide an alternative system by which the desired outcomes of the relevant authorities can be realized. Such regulations specify a level of performance, i.e., a conceptual outcome that must be achieved. In simpler terms performance regulations specify a required *result* and allow the designer flexibility as to how that result will be achieved. Any recipe may be used provided it can be demonstrated that it meets the relevant performance requirements.

In very general terms the difference between prescriptive regulations and performance regulations can be illustrated as follows: a prescriptive regulation may simply state that a designer must use a specific recipe, e.g., 5 + 5, in order to achieve an *unknown* required outcome – the inherent result 10 is not stated, no options are available and there is no flexibility.

A performance regulation, however, could simply state that the required outcome is a value of 10; the required outcome has a minimum numeric expression of performance, thus allowing the designer the flexibility to use any numeric combination to achieve the required value of 10, e.g., 8 + 2, or 6.3 + 3.7, or 5 + 5 and so on.

While the value of 10 is the mandatory outcome of the design, the method by which the designer achieves it is of no concern to the regulatory authority provided the designer can satisfy the authority that the value of 10 has been achieved. If, however, the designer uses the recipe 5 + 5, then there is no need to independently satisfy the regulatory authority as this recipe has already been identified as an acceptable means of meeting the performance requirement.

While this process brings flexibility to the processes of design and construction, it imposes a responsibility on the approval authority to ensure that the proposed recipe does achieve the required result. In many instances it is the process of determining the propriety of alternative solutions that has the potential to reduce the efficacy of performance-based regimes.

9.6 Benefits of performance-based regulations

The implementation of performance-based standards into building control systems allows flexibility in design and construction technology and provides potential for maximizing cost efficiencies to the building industry. A consequential benefit of the use of performance standards is the reduced need for dispensations to prescriptive standards.

This benefit arises from the inherent ability of the relevant approval authority to approve any alternative design that is considered to achieve compliance with the performance requirements.

In most instances, a performance-based document will provide at least one acceptable recipe for achieving compliance with the performance requirement. In some regulatory systems this acceptable recipe is referred to as a *deemed-to-satisfy* provision. A significant benefit arising from the inclusion of the deemed-to-satisfy provisions is that they accommodate the needs of practitioners who do not wish to follow the performance-based design path. It is generally expected that a majority of practitioners will use deemed-to-satisfy provisions as the principal components of their designs during the period immediately following the introduction of performance-based regimes. A second benefit of following the deemed-to-satisfy path is that an approval authority must accept a deemed-to-satisfy design as achieving compliance with mandatory performance requirements. Following this path does not ensure approval of an application, however, as approval authorities need to be generally satisfied as to the overall propriety of a design and consequent concerns may be reflected in conditions applied to the granting of approval.

The introduction of performance-based regulations should reduce the need for future amendments to building control systems with respect to matters such as changes to technology and the development of new materials, as the inherent flexibility in performance standards allows new technology to be accepted without the need for amendment of technical standards.

9.7 The structure of the performance-based building codes

The structure of performance-based building codes varies internationally and it would appear that there is no one method that is significantly better. While structures may vary it is generally accepted that the essential components of any workable structure could be presented within a three-tiered structure.

9.7.1 Tier 1 – the objectives

The top tier of a performance-based building code should present the objectives of the document. Objectives may be described as statements expressing the general intent of the building regulations, and should be a reflection of what the regulatory authority, often the government, is endeavouring to achieve through the building code. In most instances the objectives will also be a reflection of what the general community expects building codes to achieve. If this is the case it is necessary for the regulatory authority to legitimately establish the scope of these community expectations through consultation, to ensure that the content of the objectives is appropriate.

For example, the level of life safety required under the provisions of building regulations will be similar, e.g., in the event of a fire in a building it is expected that the various life safety measures in the building will perform to a standard that allows all occupants to evacuate safely. While this outcome would be a fundamental objective of both a regulatory authority and a community expectation, the objectives of building regulations extend to other aspects of minimum performance in the event of fire. For example, it is a general objective of most building codes that in the event of fire, not only will all building occupants be able to escape, but it is also expected that the building will not be a source of significant damage to other property, such as adjoining buildings. Consequently it is expected that the subject building will be constructed in such a manner that it will resist the spread of fire *from* adjoining buildings and will not be a source of spread of fire *to* adjoining buildings. If the subject building does burn down, but does so without affecting other property, that is seen as an acceptable outcome for many building codes.

It is necessary that the community fully understand that this potential outcome is accepted by the regulatory authority. If, however, a building owner wishes to construct a building in a manner whereby the subject building does not burn down in the event of a fire, that is their prerogative and that objective can be accommodated within a performance-based design. The ability to enhance the minimum objectives of building codes is a significant benefit of the application of performance-based standards.

Objectives will also need to address many other regulatory and community expectations, such as appropriate minimum standards for design and construction with regard to health and amenity issues such as ventilation, lighting and sanitary facilities.

In the development of performance-based building codes it is therefore essential that the fundamental reasons for regulating the design and construction of buildings are agreed between the regulatory authorities and the community and, subsequently, clearly expressed in the objectives of the document.

As mentioned at the outset the objectives simply express the general intent of the building regulations so that the users of the document understand what the regulations are endeavouring to achieve. They do not establish minimum standards that must be achieved and therefore, although they form an integral component of a performance-based building code, they are not mandatory requirements; the mandatory requirements are presented in the next tier.

9.7.2 Tier 2 – the performance requirements

The second tier of a performance-based building code should present the performance requirements of the document. These are the mandatory provisions that establish the minimum level of performance that must be achieved by a proposed design. They are generally written in one of two forms, being either *qualitative* or *quantitative*, and there are advantages and disadvantages attached to the use of either form.

Qualitative performance requirements

Qualitative performance requirements describe the *qualities* that must be evident in a resulting product in order for that product to comply with the regulations. Qualities could also be defined as *attributes*, *features* or *characteristics* in this context. The most significant benefit of this approach is that the performance requirements do not place any restrictions, or limits, on alternative designs, thus giving the designers total flexibility. Some practitioners, however, believe that this method is too vague to allow them to assess whether or not a resulting product actually achieves the required result, simply because the required result may not be readily evident from an initial reading.

There are also potential problems with individual designers and approval authorities having differing opinions as to the propriety of alternative solutions which can lead to inconsistencies in interpretation and application of specific provisions. Any potential design can only be viewed as an acceptable solution if both parties are satisfied that the proposed alternative solution complies with the relevant performance requirements. However, this can easily lead to debate as to the propriety of a proposed design simply because there are very few quantified criteria, or verification methods, which can be used as a standard for adjudication by an approval authority. Consequently appropriate expert judgement is essential in the development, assessment and determination of alternative solutions. Consequently, the development of appropriate independent and accessible methods of assessment is essential if industry practitioners are to be able to readily determine the propriety of alternative solutions formulated by designers.

Quantitative performance requirements

Quantitative performance requirements contain *quantified* criteria that are used as a ready means of assessing compliance of the alternative solutions with the relevant performance requirements. The benefit of writing performance requirements in quantified terms is that direct comparisons can be made to the expressed numeric or measurable values within the requirement.

While this form of regulation is more readily assessed, is does not allow total flexibility and practitioners are often unduly constrained by having to work to the expressed quantified criteria. For example, in certain instances compliance with a performance requirement that states that 'all occupants within a fire compartment must be able to be evacuated to a place of safety within three minutes' may be inappropriate; while compliance may be readily assessed, the time period stated in the performance requirement may be unsuitable in particular circumstances, such as where it can be demonstrated that untenable conditions resulting from a fire may not be present in a fire compartment for five minutes from fire initiation.

In this case it may be more appropriate to express the requirement in a qualitative manner, e.g., 'tenable conditions must be maintained within a fire compartment for sufficient time to enable all occupants to evacuate to a place of safety'. Assessment of proposed designs would require more supporting evidence, however, practitioners would have the flexibility to design appropriate systems specific to the circumstances applicable to the individual design.

In some instances the use of qualitative performance requirements in building regulations may arise from an absence of consensus among the technical experts developing the performance requirements with respect to measurable criteria that are to be included in the provisions. While this outcome is understandable the ability of some industry practitioners to agree on the propriety of proposed designs can be tested when complex designs are offered for consideration.

While performance-based building codes offer flexibility and support innovation, there are many designers who simply wish to use traditional recipes for design and construction. To accommodate these designers, performance-based building codes should incorporate a suite of recipes that have been given the approval of the regulatory authorities and are acceptable without further validation or proof of suitability. These recipes are contained within the third tier of the performance-based structure.

9.7.3 Tier 3 – building solutions

The third tier of the structure contains the optimal means that can be used to satisfy the performance requirements. Designers may formulate their own recipes to comply with the performance requirements, or they may use traditional pre-ordained recipes.

The self-formulated recipes are often referred to as alternative solutions. The onus for validating an alternative solution and demonstrating compliance with the performance requirement is placed on the person seeking approval of the design.

The traditional pre-ordained recipes are often referred to as deemed-to-satisfy provisions and these need no further validation or substantiation. In many instances a deemed-to-satisfy provision may refer to a referenced standard (i.e., a standard named in the relevant code), or other external document, that contains an acceptable means of complying with the mandatory performance requirements. For instance, structural design performance requirements can be met by the use of appropriate standards, such as ISO standards, that are simply adopted by reference within deemed-to-satisfy provisions.

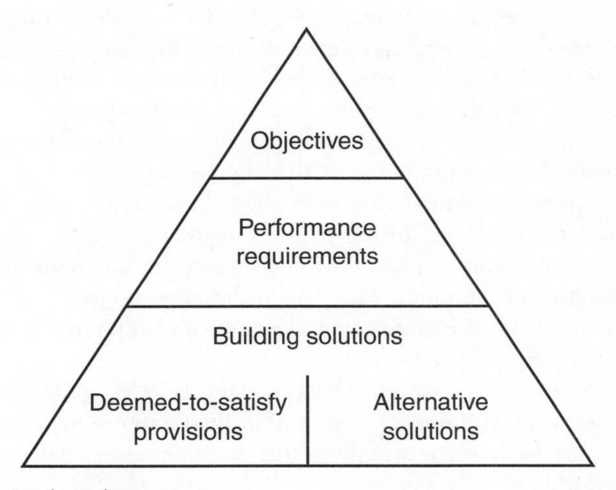

Figure 9.1 A three-tiered regulatory structure.

Designers should be able to use any combination of alternative solutions and deemed-to-satisfy provisions in order to comply with all relevant performance requirements so a performance-based regulatory regime should contain at least three tiers of provisions. The top tier provides information on what the regulatory provisions are endeavouring to achieve, the second tier specifies what is required to be achieved, and the lower tier provides ways in which those requirements may be achieved. Figure 9.1 shows a basic three tier structure.

9.8 Fire-engineered building designs

The development of complying alternative solutions may be undertaken in any manner that is satisfactory to the approval authority. The methodology utilized by the designer will generally need to involve an acceptable process which can be supported by appropriate documentation and which ultimately leaves the approval authority with no doubt as to the suitability of the alternative solution. The following discussion of fire-engineered design provides an example of a process by which certain alternative solutions can be generated and assessed.

The application of fire engineering technology is recognized throughout the world as an acceptable method of developing alternative solutions to meet relevant performance requirements for fire safety in building design. In this context, fire engineering can be described as a design methodology by which engineering principles are applied to the evaluation of hazards identified within a building and then to the design of appropriate methods of protection to deal with those hazards.

This design concept is the same as that inherent within the content of most building codes, however, traditional building codes have, through necessity, adopted a conservative approach in order to be generically appropriate across a particular class of building, whereas fire engineering allows a similar process to be applied to individual buildings within a particular class.

While appropriate fire engineering technology has been available for some time, its use in developing alternative solutions has generally been limited to applications involving objections, dispensations or appeals against the content of prescriptive provisions, largely because of the prescriptive format of most traditional building codes.

Following the introduction of performance-based building regulations in various countries throughout the world, the use of fire engineering technology as a tool for the development of alternative solutions has flourished. This increased use is due, at least in part, to the change in the content of the building regulations, as a major portion of the content of building codes address requirements for adequate standards of fire safety in the design and construction of buildings. Many major building projects throughout the world have been the subject of fire-engineered designs, including many of the buildings constructed for the Sydney Olympic Games.

Fire engineering technology facilitates the development of alternative solutions that meet the same objectives as those inherent within regulations and designers continue to explore opportunities to introduce options that accommodate their needs better than traditional deemed-to-satisfy provisions.

Another contributing factor to the recent surge in fire-engineered designs is the ongoing refinement of the technology which supports fire engineering, particularly computer technology. Advances in technology have allowed the continuing development of mathematical modelling programs that are more user friendly and more accurate than their predecessors.

Significant developments in fire engineering in Australia have been implemented through the Fire Code Reform Centre Limited (FCRCL), an independent company established to administer and direct a defined, fire engineering-related, research program. One of the significant initial outcomes of the FCRC was the publication of *Fire Engineering Guidelines* (FEG) (FCRCL, 1996). This document can be described as a process document in that it outlines procedures and methodologies for use by suitably qualified and competent practitioners undertaking fire-engineered designs. It is important to note that use of the FEG requires professional knowledge and engineering judgement and therefore it is not a document that should be used by practitioners without appropriate expertise. The document is, however, recommended reading for all practitioners as it provides a substantial base for gaining an understanding of the concepts and considerations involved in the development of an appropriate fire-engineered design.

Basically, the FEG is intended for use during the *conceptual stage* of a project and is suitable for application to both existing and new buildings. The conceptual stage can be divided into a number of steps commencing with the preparation a fire engineering design brief (FEDB) and finishing with the production of a fire safety systems report. It is during the preparation of the FEDB that relevant parties to the design and approval of the project confer and reach agreement on matters such as design objectives, acceptance criteria, hazard identification, fire scenarios to be analysed, trial concept designs and the method by which designs will be evaluated.

Following the FEDB process trial designs can be analysed and evaluated through comparison with the previously agreed acceptance criteria. In the event that the design is not appropriate, the later components of the process can be repeated with a new trial design. When a design is developed that meets the acceptance criteria, the approval process should be faster as the approval authority has been a party to the preparation of the FEDB in which the acceptance criteria were mutually agreed.

During the application of the FEDB process to more complex projects there may be instances when some of the relevant parties consider they do not have sufficient expertise to actively participate in the preparation of the brief, or there may be instances when a fire-engineered design is developed without the application of the FEDB process and the approval authority considers it has insufficient expertise to adequately assess the design. In either instance it is essential that the relevant parties are represented by an appropriately qualified person to either participate in the FEDB process, or to assess the propriety of a fire-engineered design.

Other issues that must be addressed include:

9.8.1 Selection of appropriate computer software

Approval authorities need to be satisfied that a proposed design has been developed using appropriate software and that the fire engineer is suitably experienced in the use of the software.

9.8.2 Selection of appropriate input data

If the FEBD process is not followed then both the designer and the authority must assess the validity of the input data, such as the characteristics of the 'design fire', used in producing the design.

9.8.3 Expectations of computer software

While mathematical models can provide reasonable predictions or approximations of likely outcomes, the process requires the expertise of experienced practitioners to ensure that there are adequate safety factors in the design outcomes which can accommodate the limitations of the design tools.

9.8.4 Preparation of submissions for approval of alternative solutions

Some of the most difficult assessments to be made are those regarding both the *extent* and *content* of submissions made by designers to approval authorities in support of alternative design proposals. This applies equally to documentation provided voluntarily by applicants as well as that required by the authority.

It is desirable for all relevant parties to participate in discussions at a preliminary design stage in order to prepare an agreed platform for the development and approval of design proposals. This is called *pre-lodgement consultation*. Generally, when developing building designs using prescriptive regulations, pre-lodgement consultation can be kept to a minimum as quantified design criteria are clearly set out and compliance is easily assessed. This is not the case when using performance requirements in which there may be no quantified criteria and the interpretation of the approval authority as to what is necessary to meet those requirements may vary significantly to that of the designer.

The extent of submissions

The extent of material that an applicant needs to provide in order to satisfy the approval authority of the propriety of an alternative solution will obviously be influenced by the complexity of the proposal. In the case of fire-engineered design, if the FEBD process is followed then this will usually establish the extent of the documentation that must be lodged with an application for approval, otherwise it remains the responsibility of the designer to establish what the authority requires. The final decision rests with the applicant but the approval authority may either request further information or reject an application if they feel that the documentation is inadequate.

During pre-lodgement consultation with the approval authority, the designer can explain the background to the particular design and the extent of support documentation to accompany the application can be mutually determined. This process also enables the applicant to assess the costs and benefits of alternative solutions. This decision may be influenced by a combination of factors including:

- the legislative requirements of the administrative system
- the expressed needs of the approval authority
- the expressed needs of other relevant parties to the approval process
- the likelihood of the application being approved
- the cost of the production of required support documentation
- the economic benefit of the proposed alternative solution.

If the cost of substantiating an alternative design is outweighed by the cost benefits of the design, the applicant may decide not to proceed with the initiative. This outcome may arise in more complex designs where the expertise of the approval authority may be limited and therefore extensive documentation is required.

Alternatively, at pre-lodgement discussions, the potential for the use of third party processes may be explored. A significant benefit of these processes is that they can reduce the need for approval authorities to assess proposals in areas where they lack specific expertise, and so may reduce their potential liability.

In the development of performance requirements, factors that may influence the formulation of the alternative solutions can be included within those requirements. For example, if performance requirements were developed that allowed designers to determine whether or not exit stairways should be fire-isolated, they might contain a list of issues for consideration such as:

- the height of the building
- the number of storeys connected by the stairway
- the function or use of the building
- the number of exits provided from each storey
- the evacuation time
- fire brigade intervention
- fire safety systems installed in the building

or any other issues that may be appropriate to the development of alternative solutions. Designers would be required to address each of these heads of consideration within a submission, and the approval authority would need to consider whether or not the performance requirement had been met.

Where a performance requirement does not contain specific heads of consideration, the approval authority should provide potential applicants with sufficient information during pre-lodgement discussions. Consequently, approval authorities may need to establish comprehensive listings of the fundamental issues that must be addressed in submissions prepared with respect to any performance requirement which does not list heads of consideration individually.

Pre-lodgement consultation should also determine the appropriate standards to be met depending on whether an applicant chooses to develop an alternative solution that will demonstrate compliance with either the:

- performance requirements
- equivalence to deemed-to-satisfy provisions
- a combination of the two.

The extent of support documentation will vary depending upon which of the above paths the applicant takes. The majority of applications will combine these paths as there is still a lack of credible data available for assessment of performance-based designs with regard to all fire safety, health and amenity issues.

Approval authorities generally have the ability to request additional information after an application has been lodged. This may cause a significant loss of time and incur additional costs. Even if an authority does approve an application, it may apply conditions to the approval that render a design impractical. If, however, all relevant issues can be adequately addressed prior to lodgement, and both the applicant and the authority agree on the basis of design, applications should be processed without undue delay. Pre-lodgement consultation is, therefore, fundamental to the successful development of designs using performance-based building codes. While the extent of the pre-lodgement discussions will obviously be influenced by the complexity of a particular project it should, nevertheless, be an integral component of the design process when alternative solutions are proposed, as it can provide significant benefits to both the applicant and the approval authority.

Content of a submission

Once the extent of required documentation is agreed, the content of the documentation must be considered. The basic content of any submission lodged in support of an alternative solution should include:

- identification of any particular deemed-to-satisfy provision(s) which have not been complied with in the design
- identification of the relevant performance requirement(s) which have been complied with (in lieu of the above deemed-to-satisfy provisions)
- sufficient information to satisfy the approval authority that the alternative solution meets the relevant performance requirement(s).

As performance-based design may be a relatively new process to many designers and approval authorities, it is essential that a submission establishes the basis of an alternative solution design, provides full descriptions of the proposal and describes the methods used to establish compliance. Although the basic content of any documentation will be individually tailored to a particular proposal, in general all documentation should possess the following qualities:

Clarity

A clear description of the proposal is required giving details of what the design is and why it is being proposed.

Substantiation

Applicants must demonstrate to the approval authority that compliance can be technically substantiated. The selection of a suitable means of substantiation will depend on the nature of the alternative solution.

As an established discipline, fire engineering is a popular means of developing, and subsequently substantiating, alternative solutions for fire safety performance requirements. In contrast, the substantiation of an alternative solution to the deemed-to-satisfy provisions for stair geometry, for example, may ultimately involve the construction of a prototype stair and testing its performance with simulated evacuations.

For any proposal the designer will need to demonstrate that design criteria and methods of substantiation are relevant to the particular application and that the resultant proposal can be justified. If it is appropriate, details of previous acceptance of similar proposals could be provided to the approval authority. The means by which a designer proposes to substantiate an alternative solution should be discussed prior to the preparation of detailed documentation, to ensure the propriety of the proposal and reduce the time needed for assessment by the approval authority.

Validity

The designer should state the reasons for introducing an innovative design. The validity of an option may be purely design related, however, an alternative solution may be required to satisfy specific client needs.

Economic considerations are valid reasons for adopting alternative solutions; while the objectives of most building codes relate to the achievement and maintenance of acceptable standards of safety, health and amenity, the objectives should be expressed so that compliance with the performance requirements can be achieved in a cost-effective manner.

References

An applicant must be able to show that the designers of an alternative solution are suitably qualified. Academic qualifications will need to be supplemented by appropriate experience, and supporting documents such as client references may need to be included in submissions.

The bottom line is that if the applicant cannot satisfy the approval authority that an alternative solution complies with the relevant performance requirements then approval should not be granted. In summary, when preparing documentation in support of an alternative solution, the applicant should leave no doubt that the alternative solution meets the relevant performance requirements.

9.8.5 Methods of assessing alternative solutions

For many years approval authorities throughout the world have been using a variety of administrative support systems for assessing compliance of applications with regulations. Under prescriptive codes, professional certification of structural designs and mechanical systems, for example, is a common approach. Such systems are equally relevant to performance-based alternative solutions, provided the practitioners have the ability to assess alternative solutions.

Some performance-based building codes contain provisions that detail the scope of acceptable assessment methods that can be used by approval authorities to assist in determining the propriety of alternative solutions. Administrative support mechanisms that can be used as acceptable assessment methods include:

Certificate of Accreditation

Regulatory authorities, such as agencies of state governments in Australia, may implement administrative systems whereby they, or another appropriate authority, receive submissions for accreditation and independently consider and determine whether a product or design may be suitable for accreditation. Subsequently, products or designs that gain independent accreditation must be accepted by local approval authorities (in Australia, local or municipal councils).

Under most legislative regimes, accredited products or designs cannot be refused and in this context are similar to deemed-to-satisfy provisions, however, their acceptance may be subject to specific conditions nominated on the certificate, or accreditation may be of limited duration as specified on the accreditation certificate.

Professional/private certification

An approval authority can be satisfied that a material design, product, form of construction or process complies with a performance requirement, or deemed-to-satisfy provision, or criterion for approval, through certification by a professional engineer or other appropriately qualified person. Some regulatory authorities have registers of qualified personnel or alternatively the assessment of the propriety of the certifier may be the responsibility of the local approval authority.

Comparison with deemed-to-satisfy provisions

An approval authority may assess an alternative solution against the relevant deemed-to-satisfy provisions instead of directly against performance requirements. In some instances this may benefit an applicant where the deemed-to-satisfy provisions have some form of quantified criteria to which the alternative solution can be directly compared.

Expert judgement

Any appropriately qualified person may assess the propriety of an alternative solution. While this method is similar to professional certification the expert may not be required to certify that an alternative solution is appropriate and so this process may have differing legal implications by comparison with the act of issuing a formal certificate of compliance.

All of these systems are considered suitable for assessing compliance with required standards and many of these have been accepted as a means of substantiating compliance with building regulations in the past. The continued use of support mechanisms such as these will provide valuable assistance to applicants and approval authorities alike in the future determination of alternative solutions.

9.8.6 Procedural considerations – building surveying practice

The processes by which approval authorities determine applications containing alternative solutions may differ from those applied to applications complying with deemed-to-satisfy provisions. While some authorities may be able to handle performance-based applications as easily as those relating to prescriptive regulations, there will be some that will need to review existing procedures and related operational matters such as the powers of delegated authority, and type and level of insurance cover needed due to potentially increased responsibilities arising from the assessment of alternative solutions. Local approval authorities may need to establish internal policies relating to the scope of alternative solutions that it can assess in-house.

Approval authorities may also decide which particular administrative support mechanisms it favours for alternative solutions to particular performance requirements. For example, it may be appropriate to accept professional engineer certification for structural designs, while a certificate of accreditation or expert judgement may be required for an alternative stair design.

Where performance requirements do not list particular matters for consideration, designers and approved authorities should both develop their own lists of matters that need to be addressed when either developing or assessing alternative solutions. These lists can be further developed over time as experience and knowledge levels increase.

9.9 Conclusion

It is evident that while the opportunities for introducing flexibility in design and construction in the building industry are greatly increased under performance-based building codes, flexibility in design and construction standards does not mean that industry practitioners can do as they please. It must be remembered that an approval authority should not approve the use of an alternative solution unless it is satisfied that it complies with the relevant performance requirements and the onus for demonstrating compliance with the performance requirements lies with the applicant. Therefore, while the approval authority must possess the ability to determine an application for approval, the applicant must obtain appropriate documentation to support their alternative solution.

Both parties can rely upon the various administrative support mechanisms that are in place to assist the operation of performance-based regulatory systems. Given that mechanisms for the assessment of new and innovative products have been available for many years, the question should be asked, 'What is it that actually changes?' The answer is that it is the *process* for gaining approval for the use of alternative designs that is different. It is not true to say that it is only under performance-based building codes that alternative solutions could be used; under many prescriptive regulatory regimes it is possible to have alternative solutions approved through various legislative and administrative processes, as restrictive as some of these processes may be.

What does actually change under performance-based building codes is that local approval authorities can approve alternative designs and therefore an applicant does not need to pursue acceptance of their ideas through a potentially expensive and time consuming formal or legislative system. This can be of major benefit to all building industry practitioners if the various components of the respective building control system are utilized appropriately. This is the basis of many of the challenges ahead for practitioners in the building industry.

References and bibliography

ABCB (1996a) *Building Code of Australia Class 1 and 10 Buildings – Housing Provisions* (Australian Building Codes Board).

ABCB (1996b) *Building Code of Australia Class 2 to 9 Buildings* (Australian Building Codes Board).

ABCB (1997) *National Training Program – Workshop Guide – BCA96* (Australian Building Codes Board).

Department of Local Government (1997) *BCA96 Seminars – Working With the Performance Based Building Code of Australia* (Sydney: New South Wales Department of Local Government).

FCRCL (1996) *Fire Engineering Guidelines,* First edition ISBN 0–7337–04549 (Sydney: Fire Code Reform Centre Limited).

SA, UTS and Insearch (1998) *The Operation of Performance Based Regulations in Building Design – SAA / HB126–1998* (Sydney: Standards Australia, University of Technology Sydney and Insearch Pty Ltd).

PART 2

Project procurement and management

10

Project finance and procurement

Gerard de Valence*

Editorial comment

One of the many aspects of modern construction where the role of the construction contractor has changed is in the financing of projects. Traditionally the client arranged the finance for the project and the contractor was paid for work done as the project progressed. There were a number of variations on this, such as pre-payments, accelerated payments and cash farming, and the monthly certification of work for clients before payment was an important part of the process. With the growth of the many forms of turnkey and build-and-operate projects, the task of arranging the financing has moved to those responsible for construction.

What distinguishes project finance from other forms of lending is that the project's creditors do not have recourse against assets, instead the lenders must be paid through the output and/or the cash flow that the project generates. Thus the lenders take on some of the risk of the project and will focus ways to manage and mitigate the risk they are taking on. Because the lenders accept some of the risk associated with the financial performance of the project they can be considered as investors in the project with the debt maturity defining the length of time of their involvement.

The project's revenue is the key asset that secures the finance and this is the core of project finance as distinct from conventional debt financing. Therefore the feasibility studies, demand and market analyses, and methods of guaranteeing cash flow are vitally important. Similarly, mitigation of financial exposures from fluctuations in interest rates and exchange rates are also crucial.

Putting together and structuring a project finance package is a complex process due to two factors. The first is the number of participants, the main ones being equity investors, lenders, advisers and specialized agents; the second is the broad range of measures used to provide financial security to lenders. The major risks in project financing are construction risk, market risk, economic risk and political risk.

* University of Technology Sydney

These two factors make project finance more expensive than conventional financing. However the advantages of allowing off-balance sheet financing, usually through creation of a project company, and greater risk sharing among the investors, are important. For more and more construction companies on a wider range of projects, understanding project finance and being able to structure a finance package is becoming an essential part of their ability to compete for large projects or in international markets.

The impact of this type of financing on the value of the completed project is fundamental: many projects that are the subject of project finance would be beyond the scope of any one client (public or private) and may remain forever so without the opportunities offered by these alternative financing arrangements and such projects would either never be built or would need to be substantially reduced in scope.

10.1 Introduction

One of the key investment decisions is the mixture of debt and equity in a project. A project's debt/equity mix is the capital structure. The equity component is provided by investors, often known as sponsors, while the debt comes from lenders and capital (financial) markets. Where equity is relatively straightforward, there are many forms of debt and there are many financial structures that may be used to provide debt finance for building and construction projects. The characteristic feature of project finance is that debt is not carried by investors or the value of physical assets, as with conventional debt financing, but is secured by the revenues or output produced by the project. This is non-recourse finance (i.e., not secured by assets) and as a result the concern of lenders is with the feasibility of the project, its revenue base and its sensitivity to risk factors (Tigue, 1995).

Many infrastructure projects have been financed this way, in both the advanced industrial countries and the developing economies. In some sectors, such as transport, the use of private finance to build roads, bridges and tunnels has become well established, and there are many examples of project finance in this sector. Other sectors with a lot of experience in structuring financial packages are energy (including electricity) and water supply. There are a number of examples of these types of projects given in this chapter.

Although the focus of project finance has been on infrastructure projects, it is now spreading to building projects as well. With the various forms of build and operate, design–build–finance, build and maintain, and turnkey procurement systems becoming more widespread, the ability to attract and structure the finance necessary for building projects is emerging as an important source of competitiveness for building and construction companies (de Valence and Huon, 1999). The growth of various forms of public/private partnerships in many countries has increased the importance of understanding financing options for projects, because the contractors (or consortia) that win these projects also have responsibility for financing them.

For an increasing number of commercial building projects the finance is provided only after the revenue of the finished project, or a substantial part of it, has been locked in by pre-committed tenants. This is seen in many of the business park developments that are becoming increasingly popular with many industries, such as telecommunications, software and electronics. It is common in this case for the developer to acquire the site and get a master plan approved; building is not started until a tenant commits to a long-term lease, with the building designed and customized for their use. Often the finished building will be passed on to an investment manager as part of a property portfolio.

In all these cases the project revenue is the key asset that secures the finance, and this is the core of project finance as distinct from conventional debt financing. This chapter covers the key features of project finance and identifies the characteristics that are making this an important part of the international building and construction industry. It includes a number of short case studies and a more detailed study of a power station project in China, which is used to demonstrate the large number of project agreements and risk management techniques used in project finance.

10.2 Characteristics of project finance

For most large infrastructure projects (usually worth hundreds of millions or billions of dollars, pounds, euros, etc.) a project company is created, and this company then carries out the many tasks associated with managing the design, financing, construction and operation of the project, with many of these functions contracted out to other parties. Often the members of the project company bring specialized skills to the project, such as a combination of construction contractor, financial organization, operator and maintenance manager. The consortium that built the Olympic Stadium in Sydney for the 2000 Games had such a membership. Procurement is usually through a turnkey contract to build and deliver the facilities at a fixed price, by a certain date, to specification and meeting performance warranties.

The project company secures the financing required for the project by arrangements with lenders and investors on a non-recourse or limited-recourse basis, where the main source of repayment of debt is project revenues. Where possible the project company's assets are given as collateral and debt is secured by agreements with the sponsors and other participants in the project. The actual borrower may not be the same as the project company if a separate special purpose vehicle is created.

Putting together and structuring a project finance package is a complex process (Buljevich and Park, 1999). The complexity is due not only to the number of participants, but also the range of measures used to provide financial security to lenders in a structured finance package. This makes project finance more expensive than conventional financing. The expense is due to four factors:

- the time spent by the lenders, their technical experts and their lawyers in evaluating the project and negotiating the (usually very complex) documentation
- the increased insurance cover, particularly consequential loss and political risk cover, which might be required
- the costs of monitoring technical progress and performance, and policing the loan during the life of the project
- the charges made by the lenders for assuming additional risk.

Although project finance is more expensive than conventional financing it is used because the advantages to project sponsors, investors and operators are substantial and compensate for the additional costs. The five common reasons identified by Clifford Chance (1991) for borrowers opting for project finance are:

- risk sharing, where failure of the project will not necessarily bankrupt the borrower or the investors
- management of political risks when investing in foreign countries – there is a range of political risks, such as price regulation, taxation, foreign exchange restrictions, import and/or export barriers, preferment of competitors or nationalization
- conventional loans might have an adverse effect on the borrower's or sponsor's balance sheet compared to the options available in project financing, particularly where the financing is non-recourse
- where a sponsor has borrowing restrictions, it might be necessary to use a finance structure that is not legally classified as a 'borrowing'
- tax benefits are a consideration, because tax allowances for capital expenditure and tax holidays for new enterprises might be available.

Financial closure is the point at which the principal participants (sponsors, government, lenders, suppliers) reach a formal agreement on the fundamental business structure of the project and the terms and conditions of the project's financing plan. Achieving closure can sometimes be difficult, as sponsors are not always able to raise finance for their project or sudden changes in markets conditions can affect the viability of a project.

10.3 Project agreements

The project company enters into contracts, such as a construction contract, a supply agreement, an operating and maintenance agreement and an off-take agreement, for development of the project. There can be many of these agreements in project finance deals, covering relationships (where relevant) between the project company and investors, governments or regulators, construction contractors, suppliers, the operator, customers and the financiers. These contracts improve risk management because they should allocate identified risk to the most appropriate participants. Described below are agreements on the concession period, financial completion, intercreditor agreements, construction, off-take, forward purchase and production payments and supply.

The concession, or licence, for the construction, maintenance and operation of the project is typically granted by the government to a special-purpose project company. This is particularly attractive to a government that wants to minimize the impact on its capital budget and to increase efficiency by using private sector management. Also, for developing countries, it helps promote foreign investment and the introduction of new technology.

An example of the importance of the concession is the Toll Highway Project in Latin America which secured the project debt under a concession agreement. In this project, the project company received a 20-year concession from the government to upgrade and maintain a major highway and the government retained title to the project's principal assets. Lenders were not able to create a meaningful mortgage over the assets, but, on the basis of the concession, the International Finance Corporation (IFC) made a loan on various loan securities and provided a small equity investment (Ahmed and Xinghai, 1999, p. 61).

Sponsors require assurances from government on a range of issues that might affect the project, and that adequate compensation will be paid if the government acts to the detriment of the project (e.g., through price regulation). The government is concerned that the concession company provides an adequate service, observes relevant safety and environmental protection standards, levies reasonable charges upon consumers, and maintenance and repairs are carried out. In a Build, Operate and Transfer (BOT) project, the condition of the asset on transfer is often specified in the contract.

A financial completion agreement has the major sponsors agree to provide funds if needed to meet debt service until the project achieves financial completion, which is when the debt has been repaid. After physical completion, if the company struggles to meet the conditions set for financial completion, project loans are then protected by sponsor obligations under the agreement to service loans until they are repaid.

Lenders typically enter into intercreditor agreements in order to regulate their relationship as creditors to the project company, to define the rules for the administration of their financial arrangements and, in the event of foreclosure, to share the collateral pledged in their favour. The basic purpose of these agreements is to prevent disputes among lenders that may jeopardize the interest of all creditors. The project company is not a part of an intercreditor agreement.

The construction contract is crucial, as construction costs are often around 80% of the total. The construction contract defines the duties, responsibilities and the scope of work for the parties (Turner, 1990). Any misunderstandings will affect the time and cost of the project and therefore influence the project finance package in terms of higher equity or debt needed or non-compliance of debt payment. Further, it is common to link the construction contract directly to the financial agreement; for example, the Athens Ringroad project had availability of the loans depend on the achievement of milestones in each section of the road (Esty, 2000). In this case, meeting the conditions of the construction contract is an essential part of the project finance package. A good construction contract lays the foundation of a successful project finance deal, because it helps attract potential investors and may also get a higher rating of the project, which has positive effects on the interest rates the project company is going to pay on its debts.

Often the project company enters into a long-term sale and purchase contract for project outputs, particularly in capital-intensive projects. By an off-take agreement, the off-taker agrees to buy from the project company a quantity of project output, for a set period of time and at pre-established prices. Consequently, an off-take agreement provides certainty that a project will generate sufficient cash flow to cover debt service and operating costs, and provide a reasonable return on investment. Although every project finance package has its own special features, the basic structure is that of a limited or non-recourse loan, repayable out of project cash flows with a forward purchase or a production payment agreement. Production payments are a method of financing that use the project's product and/or sales proceeds as security. Debt is serviced from the production of the project, and the lender becomes entitled to all, or an agreed proportion, of the project's production until the debt is repaid, together with the interest due. Usually the project company is required to re-purchase the product delivered to the lenders or sell it as agent for the lenders to realize cash. Under a forward purchase agreement, the lenders set up a special purpose vehicle to purchase agreed quantities of future production. The project company's obligation is to deliver the product and to service the debt. The purchase contract requires the project company to either buy back the production or to sell it on to third parties as agents for the lenders.

The Star Petroleum project in Thailand provides an example of an off-take agreement: the foreign sponsor (a US-based petroleum company) owned 64% of the project and the local sponsor (a company owned by the Thai government) owned 36%. The total project cost was approximately US$1.7 billion, to which the World Bank's financing arm, the International Finance Corporation (IFC) helped provide US$450 million in direct and syndicated loans. The project included a long-term take-or-pay agreement with the local sponsor and with a subsidiary of the foreign sponsor, covering 70% of the project's output (Ahmed and Xinghai, 1999, p. 46). The off-take agreement committed the sponsors to purchasing enough of the project's output to cover debt repayments.

The project company can have long-term supply agreements covering construction and operation of the project. Supply agreements lock in the availability and price of key supplies and generally have a term as long as that of the debt financing. The project company also may contract the operation, maintenance and management of the project to an operator, who carries most of the uninsured operating risks.

10.4 Project finance participants

The main participants in a structured finance or project finance package can be classified by the type of involvement they have; there are equity investors, lenders, advisers and specialized agents.

Equity participation begins with the major investors, or sponsors, who form the project company. The project sponsor can be one company or a consortium. The legal form and ownership of the project company are determined by a range of factors, particularly the legal framework of the host country. Equity is higher risk capital than debt because dividends can only be paid from profits, whereas interest payments are made from operating cash flow and debt is protected by various forms of security agreement.

When the capital structure of a project is agreed, the lenders will require sponsors to commit to a minimum equity contribution in the project company. While the interest burden during the construction phase is minimized if equity is contributed prior to debt financing, sponsors typically prefer to postpone equity contributions to maximize their return on equity (because they earn interest on that money by leaving it invested in other places until such time as they transfer into the project). Although the host government usually does not participate directly in a project, it might have an equity interest through an agency or be the main off-taker. In the case of a BOT project, the government will take over the project at the end of the concession period and become the owner.

On the debt side there are half a dozen roles played by banks. The scale of many projects means that financing will be syndicated among many banks, with the lending banks often coming from a range of countries. Syndication is where many banks each take a part of the debt, allowing them to spread their risk exposure over many projects. The arranger is the bank that arranges the financing and syndication of the lending. The bank that has manager status has a high level of participation in the overall facility, but will not normally assume any particular responsibilities to the borrower or other lenders. The agent bank is responsible for co-ordinating drawdowns and communications between parties, but is not responsible for credit decisions of the lenders. The engineering bank is responsible for monitoring the technical progress and performance of the project and liaising with the project engineers and independent experts. Finally, an independent trust company may be appointed as the security trustee. Where project financing involves a syndicate of lenders, security can be held by the agent bank as security trustee. However, where there are different groups of lenders an independent trustee is often used.

A number of project advisers are involved in each project. Financial advisers with expertise in the industry assist the company in determining project feasibility, in choosing the best project structure and in syndicating the debt financing. The financial adviser also prepares an information memorandum outlining the economic feasibility of the project for the lenders, and is usually a commercial or merchant bank familiar with the country where the project is located. Lawyers, usually a combination of international law firms and local lawyers, are responsible for the documentation. The legal advisers often structure the transaction. An engineering firm assists the company in all technical matters related to the design, engineering, construction, start-up and operation of the project. The technical experts often have a continuing role in monitoring the project.

There are also specialized agents, e.g., an insurance broker/risk management consultant designs, places and administers the project's insurance program. Insurance agencies are a crucial aspect of project finance and security over insurance proceeds is an important source of security for lenders, particularly when recourse to the borrower and/or sponsors is limited. Finance leasing companies acquire and lease assets to the project company in return for a rental stream calculated to cover the acquisition cost and provide a commercial return. The rating agencies such as Moodys, and Standard and Poors, give project debt a rating for quality; this affects the structure of the financing and the interest rate paid (Terry *et al.*, 2000).

Table 10.1 Financial structures of IFC infrastructure projects (Carter, 1996)

		Capital structure		**Source of debt**		**Type of debt**			
1967–1996	Number of projects	% debt	% equity	% local	% foreign	% private	% official	% IFC	% IFC syndicated
Total	115	58	42	33	67	55	10	20	15

International agencies provide funds for projects in developing countries and are responsible for arranging many project finance deals. These include the World Bank, the IFC, and regional development agencies such as the Asian Development Bank and the European Bank for Reconstruction and Development. Projects financed by these multilateral institutions are typically focused on developing the economic infrastructure of the host country. Many projects, particularly those carried out in developing countries, are arranged and co-financed by these multilateral agencies. For project sponsors based outside the host country, the involvement of the World Bank or its private sector lending arm, the IFC, might alleviate some of the concerns about the political risks that might be involved in doing business in that country.

10.5 Project risks and financial security

The objective of a risk analysis is to determine the relative importance of potential risks at the project's conceptual stage through an objective comparison and ranking of identified risks. Examples of some risk issues and risk management responses are examined below. Five risk responses identified by Baker *et al.* (1999) are:

- total avoidance of the risk
- significantly reducing the risk
- mitigating the impact of the risk
- allocating the risk to parties best able to manage it
- accepting that the risk should remain with the client.

In project finance the emphasis is on risk mitigation strategies.

Financial viability is the requirement that the project generate sufficient funds to repay loans and make a profit. There are many factors that can affect viability and the IFC identifies five major risk categories in project financing:

- construction completion risk
- market risk
- economic risk
- political risk
- force majeure risk – these are dislocations to business caused by events such as strikes, damaged equipment or government decisions (James, 1986, p. 1008).

Table 10.2 shows these risks and the risk mitigation strategies that can be used. Set against the risks are the agreements entered into by sponsors (the funds agreement or other financial support such as subordinated loans, the construction supply and operation contracts, and so on), the project assets and cash flow.

Specific financial risks relate to foreign exchange rates and interest rates. A foreign exchange exposure exists when a project's profits are affected by changes in exchange rates or when a company has future foreign currency obligations or receipts. Interest rate risk arises as a result of exposure to future movements in interest rates. To manage foreign exchange-rate risk, two options are annual adjustments by sovereign guarantees (e.g., a commodity price adjustment formula or tariff rate change), and a currency insurance program (Gavieta, 2001). Another option that can be used is a forward foreign exchange contract that fixes a future exchange rate (Terry *et al.,* 2000). For interest rates, a forward rate agreement provided by a wholesale bank allows borrowers and lenders to establish a forward interest rate for the period and amount of funding and limit risk resulting from changes in the rate of interest (Terry *et al.,* 2000).

Escrow accounts are often used in project finance. These are established to receive project revenues and are managed by the creditors or their agent, subject to a distribution agreement that covers what payments are to be made from the account and their priority. These accounts can be onshore or offshore and can be used for both local and foreign currency. In a similar way an agreement on the assignment of receivables will specify the ranking of payments to be made from project cash flows. Both of these are important if the project does not meet the project's anticipated level of output or revenue.

Table 10.2 Lender's approach to managing major project risks (Ahmed and Xinghai, 1999)

Risk to lender	Risk mitigation arrangement
Completion risk	
Delays	Turnkey contract; construction/equipment supply contracts. Specify performance obligations with penalty clauses. Project agreement to oversee construction on behalf of lenders and minority investors.
Cost overruns	Include contingency and escalation amounts in original cost estimates. Sponsor support until physical and financial completion (project funds agreement).
Site availability	Land use agreement.
Project performance	
Sponsor commitment	Strong, experienced sponsors with significant equity; share retention agreement to tie sponsors to the project.
Technology assurance	Prefer tried and tested technologies; new technologies can be used, provided the obligation to repay debt is supported by a guarantee of technological performance from the participant that owns or licenses the technology.
Equipment performance	Performance bond/guarantee from equipment suppliers. Operation agreement linking performance to compensation. Maintenance agreement.
Input activity	Supply contracts specifying quantity, quality, and pricing. Match term of supply contract to term of offtake commitment.
Management and labour performance	Experienced management team. Performance incentives and penalties. Training provided by equipment suppliers and technical advisers.
Market risk	
Demand potential	Undertake independent market assessment. Offtake contract specifying minimum quantities and prices (take or pay arrangements). Conservative financing structure. Support low-cost producers.
Payment risk	Sell output where possible to creditworthy buyers. If buyers not creditworthy, consider credit enhancements such as (1) government guarantees of contractual performance (if buyer is state-owned); (2) direct assignment of part of the buyer's revenue stream; (3) escrow account covering several months' debt service.
Economic risk	
Funds availability	Limit share of short-time financing to project; long-term finance to match project term; stand-by facility.
Interest rates	Fixed-rate financing, interest rate swaps.
Exchange rates	Match currency of project loans to project revenue, swaps and guarantees.
Inflation	Long-term supply contracts for energy and other important inputs; output prices indexed to local inflation.
Force majeure	Insurance policies and force majeure provisions.
Overall risk support	
Debt service coverage	Analysis based on pessimistic assumptions to set up-front debt/equity (D/E) ratio. Financial support until D/E ratio is reduced to safe level. Escrow accounts with debt service reserve.
Security	Mortgage and negative pledge on project assets. Assignment of concession agreement and other relevant agreements. Share pledge. Disbursement conditions and loan covenants.

There are five principal methods used by lenders to manage financial security (Ahmed and Xinghai, 1999):

- securing project assets – this involves a mortgage on the assets, where the realizable value of the assets exceeds the loan value. However, not all countries allow foreign investors to hold mortgages over domestic assets
- a weak mortgage – used in countries that prevent lenders creating adequate mortgages; e.g., Indonesia, where land cannot be mortgaged to foreigners, or countries like China and Thailand, where mortgage law is untried or still being developed. In some cases there are restrictions on project security, e.g., in Sri Lanka recourse is only available to secured fixed assets
- assignment of project receivables – determines how revenues are used for project agreements (such as concessions, supply and management agreements) and damages
- an escrow account – project cash flow is collected, and spent as agreed
- using the equity base held by the major investor to secure project loans through a pledge of shares. If a sponsor is a supplier of materials or services they can defer payments when necessary. On high-risk projects loans can be secured by insurance contracts and/or government guarantees.

10.6 Financing build, operate and transfer (BOT) projects

The BOT procurement structure is one where the host government grants a concession to a private company that will be responsible for financing, building, operating and maintaining the project to the end of the concession period, after which the project is transferred back to the host government, usually at nominal or no cost. The BOT model has become one of the most important schemes used for construction of infrastructure projects (Jones, 1991). It has become popular as it promotes the establishment of strong relationships with host governments, provides more efficient resource allocation and more effective financing, and, due to the increased confidence of investors in these projects, an improved ability to raise funds in the market. Other features contributing to the success of BOT project finance include the incentive system created by the BOT model, the ownership reversion feature, and the risk/return negotiation scheme (Buljevich and Park, 1999).

A BOT project is a method of turning over to the private sector, for a limited period, the development and initial operation of what would otherwise be a public sector project. In most BOT models, ownership reversion occurs only after the private-sector bodies have received the return of, or a satisfactory return on, the capital they have invested in the project. In return for the ownership reversion, the host government might be asked to supply credit support for project borrowings. Moreover, the BOT model consists of risk/return negotiations among the host government, the project company and the bank syndicate to develop a mutually satisfactory project structure and a fair risk/benefit sharing scheme. The bank syndicate then tailors a loan to the structure of the project. The success of this risk/return sharing scheme depends on the financial soundness of the BOT proposal (Yang-Cheng *et al.*, 2000).

10.6.1 Examples of build, operate and transfer projects

The 900 km North–South Highway project in Malaysia is an example of BOT financing where a project that was viable could not be financed by the national government. Initially the scheme was to be built by the public sector, but the Government realized that it could not afford to finance the project. In 1987 United Engineers Malaysia (UEM) was awarded a concession contract worth just over US$2 billion. Under the concession, roads already built and owned by the Malaysian Highway Authority were taken over by the new operating company. Part of the deal was that modernization and upgrading, such as improved toll booths and lighting, were carried out on existing roads. UEM attached a condition that, if traffic flows fall below an agreed level, the Government will reimburse the lost revenue. Once the concession contract was secured, UEM created a subsidiary company, Project Lebuhraya Utara-Seletum (PLUS), to act as concession holder and borrower for the project. A 30-year concession period was granted, and the Government provided a US$60 million loan drawn down over 11 years at 8%/year with a 15-year grace period (Price, 1995).

Finance for the Sydney Harbour Tunnel took 2 years to put together. The tunnel was built and is operated by Transfield Pty Ltd and Kumagai Gumi, but ownership will revert to the New South Wales Government in 2022 when the project will be debt free and will have an estimated value of AU$2 billion (Australian dollars). Most of the project finance came from AU$501 million worth of 30-year bonds, issued in eight tranches (instalments) over a 5-year period, plus AU$17 million paid by the joint venture partners for the first year's construction, a AU$223 million loan made by the state government to cover most of the second-year costs, and a AU$40 million performance bond against cost overruns and construction delays on the project lodged by the joint venture partners.

The project revenue comes from tunnel tolls limited to AU$1 (at 1986 prices) throughout the project life, and increased Harbour Bridge tolls will also be used to generate project revenue. Because the toll prices have been fixed, the NSW Government has underwritten minimum traffic flows for both the bridge and the tunnel. However, in order to prevent Transfield Pty Ltd and Kumagai Gumi enjoying excessive profits, clawback provisions have been included.

Fundraising costs for the project were substantially reduced because public equity insurance and loan syndications were both avoided, unlike the Hong Kong Tunnel and Channel Tunnel projects. The following features of the project's financial package created a favourable reaction to the 30-year securities: an extended maturity, longer than anything previously available in the Australian market, an undisclosed yield with interest payments protected by an index-linking formula, repayments of principal with quarterly interest instalments, previously available in Australia only through mortgage-backed securities, and the eight-tranche investment schedule, enabling institutions to lock in future paper pick-up (i.e., investors' agreement to provide funds) matching cash flow expectations (Price, 1995).

An example of a BOT project that failed is the Bangkok Elevated Road and Train System (BERTS) – it failed due to a number of problems and its remains stand as a witness to the risk and uncertainty inherent in the project. The concession to build and operate the system was granted in 1990 but less than 10% of the work was completed, largely because of bureaucratic delays and political upheavals (Asia Week, 1996). Begun in 1992, the concessionaire, Hopewell Holdings Limited, planned to spend 4 years completing it.

Unfortunately, the project faced protests from many parties, including bird watchers, academics, environmentalists, the disabled and a private school. Tender documents for a turnkey system were issued in March 1993 to five consortia which had expressed an interest in the project. In July, the following year, the Siemens (Germany), Adtranz (UK), Balfour Beatty (UK) and Tileman (subsidiary of Hopewell (Thailand) Ltd) consortium signed an agreement to build and operate the line. Since terminating the US$3.2 billion contract with Hopewell (Thailand) Ltd in 1998, the government has not found a way to revive the project. Tam (1999) found that there were four reasons for this project's failure: the land leasing cost, changes of government, changing construction schemes, and the difficulty of providing a tollway intersection.

10.6.2 Laibin B

Perhaps the best documented example of a high-risk international BOT project is the Chinese power plant, Laibin B. It is the second phase of the Laibin Power Project with two 350 megawatt (MW) coal-fired units and an estimated cost of US$650 million (5 billion Renminbi; RMB). The power station is scheduled to be completed in 2004, after a three and a half year construction phase, and its ownership will be transferred to the Guangxi provincial government after a 15 year concession. The operating company is a 100% foreign owned consortium of Electric de France (EdF) with 60% ownership, and GEC Alsthom (UK) with 40%. GEC Alsthom led the construction of the plant and supplied the turbine generators and the boilers; Cegelac (France) supplied instrumentation and controls and electrical equipment; EdF carried out other installations and will operate the power plant during the concession period.

The consortium will finance the project from a revenue stream based on letters of comfort (an undertaking to purchase the output) from the provincial government supporting an agreement under a fixed electricity tariff rate. Therefore, payment risks have been identified and sufficient government guarantees obtained from the provincial government, the eventual owner of the project (Ye and Tiong, 2000).

The financing of the project was arranged using foreign capital. The capital structure requirements of US$650 million had US$290 million as a syndicated loan package, with a further US$140 million provided as equity by the concession parties (EdF and Alsthom). The balance of US$220 million came from the Bank of China. Foreign investment of 66% covered equipment and technical supply. The local investment of 33% included all local materials and local sub-contracting content on the project; the debt to equity ratio was 78:22, high financial leverage. To lower the leverage and risk of insolvency of the consortium, a provision was made for local investment of 20% that converted Bank of China debt to equity and reduced the debt to equity ratio to 58:42.

The security of the debt is through an agreement with the eventual owners (the provincial government) for the power output. Supply agreements were also obtained for guaranteed delivery with extended warranty periods for the turbines, generators and coal supply. The debt service will be paid to the lenders on a semester basis during the 15-year concession period and debt repayment will commence upon the practical completion of the construction phase and commencement of the operation phase.

Laibin B used a BOT procurement method with a foreign consortium bidding for a concession from the Guangxi provincial government to build, own and operate the project throughout the concession period. The bid documents included a draft concession agreement, an agreement with the local power company and the related technical specifications and financial information. The consortium carries most risks relating to construction, such as cost overruns and delays, and operation of the plant, such as operating costs and meeting performance standards. The Chinese parties assumed most of the fluctuation in the electricity tariff, demand, political, force majeure and currency risks. The concession authority also guarantees the performance of the local power company. This scheme provided all parties with a number of advantages (Project Finance, 1999):

- clearer risk allocation – the concession authority generally accepts responsibility for many of the China risks, which have traditionally pre-occupied foreign investors and their lenders. Therefore, foreign parties require less support and comfort from Chinese government agencies
- competitive bidding – international competitive tendering led to significant advantages: transparency, fairness and a more efficient allocation of resources and expertise; this allowed bidders to focus on cost and technical factors rather than negotiation and political matters
- tariff risks – first, moving away from cost plus tariffs and guaranteed returns towards fixed tariffs promoted a more objective application of the agreed tariff formula, and second, pre-approval of the tariff formula in the agreement restricted the tariff approval authority's role to verifying that the tariff formula is being applied
- pre-packaging of approvals – the bid documents identified the package of remaining approvals, giving the successful party a regulatory road map to financial closure; the packaging of approvals reduced regulatory risks such as the continuance of approvals, tariff adjustment and foreign exchange issues.

Tiong *et al.* (2000) discuss risks associated with the scheme in detail. It was agreed that risks related to force majeure and political risks be absorbed by the Guangxi provincial government, the eventual owners of Laibin B. Other risks, such as construction risks and market risks, had appropriate measures of mitigation identified. Political risks are: change in law, corruption, delay in approval, expropriation, and reliability and creditworthiness of Chinese entities. The Guangxi provincial government has provided guarantees that tariffs will be adjusted and concession periods extended in the case of these risks eventuating. The consortium on the other hand shall maintain a good relationship with high level state and provincial officers and establish local participants in the project.

In mitigating the eight construction risks, key contract clauses have been agreed and selection criteria identified:

- land acquisition – the entire BOT scheme will commence upon the entire acquisition of the land
- compensation/restriction on equipment/materials – all equipment and materials imported for use in Laibin B will be subject to tax exemption and lesser restrictions from the Government and as such the contract sum shall not include such costs
- cost over-run – all cost over-runs are mitigated by the consortium taking all the risks by agreeing to a fixed lump-sum contract. Over-runs that are not the liability of consortium are subject to a variation that is to be agreed by the client's representative

- time and quality risks – time risks are mitigated by the consortium undertaking to pay liquidated damages for the delay of the project. As for quality, the consortium undertakes to operate the job to ISO 9002 international standards and will be subject to audit by the client's representative at various stages during construction
- default by contractor – the contractor will, prior to commencing work, submit a performance bond for 10% of the value of their contract sum to mitigate any losses due to default
- environmental damage – the environmental impact study carried out by an independent consultant resulted in a positive feasibility of the project. Project insurance will be obtained by the client for any environmental damage that may result
- input availability – the input availability will include all foreign imports and coal supply. The coal is to be supplied by the Government and the foreign imports, which mainly include the turbines and the boilers, fall under the liability of the consortium
- management performance – the success of the bid from EdF and Alsthom was based on their experience in the local Chinese market. EdF has been present in China for about 15 years as a consultant, designer and operator of large power transmission and distribution systems; Alsthom has been in China since early in the twentieth century and has supplied equipment for power stations with an aggregate capacity of more than 30,000 MW.

There are six operating risks to mitigate:

- government department default – a guarantee from the Government of the People's Republic of China has been obtained for this project
- concession company default – based on the concession companies' profiles a bank guarantee has been obtained from COFACE, the French credit export authority
- operator/labour/technology risks – the selection criteria during the tender stage has a strong emphasis on the capabilities of the concession company with regard to operating capability, provision of skilled personnel and possession of the technical expertise necessary to undertake the scheme
- prolonged downtime during operation – the concession company has undertaken to accept liability with regard to downtime
- management performance – assessment of the management performance of the consortium was undertaken as part of the selection process
- equipment performance – the supplier of the turbines and boilers (Alsthom) provided extended warranty and maintenance of 10 years beyond the concession period.

Two other risks that required market research are the demand potential and the payment risk. With the concession period being 15 years, changes in demand for project output can lead to changes in revenue and profitability problems. In hedging against market risk for Laibin B an agreement with the sponsor to purchase 80% of the project's output was agreed. The agreement uses a fixed tariff and specifies the minimum quantities and prices to be used as variables in the tariff formula. The Guangxi government will need to make specified minimum payments even if it does not take delivery. This agreement guarantees to the producer that fixed costs such as debt-service payments, fixed operation and maintenance costs and return on equity will be covered. Payment risk is mitigated by the government of China providing a guarantee of contractual performance by the Guangxi provincial government.

Two of the major financial risks associated with Laibin B are foreign exchange rate and interest rate risk. For investment in Laibin B, the consortium will receive most of their revenues in RMB. A significant portion of this revenue will need to be converted to US dollars and remitted to France (mainly) and the UK. RMB is, however, not freely convertible into US dollars; the exchange rate fluctuates continuously in the market and is subject to the approval of the State Administration for Exchange Control (Tiong *et al.*, 2000). In mitigating this risk, the fixed tariff rate and the adjustable concession period (subject to approval) is by far the most effective measure. The fixed tariff reduces the risk of the investors and transfers it to the eventual owners. It allows for an easier investment decision to be made.

The second mitigating measure to be undertaken in Laibin B for foreign currency exchange rate risk is the dual currency contract. Here, the amount to be paid, via a portion of the tariff for the turbines and boilers amounting to an estimated US$250 million, will be remitted in US dollars. The third mitigating measure involves foreign investors protecting themselves against the devaluation of the RMB – the financial package structure will use the NDF (non delivery forward) method, which is a type of hedging where no actual RMB cash is required for certain transactions. Instead, these entire transactions are conducted in US$ outside of China, with the pay-out tied to the official closing RMB rate posted each day. The NDF method requires a reference rate (the official rate) and a contract rate; the difference between the two rates then determines the gain or loss for each party in the transaction (Tiong *et al.*, 2000). Either way, transactions are automatically converted into foreign currency, so no RMB ever changes hands. This service will allow payment for the remaining foreign cost elements on the project, such as expatriate wages, control instruments and other equipment.

Fourthly, with the inclusion of the Bank of China the local component of the capital requirement will have no foreign exchange currency risk. Finally, to mitigate the risks related to interest rates, the loan from the local Chinese bank is at a fixed interest rate.

For the US$290 million of foreign investment from Agricole Indosuez an interest rate swap facility was arranged to mitigate the interest rate risk. The consortium would be able to swap from a floating interest rate to a fixed interest rate at specified stages of the concession (every 1.5 years) using an agreed swap rate, if concerned about the direction of market interest rates.

10.7 Conclusion

The methods used to finance building and construction projects is one of the most dynamic and complex areas in the modern industry. Where clients used to pay for work done, today it is increasingly common for the construction contractor or consortium to arrange the finance necessary for the projects they are responsible for. The methods first employed on infrastructure projects in the transport and energy industries are now being applied to building work.

The distinguishing feature of project finance is that the project's creditors do not have full recourse against the sponsors. The lenders must be paid through output and/or cash flow that the project generates. Although the level of recourse and the type of support given may vary from project to project, lenders in this kind of financial situation will always be taking some degree of credit risk on the project itself. Therefore the lenders will be concerned with ways to manage and mitigate the risk they are taking on. Because the lenders accept some of the risk associated with the financial performance of the project they are in a similar position to equity investors and can be considered as investors in the project with the debt maturity defining the length of time of their involvement.

Project finance is characterized by the size and complexity of the deals. These offer higher returns to lenders, because there are more risks involved, such as project feasibility, type of project, construction, financing structure, and marketing and political risks. Therefore the lenders look at the broadest range of measures possible to manage the risks they are taking on when investing in a project. This is also why most project finance is through syndicated loans, with many lenders putting in small amounts each. Such diversification is the most important risk management method available, because it allows the lenders to spread their risk over a large portfolio of projects and avoids large exposure to any single project.

There are two crucial factors common to all project finance packages; both interest rates and cash flow play significant roles, and the risks associated with them need to be managed to guarantee project success. Interest rates are primarily a payment to the lenders of funds who take on part of the risk involved with the project. As interest rates are forever changing and can mean the success or failure of a project, interest rate risk management is necessary. The common instruments used for interest rate risk includes forward rate agreements, swaps, futures and options that either eliminate the risk or hedge exposure.

The cash flow estimates are first used in the project appraisal, which then attracts the sponsors, investors and lenders involved in the project. The structure and design of a financial package is then largely built around the expected cash flow. To manage the market risk, an agreement or minimum subscriber based agreements can be used to ensure cash flow during the typical critical periods by allowing a company to secure long-term sales at a predetermined price for initial project outputs. This provides some financial security for the lenders and gives a project a chance for success.

Most of the projects that have a structured finance package will be procured through some form of a BOT contract. The period that is given for operation of the project is the time available for the investors to get loans they have made paid back, or get the return on the equity investment that was made. Again, the contrast with the traditional short-term build only contract is marked.

As construction managers become responsible for, or part of a team that undertakes, a wider range of tasks associated with their projects, such as design, financing and operation, the approach needed also changes. In effect, these types of projects internalize the traditional roles of the client, who used to provide documentation and finance for their projects, and then operate or manage the finished work. Many of the characteristics the industry has been noted for in the past, its antagonistic and disputatious nature for example, are not helpful on projects where the constructor has a stake in the long-term performance of the project.

References and bibliography

Ahmed, P.A. and Xinghai, F. (1999) *Project Finance in Developing Countries* (Washington D.C.: International Finance Corporation).

Asia Week (1996) www.asiaweek.com/asiaweek/96/0426/biz1.html

Baker, S., Ponniah, D. and Smith, S. (1999) Risk response techniques employed currently for major projects. *Journal of Construction Management and Economics*, ASCE, **17** (2), 205–13.

Buljevich, E.C. and Park, Y.S. (1999) Description of a project finance transaction. *Project Financing and the International Financial Markets* (Dordrecht: Kluwer Academic Publishers).

Carter, L. (1996) *Financing Private Infrastructure* (Washington DC: International Finance Corporation).

Clifford Chance (1991) *Project Finance* (London: IFR Publishing Ltd).

de Valence, G. and Huon, N. (1999) Procurement strategies. In: Best, R. and de Valence, G. (eds) *Building in Value: Pre Design Issues* (London: Arnold Publishers), pp. 37–61.

Esty, B.C. (2000) The Equate Project. *Journal of Project Finance*, **5** (4), 7–20.

Gavieta, R.C. (2001) Currency exchange risk and financing structure: a Southeast Asia developing country perspective. *Journal of Project Finance,* **6** (4), 49.

James, J.S. (1986) *Stroud's Judicial Dictionary of Words and Phrases*, Fifth edition (London: Sweet & Maxwell).

Jones, C.V. (1991) *Financial Risk Analysis of Infrastructure Debts* (Greenwood Publishing Group, Inc.).

Price, A.D.F. (1995) *Financing International Projects. International Construction Management Series No. 3* (Geneva: International Labour Office).

Project Finance (1999) Back to the drawing board. *Project Finance*, September. www.projectfinancemagazine.com/index110.html?/contents/publications/pf/magazine/1999/sep99/header.html&/contents/publications/pf/magazine/1999/sep99/pf-7238.html

Tam, C.M. (1999) Build-operate-transfer model for infrastructure developments in Asia: reasons for successes and failures. *Global Finance*, **17** (6), 377–82.

Terry, C., Hutcheson, T. and Hunt, B. (2000) *Introduction to the Financial System* (Melbourne: Nelson Thompson Learning).

Tigue, P. (1995) Infrastructure financing in Hong Kong: project finance leads the way. *Government Finance Review,* December, 36–7.

Tiong, R., Wang, S.Q., Ting, S.K. and Ashley, D. (2000) Evaluation and management of foreign exchange and revenue risks in China's BOT projects. *Construction Management and Economics*, **18**, 197–207.

Turner, A. (1990) *Building Procurement* (London: Macmillan Education Ltd.).

Yang-Cheng, L., Soushan, W., Dar-Hsin, C., Yun-Yung, L. (2000) BOT projects in Taiwan: financial modeling risks, term structure of net cash flows and project at risk analysis. *Journal of Project Finance*, **5** (4), 53–6.

Ye, S., Tiong, R. (2000) NPV-at-risk method in infrastructure project investment evaluation. *Journal of Construction Engineering and Management*, **126** (3) May/June, 227–33.

11

Project management in construction

Patrick Healy*

Editorial comment

The Great Pyramid of Cheops (Khufu) originally contained around 2.5 million tonnes of stone and stood almost 150 m tall. There is no universally accepted theory of exactly how such a remarkable feat of building was accomplished in an era when neither the wheel nor iron were known to the builders. The individual blocks of stone weighed around 2 tonnes each and had to be transported from distant quarries then raised to their final positions in the massive structure. The level of organization that is implied by the presence of such structures is extraordinary – no telephones, no faxes, no sources of power other than that of animals and humans, yet somehow these projects were completed, the work was done to close tolerances and produced structures that have endured for millennia.

In essence, the management of building procurement remains largely unchanged – a client decides to build (the Pharaoh or an insurance company), a building is designed (the pyramid or an office tower), materials (limestone and granite or concrete, steel and glass, and perhaps more granite) and labour (masons, labourers and roadbuilders or riggers, plasterers, tilers) are the major inputs to the construction phase, and all these processes require a framework within which they fit so that the collective effort of many people and organizations produces the result desired by the client.

In 2600 BC construction work proceeded very slowly but without the imperatives of the fiscal year, interest payments on borrowed funds, property cycles or competition from rival pyramid builders threatening to flood the market with cheap imitations. Time was less pressing than it is in the highly pressured, globalized industry that we know today. Clients today are not likely to accept the trial and error approach of the master masons of the Middle Ages who built wonderfully intricate stone structures in the gothic style that often failed structurally and had to be rebuilt. Instead, today, very complex buildings are

* University of Technology Sydney

constructed to tight deadlines and tight budgets, with amazing arrays of electronic gadgetry and sophisticated engineering services, and clients demand that they be completed on time, in budget and to high standards. The management of the procurement of such buildings is now largely directed and controlled by a new breed of industry professional – the project managers.

In the Middle Ages the architect and the builder were often the same person, usually the master mason. As the roles of designer and builder became separated the architect assumed the role of manager of the process and this remained the case until relatively recently, with the architect being responsible not only for the design of the building but also for overseeing the activities of the builder or contractor and also being the person who ultimately agreed on the amount being paid to the contractor for work completed (progress payments). Many of the functions of the architect now reside with the project manager, and in fact even the selection and hiring of the architect is often handled by the project manager.

Project managers must have a broad range of skills but probably the most important of all is the ability to manage people and to get the best performance from them. Building projects are large, costly and complex and involve many interested parties from financiers to labourers, union delegates to Chief Executive Officers (CEOs), quantity surveyors to steel fixers. For the client to achieve value for money, careful management of the processes that lead to a successful outcome is essential. The potential for disaster is ever present, with the possibility of on-site accidents, structural failure, cost overruns resulting from poor ordering procedures or unavailability of materials or specialist labour.

Project management is concerned with planning the activities, recognizing dependencies, looking ahead for potential pitfalls, keeping all concerned focused on the job at hand and monitoring and reporting progress. Successful project management is undoubtedly a major factor in the realization of value for money in building. In this chapter the fundamentals of the project management process are introduced and a number of key concepts related to the discipline discussed.

11.1 Introduction

Construction Management is the management of a building or construction site. It is recognized as a particular form of Project Management. While construction management is much older than project management, many see construction management as an industry-specific application of project management.

Project management has now emerged as a distinct discipline. With the recognition of project management as a separate discipline there emerges the opportunity to study it as an entity on its own rather than solely within the context of a specific industry. This makes it easier for lessons learned in one industry to flow to other industries. It also provides those in the building and construction industry with a new framework within which to evaluate and develop management practice.

Here the aim is to identify project management and its elements, and in doing so to differentiate between project management and process management and note the salient features of project management. It is not surprising that these features also appear in construction management, and in fact, the primary reason for studying project management in this context is to use it to improve construction management.

11.2 General management

General management is concerned with the operation of the whole enterprise. It forms a view of the future; it decides where the enterprise should go given this view of the future. It is concerned with identifying what is to be done and how it is to be done. It is concerned with the allocation of resources to the right areas. Once these decisions have been made they must be followed up and implemented.

General management includes areas such as strategic planning, marketing, financial structuring, and production, among others. As a manager one is always in general management mode. Managers are concerned with what their part of the enterprise has to do given the overall direction of the enterprise. When one acts as a general manager one will face issues and some problems that are unique, and others that are recurrent or repeating.

11.3 Project management

The issues or problems facing general management come in two different forms. There are those that can be called repeating or *process* problems, while others are one-off or *project* problems. In reality the two types of problems will come mixed together but it is useful to separate them here. Project management deals with the one-off problems, and because these one-off problems are turning up more frequently there is great interest in project management.

For example, anyone running a retail outlet (a clothes shop, say, or a food shop) will have many repeating activities, e.g., making sure that replacement stock is delivered and put out on the shelves to replace that which has been sold. In deciding how much to re-order one can gain guidance from how sales are going. If stock is clearing fast, that is, people are buying a lot and buying quickly, one has a good basis for increasing the amount of the re-order. If they reduce their buying, then orders can be adjusted accordingly. So this re-ordering part of the managing of a retail outlet is a repeating problem, and because it is repeating it has a history to help in the decision making. But every now and then the retail outlet has to decide to change the products or product line it is selling, e.g., to introduce new season clothes or new food lines – this is more of a one-off problem than the re-ordering of stock. So in managing a retail outlet one has mainly repeating problems, with one-off problems arising every so often. What has happened, however, is that these 'every so often' problems have started to arise much more frequently. New products are being introduced at a faster rate and businesses have to become more flexible so that they can cope with changing tastes. Such one-off issues have to be dealt with and it is managing these one-off issues that is the work of project management.

While builders, in many forms and guises, have been constructing one-offs for thousands of years, it is only relatively recently that the form of management they have been practising has been recognized and given the name project management. It is also known under other names: it is called construction management when applied to the specific area of site construction management, and it is called design management when dealing with the design of a building (i.e., managing the architects, engineers, quantity surveyors, etc.). The term project management is now often reserved for the management of the process of design and construction of buildings, however, the names vary from company to company. The important point is that project management has been practised in the building and construction industries for a long time and carries a number of different labels.

While project management has been used in construction for a very long time, it has, in the last half century, been recognized as something that applies to many industries. The main reason it applies to these other industries is because it applies to the management of one-off problems, and these industries have recognized that they have many one-off problems.

There is debate about the amount of industry-specific knowledge one needs to practise project management in any specific industry; one needs some but it is not clear how much. This is certainly true in the construction industry – one may not need an intimate knowledge of the process to be able to manage it successfully.

11.3.1 Defining project management

Typically, any attempt to define project management begins with a definition of a project. The Project Management Institute (PMI), based in the USA, defines a project as 'a temporary endeavor [sic] undertaken to create a unique product or service' (PMI, 1996). An updated version of this (PMI, 2000) adds 'result' to 'product or service'.

This is a perfectly valid and useful definition and it has widespread acceptance. The word 'temporary' in this context does not imply that a project occupies a short period of time, but that it has a beginning and an end. This is crucial in understanding project management; the temporary endeavour may last for a week, for a year, or for decades but it is seen as having an end point, and it ends when the product or service is delivered. While more recently there has been a push to regard the 'temporary endeavour' as extending to the whole of life of the facility or product or service there is still an end to it.

The other key word in this definition is 'unique'; a project is a one-off, it is somehow different to all that went before. It is this quality that is the source of much of the difference in how one should manage a project as against managing an ongoing operation. The lack of repetition is a key factor in determining the need for project management and in influencing how project management should be practised.

The UK Association for Project Management (APM) provides two definitions (APM, 2000):

> [A project is a] unique set of co-ordinated activities, with definite starting and finishing points, undertaken by an individual or organization to meet specific objectives within defined time, cost and performance parameters.

or

> An endeavour in which human, material and financial resources are organized in a novel way to deliver a unique scope of work of given specification, often within constraints of cost and time, and to achieve beneficial change by quantitative and qualitative objectives.

Both definitions are compatible with that of the PMI, as is that offered by the Australian Institute of Project Management (AIPM, 1996):

> A set of inter-related activities, with defined start and end dates, designed to achieve a unique and common objective.

Having defined the project, the following definitions of project management emerge; PMI provides the following definition of project management (PMI, 1996):

> [Project management is the] . . . application of knowledge, skills, tools, and techniques to project activities in order to meet or exceed stakeholder needs and expectations from a project.

So they are applying knowledge, skills, and techniques to the temporary endeavour. A later version (PMI, 2000) modifies this definition by removing reference to 'stakeholder needs and expectations' and replaces it with 'meeting project requirements'.

APM (2000) offers these definitions of project management:

> Planning, monitoring and control of all aspects of a project and the motivation of all those involved in it to achieve the project objectives on time and to the specified cost, quality and performance.

and

> The controlled implementation of defined change.

This second definition is quite succinct and telling.

The AIPM definition of project management is (AIPM, 1996):

> The planning, organizing, monitoring and controlling of all aspects of a project in a continuous process to achieve its objectives, both internal and external. It is a discipline requiring the application of skills, tools and techniques and the balancing of competing demands of product or service specification, time and cost.

All these definitions of projects and project management are essentially compatible, however, in all of them the project has been identified as the 'temporary endeavour' rather than the output at the end. There is an alternative view (Healy, 1997) that suggests that the output should be called the project and the work that leads to that output be called the project sequence. For instance, the Sydney Opera House itself, the physical building, is the project, rather than the working, the doing, on the Sydney Opera House being seen as the project. Taking the Sydney 2000 Olympic Games as another example: the project was the Games themselves, not all the work that went on before hand. Another definition now emerges (Healy, 1997):

> A project is a one-off change to be achieved by a finite, time-ordered and interrelated set of tasks. The one-off change is the project; the time-ordered set of tasks is called the project sequence. Project management is the identification of the one-off change and the management of the project sequence.

This definition has three parts: the project, the project sequence, and project management. The project is seen as a one-off change that is produced as a result of a set of tasks. Change, in this context, has a very wide meaning and can be generic in that it applies across all industries. It can mean many things: a new building, a better software package, a new way of doing things and so on. This change has to be achieved by an effort, by a set of tasks. They may be described using some qualifiers, viz. finite, time-ordered and interrelated. The tasks need to be finite in that they are expected to end, and they are time-ordered and interrelated because the tasks need to be related to one another in some time framework.

The project sequence is the set of tasks. This is commonly referred to as the project process but here we will reserve 'process' for tasks that repeat and use 'sequence' to imply a one-off and a finite effort. Sequences often have processes within them, so process management may get mixed up within project management but, for now, we have a project delivered by a project sequence.

Project management can then be defined as the identification of the project and the management of the project sequence. One of the key tasks of project management is to identify the project; not being clear about what the project is leads to all sorts of difficulties and it is up to the project manager to make sure that agreement exists as to what the project is.

In summary, project management is identifying the project (i.e., the result) and managing how to get there.

11.3.2 The elements of project management

People have been trying for a few decades now to define what project management encompasses, to identify the methods, techniques, and knowledge that constitute project management, and it is not an easy task. PMI (1996) suggests that there are nine 'knowledge areas' in project management – note the use of the word 'knowledge':

- project integration management
- project scope management
- project time management
- project cost management
- project quality management
- project resource management
- project communications management
- project risk management
- project procurement management.

Project integration management is the co-ordination of all the activities so that all is consistent, e.g., making sure that the time management is compatible with the cost management and with the scope management.

Project scope management is concerned with defining the project and defining the work that needs to be done to achieve the project. This is arguably the main driver of the project management sequence.

Project time management is ensuring completion of the project on time. Key techniques include CPM (Critical Path Method) and PERT (Program Evaluation and Review Technique).

Project cost management is concerned with finishing the project within budget – not necessarily the least expensive or cheapest or best value but simply within budget.

Project quality management is aimed at delivering the project with the required quality characteristics. There are special challenges facing project managers in this area as they are dealing with one-offs rather than repetition.

Project human resources management is concerned with getting people with the right skills working on the project sequence as long as is required.

Project communications management is about keeping people informed and meeting the information needs of various stakeholders.

Project risk management is concerned with identifying risk and deciding how to respond to it.

Project procurement management is about sub-contractor management and the acquisition of various goods and services.

PMI breaks up these nine knowledge areas into 39 processes that are embedded in the project sequence. This division is useful particularly as it is integrated with what can be loosely described as a 'plan, do, check, act' approach to managing the project sequence.

The AIPM has adopted a different approach based on competencies and defines nine units of project management competence. The names of the competencies are the same as those of PMI, that is integration, scope, time, cost, quality, human resources, communications, risk and procurement. The approach is different, however, as the AIPM wants people to display these competencies rather than simply possessing the knowledge associated with the various areas.

The APM, by contrast, has 44 topic areas grouped under seven headings: general, strategic, control, technical, commercial, organizational and people.

There are other bodies of knowledge or statements of what is the essence of project management, e.g., that put forward by the German Project Management Institute. They all are fairly similar in regard to the main sweep of topics that constitute project management, but they group them a little differently. The International Project Management Association (IPMA) has its topics arranged in a circle like a sunflower and so avoids having to attempt to group topics.

There is considerable debate about whether the professional institutions should go for knowledge or go for competencies. In rough terms we see the Americans going for knowledge, with the British, Australians and Europeans leaning towards competencies. The debate continues.

11.3.3 Examples of projects

The most famous projects are arguably the space projects, and in fact NASA has made a significant contribution to the understanding of project management. NASA often has an overall aim or outcome, for example, it wanted to get to the moon and now wants to get to Mars. The overall project, for example, getting to the moon, is broken down into a series of smaller projects. The first NASA space flight only lasted a few minutes; the object of that project was to make sure they could go up and then come down. After that they went up, went around the earth and then came down. Each project was building on a previous project until they could run a complete flight to the moon and back. Sometimes the overall project is referred to as a program, which consists of a set of projects, or one project with a set of sub-projects. This is a very powerful strategy in management and business as although you may not be able to get all of what you want now, you may be able to identify the stages you need to go through to reach the ultimate goal. There are managers who are working towards quite long-term goals, such as improving the skills in an organization, and are getting there by a series of projects. Thus you can have your own project and sub-projects and follow a project management strategy in your personal career.

A major area in which project management is employed is computing, in both hardware and software development. Putting together a large piece of software is a major endeavour; submarines, for example, need programmes with millions of lines of code, as does any jet airliner or any small fighter plane.

Battles that have had a decisive impact on the history of the world required project management: one side's project management ended in success while the other side's project management ended in failure. Victory was the desired 'project' (result) – one side achieved this while the other failed.

Allied to this there has been a long history of developing weapons for war. Weapons development continues today and is a major consumer of project management skills. Various defence departments around the world have grappled with project management and have, in the process, undertaken much research, which has brought a degree of standardization to the area. It should, therefore, come as no surprise to find project managers referring to MIL standards, which are actually US military standards.

Movies are projects; the producers are probably best described as the project managers. Conferences are projects; conference organizers are project managers. Well known examples of projects come from the construction industry: all the great buildings of antiquity were built using project management; it was not called project management then but we would recognize it now as such.

11.3.4 Separating project and process management

A 'project' has been defined very generically as a 'one-off change' that is wanted or brought about by one or more people or organizations. The one-off change can be one of a vast range of items. It can be a building, it can be a rock concert, it can be a new production process, it can be a new model car, a new software package, or a motion picture – the important characteristic is that it is a one-off. Essentially there is no second chance. It has to be right first time. To achieve this one-off, the time-ordered set of tasks, called the project sequence, is used. Thus a project sequence, while it might contain some repetition, is essentially unique.

On the other hand a process is a set of repeating activities. Examples of a process include a train service that runs to a schedule day after day, manufacturing the same product day after day, repeat performances of a play, and routine daily activities in a hotel. Essentially there is a second chance, if it goes wrong now it can be improved tomorrow or the next time round. Process management is the management of the set of repeating activities. This does not, however, include the identification of the process – setting up the process is a function of project management.

There are, then, two distinct types of management situation: process management and project management. In one, the activity is repetitive, just keeping the plant turning over producing the same product day after day; in the other, there are one-off management objectives, such as writing a new computer program.

This differentiation can be refined further: process management is, for example, the management of a chemical production process where the work is mainly directed towards keeping the process going and monitoring progress, while the installation of new machinery would be a project management.

Where an activity is repeated, there is the possibility of improving performance the next time round by identifying trends, by improving the process and so on. In quality terms, the activity is amenable to quite well recognized continuous improvement techniques. With the one-off activity, however, when it is finished, it has either worked or it hasn't. If it hasn't it is sometimes, but not always, possible to repair the damage, so it is important to get it right the first time. Examples that fall in the one-off category are constructing a building (a well-known activity often leading to lots of undesired repair work), reorganizing a company (mistakes here can be fatal) and building a ship (a bad design could result in the ship sinking).

These definitions suggest a wider separation between project sequence and process management than actually exists. Getting involved in one form of management usually implies getting involved in the other. Constructing a building involves much repetition at the trade level while running a train service involves one-offs such as introducing new services and timetables. The pure form of either process or project sequence is rare, and in fact a pure form of project sequence, involving all new and unique activities without repetition, would be a very risky business. The notion or idea of a pure project sequence or process is, however, a useful notion and helps in choosing the appropriate management response.

Construction management is not involved purely in unique activities or in the production of unique objects. This means that construction managers need to think beyond project management, even though most of their thinking will be in the project management mode. Construction managers need to be able to pick from project management what they need and also to pick from general management and process management as needed.

For clarity here 'sequence' will be used in relation to project management and 'process' in relation to process management, thus 'sequence' will imply a one-off while 'process' will imply repetition.

11.4 The players

The main participants or players in a building or construction project sequence are:

- the client
- the client's project manager
- the main supplier (the contractor)
- the main supplier's project manager
- the sub-contractor
- the sub-contractor's project manager
- the stakeholders.

Project management is not a solo performance. It involves the interaction of a number of people and organizations. The key relationship is between the client and the main supplier and what needs to be emphasized is that the role of the client is not a passive one, but one requiring considerable thought and decision.

11.4.1 Client

The client may be an individual or an organization. The project is developed to satisfy the client so it is the client who determines the need for the project. This is not an easy task and project managers need to help their clients in making this determination. When all is said and done the definition of the client boils down to this:

> The client is the entity assuming the first-line responsibility for the financial liability for the bills associated with the achievement of the project. (Healy, 1997)

'First-line' is used to distinguish between those financing the project sequence (e.g., a bank) and those who actually pay for the project sequence. In this terminology the bank would be a 'second-line'.

Often the client appoints a project manager to act on their behalf. This is usual if the client is an organization and may be so if the client is a busy person or somebody who does not wish to manage the complexities of the client role in the project sequence. There are many project management consultancies that perform the role of client project manager.

11.4.2 Main supplier

The main supplier, and on some projects there may be more than one main supplier, is that person or organization directly responsible to the client for providing the project or part of the project. The difficulty here is that there are many forms that a main supplier can take. The simplest form is one organization that will deliver the project, taking direct responsibility for all the supply and installation. In this case the builder or contractor enters into a contract as a principal together with the client. In the building and construction industry these builders or contractors are often referred to as main contractors or head contractors. Alternatively a construction manager may contract with the client to manage the supply and installation for the client but arranges matters so that the sub-contracts on the project sequence are made directly between the client and the sub-contractor. In this instance the construction manager acts in a role akin to that of an agent rather than a principal.

The main supplier organization on capital projects has to mobilize a great number of people, administer the interfaces between sub-contractors, make the payments to sub-contractors, claim and obtain payment from the client, and supply materials and locations for work. The main supplier must recruit people with relevant skills for its own organization.

The main supplier, then, may take a variety of forms. Commonly someone is nominated as the project manager for the main supplier and this overcomes many administrative problems.

11.4.3 Sub-contractors

Because no single organization has sufficient demand for all the skills needed to do all the work on a project much, if not most, of the work is given to sub-contractors. These people or organizations usually specialize in particular technologies, e.g., plumbing or electrical work. They contract to supply and/or install a part of the work required for the project. They usually enter into a contract with the main supplier (but note that they may contract directly with the client with a construction manager acting as agent). The sub-contractors are usually the repositories of the technical skills and know-how. This is often a problem for the client and main supplier in that the sub-contractor is the expert and hence knows more about the work than the client or the main supplier.

The main supplier has to acquire the services of sub-contractors and make arrangements for their accommodation within the project. This accommodation may involve providing facilities for both people and machines. Each sub-contractor usually has a project manager to manage its interests in relation to the project sequence.

11.4.4 Stakeholders

Stakeholders are those with an interest in the project, the project sequence, or the impact of either. For any large project sequence there is a wide range of stakeholders. The nature of their interest may be financial, political, or philosophical. People lending money to the project sequence are stakeholders; local government and neighbours of the proposed project are stakeholders; people concerned with the environmental impact of the project or project sequence are also stakeholders. There will be some debate about the legitimate standing of some as stakeholders, and this often takes the project manager into the political arena. Stakeholders need to be seriously considered, however, as they often have significant influence on project success.

11.4.5 Many people involved

Many people are involved with their own concerns and interests and the project manager faces a challenge in dealing with these often conflicting concerns and interests. Considerable people skills are needed if the people involved are to be kept together effectively so that the project sequence can be completed.

11.5 How project management changes process management practice

Construction management is a specific application of what is now seen in the broad sense as project management. The unique product of this exercise is a building or some other form of built structure. The unique nature of the product has important implications:

- the importance of planning activity increases
- there is a tendency towards the emergence of a single point of control
- effectiveness becomes more important that efficiency
- flatter organizational structures develop
- there is a great rise in the importance of co-ordination
- a big gap between authority and responsibility appears.

11.5.1 Planning

Because one only has one attempt at any particular project there is greater concentration on planning. Projects typically have extensive planning phases that both detail the project and work out how to achieve it.

11.5.2 Single point of control

A clear focus for decision-making reduces confusion. In a fast moving project sequence there is a strong need for clarity and co-ordination of decision-making. This leads to the need for a single point of control. This allows better tracking of the project sequence and better control of the introduction of changes.

When activities are repeated people know from practice what is needed. In fact, the information needed to manage the process moves down the hierarchy, leaving upper management to cope with more difficult or new issues. People know where to go for information because other people have built their roles around the possession of certain information.

Many companies when confronted by a project appoint a project manager to manage the project sequence independent of the organizational hierarchy. These people are then free to go up, down or across the hierarchy as needed for the benefit of the project. While project management reinforces the concept of a single point of control, it does not give absolute power; the authority of the project manager, the power of the single point of control, is limited.

11.5.3 Effectiveness versus efficiency

People who like to do everything well often find this aspect of project management hard to accept. It is much more important to be effective, i.e., to get things done, than it is to be efficient, i.e., to get things done at the lowest cost. An example might help to clarify this: because the Christmas shopping period is so important nobody developing a shopping centre will accept opening a new shopping centre after Christmas. This is the most important sales period, dominating the minds of shopkeepers. If one misses the Christmas period, one might as well miss the rest of the year. The project manager will be under enormous pressure to meet the deadline, so getting it done at a lower cost becomes less important than completing in time for the Christmas shopping period.

In practice, what happens is that one or two criteria are identified as critical. The project manager then tries to satisfy them as much as possible within the limits set by the other project constraints. This issue of effectiveness versus efficiency appears in other guises, such as that of satisfying rather than optimizing. Aiming to satisfy rather than optimize permeates project management, although the idea is anathema to many specialists. In practice, however, it is often necessary to accept sub-optimal decision-making in order to get the job done.

11.5.4 Flatter hierarchies and groups

Many of the problems to be solved in a project sequence require people to interact face to face. This is the most efficient way of deciding many project details and overcoming project sequence problems. This need for face-to-face contact does not work well with deep hierarchical management structures. As the project sequence proceeds, and people get a better understanding of the extent of the project and the project sequence, the hierarchy can deepen. At the beginning of project sequences one will have shallow hierarchies whereas deeper hierarchies can emerge later.

11.5.5 Co-ordination takes priority

The project manager is responsible for managing the interaction between the participants. The mechanical engineers will quite happily put their air conditioning ducts exactly where the electrical people want to put their cable trays – it is up to the project manager to put processes in place to deal with such situations. One of these processes might be regular co-ordination meetings; the project manager needs to ensure that there is good co-ordination between the participants, that key people know and support the project objectives, that work is properly authorized, and that priorities are properly established.

It is the responsibility of the project manager to co-ordinate the work of the specialists, not to do the specialist work. The extent of the technical knowledge that the project manager needs is really only that which is needed in order for him/her to be able to do that co-ordination.

Many see co-ordination as a shallow activity, however, as projects get bigger and more technically complex, co-ordination becomes an activity that demands a wide range of high-level personal and technical skills.

11.5.6 Authority and responsibility

There are many who believe that authority should match responsibility, that is, people should have the authority to do what is necessary in order to discharge their responsibilities. This is quite reasonable but in project management it is unlikely to happen except in the most unusual of situations. This is so because many of the decisions of project managers involve large sums of money, often in the order of millions of dollars. Very few people in organizations have this level of authority.

This is a problem in project management. Quite mundane and routine project decisions often have to go to the top of the organization for decision, because the project manager does not have the authority to make the decision. This is inefficient and slows decision-making. Some attempts have been made to match authority with responsibility, for example, by allowing people to make decisions that are within a given budget allocation, but generally the gap remains.

The issue of authority goes beyond money matters: project managers need people skills that will allow them to persuade people outside their authority and control to make appropriate contributions to the project and project sequence.

11.6 Conclusion

At the core of the preceding discussion is the crucial concept of the project as the unique product of a sequence of activities controlled by the project manager. It is the unique nature of the product that underscores the difference between process and project management and it is this difference between process and project that generates many of the problems faced by project managers. It is particularly important that the difference between process and project management is understood because it has a significant impact on the way that building and construction projects are managed.

References and bibliography

AIPM (1996) *National Competency Standards for Project Management Volume 1 – Guidelines First Approved Edition* (Sydney: Australian Institute of Project Management). www.aipm.org/

APM (2000) *Syllabus for the APMP Examination*, Second edition (High Wycombe, UK: Association for Project Management).

Healy, P.L. (1997) *Project Management: Getting the Job Done on Time and in Budget* (Port Melbourne: Butterworth-Heinemann).

PMI (1996) *A Guide to the Project Management Body of Knowledge (PMBOK®Guide).* Project Management Institute Standards Committee (Newton Square, PA: Project Management Institute). www.pmi.org/

PMI (2000) *A Guide to the Project Management Body of Knowledge (PMBOK®Guide).* Project Management Institute Standards Committee (Newton Square, PA: Project Management Institute). www.pmi.org/

12

Risk allocation in construction contracts

John Twyford*

Editorial comment

The most famous of the Laws of Edsel Murphy states 'anything that can go wrong will'. Whether Murphy was a builder is not clear but his words must resonate with many people in the construction industry. Building is a notoriously risky business and the consequences of unexpected contingencies arising can be financially devastating for many of those who are participants in the process.

Risk management is now a recognized process in the industry, although there may be some doubt as to how widely any formalized procedure is applied by any particular sector of the industry. This chapter is not, however, about assessing risk or planning for risk, but about how risk is spread between the various parties engaged in building procurement. When something does go wrong in the course of design and construction work, and, if Murphy is to be believed, it is a matter of 'when' not 'if', then there are inevitably financial consequences. Depending on how the risk is allocated under the contract for the work one or more of the parties will have to pay for the extra work required to rectify the situation. Generally the only people to profit in these circumstances are the lawyers.

Under a traditional tendered lump sum building contract, where a builder bids for the right to a job, and if successful, agrees to carry out the work for a fixed amount of money, a great deal of the risk associated with carrying on and completing the work is placed with the builder. It is this very approach, time-honoured though it may be, that is the root cause of much that troubles the industry; profit margins are low and unforeseen circumstances, often beyond the builder's control, can lead to insolvency and enormous subsequent problems with half-finished projects, unpaid sub-contractors, and completion times and budgets blown out of all proportion. Different forms of procurement assign risk in different ways and can make the difference between having a successful project or a disaster.

* University of Technology Sydney

The relationship between the question of who carries the risk and the value of the finished building is about more than just the avoidance of any additional cost that may arise. There are also concerns with delayed completion (important if the client is looking to occupy the building or wishes to lease it, which might depend on completion by a certain date, e.g., to coincide with a major event such as the Olympic Games), a perception that the building has a troubled past which may colour the view of prospective tenants or buyers, and a lack of commitment to quality if the contractor has been forced to accept what she or he believes is an unfair financial burden arising from some risk that has become an actuality.

While clients may seek to pass all risk to the contractor, and while that may be possible, the eventual costs may well outweigh the benefits. In this chapter the author discusses different contractual arrangements in the construction industry and how risk is allocated in the various forms, and looks at the consequences that may arise as a result of unbalanced or unfair allocation of risk amongst the parties.

12.1 Introduction

The idea that contracts may be used as vehicles for risk allocation in commercial transactions is by no means new or confined to the construction industry. One highly respected writer on the law of contract pointed out (Atiyah, 1981, p. 208):

> In the end, therefore, other justifications must be sought for treating agreements or mutual promises as binding. . . . The place to start, it may be suggested, is the reason for the whole arrangement. . . . Why, in other words, do people make advance arrangements in the first place? The answer surely is that they want to eliminate (or shift) risks of various kinds. Contracts for future performance are often deliberately entered for the purpose of shifting risk. . . . This is a kind of speculation, a bet, but it is likely to have a more useful social or economic purpose than a simple bet. For risk allocation between businessmen may be designed to shift risks to those who are better able to evaluate and absorb the risks, or even prevent them.

From what Professor Atiyah has said a number of questions arise. First, what is risk? Second, are there any issues of justice or commercial common sense arising from attempts by one contractual party to transfer risk to his or her contractual partner? Finally, to what extent has the law recognized or encouraged the transfer of risk between contractual parties? This chapter will attempt to answer those questions and conclude with a discussion of the manner in which a selection of standard forms of contract deal with some common risks in the construction process.

12.2 What is risk?

The Australian and New Zealand Standard AS/NZS 4360:1995 defines risk as: 'The chance of something happening that will have an impact upon objectives. It is measured in terms of consequences and likelihood.' The British Standard somewhat more obscurely says: 'A combination of the probability, or frequency, of occurrence of a defined hazard and the magnitude of the consequences of the occurrence.' In terms of a construction project this translates to a failure of a contractor, caused by random circumstances, to complete the project within the constraints of time, cost and quality contemplated by the original agreement. The construction industry is particularly susceptible to risk. This susceptibility arises from a number of reasons including the time taken to plan and execute the project, the large number of people needed to participate in the project (often from different cultures), inhospitable construction sites, specification of untried materials and the fact that construction projects are susceptible to risk cultivation by the contracting parties or those who advise them (Bunni, 1998, pp. 93–4).

The matters mentioned in the previous paragraph are all matters that the parties to a construction contract are well aware of and are addressed in both 'one-off' contracts and the standard contracts available to the industry. Usually the eventuating of a risk will manifest itself in the need for the construction period to be extended and/or the need for the contractor to execute additional work. The economic questions that arise as a consequence of such incidents are:

- should the construction period be extended because of a delay and if so, should the contractor be paid delay costs
- if an extension of time is inappropriate, should the contractor be required to pay liquidated damages
- where additional work is executed is this to be treated as a variation to the works or as part of the scope of the original works.

In terms of the texts of standard contracts such matters are dealt with under the rubric of:

- the consequences of disputes with the labour force resulting in strikes or lock-outs
- the consequences of inclement weather
- the need to execute work claimed not to be within the scope of the original transaction or to execute work arising from latent conditions.

In addition to these risks there is the risk of loss or destruction of the works by fire, flood, natural disaster, or potential liability to third parties arising from activities on the construction site. Of all risks these are usually the most carefully and extensively dealt with in construction contracts.

12.3 Justice and commercial issues arising from transfer of risk

It could be said that the outcome of a successful negotiation involves the other party to the contract taking all of the significant risk. This can result from the canny negotiation of one of the parties or, more likely, the result of one of the parties being in a dominant bargaining position. However, in terms of a final result, there would seem to be good reasons for avoiding a situation where a party undertakes a risk of which it is not aware. If the risk eventuates, the party who is prejudiced will usually defend its position by whatever legal means are available. Litigation is hardly a satisfactory outcome to a construction project; equally a dominant party forcing the risk onto a weaker party raises both economic and justice issues. As a consequence, certain so-called principles of risk allocation have evolved which are said to produce both just and economically efficient results.

The principles of risk allocation with which the construction industry will be most familiar are those proposed by the Dublin lawyer, Max Abrahamson (The Abrahamson Principles, 1979). Those principles dictate that a party to a contract should bear a risk where:

- the risk is within the party's control
- the party can transfer the risk, e.g., through insurance, and it is economically beneficial to deal with the risk in this fashion
- the preponderant economic benefit of controlling the risk lies with the party in question
- to place the risk upon the party in question is in the interests of efficiency, including planning, incentive and innovation
- if the risk eventuates, the loss falls on that party in the first instance and it is not practicable, or there is no reason under the above principles, to cause expense and uncertainty by attempting to transfer the loss to another.

These principles were adopted by a Joint Working Party comprising representatives of Government construction authorities (National Public Works Conference) and the construction industry (National Building and Construction Council) as a desirable model for use in the Australian industry (NPWC/NBCC, 1990). The definition of 'risk' adopted in the report is 'hazard, exposure to mischance, or chance of bad consequences'. The adoption of the Abrahamson theory of risk allocation has not gone unchallenged in Australia. One commentator (Davenport, 1991, p. 21) points out that it:

> . . . is misconceived and should be abandoned. The main reasons are: the theory is not capable of practical application, risks are not capable of clear definition and ambiguity results, the theory involves the introduction into contract law of principles of economic theory and equity that have failed to gain acceptance in contract law generally. [The author then proceeds to suggest a more specific solution to the problem.] The recommended alternative approach is to identify possible events and with respect to each, specify whether the contractor is entitled to: an extension of time, reimbursement of extra costs, reasonable off-site overheads and profit [or no extra time or reimbursement at all].

The economic rationale and consequences of allocating risk on the basis of the Abrahamson principles, although not referred to as such, are explained by Bunni (1998, p. 103):

> As risks associated with foreseen and identifiable hazards are calculated, it should be generally a matter of policy to determine to whom each of the risks is allocated. The most cost-effective method of allocation from the point of view of controlling the occurrence of the risk and mitigating or eliminating its adverse effects is based on the ability to exercise such control. However, risks allocated to the contractor on the basis of this method would have a cost implication if they are not his own fault, since it would be prudent for the contractor to include in his original price an element relating to this additional risk he is asked to carry. If the risk does not eventuate, the employer would have paid a larger sum than necessary. On the other hand, this may be more advantageous to the employer (principal) than to assume the risk himself and be exposed to the possibility of having to make an additional

> payment should the risk eventuate. This is particularly so where there are strict budget restrictions or where the financial considerations of the project are such that the project would not be economically viable beyond a certain limit.

It is not necessary to canvass the efficacy or otherwise of the Abrahamson formula (or any other) as this chapter is concerned primarily with the legal machinery of risk transfer. The material has been included to form a background as to why risk transfer is of such importance to contracting parties. This discussion continues with an examination of a number of standard contracts.

12.4 Legal basis of shifting construction risks from one contractual party to another

Generally the law allows the parties to a contract to make their own bargain. Accordingly, if the contract contains clear provisions doing so, then there is no legal objection to a party transferring the contractual risk to his or her contracting partner. This point is well illustrated by a decision of the New South Wales Court of Appeal in 1998. There the case concerned whether it was the consignor or the carrier that must bear the loss for the loss of cigarettes stolen from a container at a freight terminal. The consignor accepted that the contract between it and the carrier had the effect that consignor must bear all but an insignificant fraction of the loss. Accordingly, the consignor sued the owner of the freight terminal in negligence for failing to provide appropriate security services for the terminal. Mason P (in *W.D. & H.O. Wills (Australia) Limited v. State Rail Authority of New South Wales (1998)*) put the matter succinctly:

> . . . it is neither reasonable nor just in the circumstances of this case, to throw the costly burden of providing security services upon [the owners of the terminal] when they were no part of negotiated risk allocation between the parties that had primary responsibility for the safety of the shipment.

There are, however, some caveats that a party drafting such a document must bear in mind. As mentioned above, the document must clearly state the intention in this regard. The Chief Justice of the Supreme Court of South Australia (Bray CJ in *Taylor Woodrow International Ltd v. The Minister for Health* (1978), p. 9) was scathing in his description of a botched attempt to do just this:

> I must say that the departure from traditional terminology in amending the well-known variation clause so as to include, not only an addition to the work, but a change in the time provisions is not only anomalous but deplorable. It is like tipping an entirely gratuitous truck load of manure into this already sufficiently muddied stream.

Moreover, a failure by the parties to clearly state their intentions in this regard could bring into play the rules of legal interpretation. This will not always achieve the result the parties expected. How the court might read a document was spelled out by Justice Gibbs of the Australian High Court (in *Australian Broadcasting Commission v. Australian Performing Rights Association* (1973), p. 109):

> [The court must] endeavour to discover the intention of the parties from the words used in the instrument . . . the whole of the instrument has to be considered . . . every clause must if possible be rendered harmonious one with another . . . if the language is open to two constructions, that will be preferred which will avoid consequences which appear to be . . . capricious, unreasonable, inconvenient or unjust.

In addition, the *contra proferentem* rule provides that where there is an ambiguity in a deed or other instrument the document shall be construed most strongly against the maker. It should be noted that this is a rule of last resort and is to be applied only when the other rules fail to resolve an ambiguity (*Taylor Woodrow International Ltd v. The Minister for Health* (1978), p. 3).

It is proposed to continue the chapter with a discussion of how the authors of a number of standard contracts have dealt with the time/sufficiency of works dichotomy referred to earlier. In doing this the following documents will be discussed:

- AS 4000 (Standards Australia)
- JCC (Royal Australian Institute of Architects, Master Builders Australia Inc., Property Council of Australia)
- PC-1 (Property Council of Australia)
- C21 (New South Wales Department of Public Works and Services)
- NZS 3910-1998 (Standards New Zealand)
- Articles and Conditions of Building Contract (Singapore Institute of Architects, referred to hereafter as SIA)
- JCT – 1980 (Joint Contracts Tribunal, United Kingdom)
- FIDIC – 1996 (*Fédération Internationale des Ingénieurs Conseils*, the International Federation of Consulting Engineers).

The documents used in this discussion were the latest available to the author but it is recognized that some at least of the documents have been superseded by later versions. It is unlikely that any changes would be material to this discussion. The discussion of PC-1 and C21 was included not because of the intrinsic interest in the documents but rather because each represents the untrammelled views of principals as to how risk should be adjusted. It is expected that this examination will reflect the bargaining strength of the negotiating parties and their willingness to compromise. For the sake of convenience the parties to a construction contract are referred to in this discussion as principal and contractor. Different expressions including 'employer' and 'builder' are used in other contracts. Equally some contracts use the expressions 'architect' or 'engineer' for the person administering the contract; again for convenience the expression superintendent is used here.

Before dealing with the specific provisions of the contracts under discussion it is appropriate to observe that the risk that the completion of the works will be delayed is ever present and one that prudent parties to a contract must deal with. AS 4000 without completion of the schedules produces a very different risk allocation on this area of risk from the remainder of the contracts. Even so, within those other contracts there is a wide variation in how delay is dealt with. Some identify the delay events for what they are and others require the contractor to demonstrate that a delay was 'beyond his or her control'. For the parties these are serious matters. A delay will cost both parties. The principal will be deprived of the timely completion of the building with the associated loss of income and the contractor must bear the cost of remaining on site until the work is completed. The cost of protracted delays can be of such magnitude as to bring into question the principal's initial decision to commission the project or the contractor's decision to enter the contract at the contracted price.

If the contract provides that, during the delay, the contractor is not entitled to an extension of time and consequently must pay liquidated damages, then it is the contractor who is bearing the risk. If the contractor is entitled to an extension of time then both parties potentially bear the loss and it could be said that the risk is distributed equally. The principal loses from not having the building completed on time and the contractor must bear the cost of remaining on site. If the contractor is entitled to an extension of time and to recover delay costs then it is the principal who bears the risk. The point to be made is that the parties and their advisers must be aware of the risk and make an informed decision as to where the risk will ultimately lie. After the risk has been appropriately allocated it is incumbent on the parties to adjust their conduct accordingly.

12.5 Disputes with the labour force

If the Abrahamson formula were to apply to this area of risk there is little doubt that it would remain with the contractor – the contractor is, after all, the employer of the workforce and, as a matter of law, able within certain parameters to direct and control those workers. In Australia, a distinction is sometimes drawn between labour disputes that are confined to a building site or contractor where it is clear that the contractor is or should be in control of the situation, and national strikes where the presence of this control is less clear. It is noted that the Australian labour force has a propensity to withdraw its labour for reasons that might be characterized as political rather than those intended to procure a concession from a particular employer. Where a disruption to the work is caused by a labour dispute, the question arises as to whether a contractor is entitled to an extension of time to complete the works with the consequent relief from the obligation to pay liquidated damages. A second question arises as to whether or not the principal has agreed to pay delay costs to the contractor during the period of interruption of the work.

Under AS 4000 the entitlement of a contractor to an extension of time is dealt with generally under clause 34. The text of the clause invites the parties at the negotiation stage to provide if an extension of time will be granted for delays arising from a labour dispute. To trigger an extension of time for a labour dispute the parties must identify such matter as a *qualifying cause of delay* in item 23 of the schedule to the contract. For the contractor to be entitled to payment of delay costs (described in the contract as *delay damages*) under clause 34.9 the parties must take the further step of identifying labour disputes in item 26 of the schedule as a *compensable cause*. In this regard it is noted that clause 39.4 refers to *delay damages* but nowhere in the contract is the expression defined or is there a schedule item inviting the parties to agree as to what the damages should be. This means that the legal meaning of damages would apply, that is, an amount of money that would put the contractor into the same position as if the delay had not occurred. To this might be added the qualification that such damage must be a reasonably foreseeable consequence of the delay. It will be seen, that under this contract, the entitlement to both an extension of time and receipt of delay costs must be agreed to by the parties to the contract in the completion of the schedule. Neither entitlement is presumed by the text of the contract.

Clause 9.01 of the JCC document provides that the contractor is entitled to an extension of time for *any cause beyond the control* of the contractor or for matters that might be broadly characterized as defaults on the part of the principal. The question immediately arises as to whether labour disputes can be said to be beyond the control of the contractor; certainly industry-wide disputes that have a political component could be so classified. It is less certain where the contractor, by some default in his or her obligations to the labour force, causes the strike (say, by the failure to provide statutory workers' entitlements). Generally the accepted wisdom seems to be that, except in extreme cases, labour disputes are beyond the contractor's control; therefore the contractor in the event of such a dispute can expect an extension of time and relief from the obligation to pay liquidated damages. The entitlement to delay costs is dealt with in clause 10.10, which requires the parties to specifically deal with the matter in the appendix. There, under item N3, *civil commotion and industrial dispute beyond the control* of the contractor are listed with the statement that the parties must agree to a percentage (to be stated with a default position of 50%) of the *costs and expenses* of the delay that each party will bear. It is noted here that the expression *costs and expenses* is narrower than *damages* and would limit the sum that a contractor could recover. There is no provision in the document for the parties to agree to what the *costs and expenses* should be.

The combined effect of clause 10.5 and the accompanying contract particulars of PC-1 is to allow the parties to agree that the contractor may claim an extension of time in the event of the progress of the works being delayed by a labour dispute. The payment of delay costs is limited, however, to situations where the time is extended *due to a breach of the contract* by the principal (clause 10.11). Accordingly the best that the contractor can expect, based on the text of the contract, is relief from liquidated damages. It is not surprising, having regard to the antecedents of the contract, that it has a principal focus.

The fact that the C21 document was prepared for use by the NSW Government and its instrumentalities predicates that this contract should also have a client focus. The basis on which an extension of time will be granted for delay resulting from a labour dispute is perhaps more benign than might have been expected. The basis for such an extension is provided in clause 66 1.5-4: *strikes or other industrial action not caused or contributed to by the contractor and not confined to the site, the contractor or its sub-contractors*. No delay costs are payable in respect of an extension of time thus granted.

The NZS 3910, in clause 10.3, provides the basis for a contractor to claim an extension of time. Sub-clause 10.3.1(c) identifies *any strike, lockout or other industrial action* as grounds for a contractor to claim an extension. Provided the necessary conditions are met, the Engineer (superintendent) must then *determine* the appropriate extension. There are no provisions for the payment of delay costs consequent upon such determination. Accordingly, the contract favours the contractor in that an extension of time and consequent immunity from liquidated damages is provided for as a matter of course in the text of the document.

The SIA contract in clause 23(1)(e) deals with the matter in a way that is directed to the needs of the industry in Singapore. An extension of time is allowable for *industrial action by workmen, strikes lock-outs or embargoes (whether domestic or foreign)* and in this regard the workmen referred to include those involved in the manufacture or transportation of goods or materials required for the project. The extension is only allowable where the incident was not caused by *an unreasonable act or default of the contractor.* There is no provision in the contract for the contractor to be paid delay costs in respect of such a delay.

Under clause 25.3.1.2 of the JCT form of contract the architect (superintendent) *may give an extension of time* in respect of the occurrence of a *Relevant Event*. One of the *Relevant Events* referred to is defined by clause 25.4.4 as *civil commotion, local combination of workmen, strike or lock-out affecting any of the trades employed on the works* or those supplying materials and services to the works. There is no provision in the contract for the payment of delay costs in respect of such an event.

The least precise of the documents under consideration is the FIDIC contract at clause 44.1(e). There it is provided that the engineer (superintendent) may extend the time for completion for *other special circumstances which may occur, other than through a default of or breach of contract by the contractor or for which he is responsible being such as fairly to entitle the contractor to an extension of time.* The structure of the clause has the potential to cause difficulties of legal interpretation as, if it were possible to find a common thread running through sub-clauses (a) to (d), the *ejusdem generis*[1] rule might apply and limit the ambit of this provision. It is submitted that no such common thread is to be found and accordingly the rule does not apply. Even so, the expression *special circumstances* is not free from doubt. It is suggested however that the words *fairly to entitle* could cover a situation beyond the control of the contractor. In most circumstances, especially with an international contract, it could be said that a labour dispute is beyond the control of the contractor. Accordingly, the contractor could expect to be granted an extension of time with the consequent immunity from liquidated damages. There is no provision in the contract enabling the contractor to claim delay cost for the period of the extension of time.

12.6 Inclement weather

It is probably trite to make the point that the weather is beyond the control of either party to a contract. Accordingly, Abrahamson's first and second principles have no application. What can be done in the event of bad weather occurring can only be done by the contractor (thereby invoking the third principle) and equally it is the contractor who has the greatest incentive to plan for and deal with the risk (fourth principle). Add to this the fact that weather is cyclical, and that records have been kept for centuries, and it could be suggested that it would be reasonable for principals to routinely require the contractor to assume the risk for this event. Not so, and the different approaches taken by the authors of standard documents require contractors to exercise caution.

The provisions of AS 4000 relating to extensions of time for labour disputes apply equally to delays caused by inclement weather. For a contractor to be entitled to an extension of time under clause 34 for this circumstance reference would need to be made to inclement weather in item 23 of the schedule and if the contractor wished to recover *delay damages*, the matter would need also to be included in item 26 as a *compensable cause.* Some care would need to exercised here in drafting the actual words to be used in item 23 of the schedule. It is noted that the expression *inclement weather* is used in clause 1 (definition section of the document) but then only to exclude extensions of time for *inclement weather* occurring after the date *for practical completion.* The expression does not merely denote rain but might include a high wind that prevented the operation of a crane. It should also be noted that in Australia the workforce cannot be compelled to work in the rain and after a qualifying period of rain each morning the workers are allowed to go home notwithstanding the extent to which the weather improves. For a certain number of days each month the workers are paid for such absences. On this basis the cessation of work on a job might not always coincide with the accepted view of what constitutes inclement weather. Accordingly care needs to be taken with the words used in the schedule. Where a contractor is entitled to an extension of time for inclement weather he/she is thereby relieved of the obligation to pay liquidated damages.

As with AS 4000, inclement weather in the JCC document is dealt with under the general extension of time provision. Clause 9.01 allows a contractor an extension of time for a *cause beyond the control* of the contractor. There is little doubt that inclement weather is beyond the control of the contractor and accordingly the contractor is entitled to an extension of time and relief from liquidated damages. Appendix N5 allows the parties to provide who will bear the economic cost of a delay caused by inclement weather. If the parties do not complete the appendix, each will bear half of the cost of the delay. In this regard it is interesting to note that the basis of recovery is for *inclement weather or conditions resulting from inclement weather.* This extends the period of the delay for which *costs and expenses* might be recovered. A simple example would be where the period of rain was brief but the consequent flooding of the site prevented work for several days.

Clause 10.5 and the accompanying contract particulars of PC-1 allow the parties to agree that the contractor may claim an extension of time in the event of the progress of the works being delayed by inclement weather. As in the case of labour disputes, delay costs are not payable. The best that the contractor can expect is relief from liquidated damages.

Under C21 the entitlement of the contractor for an extension of time for a delay resulting from inclement weather is spelt out in some detail. The basis for such an extension is provided in clause 66 1.5-5 *inclement weather, where the aggregate number of days entitling the contractor to an extension has exceeded the allowance stated in the Contract Information*. The Contract Information is an annexure to the contract and item 25 sets out a space for the contractor to state the number of days it has allowed in its calculation of the price/construction period for interruption to the progress of the work due to inclement weather. How the parties would deal with such a provision is not clear. It is probably intended to encourage the contractor to allow for what might be the foreseeable instances of inclement weather gleaned from, say, weather records. Accordingly an extension of time would be allowed for what might be characterized as abnormal conditions. The way the item is completed would depend on the negotiating strengths of the parties. As in the case of a labour dispute the contractor is allowed only an extension of time in the circumstances described, with relief from liquidated damages but no delay costs.

The NZS 3910, in sub-clause 10.3.1(b), identifies *weather sufficiently inclement to interfere with the progress of the works* as grounds for a contractor to claim an extension of time. Whilst the text of the sub-clause is more expansive, it is doubtful if anything is gained by the additional words, as the requirement for the weather conditions to have a bearing on the progress of the works would almost certainly be implied. As in the case of labour disputes, there are no provisions for the payment of delay costs.

The SIA contract, in clause 23(1)(b), deals with the matter in a comprehensive way. An extension of time is allowable for *exceptionally adverse weather conditions (in assessing the same regard shall be had to the meteorological averages, the reasonable expectation of adverse conditions both seasonable and annual during the contract period, and to the net effect overall of any exceptionally beneficial conditions as well as the immediate effect of individual instances of exceptionally adverse conditions).* The clause requires the contractor to take into account the likely weather conditions during the contract period and it is clear from the text that no extension of time will be allowable for weather conditions that are to be expected in Singapore. The clause is even more restrictive in that favourable conditions, that may have the effect of reducing an extension of time otherwise available as a result of *exceptionally adverse* conditions, are also to be taken into account. As in the case of labour disputes, there is no provision in the contract for the contractor to be paid delay costs in respect of such a delay.

Under clause 25.3.1.2 of the JCT form of contract, the architect (superintendent) '*may give an extension of time*' in respect of the occurrence of a '*Relevant Event*'. One of the '*Relevant Events*' referred to is defined by clause 25.4.2 as '*exceptionally adverse weather conditions*'. Although the operative words are the same as those in the SIA contract this contract does not have the added words in parenthesis. It is not clear how much this omission would affect the meaning of the contract save that the contractor may be able to argue that he or she need not set off the benefit of favourable conditions against exceptionally adverse conditions. No delay costs are payable in respect of such an event.

Clause 44.1(c) of the FIDIC contract provides that the engineer (superintendent) may extend the time for completion for *exceptionally adverse climatic conditions*. As with NZS 3910, the SIA contract and JCT, this clause will require the contractor to take account of those weather conditions that might be expected. Where an extension of time is granted the contractor would be immune from the obligation to pay liquidated damages. As indicated previously, there is no provision in the contract for the contractor to claim delay costs.

12.7 Scope of works/latent conditions

A persistent source of dispute in construction contracts is the definition of the extent of the contractor's technical obligation. In general terms it is the work shown on the contract drawings and described in the specification. Contractors will want their obligation confined to the narrow limits of what these documents say whereas principals will argue that the obligation extends to work that might be inferred from the obligation to reproduce what is shown in the drawings and specification. The problem is illustrated by considering a drawing showing a building supported by appropriate foundations without any statement of the nature of the soil that needs to be excavated to construct those foundations. The contractor can fulfil its obligation by reproducing what the contract documents call for although the cost will be very different depending upon whether it is necessary to excavate in rock or earth. Where the obligation falls in a case such as that described is not necessarily answered by the standard clauses describing work or defining a variation to that work. This is particularly the case where a bill of quantities does not form part of the contract. For this reason it is common to find included in contracts clauses extending the scope of the works and specifically allocating the risk for the extra cost where adverse sub-surface soil conditions are encountered. It is proposed to examine how these issues have been dealt with in the contracts already examined.

Under AS 4000 the contractor's basic obligation is to *carry out and complete work under the Contract* (clause 2.1). There is no attempt to extend the scope of the contractor's obligation, however, care is taken to deal with the unforeseen site conditions which are dealt with under the rubric of *latent conditions*. Latent conditions are defined as *conditions on the site . . . excluding weather conditions, which differ materially from the physical conditions which should reasonably have been anticipated by a competent contractor at the time of the contractor's tender if the contractor had inspected* (clause 25.1). The effect of clause 25.3 is, after the contractor gives notice, to deem the work associated with a latent condition a variation. Where appropriate the construction time will be extended. The contractor is entitled to recover the cost of executing the additional work, delay costs, overhead expenses and profit.

The JCC document follows the same drafting policy as AS 4000. The contractor's obligation is to *proceed to execute and complete the works in accordance with this agreement* (clause 1.03.02). The works are those described in the annexed drawings and specification (clause 1.02.05). Site conditions that *differ from the conditions and characteristics shown, described or measured in this agreement or give reasonable cause for the Builder (contractor) to consider that the works require to be varied*, are treated as a variation (clause 3.02).

The text of PC-1 makes the point to the contractor that he or she must concern himself or herself with the scope of the works. Clause 7.1 under the title *The Site* provides *The Contractor warrants that it has, and will be deemed to have, done everything that would be expected of a prudent, competent and experienced contractor in: (a) assessing the risks which it has assumed under the contract, and (b) ensuring that the Contract Price contains allowances to protect it against any of these risks eventuating*. Pursuant to the 'belts and braces' principles of legal drafting, the contract continues, the contractor warrants that it has received but not relied on the site information provided by the owner (principal), clauses 7.2 and 7.7. Under the title *Construction,* a further disclaimer: *The Contractor has allowed for the provision of all plant, Equipment and Work, materials and other work necessary for the Contractor's activities, whether or not mentioned in the Works Description or any Design Documentation*, clause 8.2. These clauses are no doubt designed to protect the principal against claims for variations for items not mentioned in the contract documents. Given the detail of design documentation and the definition of variation in the Glossary of Terms annexed to the contract, this approach may not extend the contractor's obligation far beyond that in the contracts described to this point. Latent conditions are defined in the Glossary of Terms in much the same way as in the other documents. It is noted, however, that under clause 7.4 the additional work necessitated by such an event is not treated as a variation and the contractor is entitled to recover only *extra costs*. This would exclude a claim for profit related to such work.

The C21 is another document evolved from the point of view of the principal and it is therefore no surprise to find a clause that seeks to limit claims for incomplete descriptions of the work. Clause 52.2 utilizes a slightly different text, viz., *The Contractor acknowledges that: it is both experienced and expert in construction of the type and scale of the works; and it is fully aware that there are likely to be items not specifically referred to or described in the Contract which nonetheless are required to complete the works and achieve the effective and efficient operation of the works and*

. . . it has made full allowance for such items in the Contract Price. Notwithstanding the differences in text, the clause is likely to have the same legal effect as the equivalent clause in PC-1. Clause 41.1 provides that additional work directed as a result of site conditions (latent conditions) is to be valued as variation save that *no payment will be made for costs of delay or any aborted work.*

The NZS 3910 does not include any provisions purporting to extend the scope of the contractor's obligation beyond that of executing the work described in the contract documents, that is, *to construct, complete, deliver and remedy defects in the works and things described*, second schedule. Clause 9.5 defines unforeseen physical conditions (latent conditions) as has been the case in other contracts save that the definition includes weather conditions provided those conditions are the *result of weather away from the site*, clause 9.5.1. An example of such conditions would include inclement weather at the location of a quarry that prevented the quarrying of aggregate needed to batch concrete for the project. Clause 9.5.4 then provides that *the effect of the conditions notified shall be treated as if it was a variation.*

The SIA document does not contain any attempt to extend the scope of the obligation of the contractor, who agrees in the Articles of Agreement *to carry out, bring to completion, and maintain for the employer (principal) the building and other works.* Latent conditions are not dealt with as such, however, clause 14 provides that *Should any discrepancy or divergence be discovered in or between any of the Contract Documents as to the precise extent of the nature of the work to be carried out by the contractor* the architect (superintendent) is required to give a direction or instruction. Depending on what is required, the matter is then treated as a variation. As the contract requires the use of a bill of quantities in which the excavation and foundation structure is measured, the contractor would be protected by way of either variation or adjustment to the quantities.

Like the SIA document, the JCT does not purport to extend the scope of the contractor's obligation. Clause 2.1 provides that the contractor *shall upon and subject to the Conditions carry out and complete the works shown upon the Contract Drawings and described or referred to in the Contract Bills.* Equally, there is no specific reference to latent conditions. This contract also relies on the use of a bill of quantities and presumably the work associated with sub-surface conditions would be measured. On this basis any change needed to this work would be dealt with initially under clause 2.2 and then as a variation under clause 13.2.

Finally, the FIDIC takes the matter a little further, in clause 8.1 which says: *The Contractor shall, with care and diligence, design (to the extent provided for by the contract), execute and complete the Works and remedy the defects therein in accordance with the provisions of the Contract. . . . [S]o far as the necessity for providing the same [labour, materials and plant] is specified in or is reasonably inferred from the Contract.* Clause 12.2 defines latent conditions (here referred to as Not Foreseeable Physical Obstructions or Conditions) in a similar way to the other contracts discussed in this chapter. Clause 12.2 provides that the contractor may recover the cost of additional work necessitated by the discovery of such conditions and the time for completion of the project appropriately extended. The reference in clause 12.2(b) is to the recovery of *costs which may have been incurred* without reference to the variation provisions (clauses 51 and 52) which suggests that it is the cost only that is recoverable rather than the cost plus a profit margin.

12.8 Conclusion

It will be seen that there is considerable variation in the approach of the authors of the contracts discussed. In all instances save AS 4000 and PC-1 a contractor has some right to claim an extension of time for a delay arising from a labour dispute. This is possible under AS 4000 and PC-1 only if the schedule is completed appropriately. There is a potential for the contractor to claim delay costs in these circumstances under AS 4000 and JCC. This again will depend on how the schedule of AS 4000 is completed and under JCC the parties share the loss arising from a labour dispute equally unless the position is altered in the appendix. No other document provides for recovery by the contractor in such circumstances.

The treatment of inclement weather also differs from contract to contract. In a sense, if the contractor is granted an extension of time, then both parties share the burden of the delay more or less equally. The principal is deprived of his or her building during the delay without the compensation of liquidated damages and the contractor must maintain a presence on the site for the extended time without extra payment. With AS 4000 and PC-1 the parties must agree in the schedule that there will be an extension of time in the event of inclement weather. Only under AS 4000 and JCC is there a potential for the contractor to recover costs during the delay. The drafting of the inclement weather clauses in C21, SIA, JCT and FIDIC restricts the contractor's right to an extension of time to circumstances where the weather could be characterized as abnormal. The contractor must therefore allow in his or her time/cost calculations for what would be the average inclement weather during the construction period.

The authors of PC-1, C21 and FIDIC have used an expanded definition of the scope of the works, no doubt, in an attempt to head off claims for minor variations and 'claimsmanship' generally. Since PC-1 and C21 were contracts prepared on behalf of principals the intention is understandable, however, the efficacy of such legal drafting in a construction contract is by no means guaranteed. There is considerable diversity in the way the documents deal with latent conditions: AS 4000, JCC and NZS 4360 all define latent conditions in a similar manner and treat the need to execute additional work arising from such conditions as a variation. Under PC-1 and FIDIC the contractor can recover only the cost of the additional work and must forego the profit which would have been payable if the work were treated as a variation. The SIA and JCT documents do not have latent conditions clauses as such but since both documents are intended for use with a bill of quantities as a contract document, a claim for the costs associated with latent conditions could be based on adjustment of the measured quantities.

Probably the same matrix of risk allocation could be achieved using any of the documents discussed but this will not follow from the mere execution of the document. There are subtle differences and the obligations of the parties could be determined by the manner of completing the schedules to a particular document. In some instances it will be necessary to amend the document to achieve the result required. For a contractor the way the risks discussed are allocated could make the difference between a profit and a loss on a project. It should also be noted that this chapter purports to deal by way of comparison with only a sample of the risks encountered in the construction process. There, are of course, many more risks that need to be taken into account. A significant risk falling into this category is cost of a delay to the contractor whilst design problems are sorted out.

Endnote

1 This is a rule of legal interpretation to the effect that the meaning of general words in a document may be restricted to the same genus as the specific words that immediately precede them.

References and bibliography

Abrahamson, M. (1979) *Engineering Law and the I.C.E. Contracts*. Fourth edition (London, Applied Science Publishers Ltd).

Atiyah, P. (1981) *Promises Morals and the Law* (Oxford: Clarendon Press).

Australian Broadcasting Commission v. Australian Performing Rights Association (1973) 129 CLR 99.

Bunni, N. (1998) *The FIDIC Form of Contract.* Second edition (Blackwell Science).

Cremean, D. (1995) *Brooking on Building Contracts*. Third edition (Melbourne: Butterworths).

Davenport, P. (1991) Risk allocation – a new approach. *Australian Construction Law Newsletter* **19**, 21.

Dorter, J. and Sharkey, J. (1990) *Building and Construction Contracts in Australia Law and Practice* (LBC Information Services).

May, A. (1991) *Keating on Building Contracts*, Fifth edition (London: Sweet & Maxwell).

NPWC/NBCC (1990) *No Dispute, Strategies for Improvement in the Australian Building and Construction Industry.* Report by NPWC/NBCC Joint Working Party, May.

Taylor Woodrow International Ltd v. The Minister for Health (1978) 19 SASR 1.

W.D. & H.O. Wills (Australia) Limited v State Rail Authority of New South Wales; State Rail Authority of New South Wales v TNT Management Pty Limited, Matter Nos CA 40577/95; CA 40663/95 [1998] NSWSC 81 (3 April 1998).

13

Contaminated land

Grace Ding*

Editorial comment

One of the undesirable legacies of the industrial development of the past two centuries is the large amount of prime real estate in cities around the world that has been badly contaminated by industrial activity. In the years of uncontrolled expansion of industry in cities such as London and Manchester, great areas of land, often in what are now highly prized inner city or waterfront locations, were given over to all manner of factories and industrial plant: tanneries, refineries, power stations, forges, mills and foundries. Environmental controls were generally non-existent, chimneys belched thick smoke into the atmosphere, leaks and spills were allowed to simply soak away, all sorts of oils and solvents and toxic chemicals found their way into the soil and by leaching made their way into neighbouring sites, and all manner of human and industrial waste was discharged into local waterways. So bad was the situation that Engels, in 1844, described the River Irk in Manchester as '. . . a narrow, coal black, foul smelling stream full of debris and refuse . . . disgusting, blackish green slime pools . . . giving off a stench unendurable . . .' (cited in McLaren, 1983), and the average age of death around that time amongst the working classes in London was only 22, in Leeds 21, and in Manchester, an incredible 18 (Dodds, 1953, cited in McLaren, 1983).

In recent times, in many cities around the world, there has been a marked move by industry away from these inner city locations; transport has become more difficult due to congestion of streets following a change from water and rail transport to road transport, rising real estate values have made occupation of large parcels of very valuable land uneconomic, and workforces in many cases have relocated to the suburbs as inner city residential property values have priced them out of the market as traditional working class areas have been progressively 'gentrified'.

* University of Technology Sydney

Redevelopment of such land does have the potential to generate enormous profits but it can also lead to major problems due to contamination from its industrial past. The value of any development on contaminated land is naturally going to depend on finding a suitable way of dealing with the problem; if not adequately addressed at the start the consequences can be nothing short of disastrous for all concerned, owners and tenants alike. In the worst case, complete developments may have to be abandoned and demolished as residents/tenants encounter health problems such as higher-than-normal rates of cancer, asthma or other illnesses.

This chapter gives some background to both the problems of contamination and some of the possible methods of carrying out site remediation in a variety of circumstances.

13.1 Introduction

With the growth of industrialization since the eighteenth century, air and water pollution have become major issues in mankind's efforts to maintain and enhance quality of life. With the loss of fish stock in rivers adjacent to industrial centres, the destruction of trees and crops due to air pollution, and reduced life expectancies of workers in industries connected with asbestos and other chemicals, environmental degradation has moved to the top of the agenda at many national and international conferences (Meyer *et al.*, 1995).

Much research has been undertaken in order to investigate the nature, extent and impact of air and water pollutants on human health and the ecosystem. In contrast, little or no concern has been directed towards the impact of spillages and dumping of toxic wastes on industrialized land and the subsequent impact on watercourses. Contaminated land is a serious problem now confronting mankind and some developing countries pay little or no attention at all to the potentially lethal effects of contaminants in the ground. Such contaminants may pollute watercourses, which in turn provide drinking water to humans and animals or irrigation for agriculture. Contaminated land also reduces the amount of usable land for human habitation. Land scarcity is a long-term problem for the industrialized world and contaminated land has reduced this supply of usable land even further. Contaminated land is neither suitable for agriculture nor for other human activities and the effects may be such that human activity may not be viable on land adjacent to a contaminated area. The full extent of contaminated land and the hazards it could pose are still largely unknown.

Contaminated land was first identified in the late 1970s (Cairney and Hobson, 1998). Industrial countries such as the United States, The Netherlands, West Germany and the United Kingdom have moved to define contaminated land and have carried out investigations into how to treat land contamination. Other countries have so far taken little or no real action towards defining, treating or preventing land contamination. In recent times the problem of contaminated land has become more evident, and the international attention that has been given to sustainable development and controls over waste imply that land contamination is gaining higher priority and more official recognition worldwide (Cairney and Hobson, 1998).

13.2 Definition of contaminated land

There is no one internationally accepted definition of contaminated land since there is no clear demarcation of what constitutes contaminated land and what constitutes clean land. How much contamination must be present before a site is regarded as contaminated land? How clean is clean land? The attitudes of governments towards contaminated land for developed and developing countries are not alike. Some countries such as the UK, USA and the Netherlands have well-developed legislation and remedial action policies toward contaminated land whilst others have nothing or, at best, a much lower level of awareness.

The definition provided by the NATO Committee on Challenges to Modern Society (CCMS), which has been widely quoted and used, has the most international perspective of contaminated land. It defines contaminated land as:

> Land that contains substances which, when present in sufficient quantities or concentrations, are likely to cause harm, directly or indirectly, to man, to the environment, or on occasion to other targets.
>
> (cited in Harris and Herbert, 1994, p. 3)

However, the definition provided by NATO/CCMS has prompted further argument as to the definition of sufficient quantities and concentrations. Land is only considered as contaminated when substances are present in such quantities as to cause risk; if the quantities and concentrations are not sufficient to pose a risk then the land will not be considered as contaminated. Obviously this does not provide a clear and precise definition of contaminated land.

The definition provided by the Department of the Environment (DoE) in the UK is similar. The DoE have described contaminated land as 'land which represents an actual or potential hazard to health or the environment as a result of current or previous use'. (Young *et al.*, 1997, p. 1) This definition focuses on the impact on human and environmental health, with the extent of contaminants on site being of less concern.

In the UK, only land containing substances that may, or may be likely to, cause harm or risk to human health and environmental quality and which did not originate on the site would be considered as contaminated land. If land contains contaminants that are naturally present, such as radon, this land would generally be excluded even though those substances may still cause harm to the people and the environment (Hester and Harrison, 1997).

The definition provided by the Australian and New Zealand Environment Conservation Council (ANZECC) and the National Health and Medical Research Council (NH&MRC) has defined contaminated land as 'land on which hazardous substances occur at concentrations above background levels and where assessment indicates that it poses, or is likely to pose, an immediate or long-term hazard to human health or the environment' (cited in DUAP, 1995, p. 13). This definition is clearly very similar to the NATO/CCMS definition.

It is indeed hard to define land contamination. The definition of 'sufficient quantities and concentrations' is yet to be established, and such quantities and concentration levels may vary from location to location and from land use to land use. With regard to the definitions provided above, only land that contains contaminants at levels above those defined as likely to cause harm or risk, will be considered to be contaminated land. However, the purpose for which a potentially contaminated site may be used may affect the degree to which people using the site will be exposed to the risk of harm from contaminants. For instance, the presence of contaminants on site may not cause serious harm to industrial workers as they may only work in certain parts of the site. In such a situation the land may not be regarded as contaminated, however, if the land is used for residential purposes the degree of exposure to contaminants for residents may be much greater even though the same level of contamination is present.

As a result contamination may only be considered in the context of re-development and the intended use of the site. In the absence of a clear and precise definition of contaminated land it is difficult to evaluate the potential extent of its effect on human health and safety, and the necessity of its treatment.

13.3 Causes of contaminated land

Contaminated land is closely associated with past and present land use patterns and the geological composition of the land. Land contamination can arise from a range of industrial, agricultural and other activities such as mining or nuclear testing; whilst the impact of some activities will only be temporary, others will pose a long-term risk to both human health and environmental quality.

The main causes of contamination are the discharging of waste materials such as industrial and chemical waste into the ground, and the use of chemical fertilizers and pesticides in agriculture. Contamination may also be due to accidental spillages or leakages during industrial manufacturing processes, improper practices in the handling of chemicals, and leakages during storage above or below ground. The extent of the contamination may not be confined to the area directly affected but may be increased by the migration of contaminants to off-site locations by mechanisms such as dust deposition, water run-off and movement through the soil and ground water. Land contamination can be identified by polluted waters, discoloured soil and decayed vegetation. If contaminated land is not dealt with properly, it may cause significant harm or risk to humans, animals and the entire ecosystem on the site and in surrounding areas.

The following table (Table 13.1) shows some of the potential land contaminating activities and their associated contaminants:

Table 13.1 Some potentially contaminating activities and their main contaminants (DUAP, 1995)

Contaminating activities	Potential contaminants
Agriculture	arsenic, organochlorine and organophosphate-based chemicals
Airports	hydrocarbons, oil and heavy metals
Landfills	putrescible matter, paper, glass, plastics, metals, bacteriological contaminants, cytotoxic chemicals
Mining	acids, cyanides, heavy metals
Oil refineries	petroleum hydrocarbons, lead
Industry	heavy metals, oxides, lead, asbestos, cobalt, uranium, thorium
Heavy engineering	heavy metals, oxides, antifouling paints, lead, asbestos
Timber treatment works	creosote, polycyclic hydrocarbons, copper, pentachlorophenol
Termite/ant control	dieldrin, heptachlor, chlordane
Chemical and petrochemical works	tar, bitumen, solvents, fertilizers, pesticides, pharmaceuticals

13.4 Potential problems of contaminated land

Contaminated land may be the result of past or current land use, and is often caused by incorrect or improper work practices. The potential problems of contaminated land in different countries vary widely. In some countries, such as the UK, where the supply of land is at a premium, contaminated land poses a greater problem than it does in countries such as Australia where land is abundant. In such places, even if land is identified as being contaminated, it may not be as serious an issue.

The main problem of contaminated land is the reduction in the quantity of land for re-use. Contaminated land is, in principle, unsuitable for any human activity and is a particularly serious problem in urban areas where land is scarce and expensive. Built-up areas are usually densely populated and contaminated land may also pose potential risks to human and environmental health due to its toxicological or other hazardous properties. This is a potential threat not just to humans or the environment on the land itself but also to those in the vicinity due to the possible migration of pollutants through groundwater, soil and the air. With an ever-increasing global population and a growing awareness of environmental issues, the clean-up of contaminated land becomes a necessity as fewer and fewer 'green field' sites remain available for development in many areas.

Contaminants may come in three different forms: chemical, biological and physical. Chemical hazards may be caused by the present of carcinogenic substances in both land and water, biological hazards may be caused by the presence of potentially harmful pathological bacteria, and physical hazards may be caused by an accumulation of radioactive materials or a gradual build up of explosive or flammable substances such as wood dust. The existence of these hazards is a potential threat to human health, other living organisms, environmental quality and even the physical structures of buildings and services (Harris and Herbert, 1994).

Besides causing damages to the living environment, contaminated land may also be harmful to the economy of a society as a whole. Contaminated lands are not suitable for redevelopment without a substantial clean-up. The cost of the clean-up will vary greatly according to the level of contamination and the intended future usage of the land. A potential developer may need to go through a lengthy process to investigate the nature and extent of contamination on site and identify the remedial actions required. The process may take years to complete.

Contaminated lands have to be dealt with as early as possible to prevent further accumulations of hazardous substances which would further increase the time and cost required to rectify the situation.

13.5 The identification of contaminated land

Land contamination is identified through the process of site investigation. The purpose of site investigation is to identify the extent and nature of contamination and the possible damage that may be caused to humans and the environment. The information obtained from the site investigation will form the basis for risk assessment and for the decisions made regarding the design and implementation of remedial measures. Failure to identify potentially contaminated sites may result in inappropriate land use decisions and increase the risk to the health and safety of people on or adjacent to the site.

Site investigation usually starts with desk research into the history of the land. Identification of previous land uses may indicate past activities which had the potential to pollute. This in turn assists investigators as they assess the suitability of proposed future site use and the availability of suitable methods for remedial action.

A detailed site investigation may involve physical exploration on site including the excavation of trial pits and boreholes so that information about the extent of contamination to the land may be collected and analysed. The process of site investigation should be properly designed and managed to minimize potential risks to site personnel, people in the vicinity and the environment. It should also be properly executed by skilled personnel under proper supervision.

As suggested by Harris and Herbert (1994) site investigation may consist of three to five phases depending on the site situation, site coverage and the level of contamination. Table 13.2 outlines the main activities of these phases of site investigation.

Table 13.2 Five phases of site investigation

Phase	Activities
One	• make preliminary inspection of site with regard to past and present land uses • recognize, identify and interpret significant signs on site
Two	• obtain samples, e.g., vegetation, water, waste and soil for testing, analysis and reporting
Three	• make detailed investigation – including trial pits, trenches and boreholes • ascertain the characteristics of contaminants • decide on remedial action
Four	• make supplementary investigation to obtain additional information in support of risk assessment • finalize the remedial process
Five	• carry out post-treatment validation exercise to confirm the effectiveness of remedial action • investigate for compliance and performance

Phase one involves a preliminary investigation of the site. This includes a preliminary assessment to identify whether the contamination of a site presents risk of harm to people and to the environment. This normally starts with a site history review and a site inspection to look for indicators of contamination or other hazards as land contamination may result from activities that took place on or adjacent to a site. The investigation of past land uses may help to identify any previous activities on site that would help to determine the likelihood of contamination and types of contaminants that may be present. Knowledge of the previous land use is essential if hazardous materials present are to be prevented from conflicting with proposed future land use.

Historical information may be obtained by visiting the current and previous owners and occupiers, the local council archives, local libraries and through newspaper archives. The information published in any industrial or local historical literature of the district may provide further background knowledge of the site.

Besides planning and development applications, the local government may provide useful sources of information relating to the nature of previous activities on the land. Government departments such as the Environmental Protection Authority (EPA) in New

South Wales may hold information relating to any records about pollution complaints or illegal activities on site. The preliminary investigation phase provides useful background information about the site and most importantly it may identify contaminated land (areas known as 'hot spots') on site, that will provide the direction for the second phase of site investigation.

It is neither economic nor feasible to examine the entire area of a site if the site itself is large. This is a particularly serious issue for agricultural land, which is usually too large to be effectively inspected. The feedback on the first phase site investigation provides useful focus for carrying out the second phase of site investigation, which concentrates only on the 'hot spots' on the site. In this phase various samples of water, vegetation, soil and waste are collected on site. These samples are then tested, analysed and reported upon to provide information for the design of a detailed investigation.

Phase three involves a detailed investigation of the site by undertaking a comprehensive investigation of the ground conditions by using trial pits, trenches and boreholes. The purpose of this phase is to characterize the contaminants present on site and to identify the contamination pathways and target affected areas. In identifying and defining the potential pollution sources on site, the location, nature, concentration and total loading are recorded clearly for further investigation/information. The investigation as to how the contamination migrated to the final target area is an important basis for the design and implementation of any remedial strategies.

If the contamination situation on site is serious and extensive, a supplementary investigation may be required to provide further data regarding the ground condition. The fourth phase of the site investigation helps to obtain additional information to support risk assessment methodology and the confirmation of appropriate remedial strategies.

The final phase of site investigation is usually carried out following the selection of the remedial methods and continues throughout the whole remediation process. This is, therefore, regarded as a post-treatment validation exercise. The work involved in this phase is the monitoring of the remedial work and confirmation of the effectiveness of the remedial action. Investigation will also be undertaken when the process is finished on site to test the success/efficacy of the method. This will confirm whether the remedial action has removed or destroyed contaminants and if any further treatment is required. The result will determine whether the land can be re-used for its intended purpose.

13.6 Risk assessment and estimation

Young *et al.* (1997) define risk assessment as the work concerned with the gathering and interpretation of information on the characteristics of contamination sources, pathways and receptors at a specific site and understanding the uncertainties inherent in the ensuing assessment of risk. The purpose of risk assessment is to determine the acceptable level of risk associated with the site such that there is not an unacceptably high risk of harm to public health and the ecosystem. The information obtained through risk assessment will be fundamental to the determination of remedial actions in order to reduce or control risks at an acceptable level. The process of risk assessment has four key components (Young *et al.*, 1997):

- identification of contaminants
- exposure assessment

- toxicity assessment
- risk characterization

Hazard identification involves the collection of information about the contaminants, the site and the surrounding environment, in order to identify the sources, exposure routes and targets. The type and amount of contaminants present must be known in order to identify their intrinsic toxicity and any associated health risks. This type of information mainly comes from the site investigation. The observed levels of contamination are compared with published data on natural 'background' levels of contaminants and local background concentrations (Young *et al.*, 1997). The comparison between observed levels and the background concentration relies on the development of environmental assessment criteria for soils and groundwaters against which analytically determined site concentrations can be assessed. There are different approaches that assist in the determination of the significance of contamination of sites intended for development:

- The Inter-departmental Committee on the Redevelopment of Contaminated Land (ICRCL) is a compendium/directory of trigger concentration values developed in the UK. This guideline provides two sets of values, the 'trigger' value which indicates when further investigation is necessary, and the 'action' value which indicates when some form of remedial action is likely for the proposed end use of the site (Harris and Herbert, 1994).
- The Dutch Government and the Ontario Ministry of Environment have adopted an approach which provides guideline values of background, intermediate and heavy levels of contamination for soil and waters (Young *et al.*, 1997).
- The development of environmental criteria called risk-based environmental assessment criteria for soils and groundwaters (Young *et al.*, 1997).

Exposure assessment is the assessment of probable human exposures; this is the key to the choice of a remediation method. Contaminants can be transported by wind and groundwater, and through the air by ingestion, inhalation or direct contact, to reach the targets. Through the possible exposure routes individuals may be injured by contaminants. It is, therefore, necessary to identify and assess each of these routes to determine the likely amount, frequency and duration of exposure to contaminants.

The assessment of toxicity of contaminants to human health is an area that requires attention and it involves the prediction of health effects associated with the contaminants. It is necessary to describe the variation of health effects with the variation in dosage. Every contaminant will have a different minimum dose below which it will have no harmful effect on human health; above this minimum level its effects will become more evident. The purpose of toxicity assessment (Harris and Herbert, 1994) is to determine the effect of the hazard on the target under the conditions of exposure. Effects of hazards on individuals involve a consideration of dose–response relationships, biological mechanisms regulating responses to different types of substances and other factors affecting the response of targets.

Combined information on the identification of contaminants, exposure and toxicity assessment constitutes the risk characterization for the site. These details should be prepared for each of the remedial alternatives being considered so that the health benefits of each can be compared (Young *et al.*, 1997; Cairney and Hobson, 1998).

13.7 Planning, design and implementation of remediation

Following the assessment of risk on site, if the land poses an unacceptable risk to public health or the biophysical environment, remedial action is required to minimize or eliminate present or future risk and thus protect human health and the biosphere. The ultimate aim of remediation is to eliminate the hazard by completely removing or treating contaminated soil/material, on- or off-site.

13.7.1 Remedial strategies

Remedial work should be well planned, and designed to avoid inconvenience or danger for people working or living on or near the site. The possible impact generated by remedial work, such as noise, vibration, smell, emissions or pollution of underground water should also be minimized. Possible exposure of residents and on-site workers to contaminants in air, soil and water, as well as possible ecological impacts, should be given consideration prior to designing and choosing the remediation technology to be used.

13.7.2 Remedial methods

There are numerous technologies available nowadays for dealing with land contamination. The decision as to which technology is used depends on the nature of contaminants, the level of risk and the degree of exposure to the general public, in addition to the considerations of the total cost and any associated legislation. Thus a technology may be suitable for a certain form of contamination but be simply too expensive to carry out.

Remedial methods are broadly divided into three categories: risk avoidance methods, engineering methods and process-based methods as detailed in Figure 13.1 (American Society of Civil Engineers, 1996; Hester and Harrison, 1997; Cairney and Hobson, 1998).

Risk avoidance methods involve changing land use, site layout, location of services and other related infrastructure, so that the exposure of people to contaminants is minimized. These methods are particularly useful on vacant lands since land use and layout are freely and cheaply redesigned. Changing land use and layout for occupied land may, however, be impossible or uneconomic.

The engineering methods used in remediation are traditional civil engineering techniques that involve excavation and removal of contaminated soils from site. The excavated soils may be disposed of on-site at a designated location or off-site in a controlled landfill. These methods do not change the state of contamination as the contaminants are neither removed nor destroyed, merely relocated. In some situations excavated soil is treated using process-based systems prior to disposal in order to minimize hazardous effects. This is the most utilized strategy since the cost is comparatively low and it is a rapid method of dealing with a contaminated site. However, with the shortage of landfill sites and the environmental problems associated with landfills, the dumping of contaminated soil may not be a feasible method in the near future.

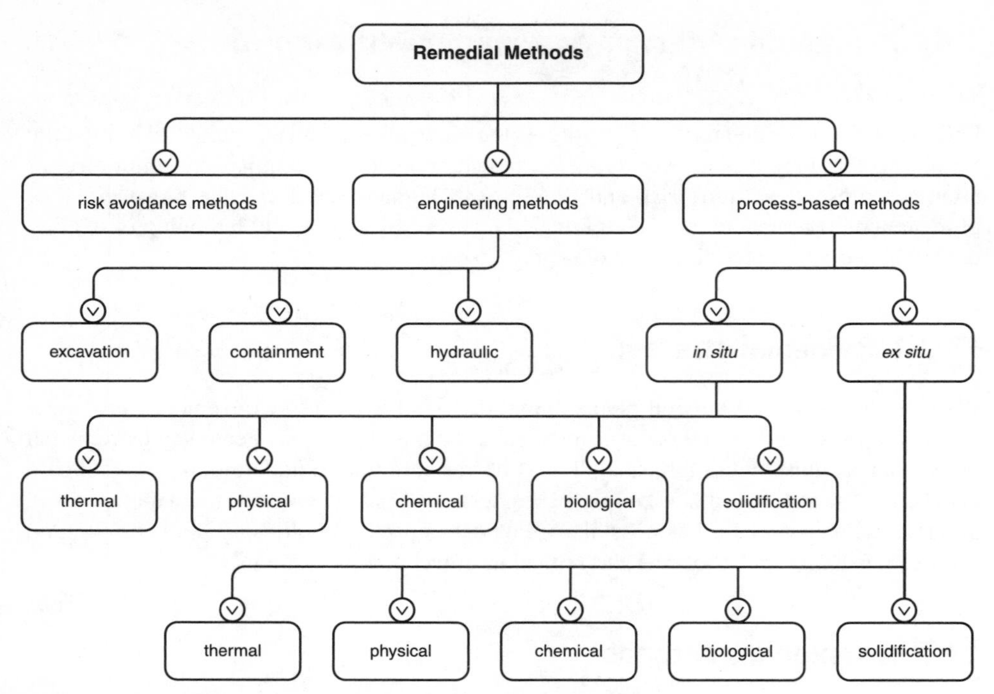

Figure 13.1 Classification of remedial methods (Harris and Herbert, 1994).

Apart from removing contaminated materials, contaminants can be isolated from the surrounding environment. Isolation may be achieved by inserting vertical low permeability barriers or liners. These methods help to prevent or limit the lateral or vertical migration of contaminants through ground water or sub-soil movement. These methods include displacement (e.g., sheet steel piles), excavation (e.g., clay barriers, slurry trenching) and injection (e.g., jet grouting). They are particularly useful for the control of pollution in landfill areas. Horizontal barriers may also be used together with vertical barriers to completely encapsulate a contaminant source. Once again, however, these methods do not practically treat or destroy contaminants and these contaminants may have adverse reactions with the barrier materials in the long term.

If the contamination is concentrated on the surface, a 'cap and cover' system may be adequate. A cover system consists of covering the contaminated area with a single layer or a succession of layers of non-contaminated materials. The purpose of a cover system is to provide a barrier with specific physical and chemical properties that will prevent humans and the ecosystem being exposed to harmful substances. Again these methods do not treat or destroy contaminants and these may still restrict the future use of the site.

The process-based methods involve using various process methods to treat contaminated soil *in situ* or *ex situ*. These methods can be used together with engineering methods to treat excavated soil at a designated location or in a laboratory. The purpose of these methods is to remove, stabilize or destroy contaminants in the soil. The most popular process-based methods are biological, chemical, solidification, thermal and physical methods.

Biological methods use naturally occurring micro-organisms, such as bacteria or fungi, to degrade the chemical pollutants into non-toxic compounds such as carbon dioxide, water and simple inorganic compounds. These methods are suitable for treating large areas of land that have been contaminated by petroleum-based products over a wide range of concentrations. These treatments are applicable for both contaminated soil and groundwater and potentially can be combined with other processes in a treatment regime.

Chemical methods use chemicals to destroy or neutralize hazardous compounds. Chemicals are added to the soil to stimulate the appropriate chemical reaction. Some of the added chemicals, by-products and waste may, however, remain in the soil and become hazardous in the long-term resulting in the need for further treatment. The common types of chemical processes are oxidation-reduction, dechlorination, extraction, hydrolysis and pH adjustment.

Solidification helps to solidify contaminated materials into less mobile chemical forms by binding them into an insoluble matrix. These processes can be used to treat soils, wastes and water. Common formulations include those based on cement, silicates, lime, thermoplastics and polymers. The long-term performance of these methods is uncertain, however, and may have potential health and environmental impacts.

Thermal methods use heat to remove or destroy contaminants by incineration. These processes have the potential for complete destruction of contaminants. The three main types of process are thermal desorption, incineration and vitrification. Hot air or steam is introduced directly into the ground to strip away contaminants, or electrical energy may be delivered through an array of electrodes inserted into the ground to destroy contaminants. These methods are energy intensive processes that may produce waste by-products with potential public and environmental impacts.

Physical methods remove contaminants from the soil matrix by exploiting differences in their physical properties. Some external force is applied that alters some physical characteristic thus enabling separation to occur. Specific methods include soil washing, solvent extraction, electrokinetics and soil vapour extraction. These methods may also produce waste streams that are harmful to the environment.

13.7.3 Selection of remedial methods

In some situations a single remedial method may be sufficient to handle all the risks presented by the site. More commonly, however, more than one measure or a combination of measures will be required to completely address all the risks. This situation will occur, for instance, if different types of contaminants are found on the same site. While there are different types of remedial methods according to the nature of the contaminants in the soil or water, any remedial measures undertaken should be practical and meet cost requirements.

Harris and Herbert (1994) suggest that the process of selecting remedial action should follow a structured framework, from identification, to evaluation through to screening, in order to choose the most appropriate methods. Any information obtained from site investigation and risk assessment is crucial as it helps those involved anticipate the potential benefits and limitations of the different approaches that are appropriate for the particular types of contaminants on the site.

Table 13.3 Summary of remedial action selection criteria (Harris and Herbert, 1994)

Criterion	Considerations
Applicability	• action must be suitable for both the contaminants and the contaminated medium • method must be able to reach the contaminants • some contaminants may be treated by certain types of approach only
Effectiveness	• must be effective in the long term • must be capable of achieving the level of treatment and risk reduction desired
Limitation	• excessive time constraints for some biodegradation processes • control of atmospheric emissions from thermal treatments • site location and size may restrict the methods to be used • the site must remain operational during the remediation
Cost	• indicative cost is required for each method to allow comparison with the budget • sometimes difficult to predict due to unknown factors and innumerable variables • the cost of process-based treatments is generally substantially higher than for other methods
Track record	• methods that have been used successfully before may be more desirable
Availability	• the outcome of previous case studies may determine the suitability of a method • the availability of certain methods may be limited • some methods may be available overseas only
Information requirements	• the selection of appropriate methods will rely on sufficient information and risk assessment from site investigation
Environmental	• choose the method that has the least adverse effects on impact the environment
Health and safety needs	• method chosen must comply with health and safety provisions during implementation
Post-treatment management needs	• to verify the success of the remediation process • to monitor the long-term integrity of the remediation

Selection criteria can be used to help in the initial screening and evaluation of remedial action methods. Factors to be considered at this stage are summarized in Table 13.3.

Based on the consideration of these selection criteria, a range of remedial methods for treating particular contaminants can be established. A number of alternatives should then be developed to allow more choice during the final selection. A 'no action' base case should also be included as one of the alternatives for comparison with other treatment processes. At this stage all the alternatives are analysed in detail to determine their respective strengths and weaknesses in order to arrive at a final ranking of these alternatives and a decision can then be made accordingly.

13.8 Case studies

The following examples relate to two large former industrial sites in Sydney. In recent years many large sites quite close to the commercial centre of Sydney, that were previously used for a range of industrial purposes, have been vacated and redeveloped as land costs have pushed industry to relocate to outer suburban locations or even to regional centres. Many of these sites were extensively contaminated as a result of their previous use and have required detailed and expensive remediation work.

13.8.1 Homebush Bay Olympic Site

The Homebush Bay Olympic Site is located 16 km west of Sydney's central business district (CBD) and covers approximately 660 ha of what was seriously contaminated land. Before European settlement the area consisted of extensive tidal wetlands and open woodlands with significant ecological value. Homebush Bay contains the largest mangrove forests and second largest saltmarsh community in the Sydney Harbour region and maintains more than 100 different plant species, and supports a diversity of birds, mammals, reptiles and amphibians. Over a period of 100 years the area in Homebush Bay was degraded and underwent dramatic transformation by land reclamation carried out from 1948 to the early 1960s, and the illegal dumping of waste. The rehabilitation program for Homebush Bay was one of the biggest land clean-up projects in Australia and of great importance to the urban environment in the neighbourhood. The Olympic Co-ordination Authority (OCA, 2000) was faced with the very difficult task of providing space that was suitable for international sporting facilities and venues, as well as parkland and open space for the people in the vicinity.

The Homebush Bay project was carried out by the OCA which was established in 1995 to plan and manage the rehabilitation program of the site as part of the preparation for the Olympic and Paralympic Games in 2000. The OCA, together with other environmental specialists and community representatives, developed the environmental guidelines for the planning and management of the Sydney Olympics in line with the ecologically sustainable development principles of the International Olympic Committee. The site also provides a new suburb with a mixed usage of land for residential, commercial, sporting facilities and parkland.

In the past Homebush Bay had been used for many industrial activities. It was the site of an abattoir, a brickworks, an armaments depot and was also used extensively as an illegal dumping ground for municipal and industrial waste. The area had been seriously contaminated by leachate from waste disposal and the seepage had contaminated the surface and ground water systems. Extensive soil investigation was carried out in the area and samples of soil, sediment and surface water were obtained for analysis. It was discovered that the site contained putrescible waste, heavy metals, waste oil products, asbestos, pesticides and other industrial chemicals that are highly unsuitable in areas used for human activities and natural habitat.

The remedial strategy to restore the Homebush Bay included the excavation of 400 tonnes of contaminated soil containing chemical wastes such as chlorinated benzenes and chlorophenols, pesticides, polycyclicaromatic hydrocarbons, dioxins and furans for treatment. The treatment of contaminated soil was undertaken in two stages: the first stage was carried out to separate the contaminants from the soil and the second was conducted to destroy all the contaminants.

During the first stage chemical waste was separated from the soil using an indirectly-heated thermal desorption plant and the contaminants were concentrated in a waste sludge. The treated soil was removed and safely contained in a specially designed containment area on site. These landfills were capped by a minimum of 0.5 m of clay to prevent leachates from leaking into the surrounding environment and polluting the watercourses. Sub-surface leachate drains were constructed to capture any groundwater seepage from the landfills and subsequently leachates will be treated at a nearby liquid waste treatment plant before discharging back into the sewer. Artificial topsoil has been placed on the clay topping to enable extensive landscaping to take place. The landfill areas have been landscaped with native plants and trees to reduce the amount of stormwater entering the landfills and to create semi-natural habitats for flora and fauna.

In the second stage of the remediation programme the separated chemical waste in the waste sludge was destroyed by a decomposition process that broke down the contaminants by reversing the chemical process. The residues were transported to a waste treatment plant at Lidcombe (a nearby suburb) for further treatment (EPA, 1997).

13.8.2 ICI Dulux paint manufacturing factory

In 1996, after two years study and investigation, ICI Dulux announced the world's largest soil-washing decontamination project on one of their former industrial sites at Cabarita, not far from the Olympic site at Homebush. ICI Dulux had been manufacturing paint products on the 9 ha site on the Parramatta River for more than 75 years. During that time ICI buried much of its waste on site which seriously contaminated the soil and polluted adjacent watercourses. Paint production on the site ended in December 1994. During the following two years of soil investigation lead levels in the soil were found to be, on average, between seven and ten times the Australian maximum allowable level.

The clean-up project included the demolition of more than thirty existing buildings on site with most of the building materials such as timbers, bricks, concrete and steel being decontaminated and recycled for reuse in other projects. The world's largest soil decontamination plant was installed to carry out soil-washing on site. The plant was able to clean and re-use up to 80% of the 140 000 tonnes of polluted soil. It was able to reduce the lead levels in the soil from levels of between 2000 and 3000 parts per million (ppm) to levels of between 150 and 200 ppm, which was well below the 300 ppm maximum allowed under Australian regulations. The contaminated residues were stabilized in concrete and dumped in landfills. The overall project cost was approximately AU$20 million with about $5 million spent on the decontamination plant alone. The project was successfully completed and the former industrial site transformed into a medium density housing development (Healey, 1996).

13.9 Conclusion

Land contamination is a major environmental problem that must be confronted both now and in the future. There is, however, a lack of any clear or precise definition of contaminated land and governments from the developed and developing countries have substantially different attitudes towards contaminated land. These factors add to the problems associated with the management of contaminated land.

The nature and extent of contamination of land and its associated risk to the general public and the environment is largely unknown and it is an area requiring further research.

The potential risks of contaminated lands can only be removed if they are treated effectively. The primary objective of remediation of a contaminated site is to reduce the actual or potential environmental threat, or to reduce unacceptable risk levels to levels that are acceptable. A significant amount of research has been undertaken into the area of remedial treatment for contaminated ground, however, most remedial treatments are both time consuming and costly which can prove uneconomical for some developments. Future research into treatment processes may have to be directed at efficiency and cost effectiveness in order to make the clean-up of contaminated sites a more attractive proposition. Not until contaminated sites are dealt with efficiently will the environmental problems be resolved.

References and bibliography

American Society of Civil Engineers (1996) *Monitor Well Design, Installation, and Documentation and/or Toxic Waste Sites* (New York; ASCE Press).

Boulding, J.R. (1995) *Practical Handbook of Soil, Vadose Zone and Groundwater Contamination: Assessment, Prevention and Remediation* (Boca Raton: Lewis Publishers).

Cairney, T. and Hobson, D.M. (1998) *Contaminated Land Problems and Solutions* (London: E. & F.N. Spon).

Dodds, J.W. (1953) *The Age of Paradox – A Biography of England, 1841–1851* (New York: Rinehart).

DUAP (1995) *Contaminated Land* (Sydney: Department of Urban Affairs and Planning).

Engels, F. (1845) *The Condition of the Working Class in England*, cited in McLaren (1983).

EPA (1997) *New South Wales State of the Environment 1997* (Sydney: NSW Environment Protection Authority).

Hadley, M. (1996) New systems for evaluating the risk or blight associated with contaminated land. In: *Proceeding of RICS Research Conference: The Cutting Edge*. The Royal Institution of Chartered Surveyors.

Harris, M. and Herbert, S. (1994) *ICE Design and Practice Guide: Contaminated Land Investigation, Assessment and Remediation* (London: Thomas Telford).

Healey, K. (1996) Waste management. *The Spinney Press*, **66**, 14.

Hester, R.E. and Harrison, R.M. (1997) *Contaminated Land and its Reclamation* (London: Thomas Telford).

Janikowski, R., Kucharski, R. and Sas-Nowosielska, A. (2000) Multi-criteria and multi-perspective analysis of contaminated land management methods. *Environmental Monitoring and Assessment*, **60** (1), 89–102.

Lewry, A. and Garvin, S. (1995) Construction on contaminated land. In: *Proceeding of RICS Research Conference: COBRA 1995.* The Royal Institution of Chartered Surveyors.

McLaren, J.P. (1983) Nuisance law and the industrial revolution – some lessons from social history, 3, *Oxford Journal of Legal Studies* 155.

Meyer, P.B., Williams, R.H. and Yount, K.R. (1995) *Contaminated Land: Reclamation, Redevelopment and Reuse in the United States and the European Union* (Aldershot: Edward Elgar).

McGarty, J. and Sturge, K. (1999) Waste not, want not? Barriers to residential brownfield development. In: *Proceeding of RICS Research Conference: The Cutting Edge.* The Royal Institution of Chartered Surveyors.

Olympic Co-ordination Authority (2000) *Environment Report 1999* (Sydney: Olympic Co-ordination Authority).

Page, G.W. (1997) *Contaminated Sites and Environmental Cleanup: International Approaches to Prevention, Remediation and Reuse* (Academic Press).

Petts, J., Cairney, T. and Smith, M. (1997) *Risk-Based Contaminated Land Investigation and Assessment* (John Wiley and Son).

Powell, J.C. (1996) The evaluation of waste management options. *Waste Management and Research*, **14**, 515–26.

Richards, T. (1996) Contaminated land: more certainty less chance. *CSM*, February, **5** (5), 25.

Samuels, R. and Prasad, D.K. (1994) *Global Warming and the Built Environment* (London: E. & F.N. Spon).

Sarsby, R.W. (1998) *Contaminated and Derelict Land* (London: Thomas Telford).

The Royal Institution of Chartered Surveyors (1996) Contaminated land, *CSM Supplement*, April.

Wood, G. (1995) Contaminated land and the surveyor. *CSM*, April, **4** (6), 25.

Yost, P.A. and Halstead, J.M. (1996) A methodology for quantifying the volume of construction waste. *Waste Management & Research*, **14**, 453–61.

Young, P.J., Pollard, S. and Crowcroft, P. (1997) Overview: context, calculating risk and using consultants. In: Hester R. and Harrison R. (eds) *Contaminated Land and its Reclamation* (London: Thomas Telford).

14

The foundations of lean construction

Lauri Koskela*, Greg Howell†, Glenn Ballard† and Iris Tommelein‡

Editorial comment – Chapters 14 and 15

The creation of value in building and construction projects has a particularly strong place in the lean construction philosophy. 'Lean' is a way to design production systems to minimize waste of materials, time, and effort in order to generate the maximum possible amount of value. This chapter and Chapter 15 introduce the ideas and techniques of lean construction. These chapters do not give detailed instructions for implementing lean construction, but they give a comprehensive overview of the philosophy and practice of lean as its applies to construction. Chapter 14 provides an overview of lean construction as a theory-based approach to project management, which is compared to current project management, and outlines the lean-based project delivery system and its implementation. Chapter 15 describes several tools and techniques that support this new approach.

This chapter starts with a discussion of a theory of production. Our understanding of systems of production and associated production theory and related tools can be classified into the transformation, flow, and value concepts. Lean production attempts to integrate these three concepts of production. The authors argue that current project management attempts to manage by scheduling, cost and output measures, but these are often not effective. By contrast, lean construction attempts to manage the value created by all the work processes used between project conception and delivery. Next, the phases of the Lean Project Delivery System[1] are explained, and how the inter-relationships between these phases can be managed. The chapter finishes with a discussion on organizational change and culture.

* VTT Building and Transport, Finland
† Lean Construction Institute
‡ University of California, Berkeley

Chapter 15 looks at the areas of production management, lean design, lean supply, and lean assembly. This is a thorough introduction to some of the techniques that distinguish lean construction from traditional project management. Importantly, the Last Planner[1] system of production control is clearly explained and the three components of this system are outlined. The Last Planner is one of the core ideas in lean construction. This is followed by a discussion that shows how the philosophy of value generation and waste reduction can be applied to design. The section on lean supply shows how lean thinking brings together the product design, detailed engineering, and fabrication and logistics aspects of construction projects. Finally, the tools and techniques used in addition to the Last Planner for lean assembly are described.

14.1 Introduction

Since the mid-1990s lean construction has emerged as a new concept, both in the discipline of construction management and the practical sphere of construction. There are two slightly differing interpretations of lean construction. One interpretation holds that the question is about the application of the methods of lean production to construction[2]. In contrast, the other interpretation views lean production as a theoretical inspiration for the formulation of a new, theory-based methodology for construction, called lean construction. The latter interpretation has been dominant in the work of the International Group for Lean Construction, founded in 1993.

Here, the view of lean construction as a novel theory-based approach to construction is adopted. This does not mean, however, that the view of lean construction as a kit of methods is totally rejected; rather, methods and tools from lean production are introduced when justified.

14.2 Theoretical considerations

Let us first clarify the basic issues. What do we do with a theory of production? What do we require from it?

14.2.1 What is a theory of production?

An explicit theory of production will serve various functions (Koskela, 2000). A theory provides an explanation of observed behaviour and it thereby contributes to understanding. A theory provides a prediction of future behaviour. On the basis of the theory, tools for analysing, designing and controlling can be built. A theory, when shared, provides a common language or framework, through which the co-operation of people in collective undertakings (such as a project or a firm) is facilitated and enabled. A theory gives direction in pinpointing the sources of further progress. A theory can be seen as a condensed piece of knowledge: it empowers novices to do the things that formerly only experts could do. It is thus instrumental in learning. Once a theory has been made explicit, it is possible to constantly test its validity. Innovative practices can be transferred to other settings by first abstracting a theory from that practice and then applying it in target conditions.

The primary characteristic of a theory of production is that it should be prescriptive: it should reveal how action contributes to the goals set for production. On the most general level, there are three possible actions:

- design of the production system,
- control of the production system in order to realize the production intended, and
- improvement of the production system.

Production has three kinds of goal. First, there is the goal of getting intended products produced in general (this may seem so self-evident that it is often not explicitly mentioned). Second, there are goals related to the characteristics of the production itself, such as cost minimization and level of utilization (internal goals). Third, there are goals related to the needs of the customer, such as quality, dependability and flexibility (external goals). Furthermore, the theory of production should cover all essential areas of production, especially production proper and product design.

From the point of view of practice of production management, the significance of the theory is crucial; the application of the theory should lead to improved performance. Conversely, the lack of the application of the theory should result in inferior performance. Here is the power and significance of a theory from a practical point of view: it provides an ultimate benchmark for practice.

14.2.2 What theories regarding production do we have?

What have scientists forwarded as theories? What theories have actually been used in practice? Throughout the twentieth century, the transformation view of production has been dominant. In the transformation view, production is conceptualized as a transformation of inputs to outputs. There are a number of principles by which production is managed. These principles suggest, for example, decomposing the total transformation hierarchically into smaller transformations, called tasks, and minimizing the cost of each task independently of the others. The conventional template of production has been based on this transformation view, as well as the doctrine of operations management.

The transformation view has its intellectual origins in economics, where it has remained unchallenged to this day. The popular value chain theory, proposed by Porter (1985), is one approach embodying the transformation view. A production theory based directly on the original view on production in economics has been proposed by a group of scholars led by Wortmann (1992). However, this foundation of production is an idealization, and in complex production settings the associated idealization error becomes unacceptably large. The transformation view of production has two main deficiencies: first, it fails to recognize that there are phenomena in production other than transformations, and second, it fails to recognize that it is not the transformation itself that makes the output valuable, but, instead, that there is value in having the output conform to the customer's requirements. The transformation view is instrumental in discovering which tasks are needed in a production undertaking and in getting them realized, however, it is not especially helpful in figuring out how to avoid wasting resources or how to ensure that customer requirements are met in the best possible manner. Production managed in the conventional manner therefore tends to become inefficient and ineffective.

The early framework of industrial engineering introduced another view on production, namely that of production as flow. The flow view of production, first described in scientific terms by Gilbreth and Gilbreth (1922), has provided the basis for just-in-time (JIT) and lean production. This view was translated into practice by Henry Ford, however, his implementation was misunderstood. The flow view of production was further developed only from the 1940s onwards in Japan, first as part of war production and then in automobile manufacturing at Toyota.

The flow view is embodied in 'lean production,' a term coined in the 1980s by researcher John Krafcik to characterize Toyota's manufacturing practices. In the flow view, the basic thrust is to eliminate waste from flow processes. Thus, such principles as lead time reduction, variability reduction, and simplification are promoted. In a breakthrough book, Hopp and Spearman (1996) show that by means of queuing theory, various insights that have been used as heuristics in the framework of JIT can be mathematically proven.

A third view on production was articulated in the 1930s, namely that of production as value generation. In the value generation view, the basic goal is to reach the best possible value from the point of the customer. The value generation view was initiated by Shewhart (1931). It was further refined in the framework of the quality movement but also in other circles. Principles related to rigorous requirements analysis and systematized flowdown of requirements[3], for example, are forwarded. Cook (1997) recently presented a synthesis of a production theory based on this view.

Thus, there are three major views on production. Each of them has introduced practical methods, tools, and production templates. Nevertheless, except for a few isolated endeavours, these views – as candidate theories of production – have raised little interest in the discipline of operations management. As stated earlier, there has not been any explicit theory of production. Consequently, the important functions of a theory, as outlined, have not been realized either from the viewpoint of research or from the viewpoint of practice.

These three views do not present alternative, competing theories of production, but rather theories that are partial and complementary. What is needed is a production theory and related tools that fully integrate the transformation, flow, and value concepts. As a first step towards such integration, we can conceptualize production simultaneously from these three points of view: transformation, flow, and value. A number of first principles stemming from each view can be induced from practice or derived from theory. An overview of this integrated view, called the TFV theory of production, is presented in Table 14.1. These three conceptualizations remain partial, however, the ultimate goal should be to create a unified conceptualization of production instead.

The crucial contribution of the TFV theory of production lies in calling attention to modelling, structuring, controlling, and improving production from these three points of view combined. In production management, management needs arising from the three views should be integrated and balanced. In practice, the domains of management corresponding to the three views may be called task management, flow management, and value management. The constituents of the TFV theory of production are not new,

Table 14.1 TFV theory of production (Koskela, 2000)

	Transformation view	Flow view	Value generation view
Conceptualization of production	As a transformation of inputs into outputs	As a flow of material, composed of transformation, inspection, moving and waiting	As a process where value for the customer is created through fulfilment of his/her requirements
Main principle	Getting production realized efficiently	Elimination of waste (non-value-adding activities)	Elimination of value loss (achieved value in relation to best possible value)
Methods and practices	Work breakdown structure, MRP, organizational responsibility chart	Continuous flow, pull production control, continuous improvement	Methods for requirement capture, quality function deployment
Practical contribution	Taking care of what has to be done	Making sure that unnecessary things are done as little as possible	Taking care that customer requirements are met in the best possible manner
Suggested name of practical application of the view	Task management	Flow management	Value management

however, the TFV theory supports the new insight, that there are three fundamental phenomena in production that should be managed simultaneously.

14.3 Why conventional construction project management fails

Conventional project management in construction is inadequate because it does not rest on a TFV theoretical framework (Johnston and Brennan, 1996; Howell and Koskela, 2000; Koskela and Howell, 2001). From the first moments, construction projects are managed today by breaking them into pieces or activities, estimating the time and money to complete each, applying the critical-path method (CPM) to identify a logical order, and then either contracting externally or assigning internally to establish responsibility. In either case the pieces or activities are treated much the same. Project managers use the schedule to determine when each activity should start and push for work to begin on the earliest start date. Control begins with tracking and rests on the thermostat model[4]. Project controls determine if each activity and the total project are within their cost and schedule limits. Action is taken either to speed or re-sequence activities if delays threaten required completion. In many cases, additional workers are mobilized to speed completion but this then reduces productivity. Hard choices are made and risks shifted among participating organizations depending on commercial terms and other factors. While the project manager is struggling to achieve project objectives, those responsible for each activity work towards assuring or improving their estimated performance.

Why is it that this approach, which sounds reasonable, so often fails in practice? From the lean construction perspective, current practice rests on a defective model of the project, the work involved, and its control. Simply put, current project management attempts to manage activities by centrally applied scheduling and to control them using output measures. It fails even in the attempt to manage activities and misses entirely the management of work flow and the creation and delivery of value.

Projects today are complex, uncertain, and quick (CUQ) (Shenhar and Laufer, 1995). The pressure for ever-shorter durations will always be with us. Complexity and uncertainty arise from multiple contending and changing demands of clients, the market place and technology. The pressure for speed adds to the burden. In this dynamic environment, activities are rarely linked together in simple sequential chains; rather work within and between tasks is linked to work in others by shared resources and/or depends on work underway in others.

Co-ordinating work on CUQ projects cannot be assured even with highly detailed CPM schedules. These schedules portray the project as a series of activities and ignore the flow of work within and between them. The reliable release of work from one crew to the next is assumed or ignored. Project managers who rely on these schedules struggle with uncertainty but rarely see it arising within the project from their reliance on project level scheduling and control of activities (Tommelein *et al.*, 1999).

Controlling by tracking activity completion and cost fails to assure reliable work flow (Howell and Ballard, 1996) because this type of control rests on the thermostat model applied to output measures. The thermostat model triggers when a variance is detected and it assumes that there are direct links to the cause of the variance. If the room temperature is above the set point, the thermostat turns off the furnace. Output measures based on estimated expenditures of work hours and duration are not linked to the 'furnace' on a project. At best, variances on a project trigger some investigation by supervisors but this is often aimed at justifying why the standard is incorrect for the circumstance. Given the circumstance, there is often little that supervisors can do to increase the production rate and/or reduce costs in the face of unpredictable release from upstream and poor co-ordination with adjacent crews or design squads.

Too often, steps taken by one supervisor to improve performance of the activity in his charge reduce total project performance by further reducing the reliable release of work to the next team. For example, this happens when crews choose to install the easier work first in order to improve their performance numbers; pipefitters call this 'showpipe'. Here we see a deeper problem in the way projects are managed: the attempt to optimize each activity inevitably leads to sub-optimal outcomes for the project. Despite efforts to build teamwork (e.g., through partnering) commercial contracts and cost/schedule controls lead to adversarial relations as managers responsible for interacting activities struggle to advance their interests by optimizing their activity with little concern for the problems this causes others. The business objectives of project-based producers and the client seem inevitably opposed as the project manager tries to complete the project.

Value to the client in this situation is understood as meeting the original design within cost, schedule, and quality limits – change is the enemy. Current project management certainly tries to deliver value to the client present at the beginning of the project. In the CUQ world, delivering value means increasing the ability of the customer at the end of the project to achieve their purposes. Circumstances change quickly when projects are CUQ, so completing a project that does not increase capability within schedule and budget limits set at the beginning is of little use. Change is certainly difficult to manage with current techniques that push for early decisions and local optimization.

The failures of current project management help define the requirements for a new approach. This new approach must rest on the expanded TFV foundation. In practice this means the management system must optimize performance at the project level in a complex and uncertain setting, always pressed for speed.

14.4 Lean project delivery in construction

The phrase 'project delivery system' has traditionally been used to indicate the contractual structure of the project, e.g., design–bid–build or design–build. 'Delivery' in this context is understood to be a type of transaction and a key question is how to structure the transaction. Design–build is seen as a means for providing a client with a single contracting entity with which to interact, as opposed to holding contracts with multiple players and thus inheriting the task and risk of co-ordinating their actions. By contrast, the lean construction community understands 'delivery' in terms of the actual work processes used to move a facility from concept to customer (Ballard and Zabelle, 2000a, b).

In the realm of construction, delivery involves designing and making capital facilities – buildings, bridges, factories, and so on. Construction differs from other types of project-based production systems by the type of products it produces, the differentiating characteristic of which is that they eventually become rooted in place. Construction's products share with airplanes and ships the characteristic that, in the process of assembly, they become too large to move through workstations, so workstations must be moved through the products. Consequently, buildings, airplanes, and ships are made using fixed position manufacturing. Unlike airplanes and ships, however, buildings and bridges are rooted in place and are designed for a specific location, often both technically and aesthetically.

Traditional project delivery systems pursue the 'task' of project delivery and neglect both value maximization and waste minimization. This approach confuses the 'task' view with managing the project. A lean project delivery system is one that is structured, controlled, and improved in pursuit of all three goals, i.e., the transformation/flow/value goals proposed by Koskela (2000). While techniques are important, and such techniques as 'kanban'[5] have become identified with lean production systems, all systems that pursue the TFV goals are, in principle, lean delivery systems, though some will be 'leaner' than others. Since it is impossible to achieve simultaneously the elements of the lean ideal (which is to provide a unique product to each customer, in zero time, with nothing in stores or any other kind of waste), techniques will come and go, but the goals will be pursued perpetually. The lean project delivery system[6], as currently conceived, incorporates many elements from advanced practice in construction today. However, they are integrated into a complete delivery system, rather than occurring in isolation. In addition, many similarities between lean and traditional practice prove, on examination, to be superficial. For example, design–build modes of structuring contractual relations might seem to share with the lean system characteristics such as cross-functional teams and integrated design of product and process. Design–build as such has nothing, however, to do with how things are designed and built, only with how a client procures its capital facilities. Design–build modes of delivery only pursue the transformation goal of production systems, and do not as such pursue the value or flow goals.

14.4.1 Lean Project Delivery System (LPDS[1]) Model

Projects have long been understood in terms of phases, e.g., pre-design, design, procurement, and installation. One of the key differences between traditional and lean project delivery concerns the relationship between phases and the participants in each phase. The model in Figure 14.1 represents those phases in overlapping triangles, the first of which is Project Definition, which has the job of generating and aligning customer and stakeholder values, design concepts, and design criteria. Those three elements are determined recursively. In other words, each may influence the other, so a conversation is necessary among the various stakeholders. Typically, like a good conversation, every person leaves with a different and better understanding than they brought with them. Traditionally, project definition has been done by the architect (or engineer, for non-building projects) working alone with the client. In Lean Project Definition, representatives of every stage in the life cycle of the facility are involved, including members of the production team that is to design and build it.

Alignment of values, concepts, and criteria allows transition to the Lean Design phase, in which a similar conversation occurs, this time dedicated to developing and aligning product and process design at the level of functional systems. During this phase, the project team stays alert for opportunities to increase value. Consequently, the project may revert to Project Definition. Further, design decisions are systematically deferred to allow more time for developing and exploring alternatives. By contrast, traditional design

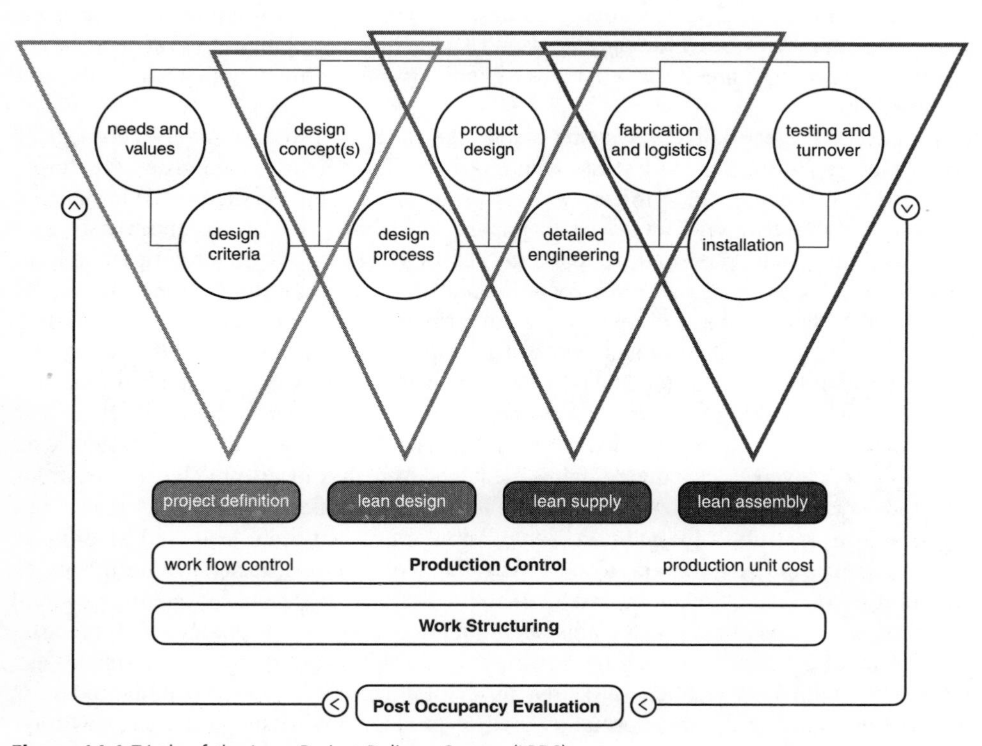

Figure 14.1 Triads of the Lean Project Delivery System (LPDS).

management is characterized by demands for a freeze of design and by a tendency to narrow a set of alternatives to a single selection very quickly. Although done in the name of speed (and often abetted by limited design fees), this causes rework and turmoil, as a design decision made by one specialist conflicts with the design criteria of another. The 'set-based' strategy employed in Lean Design allows interdependent specialists to move forward within the limits of the set of alternatives currently under consideration. Obviously, time is rarely unlimited on construction projects, so selection from alternatives must eventually be made. The practice in lean design is to select those alternatives at the *last responsible moment*, which is a function of the lead time required for realizing each alternative. Reducing those lead times by restructuring and streamlining supply chains allows later selection so that more time can be invested in designing and value generation.

The transition to detailed engineering occurs once the product and process design for a specific system has been completed and released for detailing, fabrication, and delivery. At least the latter two functions occur repetitively over the life of a project, hence the model shows Fabrication and Logistics as the hinge between Supply and Assembly.

Assembly completes when the client has beneficial use of the facility, which typically occurs after commissioning and start-up. The management of production throughout the project is indicated by the horizontal bars labelled Production Control and Work Structuring, and the systematic use of feedback loops between supplier and customer processes is symbolized by the inclusion of post-occupancy evaluations.

14.4.2 How is the LPDS structured, controlled and improved for achieving the TFV goals?

Management of a production system consists of structuring the system to achieve its goals, controlling the system for goal achievement during execution, and improving both structure and control during execution and between projects (Koskela, 2000). Projects are *structured* to pursue the TFV goals by the application of many principles and techniques. Ballard *et al.* (2001) present a more fully developed hierarchy of ends and means. Techniques include:

- involving downstream players in upstream decisions
- deferring commitments to the last responsible moment
- aligning the interests of participants, e.g., so that it is always in the interest of the producer to maximize value for the customer
- selecting, sizing, and locating buffers to absorb variability and match the value of time versus cost for each customer.

The essence of traditional project control is in monitoring actual performance, comparing it to planned or intended performance, and identifying negative variances on which management should act. In other words, it is like trying to steer a car by looking in the rear view mirror. Lean production *control* is achieved through a systematic process for making assignments ready to be performed, combined with explicit commitment by people at the production level to what work will be released to their 'customer' processes in the next plan period, which is typically 1 week, and ongoing identification and action on root causes for plan failures.

Improvement is accomplished between projects primarily through post-occupancy evaluations, which examine both product and process. To what extent was design and construction based on a correct determination of customer and stakeholder values? To what extent was the facility designed and delivered so as to allow realization of customer and stakeholder purposes?

Within projects, improvement is closely linked to control. For example, the Last Planner system of production control[7] tracks plan reliability through its Percent Plan Complete measurement and also identifies reasons for plan failure so they can be acted upon (Ballard and Howell, 1998).

14.4.3 Linking the LPDS Upstream and Downstream

The starting point for project delivery varies widely. Clients with on-going capital facilities programs typically perform a business analysis and feasibility assessment prior to engaging a delivery team. In other cases, analysis and feasibility may occur only after engaging the team. Generally, it has been found to be preferable for the delivery team to be involved earlier in business analysis and feasibility assessment. When a team is engaged after those functions have been performed, the first task should be to review previous planning, so they can at least thoroughly understand the business case of the client, and may be able to make valuable contributions regarding alternative possibilities, previously unconsidered options, the cost or time of options, and so on.

The Lean Project Delivery System produces a facility for a customer to use. Customer use can be represented by a fifth triad, containing Commissioning, Operations and

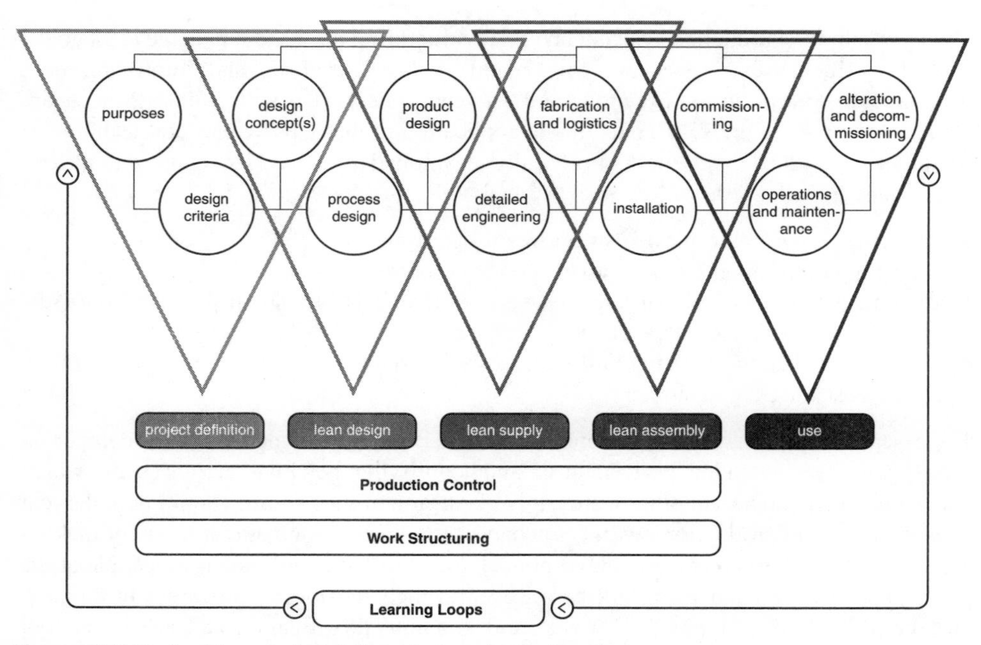

Figure 14.2 Triads of Lean Project Delivery System plus Facility Use.

Maintenance, Alteration and Decommissioning, all of which are anticipated in the previous phases (Figure 14.2).

The hand-off from the delivery team to the operations and maintenance team is typically done during commissioning and start-up of the facility. However, strictly speaking, hand-off occurs once the facility is operating to targeted performance. Consequently, that ramp up time should be included in measurements of project duration, and also be included in efforts to reduce project duration.

14.4.4 Summary of the LPDS

The comparison of LPDS to more traditional systems shows that this new approach is a radical departure from current practice (Table 14.2).

Table 14.2 Comparison of traditional and lean project delivery systems

Lean	Traditional
Focus is on the production system	Focus is on transactions and contracts
TFV goal	T goal
Downstream players are involved in upstream decisions	Decisions are made sequentially by specialists and 'thrown over the wall'
Product and process are designed together	Product design is completed, then process design begins
All product life cycle stages are considered in design	Not all product life cycle stages are considered in design
Activities are performed at the last responsible moment	Activities are performed as soon as possible
Systematic efforts are made to reduce supply chain lead times	Separate organizations link together through the market, and take what the market offers
Learning is incorporated into project, firm, and supply chain management	Learning occurs sporadically
Stakeholder interests are aligned	Stakeholder interests are not aligned
Buffers are sized and located to perform their function of absorbing system variability	Participants build up large inventories to protect their own interests

14.5 Implementation

Implementing this approach in existing organizations or with people schooled in current practice is hardly automatic. Lean-based project management requires changes in individual behaviour and larger organizational development efforts to overcome the ways current practice contradicts the new. Implementing Lean Construction requires the progressive application of a new way to design project-based production systems. The change required is both conceptual and practical. Changing long-held ways of thinking and acting is hard but rewarding work. Changing procedures, techniques and corporate systems is the easy part; changing minds is the real challenge. The lean literature, books like 'Lean Thinking' (Womack and Jones, 1996) are full of stories about companies and people making the transition. Urgency, leadership, focus, structure, discipline, and trajectory themes are apparent in both these stories and in construction and some patterns can now be perceived.

14.5.1 Urgency

Companies that are more likely to implement lean processes will view themselves as being in the business of execution of production (or services). This is in contrast to companies that see themselves as being in the business of brokering projects by contracting work from others, while minimizing their own involvement in the actual execution of the work. Among these companies, declining performance creates the urgency for action. Fear seems a better stimulus than greed for driving change. In any case, the leadership of a company going lean must first explain why this change is needed so that people understand the context for the effort.

14.5.2 Leadership

While a steady purpose must be communicated, transformational leadership requires getting change started and sustaining it. Perhaps the best metaphor is teaching a child to ride a bicycle; lectures and explanations help but the only place to learn is in the saddle, so demonstrating the new behaviour and causing action, getting people involved in doing different things, is vital. Continuing with the bike metaphor, leaders must expect some falls and scraped knees. Leaders often miss the fact that others throughout the organization have been trying for some time to ride the bike and have been criticized for both their effort and mistakes by those holding fast to current practice. Leadership becomes a matter of putting new people in the saddle and finding and encouraging those already trying the ideas on their own. Catching people doing it right, praising them and rewarding them, builds credibility and confidence that this is more than another passing fad.

14.5.3 Focus

Making quality assignments by applying the Last Planner System of Production Control is the place to start. This system brings real change at all levels, produces measurable results, and once in place leads to wider change in the way projects are designed, supplied and controlled. Pilot projects can be established to both prove the ideas in practice and to make apparent the differences between the new and current practices. These projects should not be seen as experiments or tests because people will sense that the commitment to lean is not yet firm. One caution: companies often incorporate some ideas and practices from lean into their current planning and management approach – the result is better performance and a reduced sense of urgency. The company then claims they are 'doing it' even though no real change has occurred.

While there are many differences between the lean approach and current practice, an important implementation milestone occurs when the project organization shifts from only measuring the performance of each activity (the task view) to actively improving the predictable release of work from one specialist to the next (the flow view). Since planning at the assignment level is what finally causes work to be done, the ability of the planning system to predict, indeed cause, a certain time when specific tasks will be completed can be measured[8]. This milestone matters because it indicates that the organization, by the controls it employs, is shifting from trying to optimize the performance of each activity to optimizing at the project level.

14.5.4 Structure

Kotter (1996) speaks of developing and expanding the 'guiding coalition' as a key to transformation. In construction companies we see more or less formal steering committees made of executives, key training and coaching staff, and leaders from throughout the company. These groups plan and carry out the implementation activities, develop materials and collect and tell success stories. This forum is also where contradictions between lean and current practice are identified and resolved.

14.5.5 Discipline

Lean is not a programmatic patch or a one-time problem to solve. It is a different way to think and act that must be learned through disciplined practice. Keep at it and keep the effort to perform better against the lean ideal visible. Some companies attempt to manage the transformation on their own or with only modest help. Training is required but it alone is insufficient to assure success. Significant coaching is also required; this means having people work on projects with the management team to assure the system is installed and running. Project managers and superintendents are not the kind of people who ask for help so the coaches need to be proactive and engaged.

14.5.6 Trajectory

Most companies start with pilot implementation of the Last Planner System. This system is designed to assure the reliable release of work from one station to the next. It is not uncommon for those leading this effort to come to the startling realization of the power of this idea, as in 'This reliability stuff is really important'. (This is an interesting moment, much like when a child realizes the tremendous freedom, speed, and range made possible by learning to pedal a bike.) In construction, this realization usually means that the practitioner understands that new levels of performance are really possible and that changes in design, supply, assembly and control will lead to even better results.

Two models of organizational change are now apparent. The first is the more classic, larger organizational change model that includes developing vision and values, aligning interests, re-examining practices, and taking first steps. These efforts involve multiple activities on many fronts. They stress immediate action, getting people on the bike, in parallel with other efforts. Another model for change is emerging and, while it is relatively new, it appears to offer great promise. This approach implements the Last Planner System in conjunction with focus and training on making and keeping reliable promises (Winograd and Flores, 1986). These skills provide an immediate link between the design of the planning system and the human and organizational issues required for its implementation. Just as the focus on reliable work flow creates a line for continuous action linked to improved system performance against the lean ideal, the pressure for making and keeping reliable promises progressively reveals contradictory organizational policies and practices. This is not to suggest that a company cannot successfully implement lean construction without installing the Last Planner System and Reliable Promises, but it does suggest that a more direct route to implementation may be more effective.

14.6 Conclusion

Lean construction is still, to a considerable extent, 'work-in-progress'. However, its development to date supports two major claims: first, lean construction is based on a better theory than conventional construction; second, lean construction is more effective than conventional construction. Thus, lean construction is not just another specific approach to construction, but rather a challenger of the conventional understanding and practice of construction. In consequence, it is in the interest of every player in the construction sector to assess this new thinking and practice.

The future development of lean construction will have two directions: breadth and depth. On one hand, the seminal ideas of lean construction were related to the management of site operations. After that, new methods were developed for supply chain management, design management, cost management, and for total project delivery. This process of increasing breadth will eventually lead to the situation where all issues of construction project delivery have a methodical solution based on the new theoretical framework.

On the other hand, this new theoretical framework is – and should be – constantly moving, leading to increasing depth. Up until now, the main focus of theoretical development has been on the theory of production and its application to the specific characteristics of construction. Next, the theory of management and the theory of communication need to be clarified and integrated into the existing body of theoretical knowledge.

Among managerial sciences, the quest for a theory is not a phenomenon restricted to construction management. Rather, a similar movement is emerging in the wider fields of operations research and management science (Saaty, 1998). The characterization of a shift of focus, from individual problems to a theory of the system where the problems are embedded, presented in this wider context is perfectly adequate also regarding construction management (Saaty, 1998):

> After more than a half century of tinkering with and solving problems, we need to characterize the system underlying our activity, classify, and generalize its problems.

Endnotes

1 Lean Project Delivery System (LPDS) and Last Planner are both Trademarks.
2 This conception is common especially in the UK. The attacks by Green (1999) on lean construction seem to address this 'tool cocktail' conception.
3 Flowdown of requirements refers to the stagewise decomposition and conversion of high level requirements to requirements for part design, fabrication and assembly.
4 In the thermostat model (Hofstede, 1978), there is a standard of performance, and performance is measured at the output of the controlled process. The possible variance between the standard and the measured value is used for correcting the process so that the standard can be reached.
5 The Japanese word 'kanban' means card or sign board. In the Toyota production system, cards are often used for controlling the flow of materials through the factory. The basic concept is that a supplier or the warehouse only delivers components to the production line as and when they are needed eliminating the need for storage in the production area. Supply points along the production line only forward desired components when they receive a card and an empty container, indicating that more parts are needed in the production line (Hopp and Spearman, 1996; Olson, 2001).
6 See Ballard (2000).
7 The Last Planner system is described in detail in Chapter 15.
8 The test question to determine if the organization is serious about managing work flow is: 'Are you measuring the performance of your planning system with PPC (Percent Plan Complete) and acting on reasons?' Even here we occasionally find companies who use the terms but modify the measurement criteria to measure the amount of work completed rather than the ability of the planning system to assure release of work from one crew to the next. The focus should remain on doing it right. Both these methods are presented in more detail in Chapter 15.

References and bibliography

Ballard, G. (2000). Lean Project Delivery System. Lean Construction Institute White Paper No. 7, September. www.leanconstruction.org.

Ballard, G. and Howell, G. (1998) Shielding production: essential step in production control. *Journal of Construction Engineering and Management*, ASCE, **124** (1), 11–17.

Ballard, G., Koskela, L., Howell, G. and Zabelle, T. (2001) Production system design in construction. In: *Proceedings 9th Annual Conference of the International Group for Lean Construction*, 6–8 August, National University of Singapore.

Ballard, G. and Zabelle, T. (2000a) Project Definition. *Lean Construction Institute White Paper No. 9*, October. www.leanconstruction.org

Ballard, G. and Zabelle, T. (2000b) Lean Design: Process, Tools & Techniques. *Lean Construction Institute White Paper No. 10*, October. www.leanconstruction.org

Construction Task Force (1998) *Rethinking Construction (The Egan Report)* (London: Department of the Environment, Transport and the Regions).

Cook, H. (1997) *Product Management – Value, Quality, Cost, Price, Profit and Organization* (London: Chapman & Hall).

Gilbreth, F.B. and Gilbreth, L.M. (1922) Process charts and their place in management. *Mechanical Engineering*, January, 38–41.

Green, S.D. (1999) The dark side of lean construction: exploitation and ideology. In: *Proceedings 7th Annual Conference of the International Group for Lean Construction* (IGLC-7), Berkeley, CA, 26–8 July, pp. 21–32.

Hofstede, G. (1978) The poverty of management control philosophy. *Academy of Management Review*, July, pp. 450–61.

Hopp, W. and Spearman, M. (1996) *Factory Physics: Foundations of Manufacturing Management* (Boston: Irwin/McGraw-Hill).

Howell, G. and Ballard, G. (1996) Can project controls do its job? In: *Proceedings 4th Annual Conference of the International Group for Lean Construction*, Birmingham, England.

Howell, G. and Koskela, L. (2000) Reforming project management: the role of lean construction. In: *Proceedings 8th Annual Conference of the International Group for Lean Construction* (IGLC-8), 17–19 July, Brighton.

Johnston, R.B. and Brennan, M. (1996) Planning or organizing: the implications of theories of activity for management of operations. *Omega, International Journal of Management Science,* **24** (4), 367–84.

Koskela, L. (2000) *An exploration towards a production theory and its application to construction* (Espoo: VTT Publications). www.inf.vtt.fi/pdf/publications/2000/P408.pdf

Koskela, L. and Howell, G. (2001) Reforming project management: the role of planning execution and control. In: *Proceedings 9th Annual Conference of the International Group for Lean Construction* (IGLC-9), 6–8 August, Singapore.

Kotter, J. (1996) *Leading Change* (Harvard: Harvard Business School Press).

Olson, J. (2001) *Kanban – An Integrated JIT System.* www.geocities.com/TimesSquare/1848/japan21.html

Porter, M. (1985) *Competitive Advantage* (New York: The Free Press).

Saaty, T.L. (1998) Reflection and projections on creativity in operations research and management science: a pressing need for a shift in paradigm. *Operations Research*, **46** (1), 9–16.

Shenhar, A.J. and Laufer, A. (1995) Integrating product and project management – a new synergistic approach. *Engineering Management Journal*, **7**(3), 11–15.

Shewhart, W.A. (1931) *Economic Control of Quality of Manufactured Product* (New York: Van Nostrand).

Tommelein, I.D., Riley, D. and Howell, G.A. (1999) Parade game: impact of work flow variability on trade performance. *Journal of Construction Engineering and Management,* ASCE, **125** (5), 304–10.

Winograd, T. and Flores, F. (1986) *Understanding Computers and Cognition: A New Foundation for Design* (Norwood, NJ: Ablex Publishing).

Womack, J.P. and Jones, D.T. (1996) *Lean Thinking: Banish Waste and Create Wealth in Your Corporation* (New York: Simon & Schuster).

Wortmann, J.C. (1992) Factory of the future: towards an integrated theory for one-of-a-kind production. In: Hirsch, B.E. and Thoben, K.-D. (eds) *'One-of-a-kind Production': New Approaches* (Amsterdam: Elsevier Science), 37–74.

15

Lean construction tools and techniques

Glenn Ballard*, Iris Tommelein†, Lauri Koskela‡ and Greg Howell*

15.1 Introduction

Various tools and techniques have been developed to implement the Lean Project Delivery System (LPDS[1]) described in the preceding chapter. No list will be accurate for long, as innovation is very much underway and new tools and techniques emerge all the time.

15.2 Lean production management

Production management is at the heart of lean construction and runs from the very beginning of a project to handover of a facility to the client. Lean production management consists of Work Structuring and Production Control.

15.2.1 Lean work structuring

Lean work structuring is process design integrated with product design and extends in scope from an entire production system down to the operations performed on materials and information within that system. Lean work structuring differs from work breakdown structure (a technique of traditional, non-lean project management) in the functions it performs and the questions it answers, which include (also see Ballard, 1999):

- in what chunks will work be assigned to specialists?[2]
- how will work chunks be sequenced?
- when will different chunks of work be done?

* Lean Construction Institute
† University of California, Berkeley
‡ VTT Building and Transport, Finland

- how will work be released from one production unit to the next?
- will consecutive production units execute work in a continuous flow process or will their work be de-coupled?
- where will de-coupling buffers be needed and how should they be sized?
- how will tolerances be managed?

Lean work structuring produces a range of outputs including:

- project execution strategies
- project organizational structures, including configuration of supply chains
- operations designs (Howell and Ballard, 1999)
- master schedules
- phase schedules.

Collectively, these amount to a design of the 'temporary' production system (Figure 15.1, a partial ends–means hierarchy) and its links with the 'permanent' production systems

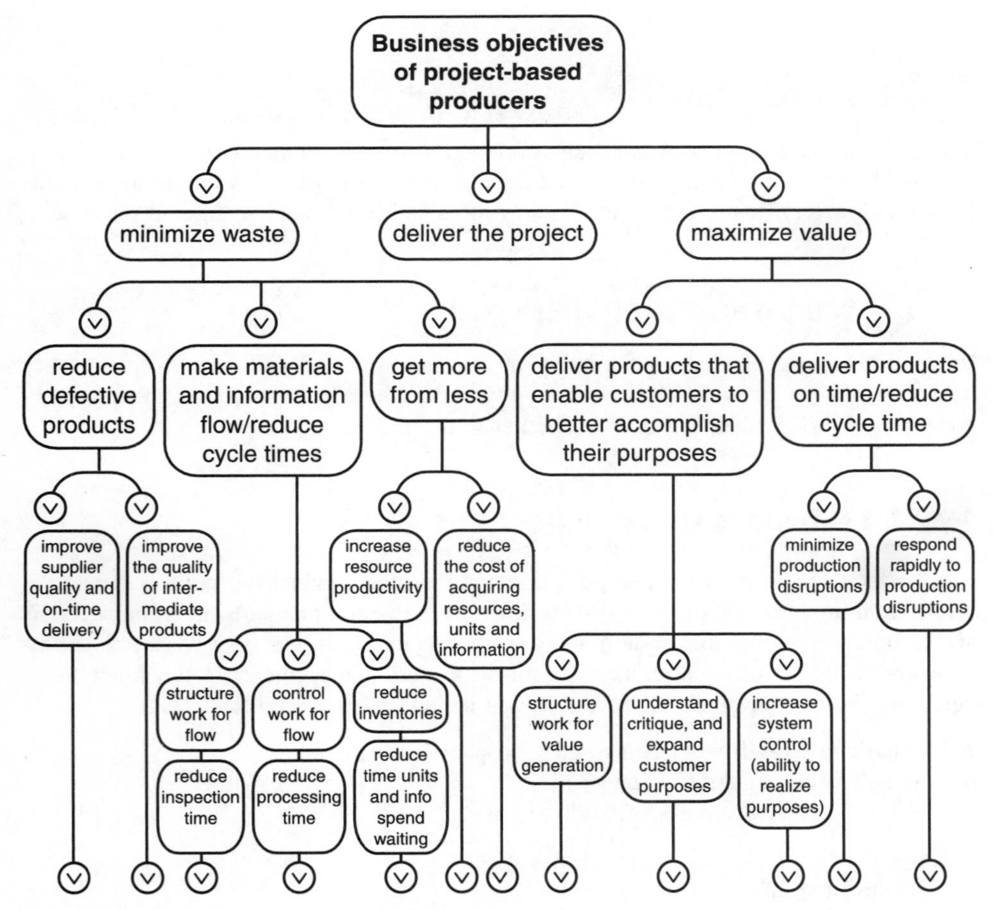

Figure 15.1 Production system design (Ballard et al., 2001).

that exist independently of the project. Production system design is said to be 'lean' when it is done in pursuit of the transformation/flow/value (TFV) goals.

15.2.2 Schedules

Schedules are those outputs of work structuring that link directly with production control. Schedules and budgets specify *should*, while production control translates *should* into *can*, by making scheduled activities ready for assignment, and eliciting a specific commitment to what *will* be done during the next week (or other near-term plan period). The lean construction principles applicable to schedules are:

- limit master schedules to phase milestones, special milestones, and long lead items
- produce phase schedules with the team that will do the work, using a backward pass, making float explicit, and deciding as a group how to use float to buffer uncertain activities.

Contrary to traditional construction management wisdom, detailed schedules produced at the beginning of a project do not assure project control. With few exceptions, the only thing known for certain at that time is that the project will *not* be executed in accordance with that schedule. What is needed, instead, is a hierarchical planning system that progressively develops detail as time for action approaches. Project control is to be achieved by continuously making adjustments in steering as we move through time, rather than by developing a network of detailed orders in advance, then monitoring and enforcing conformance to those (often unrealistic) orders.

The appropriate functions of master schedules are to:

- give us confidence that an end date and milestone dates are feasible
- develop and display execution strategies
- identify and schedule long lead items (defined as anything that cannot be 'pulled' to the project within the lookahead window)
- divide the project into phases, identifying any special milestones of importance to the client or other stakeholders.

Phase schedules, sometimes called 'pull schedules' in the lean construction community, are produced by those who will do the work in that phase, beginning with a backward pass from a target milestone. The best way yet discovered for producing phase schedules is to have a team of different specialists write down their tasks on cards and stick them to a wall, thus creating a logic network, which can be revisited by the team until sufficient float is generated to buffer uncertain activities (Ballard, 2000a). Phase scheduling through this collaborative approach assures the selection of value adding tasks that release other work.

15.2.3 Last Planner[1] system of production control

Planning is followed by control, i.e., by management processes governing execution so that project objectives are best achieved. The Last Planner refers to that individual or group that commits to near-term (often weekly) tasks, usually the front line supervisor,

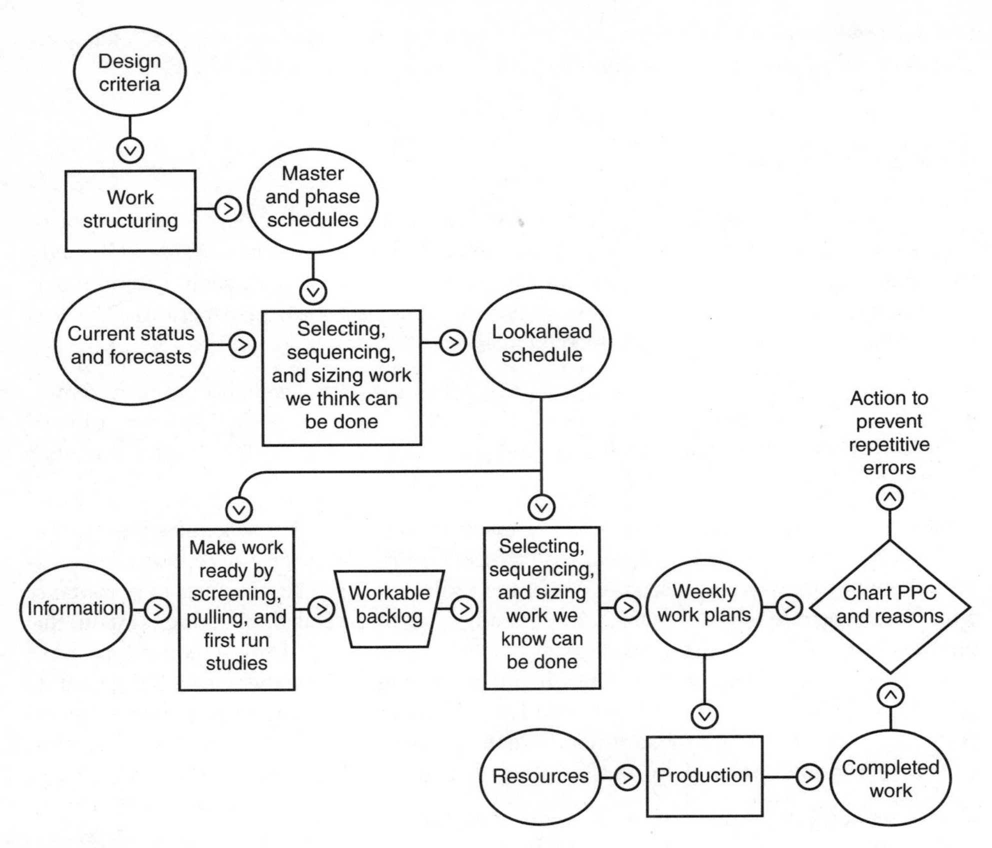

Figure 15.2 Last Planner System of production control.

such as a construction foreman, a shop foreman, or a design team boss. The Last Planner system of production control (Figure 15.2; Ballard and Howell, 1998; Ballard, 2000b) has three components:

- lookahead planning
- commitment planning
- learning[3].

The primary rules or principles for production control are:

- drop activities from the phase schedule into a 6-week (typical) lookahead window, screen for constraints, and advance only if constraints can be removed in time
- try to make only quality assignments – require that defective assignments be rejected
- track the percentage of assignments completed each plan period (PPC or 'per cent plan complete') and act on reasons for plan failure.

Lookahead planning

The functions of lookahead planning are to:

- shape work flow sequence and rate
- match work flow and capacity
- maintain a backlog of ready work (workable backlog)
- develop detailed plans for how work is to be done (operations designs).

Tools and techniques include constraints analysis, the activity definition model, and first run studies (first run studies are described in section 15.5 on Lean assembly).

Constraints analysis is done by examining each activity that is scheduled to start within the next 6 weeks or so. Six weeks is typical, but lookahead windows may be shorter or longer, depending on the rapidity of the project and the lead times for information, materials and services. On one hand, since long lead items are items that cannot be pulled to a project within the lookahead window, extending that window offers the possibility of greater control over work flow. On the other hand, attempting to pull too far in advance can affect one's ability to control work flow on site. Consequently, sizing of the lookahead window is a matter of local conditions and judgment.

The rule governing constraints analysis (Table 15.1) is that no activity is to be allowed to retain its scheduled date unless the planners are confident that constraints can be removed in time. Following this rule assures that problems will be identified earlier and that problems that cannot be resolved in the lookahead scheduling process will not be imposed on the production level of the project, whether that be design, fabrication, or construction.

The activity definition model (ADM, Figure 15.3) provides the primary categories of constraints: directives, prerequisite work, and resources. Directives provide guidance according to which output is to be produced or assessed; examples are assignments, design criteria and specifications. Prerequisite work is the substrate on which work is done or to

Table 15.1 Illustration of Constraints Analysis

Project: Mega Building
Report date: 3 Nov

Activity	Responsible party	Scheduled duration	Directives	Prerequisites	Resources	Comments	Ready?
			Constraints				
Design slab	Structural engineer	15 Nov to 27 Nov	Code 98 Finish? Levelness?	Soils report	10 hours labour, 1 hour plotter		No
Get information from client re floor finish and level	*Structural engineer's gofer*	*3 Nov to 9 Nov*	OK	OK	OK		Yes
Get soils report from Civil	*Structural engineer*	*By 9 Nov*	OK	OK	OK		Yes
Layout for tool install	Mechanical engineer	15 Nov to 27 Nov	OK	Tool Configurations from mfr	OK	May need to coordinate with HVAC	No

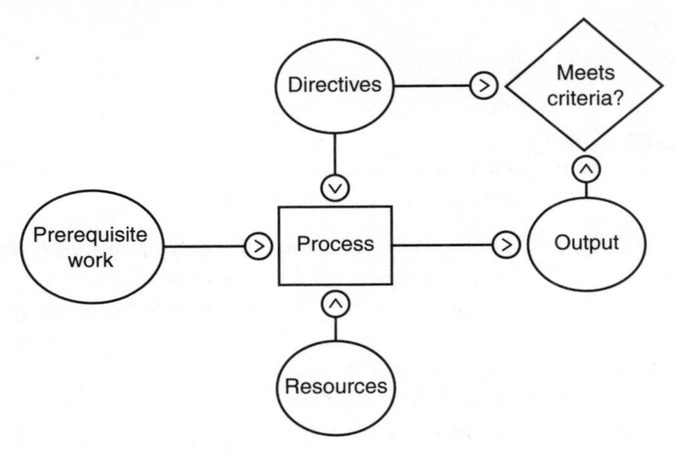

Figure 15.3 Activity Definition Model (ADM).

which work is added. Examples include materials, whether 'raw' or work-in-process, and information that is input to a calculation or decision. Resources are either labour, instruments of labour, or conditions in which labour is exercised. Resources can bear load and have finite capacities, consequently, labour, tools, equipment and space are resources.

ADM is a tool for exploding phase schedule activities into greater detail. Explosion occurs through specification of constraints and through further detailing of processes.

Commitment planning

The Last Planner presents a methodology to define criteria for making quality assignments (Ballard and Howell, 1994). The quality criteria proposed are:

- definition
- soundness
- sequence
- size
- learning (not, strictly speaking, a criterion for assignments, but rather for the design and functioning of the entire system)

The Last Planner considers those quality criteria in advance of committing production units to doing work in order to shield these units from uncertainty. The plan's success at reliably forecasting what work will get accomplished by the end of the week is measured in terms of PPC (Figure 15.4). Root causes for plan failure are then identified and attacked, so that future problems may be avoided.

Increasing PPC leads to increased performance, not only of the production unit that executes the Weekly Work Plan (Table 15.2), but also of production units downstream as they can better plan when work is reliably released to them. Moreover, when a production unit gets better at determining its upcoming resource needs, it can pull those resources from its upstream supply so they will be available when needed. Implementation of the Last Planner therefore results in more reliable flow and higher throughput of the production system.

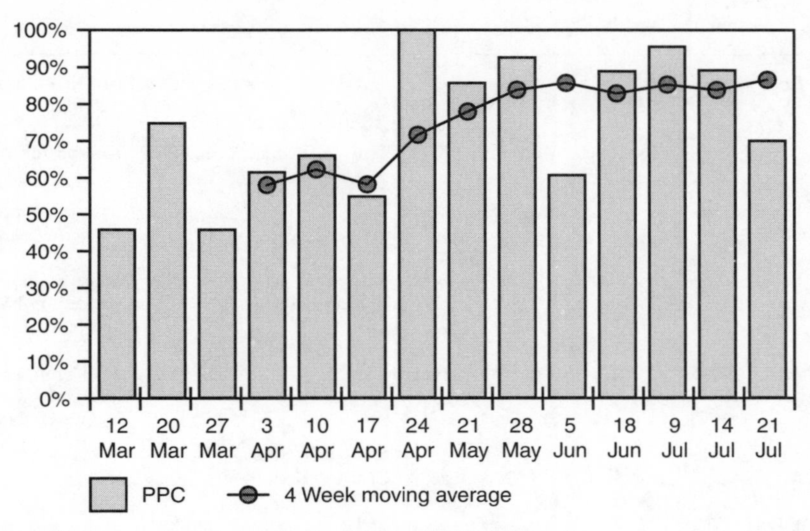

Figure 15.4 PPC chart – electrical contractor (1995 project in Venezuela).

Learning (reasons analysis and action)

Each week, last week's weekly work plan is reviewed to determine what assignments (commitments) were completed. If a commitment has not been kept, then a reason is provided (Figure 15.5). Reasons are periodically analysed to root causes and action taken to prevent repetition. Obviously, failure to remove constraints can result in lack of materials or prerequisite work or clear directives. Such causes of failure direct us back to the lookahead process to seek improvements in our planning system. Some failures may result from the last planner not understanding the language and procedures of making commitments or from poor judgement in assessment of capacity or risk. In these cases, the individual planner is the focus of improvement. Plan failures may also result from more fundamental problems – such as those to do with management philosophy, policy, or conflicting signals.

Whatever the cause, continued monitoring of reasons for plan failure will measure the effectiveness of remedial actions. If action has been taken to eradicate the root causes of materials-related failures, yet materials continue to be identified as the reason for failing to complete assignments on weekly work plans, then different action is required.

15.2.4 Benefits of lean production management

Lean production management is dedicated to reducing and managing variability and uncertainty in the execution of project plans. A starting point is the recognition that much uncertainty on construction projects is a consequence of the way projects are managed, rather than stemming from uncontrollable external sources. Making assignments ready by removing constraints within the lookahead process eliminates potential variability. Shielding production from work flow variability is often the place to start in implementation of the entire LPDS. Work structuring addresses the problem of managing

Table 15.2 Construction Weekly Work Plan

Project: Pilot **ACTIVITY**					**1 Week plan**					**FOREMAN: Phillip** **DATE: 20/9/96**	
	Est	**Act**	**Mon**	**Tue**	**Wed**	**Thu**	**Fri**	**Sat**	**Sun**	**PPC**	**Reason for variances**
Gas/F.O. hangers O/H 'K' (48 hangers)			xxx Sylvano, Mario, Terry	xxx						No	Owner stopped work (changing elevations)
Gas/F.O. hangers O/H 'K' (3 risers)					xxx Sylvano, Mario, Terry	xxx	xxx	xxx		No	Same as above-worked on backlog and boiler breakdown
36" cond water 'K' 42' 2–45 deg 1–90 deg			xxx Charlie, Rick, Ben	xxx	xxx					Yes	Matl. from shop rcd. late Thurs. Grooved couplings shipped late
Chiller risers (2 chillers per week)						xxx Charlie, Rick, Ben	xxx	xxx		No	
Hang H/W O/H 'J' (240'-14")			xxx Mark M, Mike	xxx	xxx	xxx	xxx	xxx		Yes	
Cooling tower 10" tie-ins (steel) (2 towers per day)			xxx Steve, Chris, Mark W.	xxx	xxx	xxx	xxx	xxx		Yes	
Weld out CHW pump headers 'J' mezz. (18)			xxx Luke	xxx	xxx	xxx	xxx	xxx		Yes	
Weld out cooling towers (12 towers)			xxx Jeff	xxx	xxx	xxx	xxx	xxx		No	Eye injury. Lost 2 days welding time
F.R.P. tie-in to E.T. (9 towers) 50%			xxx Pat, Jacky, Tom	xxx	xxx	xxx	xxx	xxx		Yes	
WORKABLE BACKLOG Boiler blowdown–gas vents–rupture disks											

remaining variability through thoughtful location and sizing of inventory and capacity buffers. Production management in its entirety assures as far as possible that each 'workstation' (specialist production unit) does work in the right sequence and rate for reliable release to its 'customer'.

Implementation of lean production management has been shown to substantially improve productivity. Figure 15.6 shows the change in PPC on a Chilean project after implementation of the Last Planner system. Figure 15.7 shows that project productivity improved by 86% in consequence of this improvement in work flow reliability.

This impact of PPC on productivity can be explained by reference to Figure 15.8, which is a somewhat peculiar presentation of the time/cost trade-off. From queuing theory, we know that the wait time of a work item accelerates as a processor approaches 100% utilization, assuming some variability in the system; the greater the variability, the steeper the acceleration. In traditional project management, it is assumed that variability is independent of management action and consequently that the trade-off between time and

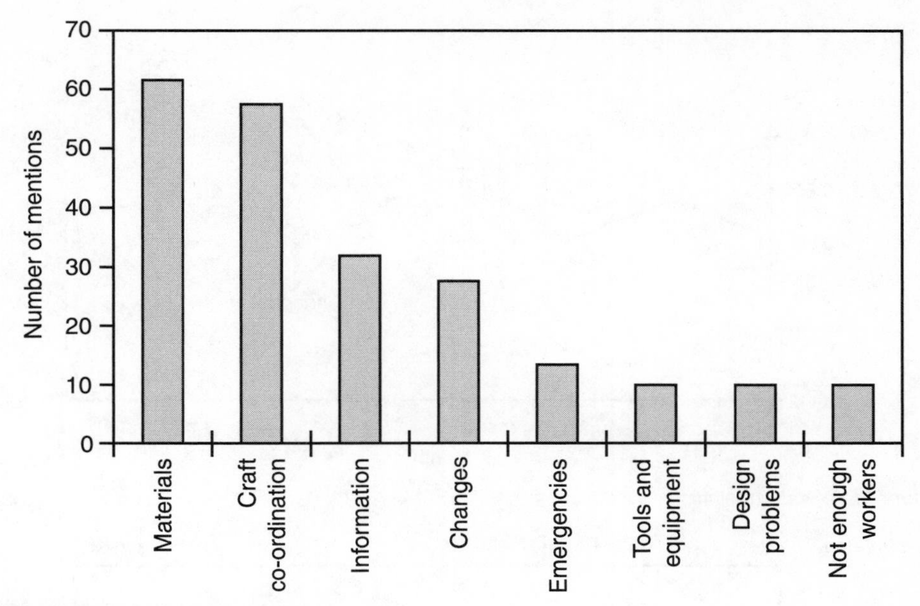

Figure 15.5 Reasons for plan failure.

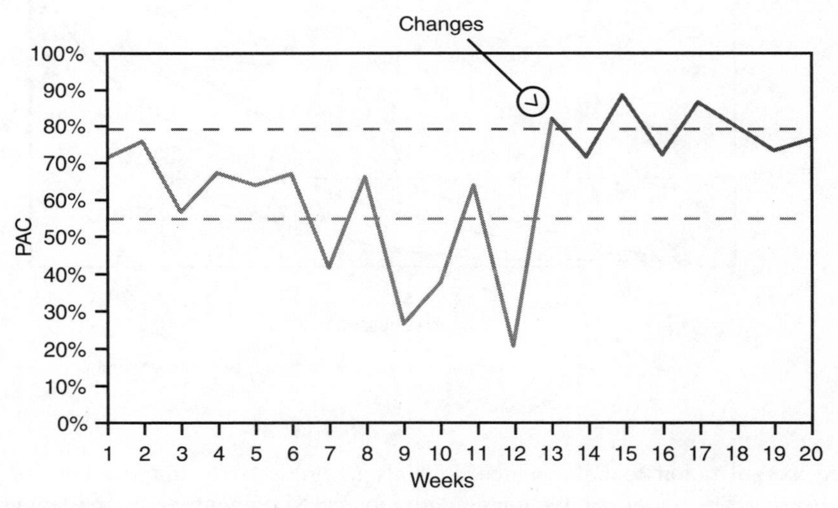

Figure 15.6 PPC improvement with Last Planner implementation (project in Chile, 1998).

cost is fixed, the only discretion for decision making is in finding the exact point where the trade-off can best be made.

The facts are quite otherwise. As demonstrated by applications of lean production management, variability in a production system (e.g., a project) can be reduced through management intervention, thereby changing the trade-off to be made between time and cost. In terms of Figure 15.8, a PPC of 50% might correspond to a utilization rate of 50%. To maintain the same pace of completion, increasing PPC to 70% might allow an increase

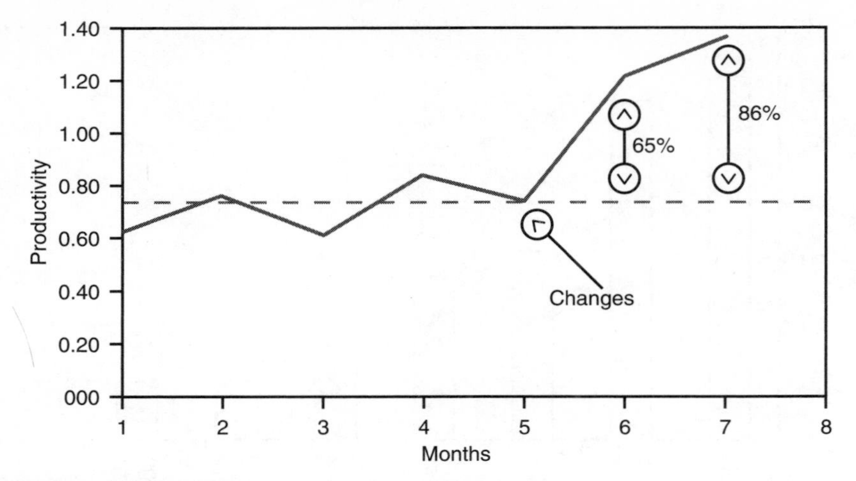

Figure 15.7 Productivity improvement with rising PPC.

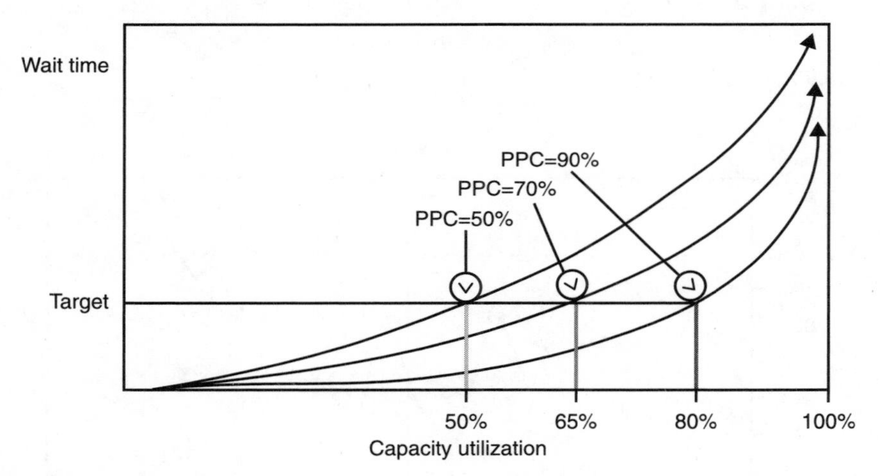

Figure 15.8 Improving the trade-off between time and cost by reducing work flow variability.

in capacity utilization to 65%, which amounts to productivity improvement of 30%. Alternatively, the pace of completion could be increased without increasing labour cost. Numerous instances of 30% or more improvement in productivity have been recorded in tandem with increases in PPC from roughly 50% to 70%.

15.2.5 Summary

Lean production management begins with work structuring and ends with production control, but these are related like the sides of a spinning coin. The coin keeps spinning throughout a project, so that production control takes on the job of executing schedules produced by work structuring, but work structuring has the job of buffering work flow against remaining variability, and continually adjusting plans as the future unfolds.

Structuring production systems, whether large or small, for the TFV goals increases value generation and reduces waste. Production control reduces variability in work flow and significantly elevates the level at which time/cost trade-offs are made. With less variability and uncertainty, inventory and capacity buffers can be reduced, further improving both time and cost performance. Even more important, project participants at every level become actively involved in the management of the project, achieving a transition from a centrally controlled, order-giving form of organization to a distributed control form of organization.

15.3 Lean design

Design is understood as encompassing not only product design, the traditional sphere of architects and engineers, but also process design. Product design determines what is to be produced and used, while process design determines how to produce it or use it. Process design includes structuring the project organization, deciding how to perform specific design and construction operations, deciding how to operate or maintain a facility, and deciding how to decommission and 'un-assemble' a facility. As previously discussed, design of both product and process are understood to be undertaken in pursuit of the TFV goals.

Designing can be likened to a good conversation, from which everyone leaves with a different and better understanding than anyone brought with them. How to promote that conversation (iteration), how to differentiate between positive (value generating) and negative (wasteful) iteration, and how to minimize negative iteration are all critical design management skills.

Some claim that purposes and design criteria can be defined prior to the design process, but iteration – a conversation – is necessary between purpose, concept and criteria. This necessitates exploration of alternative futures, which can often best be represented by sketches and models. Consequently, the project definition (Ballard and Zabelle, 2000a) triad of the LPDS (Figure 14.2) includes design concepts. The challenges of lean design (Ballard and Zabelle, 2000b) include:

- controlling project objectives of time and cost, and the goal of waste minimization without decreasing value
- generating, evaluating, and identifying alignment between purposes, concepts and criteria; the requirement for transitioning a project from project definition into design proper
- capturing and making accessible the design rationale of a facility, i.e., the decisions that were made, the alternatives that were considered, the criteria by which alternatives were evaluated and so on, throughout its life
- minimizing value loss as a project moves through its phases.

15.3.1 Practical methods

Figure 15.9 presents an overview of tools and techniques for lean design.

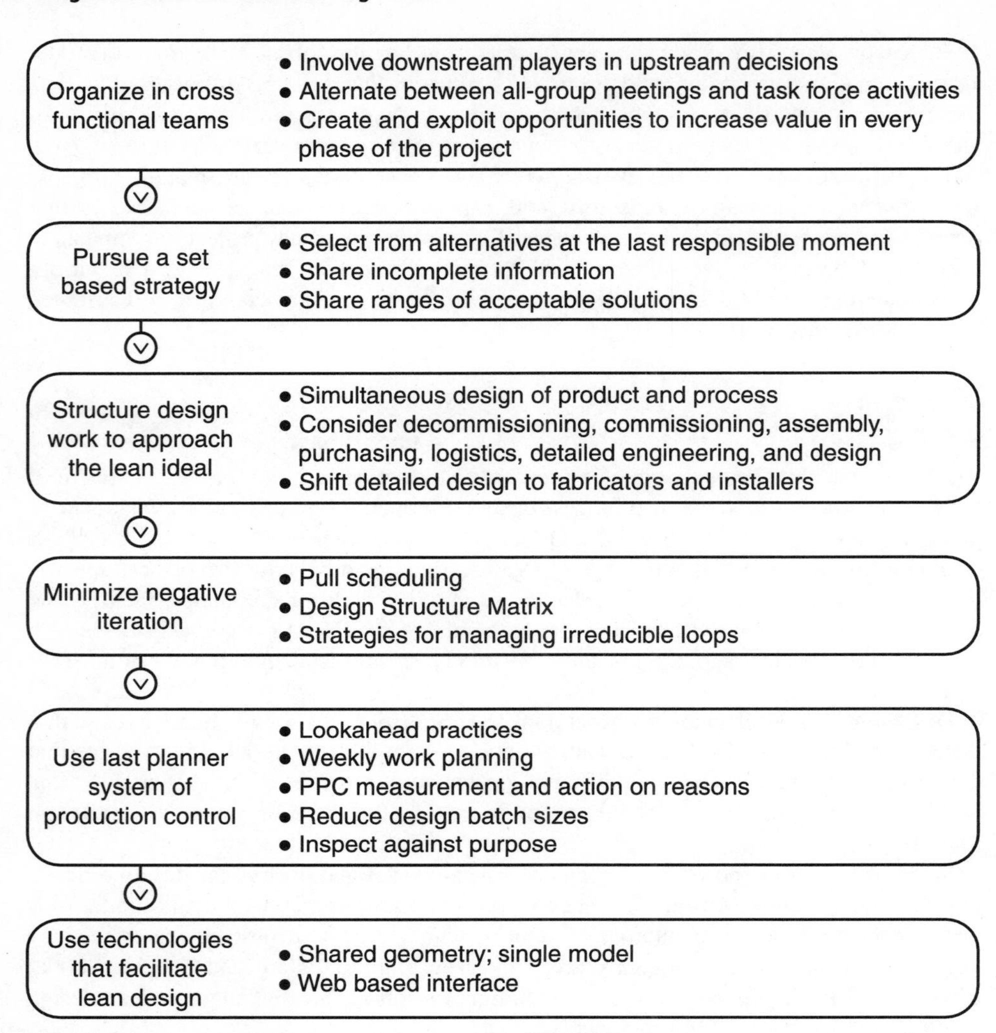

Figure 15.9 Overview of lean design (Ballard and Zabelle, 2000b).

Organize in cross-functional teams

Cross functional teams are the organizational unit for all phases of the LPDS. All stakeholders need to understand and participate in key decisions. It is not possible, however, for everyone to meet continuously and simultaneously; some division of labour is required. In general, the appropriate pattern to follow is alternation between bigger alignment meetings and work by individuals or by smaller teams on the tasks identified and agreed in those meetings.

Facility systems and sub-systems offer natural groupings for the formation of cross-functional teams. For example, a Foundation team might consist of the structural engineer, foundation contractor, key suppliers, etc. In addition, representatives from Superstructure, Mechanical/Electrical/Plumbing/Fire Protection, Interior Finishes, and other teams would participate.

In the design phase, the natural division is between product and process design, but the trick is to counteract the developed tradition of producing them separately and sequentially. Information technology can be helpful by making the state of both more visible, e.g., through shared, integrated models. Nevertheless, having representatives of each relevant specialty assigned to each team will always be essential.

Pursue a set-based design strategy

'Set-based engineering' has been used to describe Toyota's application of a least commitment strategy in its product development projects (Ward *et al.*, 1995; Sobek *et al.*, 1999). That strategy could not be more at odds with current practice, which seeks to narrow alternatives to a single point solution rapidly, but at the risk of enormous rework and wasted effort. It is not far wrong to say that standard design practice currently requires each design discipline to start as soon as possible and co-ordinate only when collisions occur. This has become even more common with increasing time pressure on projects, which would be better handled by sharing incomplete information and working within understood sets of alternatives or values at each level of design decision making, e.g., design concepts, facility systems, facility sub-systems, components, and parts.

Toyota's product development process is structured and managed quite differently even compared to those of other Japanese automobile manufacturers. Toyota's product development develops multiple design alternatives, produces five or more times the number of physical prototypes than their competitors and puts new products on the market faster than their competitors and at less cost.

Toyota's superior performance is probably a result of reducing negative iteration, with that reduction being more than sufficient to offset time 'wasted' on unused alternatives. Negative iteration occurs as a result of each design discipline rushing to a point solution, then handing off that solution to downstream disciplines in a sequential processing mode.

Whether or not one has the time to carry alternatives forward, would seem to be a function of understanding when decisions must be made lest the opportunity to select a given alternative is lost. We need to know how long it takes to actually create or realize an alternative. If the variability of the delivery process is understood, safety-time can be added to that lead-time in order to determine the last responsible moment[4]. Choosing to carry forward multiple alternatives gives more time for analysis and thus can contribute to better design conditions.

Structure design work for value generation and flow

In the LPDS the intent is to structure work in pursuit of the lean ideals, i.e., to deliver what the customer needs, to deliver it instantly and to deliver it without waste. Such a work structure must anticipate every delivery phase, i.e., how the product is to be decommissioned at the end of its life, how it is to be altered to meet changing needs, how it is to be operated and maintained, how it is to be commissioned for operation and use, how the product is to be assembled, how the components are to be procured and fabricated, how the supply chains providing those components are to be configured, and how to structure commercial arrangements so that the relevant stakeholders and experts can be involved in making those decisions. All these 'process' considerations (and more) have to be determined in intimate conjunction with product design.

Integrating design of product and process means considering and deciding *how* to build and use something at the same time as we consider *what* to build. The challenge is to overcome the tradition of first designing the product, then throwing it over the wall to someone else to decide how or if it can be built, operated, altered, etc. Besides old habits, we have to overcome the centrifugal force of specialization and inadequate commercial models. Specialization is unavoidable, but we can do better at educating designers regarding process design criteria and at educating builders regarding product design criteria. As regards commercial models, even in projects executed under design–build forms of contract, it is now common to use design–bid–build with specialty contractors, which makes it impossible to involve them in the design process proper.

Minimize negative iteration

By definition, negative iteration does not add value. If conversation is the image for positive iteration, a bar-room brawl represents negative iteration. Several strategies have been identified for minimizing negative iteration. The first is the use of phase scheduling[5], plus re-organizing the design process as indicated below. Design naturally involves some irreducible loops, e.g., the mutual determination of structural and mechanical loads. Once identified, looped tasks are jointly assigned to the relevant teams of specialists. Those teams must then decide how to manage their interdependent tasks. General strategies that govern all design activities include use of the Last Planner system of production control and practicing set-based design. Specific strategies for managing iterative loops that have thus far been identified include:

- holding team meetings to accelerate iteration
- designing to the upper end of an interval estimate, e.g., design a structure so it will hold the maximum load that might be placed on it
- shifting early design decisions where they can best be made, perhaps outside the looped tasks
- sharing incomplete information.

Functional specialization, sequential processing and fear of liability drive designers and engineers to share work only when it is completed. Concurrent design requires just the opposite: frequent and open sharing of incomplete information so each player can make better judgements about what to do now. Information technology will go some way towards enabling such sharing, but the key obstacles will be old habits of thought and action. Education, frequent reminders, and ultimately successful experiences will be necessary in order to promote open sharing. Of course, it will remain necessary to properly qualify the status of information that is released. If you are considering a design concept, that is very different from releasing a model or sketch which is to be the basis of others' work. Until the new culture is developed, it seems prudent to attack that development within the optimum conditions provided by collocated teams dedicated to working together over multiple projects.

Design production control

In the LPDS, production control is applied in all phases, as soon as any type of plan is created. The Last Planner system of production control (Ballard, 2000b) is used throughout. Here we discuss only two techniques: reducing design batch sizes and a control technique that was developed in response to the demands of design processes, namely, evaluation against purpose.

All too often specialists transmit completed design information with little regard for the needs of other team members and downstream customers (Zabelle and Fischer, 1999). Design decisions and outputs are grouped (batched) in traditional ways that were developed when designing and building were not integrated, and when each discipline tended to practice throw-it-over-the-wall, sequential processing of project information. What is needed is to divide design outputs and communicate them more frequently to release other design work. Typically, this produces smaller batches and speeds up the design process. Since the set-up time[6] to review design information is short for engineers who are up-to-date with the status of a design, the penalty for batch size reductions is very small and does not negatively affect the design processing capacity of a team. Rather the opposite appears to be true: the set-up time and the work-in-process inventory become quite large if the batch size of design information is increased. In addition to dividing design decisions and outputs into smaller batches, it is also necessary for the various specialists to learn how to communicate incomplete information without misleading their co-workers. The mechanical engineer may be very happy to hear that heat loads may change, (s)he can decide to defer other decisions or perhaps arrange redundant capacity if there is no time for waiting. Currently, both tradition and fear of liability constrain the free flow of information among designers and engineers.

Project production can be abbreviated to the steps:

- determine customer (and other stakeholder) purposes and needs
- translate those purposes and needs into design criteria
- apply those criteria to the design of product and process
- purchase, fabricate, deliver and install materials and components in accordance with that design.

Value is maximized when needs are accurately determined and when those needs are maximally satisfied by the product produced and the process employed to produce it. The industry habit, however, is to inspect outputs in that sequence of steps against standards, assuming that those standards themselves have been properly established. For example, fabricated items are inspected against the detailed fabrication drawings and relevant specifications; what is missing is re-evaluation of the drawings and specifications against stakeholder needs, on the chance that conditions or knowledge have changed, or that a better idea has emerged.

Any opportunity to improve understanding of purposes and needs, design criteria, or design should be taken whenever doing so adds value. The operating assumption in current practice is that the process only flows one way, so it is almost impossible to make an improvement downstream once a 'standard' has previously been established.

15.3.2 Technologies facilitating lean design

A key support tool for simultaneous product and process design (and for work structuring in general) will be integrated product and process models, i.e., complex databases capable of representing product design in 3D and also capable of modelling the manufacturing, logistics, assembly, commissioning (start-up), operations, maintenance, alteration, and decommissioning of that product or its components.

Designing within a single model has obvious advantages, e.g., minimizing interferences, visualization and exploration of alternatives, and the creation of a tool for use during post-construction operations, maintenance, and future adaptations. Even better, and certainly more practical, is accessing the data generated on different platforms and using that data to produce models, do trade-off analyses, and so on. Three-dimensional modelling can be useful in project definition, design production, and detailed engineering. In project definition, models can be used like sketches to display alternative concepts. In design proper, models can be used to ensure that the design of systems, sub-systems and components are adhering to interface specifications. In detailed engineering, the product can be built in the computer before being built in physical space and time.

> Once specialists are involved early and organized into cross-functional teams by facility systems, they can use the computer model as a tool to integrate and test components within their systems and to verify the compatibility of various system architectures. The computer model then becomes a value generation tool that supports the real-time consideration of multiple concepts for each system and component. Rather than a more sophisticated form of drafting, the model becomes a tool for simulating the product (and increasingly the process as well), so that better decisions can be made. Ultimately, and perhaps not in the far distant future, we will learn how to design and build the project, process, and team in the course of modelling. (Zabelle and Fischer, 1999)

15.3.3 Benefits of lean design

Adopting a lean approach to design improves both value generation and waste reduction. Value is generated through more methodical and thorough processes for identifying, challenging and clarifying customer and stakeholder purposes, through pursuit of a set-based strategy and the additional time provided for exploration of alternatives and analysis of trade-offs, and through practices such as evaluating outputs against purposes rather than only against immediate requirements. Waste is reduced through the superior product/process design that results from the lean approach. Product designs that are more easily, safely and rapidly built, product designs that are more economically and efficiently operated and maintained, product designs that are more easily altered to changed needs and that cause less environmental damage during realization and on demolition – all these are the result of integrated product and process design, the hallmark of lean design.

15.4 Lean supply

15.4.1 What is lean supply and who is involved in it?

Generally speaking, supply refers to the hand-off between a supplier and a customer. To stress the F in the TFV perspective, supply comprises three flows:

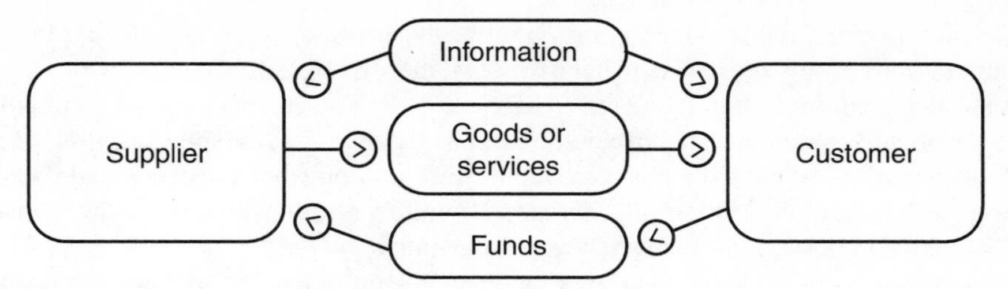

Figure 15.10 Supply flows of information, physical goods or services and funds.

- a two-way exchange of information
- a one-way flow of goods (temporary or permanent materials) or services (people and equipment) from the supplier to the customer
- a one-way flow of funds from the customer to the supplier.

Figure 15.10 illustrates these flows as going directly from a supplier to a customer and back, however, flows may also occur in an indirect fashion. More specifically, supply refers to the processes that result in delivery of goods and services to the site, which is the ultimate location of assembly of construction components. Accordingly, Lean Supply is the third triad in the LPDS (Figure 14.2).

Lean supply is achieved by implementing lean production principles and techniques (such as transparency, pull, load levelling, and just-in-time delivery) to supplier–customer relationships, across organizational boundaries. In the process of implementing Lean Supply, companies may find that impediments to effective implementation exist not only due to organizational boundaries, but also due to the formation of functional silos within their own organization.

Suppliers and customers linked by various flows are also referred to as a 'supply chain'. The lean production techniques so appropriate for streamlining production processes within a single shop, a department, or an organization equally apply to supplier–customer relationships across organizational boundaries. The field of supply-chain management (SCM) is, in fact, an outgrowth of efforts to integrate the functions, goals and objectives of logistics (by nature a flow-oriented task), procurement and production. These developed as independent fields of study after World War II. Many of today's best practices in SCM mirror the goals, objectives and techniques of lean production.

15.4.2 Supply chain challenges

Today's supply chains in construction are long and complex. They include not only owners, designers, engineering specialists, contractors and sub-contractors, but also manufacturers, shipping agents, and other suppliers of goods and services, ranging from commodities to highly specialized made-to-order products. On any specific project, hundreds if not thousands of such companies may constitute the supply chain. A holistic approach towards project delivery as proposed by the LPDS therefore includes the configuration of an extremely complex system.

Many relationships in project supply chains are forged on a one-on-one basis in order to meet the business objectives of the companies involved; they are rarely established with overall supply chain performance in mind. In fact, metrics to gauge overall supply chain performance are sorely lacking in the construction industry. Relationships may or may not pre-date any one project, but irrespective of this, specific agreements are typically invoked opportunistically, based on capabilities and estimated capacity at the time a project's needs arise. Construction supply chains require rapid configuration when those demands become known. Regrettably, forecasting such demand accurately any considerable period ahead of time is nearly impossible in many sectors of our industry.

Construction supply chains also are fraught with inefficiencies. Buyers, like sellers, make unclear and unreliable commitments, unintentionally or intentionally, in terms of specifying exactly when and what goods or services will be needed, and how and when deliveries or hand-offs will take place. They do so in part because it is easier to be vague than to be precise (Vrijhoef *et al.*, 2001), and few management systems are designed to promote reliability (the Last Planner is an exception). Moreover, supply agreements rarely spell out penalties for unreliable performance; if they do, such penalties are seldom enforced or end up being unenforceable. Increasing unreliability leads to deteriorating performance (e.g., Tommelein *et al.*, 1999), but today's contracts and legal practice do not provide incentives for supply chain participants to behave differently. Complex and unreliable systems tend to be wasteful and this is all too true of construction supply chains.

15.4.3 Improving lean supply

Lean supply (Figure 14.2) spans Product Design, Detailed Engineering, and Fabrication and Logistics.

Product design for lean supply

Lean supply in the architecture–engineering–construction industry deliberately includes lean product design, which is the counterpart of product development for manufacturing. Accordingly, lean supply can leverage the tight interface with lean design upstream, in addition to taking advantage of lean assembly downstream.

Organize in cross functional teams. The value of organizing in cross-functional teams was already pointed out in Lean Design, however, it is worth stressing the involvement of non-traditional players in these teams. Apart from the traditional players – architects, engineers and contractors – the construction supply chain comprises materials suppliers and fabricators, as well as procurement, materials management and logistics providers. The latter contribute significantly to supply-chain performance by linking design to execution and should therefore be included early on as integral players in a lean team. (Note that our educational systems separate architecture from engineering and construction students, and that few such students are apprised of functions performed and value provided by this 'missing' link.) Specialty contractors and key component suppliers must sit at the table and engage in conceptual design discussions so that they can be informed early of emerging requirements and reveal their production constraints.

Lean design means design for procurability, design for constructability, design for maintainability, in short, Design for X (DFX) where X stands for an ability downstream in the supply chain. Similarly, lean supply recognizes that the design and design detailing functions must be fulfilled with procurement (and other downstream processes) in mind. Detailing commits to a specific product, and when combined with procurement, it also commits to a specific vendor and the associated production process (e.g., information required by that vendor and that vendor's production lead times and quality) (Sadonio *et al.*, 1998).

Long-term supplier relationships. Relationships between buyers and sellers have traditionally been transactional and established on a one-on-one basis to meet the requirements of a single project. In contrast, lean production advocates longer-term, multi-project, relational agreements. In order to leverage the benefits to all involved, those agreements are best established with a select few suppliers, where the actual number in 'few' must reflect the nature of the product or service that is supplied. Longer-term agreements provide an incentive for the parties involved to streamline their various flow processes (e.g., by setting up standard work processes and unambiguously defining deliverables) and they provide greater opportunity for feedback and learning.

Early identification of suppliers and their involvement in design can prevent the design–bid–*redesign*–build work that is all too common today – the consequence of designers not understanding the possibilities, requirements, criteria and constraints faced by those detailing and then executing a design. Lean suppliers will further avoid waste by, for example, providing quality products and making reliable deliveries, thereby obviating the need for customers to inspect, expedite or reschedule activities. The LPDS treats suppliers as undiminished participants in the project delivery system and structures rewards commensurately.

Detailed engineering for lean supply

The detailed engineering, fabrication and logistics tasks interact closely with product design, since the ease of making and transporting goods, and ultimately the ease of installation, are but some lean design criteria. The benefits of shifting detailed design to fabricators and installers (including specialty contractors) have been stressed already.

Detailed engineering of engineered and made-to-order products combined with the approval process frequently has a lead time that is significantly greater than that of fabrication itself and is therefore a prime target for efforts to reduce cycle time. Batching as well as multi-tasking practices in detailing and fabrication, like those in design, are detrimental to throughput performance. In contrast, standardization of products and processes would improve performance. Note that such standardization does not need to stifle innovation; in the computer industry, for example, new standard interfaces between components are regularly created to keep up with the lightning-fast evolution of technologies.

Fabrication and logistics for lean supply

The ideal one-piece flow of 'lean' translates into small and frequent hand-offs in a continuous-flow system driven by customer pull. Pull means that goods and services are delivered only upon demand in response to a real need, as opposed to being delivered to stock so as to meet a forecast need. Forecast construction needs are typically captured in master schedules, which are critical path method (CPM)-based push schedules. Systems fraught with uncertainty perform significantly better when such schedules are augmented by real-time pull (Tommelein, 1998). Today's construction supply practices face major opportunities in this regard. All too often, uncertain forecasts and long lead times, combined with poorly-chosen financial incentives, lead to the build-up of large on-site inventories of materials not needed immediately for installation. More accurate short-term planning (e.g., the Last Planner) combined with closer-to-real-time communication from customer to supplier (e.g., web-based systems that provide information transparency) and the ability of suppliers to deliver quickly to real demand (short lead times), are all needed to implement pull in lean supply.

Admittedly, lead times in supply will continue to be significant especially when transactions are conducted in a global marketplace. Few manufacturers that serve the construction industry already have become lean. Others are making the transition but do not yet know how (or prefer not) to let supply chain partners take advantage of their 'leaning' and leverage the lessons learned by applying them across organizational boundaries. The move towards lean is especially notable in companies that span multiple echelons in the supply chain, such as specialty contractors that have their own off-site fabrication shops and field-install their fabricated materials. The benefits of lean increase with the scope of application of lean.

Transportation. Transportation of construction goods is largely done on a supplier-by-supplier and material-by-material basis. The construction industry has not taken advantage of opportunities offered by other supply arrangements, such as earlier consolidation of various materials that will be needed later at the same time (examples are kitting of parts and prefabrication of modules). Like traditional designers, manufacturers, fabricators and other suppliers batch their deliverables in order to minimize set-up costs and take advantage of transportation costs. Even though faster and more economical means of transportation are being developed, transportation will remain a source of lead time with uncertainty. The supply chain can be made leaner by procuring from suppliers that are located in closer geographic proximity.

Purchasing vs. releasing for delivery. The distinction between design and execution equally applies to purchasing and release for delivery. For example, Tommelein and Li (1999) describe the process for ordering ready-mix concrete as two-fold; first, placing an order with the ready-mix batch plant one or several weeks prior to site need in order to reserve production capacity and allow time for the plant to order and receive concrete ingredients; second, calling in a few hours or a day in advance of the placement to confirm the time at which the site will be ready to receive the truck, with concrete mixed at most half an hour or so prior to delivery. Purchasing and releasing are often thought of in combination. By recognizing that they are separate (Ballard, 1998), one gains an additional degree of freedom in configuring and managing the supply chain, which serves the implementation of lean supply.

15.4.4 Supply chain design and control

Lean supply chains must be designed with the following value-adding tasks in mind:

- physical movement of products
- change in unit of hand-off

- temporary storage or velocity adjustment to allow for synchronization
- providing timely information.

The focus of lean supply is on the physical movement of products or delivery of services to the customer. While 'transportation' is often regarded as a wasteful activity, particularly when it involves re-handling of materials, at least some part of it is value adding because construction sites are located some distance away from raw material quarries.

Lean supply takes advantage of the opportunity to change the unit of hand-off, when batch sizes of outputs released in upstream processes do not match batch sizes of inputs required in downstream processes. Changes in units of hand-off are needed only because one-piece flow was not possible.

Lean supply purposefully seeks out opportunities for early assembly, modularization, use of standard materials and components, kitting or 'bagging and tagging,' and other kinds of (re)packaging in order to avoid matching problems and the effect of merge bias downstream. Matching problems occur when several items are needed at the same time for final assembly, but one or several are missing (e.g., Tommelein, 1998) so that no work can be completed. Merge bias means that any one matching item being late will delay final assembly. Synchronization of the supply chains that merge at final assembly, and tight control on their duration are therefore key factors in achieving lean supply.

Lean supply also aims to synchronize supplier and customer production, adjusting transportation speeds or providing for intermediate staging (buffering) because supply and demand could not otherwise be balanced. Fabricators aim to level job-shop schedules whereas installers face surging demands to meet project schedules. The two are therefore often in tension with each other (Akel *et al.*, 2001).

Information transparency

Finally, lean supply requires the judicious creation of information transparency to alleviate demand uncertainties in the various supply organizations and to enable the pulling of resources through those parts of the supply chain where it can be achieved.

15.4.5 Roles of SCM in construction

Vrijhoef and Koskela (2000) identified four roles of supply SCM in construction. The alternatives are numbered in Figure 15.11 and the descriptions below paraphrase their text.

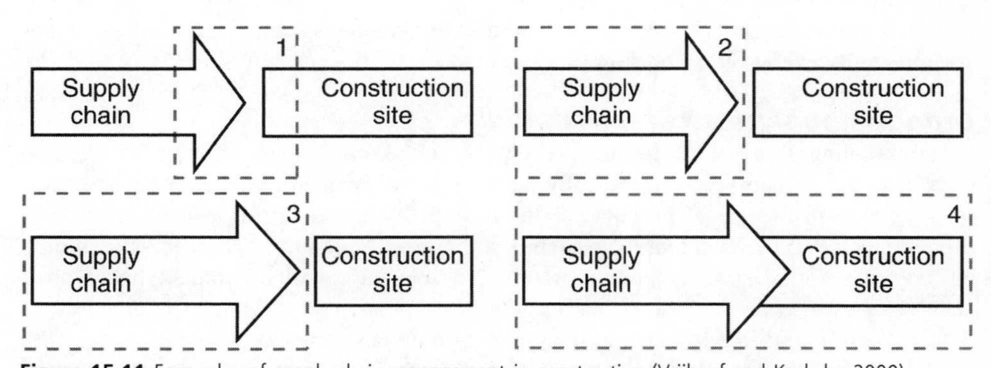

Figure 15.11 Four roles of supply chain management in construction (Vrijhoef and Koskela, 2000).

Each role could be played by one or several supply chain participants. Roles are not mutually exclusive and in fact are often pursued jointly.

- SCM focuses on the impact of the SC on construction site activities and aims to reduce the cost and duration of those activities. The primary concern therefore is to establish a reliable flow of materials and labour to the site. Improvements may be achieved by focusing on the relationship between the site and direct suppliers. The contractor, whose main interest is in site activities, is in the best position to adopt this role.
- SCM focuses on the SC itself and aims to reduce costs, especially those related to logistics, lead time and inventory. The contractor, but also materials suppliers and component manufacturers, may adopt this role.
- SCM focuses on transferring activities from the site to earlier stages in the SC and aims to reduce total installed cost and duration by avoiding inferior conditions on site or achieving wider concurrency between activities, which is not possible due to technical dependencies on site. Designers, suppliers or contractors may adopt this role. Owners play a key role in this regard, by selecting SC participants and providing them with incentives to perform.
- SCM focuses on the integrated management and improvement of the SC and site production, that is, site production is subsumed by SCM. Owners, together with designers, suppliers, and contractors, may play this role. For example, a light-fixture manufacturer adopted this role (Tsao and Tommelein, 2001) for one part of the construction supply chain.

The challenge of lean supply therefore is not on any one supply chain participant, but on all participants.

15.5 Lean assembly

Lean Assembly is the fourth triad in the LPDS (Figure 14.2). It includes Fabrication and Logistics, Installation, and Commissioning. In construction, the final installation of components (modular or not) takes place *in situ*. 'Lean' aims to minimize this effort while also expediting the entire delivery process.

15.5.1 Practical methods

In addition to using the Last Planner to control the planning system, lean assembly takes advantage of several other tools and techniques.

Lean assembly for commissioning

Commissioning is a set of formal procedures for assuring that what is delivered to customers meets their needs. It typically includes some means for assessing the adequacy of design, conformance of products to the design (including testing and integration of subsystems into functional facility systems), and preparation of the customer for assuming custody and control, as in operator training. It may also include some type of post-occupancy evaluation.

Commissioning is the last step in the production process prior to turn-over of a facility to the owner. Turn-over defines the pull of the customer, so it is with turn-over in mind that all tasks in the delivery system must be sequenced and executed. To reflect this pull, commissioning and the other elements of lean assembly are discussed in reverse order of their appearance in the triad. The work of different trades must thus be co-ordinated and completed at set times in order to allow for entire systems to be turned on, calibrated and tested. In traditional scheduling practice other network logic may prevail over this, thereby jeopardizing the turn-over date.

Lean installation

First run studies for operations design (Howell and Ballard, 1999). Construction operations are the ways crews use what they have to do work. Work methods appear simple enough when represented in the estimate, but that design is seldom detailed or explicit at the step or sub-cycle level. Under lean construction, the design of the product and the process occur at the same time so factors affecting operations are considered from the first. Ultimately, operations design reaches all the way through the delivery system, as it is part of work structuring.

In product design, decisions about the selection of materials, their joining and configuration on site, constrain the work method. Designers thus constrain the range of solutions left. That range will narrow further as time advances and downstream planners make additional assumptions; downstream planners are thereby progressively constrained while hopefully being able to find at least one acceptable method.

All details are rarely resolved before an activity begins and most operations continue to evolve once underway. This image contradicts the view that there is one right or standard way to work. It suggests two strategies in the extreme: one is to leave as much flexibility as possible for the last planner, while the opposite is to completely prescribe all details in advance and then assure that the planned circumstance happens.

Designing operations under either strategy has obvious problems. In the first, total flexibility makes planning and co-ordination with others difficult, and projections of cost or completion unreliable. In the second, early prescription ignores late developments. Insofar as construction is a prototyping process, expecting to prescribe all details is unrealistic, nevertheless, establishing procedures early and supporting them by intermediate planning might improve the reliability of work flow. Just as with product design options, process design options progressively disappear as time passes because we have hit the lead time of suppliers (last responsible moments). Certain options may also be eliminated by examination.

The design of an operation may be specified in front-end-planning, but more design work will remain to be done in the engineering phase and within the lookahead process when the work package is released to the crew. First run studies must be a routine part of planning, conducted preferably 3 to 6 weeks prior to the start of a new operation (Howell and Ballard, 1999). They include actually performing the operation in as realistic a manner as possible, in order to try out and learn how to best perform the work involved, identify skills and tools available or needed, interaction of the operation with other processes (e.g., Howell *et al.* 1993), and so on.

The interdependence between product and process design can be explored using computer models of the design (such as discrete-event simulation and 3D computer-assisted design; CAD) so that work can be structured to best meet project objectives. Issues to be considered include:

- design of the product itself
- available technology and equipment
- site layout and logistics
- size of work packages released to the crews
- size of work packages released to downstream crews
- potential site environment (temperature, precipitation, wind, etc.)
- safety
- expected experience and skills of craft workers and supervisors
- craft traditions or union work rules.

First run studies and operations design are not limited to repetitive operations. Indeed, all operations should be subjected to a design study on each project. Typical studies include process, crew balance and flow charts, as well as space schedules that show how resources move through space and work progresses. It is of utmost importance to measure and understand variability in arrival rates of inputs and processing durations. Construction operations usually begin with a significant uncertainty but first run studies will reduce it.

First run studies result in identifying a good (not the best, although this is the lean ideal) way to do work, thereby setting a standard against which all those conducting the work can gauge performance. Standardized work is a hallmark of lean production, but such standards should not be viewed as rigid in any way. They are subject to examination and improvement (e.g., 'kaizen', a Japanese term used in the Toyota production system to refer to the search for continuous improvement) to result in a new and better standard when appropriate. Standards are very important, however, in that they make it easy to delegate responsibility for execution and control to those conducting the work, and they facilitate learning by clearly defining a process that can be mutually agreed upon and critiqued.

Aiming for continuous flow (Ballard and Tommelein, 1999). A continuous flow process (CFP) is a type of production line through which work is advanced from station to station on a first-in–first-out basis (Ballard and Tommelein, 1999). The idea is to balance, approximately, the processing rates of the different stations so that all crews and equipment can perform productive work nearly uninterruptedly while only a modest amount of work-in-process (WIP) builds up in between stations.

The objective of achieving continuous flow is maximizing the throughput of that part of the system while minimizing resource idle time and WIP. Just as pull techniques are limited by the relative size of supplier lead times and windows of reliability, not all work can be structured in CFPs. However, doing so where possible reduces the co-ordination burden on the 'central mind' and provides 'bubbles' of reliable work flow around which other work can be planned.

Examples of work that can be executed as a CFP include excavating footings, placing formwork and rebar, then inspecting prior to placing concrete, and subsequently curing, stripping and finishing; or finishing rooms of a hotel or hospital (painting, carpeting, etc.) one after the other. The key to CFPs is that work gets done in small chunks, each chunk is involved in one production task (or operation) and, once processed, is worked on in subsequent production tasks. In the mean time, the first task gets repeated, and so the process continues.

In order to assess whether or not continuous flow is appropriate, and then to achieve it, a number of steps must be taken. The steps in CFP design are:

- data collection
- definition
- rough balancing
- team agreements
- fine balancing
- change guidelines.

While identifying the characteristics of individual operations, e.g., in terms of work content, method design (though this may change), set-up time, minimum resource unit, minimum process batch size, capacity, space and access needs, and more, one also needs to pay attention to available skill sets and equipment capabilities. One should not be misled, however, by contractual or union boundaries that may constrain the view on operations.

Once descriptive data on individual production tasks and their alternatives is available, different tasks can be put together into a system and the potential for it to be made into a CFP identified. The team must decide which parts will be made into CFPs and which parts will be decoupled by means of buffers. This decision is driven in part by the amount of flexibility that exists in the operation's design and the required resources; technology might also be a driving factor.

In the rough balancing stage, specific site constraints must be considered as they define the pace of the operations as work progresses to complete the project at hand. Balancing a system to achieve continuous flow is done by a combination of techniques, including assigning capacities, mutual adjustment, inventory buffers and capacity buffers.

For self-governance, the specialists operating at different stations in the line must agree on a division of operation, pacing or production rate, the size and quality characteristics of transfer batches, balancing techniques such as multi-skilling or rate adjustment, and strategies for adjusting to differences in load over time and other variability if unforeseen needs arise.

Multi-skilling. Lean production promotes multi-skilling of teams of workers so they will be able to perform more than just a few specialist tasks and assemble a multitude of systems, thereby avoiding process fragmentation otherwise imposed by tradition or trade boundaries. Multi-skilled workers can better support and maintain CFPs by being able to do a broader range of work, which is especially important when work flows are variable.

Fabrication and logistics for lean assembly

Preassembly. The greater number of components that are pre-assembled prior to their final installation, the more straightforward the final assembly process becomes, provided, of course, that assemblies can be managed logistically.

Standardized and interchangeable parts. The repeated use of standardized parts greatly eases assembly; not only will crews be familiar with the parts, they will also be able to learn from their repeated use. In addition, the use of a limited number of parts keeps matching problems at a minimum.

Just-in-time deliveries. Lean assembly must, of course, be closely co-ordinated with lean supply. Ideally, materials will be received just-in-time (JIT) and strategically located on site. JIT does not mean that everything is delivered at the last minute – a better translation from the corresponding Japanese concept is 'at the appropriate time'. Materials must be buffered as needed to match work flow requirements both upstream and downstream in the process. In a CFP, WIP will thus be minimal.

One-touch handling. One-touch handling is a lean ideal that provides a good metric for otherwise numerous re-handling steps from receipt on site to laydown, and from issuing to staging of materials prior to their final installation. Some materials can be directly installed, whereas others are parts or components for sub-assemblies yet to be produced. Of those items that are ready for installation, some can be directly installed from the delivery vehicle. Three rules-of-thumb for one-touch material handling are:

- off-load directly from delivery vehicle into final position when possible (e.g., pipe spools, most equipment)
- if direct off-loading is not possible off-load within 'crane reach' of final position (e.g., structural steel requiring pre-assembly at the site)
- deliver consumables (e.g., grinding disks, gloves) and commodity materials (e.g., fittings, gaskets, bolts) directly into the hands of users, as opposed to warehousing and issuing them based on requisitions. Minimum–maximum inventory rules can be followed to conform to this rule while matching work place characteristics.

Distributed planning. Finally, recognizing that there are numerous Last Planners on any one project – planning is inherently a distributed task – lean assembly relies on information flows that support distributed co-ordination of shared resources including space (Choo and Tommelein, 2000). A key to effective distributed planning is to recognize that each planner needs to plan with significant detail but not all that detail is to be revealed to everyone else with whom work must be co-ordinated. Some others will want certain details, whereas others do not. Detail must be selectively revealed as and when needed, depending on the circumstances. This thinking, like so many other concepts in the LPDS, requires a paradigm shift from current practice, which is dominated by the central control paradigm.

15.5.2 Summary

Many of the tools and techniques described under lean assembly are equally applicable in and across other triads of the LPDS. We have described them here because many of our experiences with lean production were gained in construction, before we considered lean project delivery with a broader scope. Practitioners may find quick rewards by applying our lean construction tools and techniques at the site level, prior to covering more scope.

15.6 Conclusion

Many powerful tools and techniques have been developed to manage the LPDS. Some of these are conceptual, some are procedural, some are embedded in software. Whereas several tools are simple, others are more complex, for example, the Last Planner system is a complex tool, itself including multiple rules and techniques, namely the Activity Definition Model, constraints analysis, and PPC. One-touch material handling is a conceptual and simpler tool, an ideal to be pursued with rules to be followed in its pursuit.

This varied tool set is very powerful in the hands of managers inspired by the lean conceptualization of projects and of project management. Bertelsen *et al.* (2001) reported at the Third Annual Lean Construction Congress that Danish contractors had reduced project durations by 10%, increased productivity by 20%, and improved profitability 20–40% on projects where they applied lean principles. Like everyone else, they are still in the early stages of their lean revolution, and have not yet applied all elements of the lean system nor yet applied all of its tools and techniques.

A true revolution in construction management is underway and it is as yet far from achieving its full potential. Indeed, the lean ideal suggests that 'full potential' will never be reached, as pursuit of the ideal eclipses all previous performance benchmarks. New tools and techniques will undoubtedly be developed in the never-ending pursuit of perfection.

Endnotes

1 Lean Project Delivery System (LPDS) and Last Planner are both Trademarks.
2 For an excellent example of chunking strategies, see Tsao *et al.,* 2000.
3 Extension of commitment planning and learning to direct workers is a likely future step in the evolution of lean construction.
4 There appears to be an opportunity for alternatives such as concrete and steel superstructures to compete on lead time. Shorter, less variable lead times would allow delaying design decisions to accommodate customer needs for late-breaking information, or could be used to shorten overall project durations.
5 Other tools can also be useful; e.g., the design structure matrix, which is used to sequence design tasks to eliminate avoidable looping – see www.mit.dsm.
6 Set-up time is a term from manufacturing indicating the time required to switch from producing one product to producing another product or to producing to a different set of specifications. An example from construction is changing crane booms. An example from fabrication is changing heads on the machine that produces different sizes of round sheet metal duct. An example from design is the time required to collect and focus one's thoughts when interrupted from a complex intellectual task.

References and bibliography

Akel, N.G., Boyers, J.C., Tommelein, I.D., Walsh, K.D. and Hershauer, J.C. (2001) Considerations for streamlining a vertically integrated company: a case study. In: *Proceedings 9th Annual Conference of the International Group for Lean Construction* (IGLC-9), 6–8 August, National University of Singapore.

Alarcon, L. (ed.) (1997) *Lean Construction* (Rotterdam: A.A. Balkema).

Ballard, G. (1998) Implementing pull strategies in the AEC industry. *Lean Construction Institute White Paper No. 1.* www.leanconstruction.org

Ballard, G. (1999) Work structuring. *Lean Construction Institute White Paper No. 5.* www.leanconstruction.org

Ballard, G. (2000a) Phase scheduling. *Lean Construction Institute White Paper No. 7.* www.leanconstruction.org

Ballard, G. (2000b) *Last Planner System of Production Control.* PhD Dissertation, Dept. of Civil Engineering, University of Birmingham, U.K. www.leanconstruction.org

Ballard, G. and Howell, G. (1994) Implementing lean construction: stabilizing work flow. In: *Proceedings 2nd Annual Conference on Lean Construction.* Pontificia University Catolica de Chile, Santiago, September. www.vtt.fi/rte/lean/santiago.htm

Ballard, G. and Howell, G. (1998) Shielding production: essential step in production control. *Journal of Construction Engineering and Management,* ASCE, **124** (1), 11–17.

Ballard, G., Koskela, L., Howell, G. and Zabelle, T. (2001) Production system design in construction. In: *Proceedings 9th Annual Conference of the International Group for Lean Construction*, 6–8 August, National University of Singapore. www.leanconstruction.org

Ballard, G. and Tommelein, I.D. (1999) Aiming for continuous flow. *Lean Construction Institute White Paper No. 3.* www.leanconstruction.org

Ballard, G. and Zabelle, T. (2000a) Project definition. *Lean Construction Institute White Paper No. 9.* www.leanconstruction.org

Ballard, G. and Zabelle, T. (2000b) Lean design: process, tools & techniques. *Lean Construction Institute White Paper No. 10.* www.leanconstruction.org

Bertelsen, S., Christoffersen, A.K., Jensen, L.B. and Sander, D. (2001) Studies, standards and strategies in the Danish construction industry implementation of the lean principles. In: *Getting it Started Keeping it Going, Proceedings of the 3rd Annual Lean Construction Congress*, Lean Construction Institute, August, Berkeley, CA.

Choo, H.J. and Tommelein, I.D. (2000) WorkMovePlan: database for distributed planning and coordination. In: *Proceedings 8th Annual Conference of the International Group for Lean Construction* (IGLC-8), 17–19 July, Brighton, UK. www.sussex.ac.uk/spru/imichair/iglc8/08.pdf

Howell, G. and Ballard, G. (1999) Design of construction operations. *Lean Construction Institute White Paper No. 4.* www.leanconstruction.org

Howell, G., Laufer, A. and Ballard, G. (1993) Interaction between subcycles: one key to improved methods. *Journal of Construction Engineering and Management*, ASCE, **119** (4), 714–28.

Sadonio, M., Tommelein, I.D. and Zabelle, T.R. (1998) The LAST DESIGNER'S Database-CAD for sourcing, procurement and planning. In: *Proceedings of Computing Congress '98*, ASCE, pp. 364–75.

Sobek, D.K., Ward, A.C. and Liker, J.K. (1999) Toyota's principles of set-based concurrent engineering. *Sloan Management Review,* Winter, **40** (2), 67–83.

Tommelein, I.D. (1998) Pull-driven scheduling for pipe-spool installation: simulation of lean construction technique. *Journal of Construction Engineering and Management*, ASCE, **124** (4), 279–88.

Tommelein, I.D. and Li, A.E.Y. (1999) Just-in-time concrete delivery: mapping alternatives for vertical supply chain integration. *Proceedings 7th Annual Conference of the International Group for Lean Construction* (IGLC-7), 26–8 July, Berkeley, CA, pp. 97–108.

Tommelein, I.D., Riley, D. and Howell, G.A. (1999) Parade game: impact of work flow variability on trade performance. *Journal of Construction Engineering and Management*, ASCE, **125** (5), 304–10.

Tsao, C.C.Y. and Tommelein, I.D. (2001) Integrated product/process design by a light fixture manufacturer. In: *Proceedings 9th Annual Conference of the International Group for Lean Construction* (IGLC-9), 6–8 August, National University Singapore.

Tsao, C.C.Y., Tommelein, I.D. and Howell, G. (2000) Case study for work structuring: installation of metal door frames. In: *Proceedings 8th Annual Conference of the International Group for Lean Construction*, University of Sussex, Brighton, UK.

Vrijhoef, R. and Koskela, L. (2000) The four roles of supply chain management in construction. *European Journal of Purchasing and Supply Management,* **6**, 169–78.

Vrijhoef, R., Koskela, L. and Howell, G. (2001) Understanding construction supply chains: an alternative interpretation. In: *Proceedings 9th Annual Conference of the International Group for Lean Construction* (IGLC-9), 6–8 August, National University of Singapore.

Ward, A., Liker, J.K., Cristiano, J.J. and Sobek, D.K. II (1995). The second Toyota paradox: how delaying decisions can make better cars faster. *Sloan Management Review,* Spring, **36** (3), 43–61.

Zabelle, T.R. and Fischer, M.A. (1999) Delivering value through the use of three dimensional computer modeling. In: *Procedings 2nd Conference on Concurrent Engineering in Construction*, Espoo, Finland.

16

Waste management in the construction industry

Martin Loosemore*, Helen Lingard† and Melissa Teo‡

Editorial comment

We live in an age of conspicuous consumption – an age where one day's marvel is tomorrow's landfill, and all too often a toxic problem a few days later. Take personal computers as an example: almost as soon as we upgrade to the latest model our new unit is obsolete, and disposal of computers perhaps only 2 to 3 years old is becoming a major headache in developed countries. Monitors contain a variety of harmful substances that preclude us from simply sending them to landfill. Keyboards and CPUs contain quantities of materials such as copper and plastic that are thrown away, necessitating the extraction and use of ever greater quantities of the raw materials used to produce them.

If, as the highly influential twentieth century architect, Philip Johnson (1964), suggested, 'architecture is the art of how to waste space', then construction might be described as the art of generating waste while enclosing space. The construction industry worldwide is a conspicuous consumer of raw materials of many types and it does not have an enviable record in its attitude to managing the waste that is produced both during construction and as a result of the demolition of buildings. Typically construction waste has been indiscriminately loaded into containers and taken to landfill with no attempt to salvage anything that might be of use. Estimators have routinely allowed for substantial wastage of materials during construction and have passed the additional cost on to clients as part of their tender price, just as they have done with the dumping fees associated with sending waste to landfill. Some demolition materials, where easily removed from the building and where a ready market for such materials existed, were recycled, but this tended to be limited to components and assemblies such as doors and windows, or unit products such as bricks.

* University of New South Wales, Sydney
† University of Melbourne
‡ Building Construction Authority, Singapore

We are now in a transition period, however, where the combined forces of tightening environmental legislation and economic pressures, largely the result of rapidly increasing costs for dumping waste due to a growing shortage of available landfill sites, are tipping the scales in favour of careful waste management at all stages of a building's life, from design through to demolition. This is leading to the establishment of markets that trade in recycled materials, whether reprocessed or not, and the economic benefits of saving useful materials and specifying recycled materials and products are prompting all parts of the industry to look for ways to cash in, in this area.

The case studies in this chapter amply illustrate that while waste can be reduced substantially, and that such reductions can improve contractors' profitability, there is still some way to go before all the participants in the process are wholehearted in their support of waste management. Certainly waste from all sectors of the industry is being reduced and better managed but the sheer volume of waste produced means that much more remains to be done.

16.1 Introduction

The issue of waste management is critically important for two main reasons. Firstly, it is estimated that we only have 100 years of cheap oil remaining, 300 years of cheap coal and far fewer years of certain ores and chemicals (Baird *et al.*, 1990). Secondly, in addition to depleting the world's natural resources, our pollution may be irreparably damaging its natural eco-system. Evidence of this is provided by a recent United Nations survey which is the most extensive of our environment so far (UN, 2000). This report indicates that the earth's climate is changing at such a pace that eventually the planet might be uninhabitable without special protection (Lemonick, 2000; Linden, 2000).

The construction industry can play a major role in solving these environmental problems. For example, it has been estimated that its products and processes consume up to 50% of all resources taken from nature and that it contributes approximately 50% of all waste going to landfills (Ferguson *et al.*, 1995; Anink *et al.*, 1996; Mak, 1999). The industry also consumes land at an alarming rate; in England, for example, an area of land the size of Wales has been developed since 1960 (Thompson, 2000).

As it is one of the biggest 'consumers' of nature, it is not surprising that the construction industry has become the target of environmentalists and, more recently, of government attempts to manage the environment more effectively. For example, the UK's Environment Agency has targeted the construction industry as a relatively poor environmental performer in relation to other industries, publicly naming its worst offenders and imposing fines on polluters. In Australia, the Environmental Protection Agency has gone further by identifying construction as a priority area for waste reduction and setting an objective of 60% reduction in waste between 1990 and 2000. As Faniran and Caban (1998) argue, landfill disposal can no longer be seen as a convenient solution to the construction industry's inability to manage its waste. Little is known, however, about the waste management process in construction and the purpose of this chapter is to discuss the complexities of this process, particularly from an attitudinal perspective. This is important because changing attitudes towards waste management is a fundamental step in changing the practices of a traditionally cost-driven industry that is cynical of attempts to protect an environment upon which it is so dependent. The challenge for the construction industry is to make sustainability a way of life rather than a philosophy promoted by a few passionate enthusiasts.

16.2 The problem of resource depletion

Before progressing to a detailed discussion of waste management practice in the construction industry, it is worthwhile considering the reasons why a sustainable construction industry is so urgently needed. In doing so, it is appropriate to revisit the problem of resource depletion highlighted in the introduction because it is far more serious when viewed from the perspective of future demographic changes, increasing affluence and technological advances. Collectively these factors are causing pollution levels and resource depletion to increase at a far greater pace than the expansion of the world's population.

16.2.1 Demographics

While the world's population took millions of years to reach 1 billion in 1830, it took only a century to reach 2 billion, a further 30 years to reach 3 billion and then 15 years to reach 4 billion (Rita, 1985). With a net growth-rate of 130 million per year (equal to Germany's population), it is estimated that it could be as high as 10.7 billion by the year 2050 (Kluger, 2000). In terms of resource consumption, this exponential population growth is worrying; when one considers the demographics of this growth, the situation is even more concerning. The main problem is that population growth is mainly taking place within developing regions such as Africa and South-East Asia where the potential for growth in resource consumption per capita far exceeds that of developed nations. One reason for this is related to anticipated birth rates in these countries: for example, while 45% of the world's population lived in developing countries in 1940, by the early 1980s it was 65%, and by 2034 it is predicted to be 93%. More importantly, while the average annual per capita use of energy is equivalent to about 10 000 kg of coal in the US it is as low as 45 kg in the world's poorest countries, which provides a frightening indication of the potential for future growth in energy consumption (Ehrlich and Ehrlich, 1986).

16.2.2 Technology

The main driving force behind resource depletion and pollution has been a self-perpetuating technological revolution. Wildavsky (1988) has argued that the main threat associated with technology is our tendency to over-estimate our capabilities, causing us to take ever-greater risks with the environment. Indeed, it is not difficult to find examples to support Wildavsky's theory since there have been many environmental disasters which have been the consequences of peoples' inability to control high technologies. Probably the most significant was the gas leak at the Union Carbide plant in Bhopal in India in 1982 which claimed 3500 lives and over 10 000 casualties (Shrivastava, 1992). Another prominent example was the grounding of the Exxon Valdez oil tanker in 1989 which spilt 11 million gallons of oil, causing devastating damage to the marine life and fishing industry of Alaska (Goldberg and Harzog, 1996).

16.2.3 Affluence

Another damaging consequence of the technological revolution has been the growth of 'the convenience society'. Commoner (1972) reported that in the 25 years following the Second World War, the production of non-returnable milk bottles had increased by 53 000%, synthetic fibres by 5980% and mercury used for chlorine detergents by 3930%. In contrast, the production of fibre, wool and traditional soap products had decreased. The implications of these changes become clear when the differences in energy needed to produce these different products and their differing impact upon the environment is considered. For example, detergents are often derived from non-renewable resources such as oil, consume three times the energy of soap in manufacture and put many more non-biodegradable pollutants into the environment. The same applies to synthetic fibres when compared to natural fibres.

16.2.4 Management trends

In response to technological advances there have been advances in management techniques and there is a growing recognition that they also have a major role to play in the demise of a sustainable society (Richardson, 1996). Richardson argues that modern business techniques such as lean production and business process re-engineering increase the vulnerability of organizations to crises by their ruthless focus upon efficiency. This places people under untenable pressures, resulting in more mistakes and crises which can seriously impact upon the environment. For instance, Richardson (1996) cites an example, in the early 1990s, of a chemical company in the UK which was fined £5000 for polluting a river 21 times in 1 year. A court found that commercial pressures were to blame for the pollution incidents, causing people to overlook the need for stringent but costly checks on the company's waste disposal methods.

16.3 The sustainability agenda

Concern with environmental degradation is not new: Durham (1999), for example, points to pollution legislation in thirteenth century Britain under which coal fires were banned in some cities due to their harmful effects on peoples' health. It was not until the late twentieth century, however, that these parochial concerns grew into a global agenda for environmental protection because of growing disquiet about the health impact of increased pollution levels, resource depletion and ozone depletion. The birth of this movement can be traced back to *The Limits of Growth* report and the first United Nations Conference on the environment in 1972 (Meadows *et al.*, 1972). This event first exposed the world as a place of limited resources with a finite capacity to support human life, and as a delicate eco-system that could be irreparably damaged by unabated industrialization. Although this led to the first international treaty to protect the environment, it was not until *The Brundtland Report* of 1987 (WCED, 1987) that the environmental movement began to gain serious support. The important contribution of *The Brundtland Report* was to interest the international business community in environmental issues by highlighting the threat of environmental damage to economic systems and wealth. According to Brundtland, it was no longer correct to speak of energy crises, environmental crises, economic crises and development crises – they were merely different aspects of the same ecological crisis. Furthermore, ecological and economic systems are connected to each other and nations are bound together by both economic and ecological systems (Mannermaa, 1995) and the fact that we are facing ever greater threats of economic catastrophe is a reflection of the fact that the developed world is already reaching the limits of its natural resources. The other important contribution of *The Brundtland Report* was to endorse the idea of sustainable development. This was defined as 'development which meets the needs of the present without compromising the ability of future generations to meet their needs'. Although this definition was seen as a compromise by environmentalists, it was popular with the business community because it stressed evolution rather than revolution. This comforted business leaders by reassuring them that radical change was not needed and that it was enough to make modifications to existing work-practices to enable economic activities to be sustained into the future.

It has since been recognized, however, that the business community's public support for sustainable development was largely rhetoric and that there were powerful interest groups driving the environmental agenda for economic rather than environmental reasons. As McCully (1991) pointed out, the US is the world's most polluting country, not because of a lack of efficient technologies but because of the political might of its oil, coal and automobile industries. The consequences of this were revealed by Meadows *et al.* (1992) in a replication of their 1972 study, *The Limits of Growth.* In their new report entitled *Beyond the Limits,* they argued that our use of many resources had already passed rates that were physically sustainable and that without significant reductions in material usage and energy flows there would be an uncontrolled decline in per capita food output. To them, a sustainable world was still possible but only with radical changes in the way we work and live.

16.4 Unsustainability in the construction industry

While the concept of sustainability has been in existence since the appearance of *The Brundtland Report* in 1987, it is relatively new to the construction industry. Indeed, the construction industry has an atrocious record in following the green agenda, has had a lacklustre approach to environmental targets and has demonstrated a resilient ability to ignore the environmental writing on the wall (Barrie, 1999). In construction, sustainability seems to be treated as another buzz-word to the extent that Barrie (1999, p. 3) found that managers in the construction industry were not sure of what the term meant:

> It seems to cover everything from committing the whole supply chain to making buildings more sustainable, to convincing individual firms that going green will not push them into the red.

This statement exposes one of the greatest barriers to sustainable practices in the construction industry, namely, who is to pay for green measures? Even if developers are keen to build green offices, they are often thwarted by the hostility of their investors or tenants. As Barrie (1999, p. 3) points out, 'if a tenant wants an office space for 5 years why would they worry whether a green building is cheaper to run over 25?' A further problem is the structure of the construction industry, which relies heavily upon sub-contracted labour from very small, highly geared companies who are handed relatively high levels of risk and who cannot afford to keep up with the latest management thinking. Not only does this make the issue of monitoring environmental performance difficult but it creates an attitudinal problem revolving around short-term, cost-driven values.

Yet, even in this difficult context, leading construction firms have developed environmental management systems (EMSs), often operating on the back of existing quality management processes. Other smaller businesses have also made a start by addressing specific environmental problems, such as energy efficiency, managing erosion or sedimentation, or waste management. These changes have largely been driven by increasing public awareness concerning environmental issues, government intervention (in the form of legislation) and indirect pressure, through tendering requirements.

Public awareness of environmental issues has been raised to such a degree that the community now expects business enterprises to demonstrate exemplary environmental performance. These expectations are often expressed through membership of pressure groups, which can exert considerable influence through the political process, lobbying decision-makers or sometimes through direct actions, such as demonstrations. Governments have responded to this by incorporating elements of environmental management into their own operations, purchasing requirements and tendering conditions. For example, in Australia, it is the policy of the government of New South Wales to require a corporate environmental management system or a project environmental management plan to be submitted with tender documents. In the case of projects of Aust$10 million or more, or projects deemed to be environmentally sensitive, contractors are also required to have their EMS accredited by a government agency (Commonwealth Department of the Environment and Heritage, 2000).

Legislation has also been enacted to establish environmental standards, and monitoring and reporting requirements. The legislative approach adopted is not prescriptive – instead, governments are seeking collaboration with industry in solving environmental problems and prefer incentives and co-operative agreements to a punitive approach. The resulting need for industry to be proactive in managing environmental issues is reflected in the way that courts rule on environmental matters. In the recent *Australian Environmental Protection Agency v. Ampol* (1995) case, which was heard in the New South Wales Land and Environment Court in 1995, it was stated:

> Ampol's obligation under the Environmental Offences and Penalties Act did not rest on any of the standards; it was rather a general obligation to avoid or minimize environmental harm. (Smith, 1995)

The judge's statement clearly shows that community and government expectations exceed compliance with minimum requirements laid down in standards. Proactive steps to manage environmental issues are essential to demonstrate compliance with legal determinations like 'environmental duty', exercising 'due diligence' and taking 'all reasonable and practical measures'.

Furthermore, in many large industrial enterprises, environmental management is recognized, at corporate level, not purely as an environment tool, but as providing the benefit of a competitive edge for its users in gaining preferred supplier status. In many markets, including construction, good environmental credentials may soon become decisive determinants of a firm's competitiveness. An organization's performance in environmental matters and other areas of social responsibility is increasingly a factor in the decision-making of discerning customers, job seekers and investors. Evidence also suggests that substantial cost savings are to be gained from implementing a cleaner production approach aimed at preventing or minimizing waste, conserving energy and other resources, and improving environmental management. In this context, the need for firms to address their business roles and responsibilities within the bigger picture of sustainability is recognized in mandatory environmental reporting requirements. For example, in Australia, s.299(1)(f) of Corporations Law now requires companies to demonstrate that they have fulfilled their environmental obligations to shareholders each year.

It is recognized that solid waste management is one area in which construction industry participants can improve their environmental performance considerably. Langston and Ding (2001) argue that effective waste management and the minimization of site waste is essential if a sustainable construction industry is to be achieved. Furthermore, these improvements can be made without necessarily implementing a full EMS, making effective waste management practice achievable, even for the small- and medium-sized enterprises that make up the vast majority of construction firms. Waste management is one of the more pressing issues in managing the environmental consequences of construction because the increasing scarcity of new, virgin materials has made the reliance on non-renewable resources untenable. In Australia, around one tonne of waste per person is sent to landfill each year and it is estimated that 14 million tonnes of waste is disposed of in Australian landfills each year. There is rising concern about the amount of waste generated and disposed of by the construction industry.

Of the total waste stream, approximately 15% is construction waste (Environment Australia, 2000). This figure is a national average and considerable variation occurs between states; for example, it is reported that building rubble alone has contributed as much as 27.4% of the total waste stream in Perth (McDonald and Smithers, 1998) and makes up 44% of landfill, by mass, in Victoria (McDonald, 1996). With these high proportions, a substantial reduction in construction waste will inevitably lead to a considerable reduction in the overall waste stream, prolonging the life of existing landfill sites. There is also evidence that construction and demolition waste poses other environmental problems; the growing use of organic polymers and chemical additives in construction renders construction waste increasingly liable to contamination by hazardous substances (Lahner and Brunner, 1994). It is also now understood that much construction waste, which was previously regarded as inert, generates harmful leachate (Apotheker, 1992). In the UK, toxins arising from the burning of treated waste wood appear to be accumulating in the food chain (CIRIA, 1993).

16.5 Good practice in solid waste management

The practice of solid waste management is underpinned by a hierarchy of waste control measures (Peng *et al.*, 1997). The principle of this hierarchy, depicted in Figure 16.1, is that control measures that aim to reduce waste at source (reduction and re-use) are preferable to recycling, which requires that waste be re-processed before it can be re-used. In turn, recycling is preferable to the disposal of waste either by incineration or as landfill.

Implicit in this hierarchy is the need to re-examine the life cycle of building materials and components. Life cycle assessment is the framework within which a material or

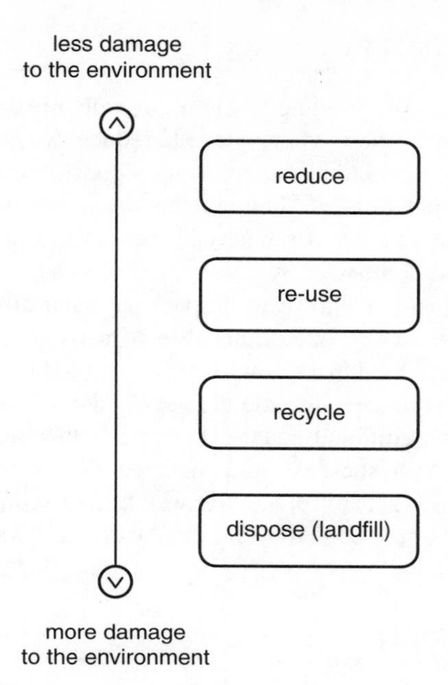

Figure 16.1 The waste management hierarchy.

product is examined through its entire existence from raw material extraction, through manufacture, construction, use, maintenance and disposal. It is imperative that the emphasis must shift from a 'cradle-to-grave' analysis to a 'cradle-to-cradle' approach in which materials and components do not flow in one direction from extraction to disposal but instead follow a circular path from 'birth' to 're-birth'. In order to achieve this circular flow, however, an integrated effort must be made by all participants in the construction process. For example, there must be a secondary market for re-used or recycled materials if it is to be feasible for demolition or construction operators to embark upon salvage or recycling processes. This requires that designers be willing to specify products that are salvaged or whose content is partially or wholly recycled, that there is a processing, supply and distribution network for these materials and that they are recognized to be compliant with the requirements of building regulations and environmental health standards (Neale, 1994). Effective waste management therefore requires a collaborative effort through the construction supply chain in which the client, designer, materials manufacturers and suppliers, construction and demolition contractors and waste disposal contractors all have an important role to play.

In Australia, a group of construction firms and industry bodies have committed to working, with Government, towards best practice in waste minimization. This programme, known as the WasteWise Construction Programme, is in its second phase. Major construction firms, client bodies and designers are represented among the WasteWise partners. The achievements of the WasteWise Programme to date, some of which are described below, are impressive and demonstrate the effectiveness of adopting a collaborative approach to waste management.

16.5.1 Innovative design

Innovations in the design of buildings which provide greater flexibility and permit refurbishment and fit-out as and when needed, reduce waste. Buildings can also be designed to facilitate de-construction (the opposite of construction), which is the preferred method of demolition if materials and components are to be salvaged for re-use. Design decisions also determine the volume of materials incorporated into a structure and the use of materials with recycled content.

Innovative design can have a significant impact on material wastage, especially when this is considered over the entire operational life of a facility. The roof of the Sydney Olympic Main Arena Grandstand, for example, was designed in such a way that it required 22% less steel than alternative designs, and the nearby domed Showground Hall used less building materials than a traditional square structure. Other buildings were constructed using recycled concrete as a sub-base and recycled timber was incorporated into the Clydesdale Pavilions. Waste generation on-site was further reduced by the extensive pre-fabrication of components off-site (Environment Australia, 2000).

16.5.2 Material supply

Suppliers also have a role to play in promoting waste minimization: they can rationalize production by standardizing components and reducing packaging, and they can participate in the manufacture of products with recycled content. BHP, a leading Australian mining and resources company, manufactures 100% recycled steel products in its Sydney mini-mill which uses up to 300 000 tonnes of scrap steel each year. Innovative supply arrangements can also promote better waste management and it is reported that some producers of construction materials are beginning to take responsibility for the entire life cycle of their products, adopting practices like 'taking-back' surplus or recyclable materials (Environment Australia, 2000). For example, it is reported that Interface Australia Pty Ltd has refocused its supply arrangements to achieve less wasteful practices. The company, which supplies modular carpet tiles, has adopted a system through which customers lease rather than buy the carpet. Customers pay a monthly fee for the value and use of the carpet supplied. This monthly fee includes warranty, design, manufacture, installation, maintenance and final reclamation of the product (Environment Australia, 2000).

16.5.3 Construction

Construction contractors can assist with waste reduction on-site through the implementation of project-specific waste management plans, and good site management and organization. Accurate estimating and ordering with nil allowance for waste reduces the volume of waste, the cost of materials and the cost of waste disposal. The identification of recyclable waste and identification of disposal options available in the geographical area of the site are also essential to effective management of site waste. The decision to recycle certain waste streams requires that the contractor establish a system for collection of recyclable materials. These materials may be separated from general waste on-site or off-site, but if on-site separation is to be used, waste bins should be clearly labelled and placed in sufficient numbers in convenient locations around the site. It is reported that through its waste minimization measures, Fletcher Construction reduced waste disposal costs by 55% and increased profits compared to another project, of very similar design and construction, in which waste minimization measures were not implemented (McDonald, 1996).

In the construction industry, in which a large proportion of the work is sub-contracted, ensuring supplier and sub-contractor compliance with waste management plans is crucial. This can be achieved through up-front contractual requirements, e.g., in New South Wales and Victoria, Multiplex Construction negotiates with suppliers and sub-contractors to reduce packaging at source. On the Stadium Australia project, 7000 fire doors were delivered direct from the factory without packaging and sub-contracts required compliance with project waste management plans and specified that packaging materials had to be returned to the original suppliers (Commonwealth Department of the Environment and Heritage, 2000).

16.5.4 Demolition

The demolition process generates waste in greater volumes than construction, however, reduction of waste can be achieved through good planning. Deconstruction, through which buildings are taken apart piece by piece, can provide opportunities for the salvage and re-use of building components. During the demolition of Phillipines House in inner Sydney, Metropolitan Demolition Pty Ltd set a goal of 83% diversion from landfill. Site separation of materials was conducted with re-usable materials being salvaged and sold, and recyclable materials removed in stages and recycled (Commonwealth Department of the Environment and Heritage, 2000).

Salvage and re-use of building components requires a network through which materials and purchasers can be matched. There are reported to be very successful material/waste exchanges operating in the United States, e.g., Mason Brothers in Vermont, who use a barter system to trade used items from buildings and are reportedly making a profit (Witten, 1992). Mason Brothers do not specialize in historical treasures but also supply ‘low-end’ materials, such as US$10 doors and chipped bathtubs. Another successful broker of salvaged items is Urban Ore in Berkeley, California. The company operates a 2.2 acre (0.89 ha) warehouse, employs 16 people and reported expected gross sales of US$1 000 000 in 1992 (Hazen and Sawyer PC, 1993). At present, such exchanges are in their infancy in Australia although some government agencies have created exchange web sites. For example, in 1999 the New South Wales Waste Boards (www.wasteboard.nsw.gov.au) established the Australian Reusable Resources Network (www.arrnetwork.com.au). Membership of this network is free and allows members to request or seek potential buyers for re-usable materials on the Internet.

Where salvage and re-use are not possible, due for instance to the use of composite materials, demolition waste can still be recycled. For example, during the demolition of the Balmain Power Station in Sydney, around 18 000 tonnes of concrete and brick and 2000 tonnes of steel were recycled (Commonwealth Department of the Environment and Heritage, 2000).

16.5.5 Waste disposal

Waste disposal contractors can also assist in the waste management process through offering services, additional to the provision and emptying of skips, such as site planning, materials management and transport.

As the above examples of good practice illustrate, technological barriers to the management of solid construction waste are rapidly being overcome. Construction waste has a high potential for recovery and re-use (Cosper *et al.*, 1993; Schlauder and Brickner, 1993) and recycling options for solid construction waste are also increasing (Merry, 1990; von Stein and Savage, 1994). However, the extent to which reduction, re-use and recycling of solid construction waste can be achieved depends, to a large extent, on attitudinal and motivational influences on the behaviour of participants at all levels of the construction process, from site operatives, to foremen, managers, clients, designers and suppliers; pressures to complete work quickly, for example, might lead a tradesman to cut components from new material rather than spending time locating suitable previously cut pieces (Federle, 1993).

In the construction industry, Skoyles and Skoyles (1987) were the first to recognize that the problem of material wastage was more dependent upon the attitudes and behavioural tendencies of individuals involved in the construction process than upon the technical processes it employed. Since then other studies by Heino (1994), Soibelman *et al.* (1994), Guthrie and Mallett (1995), Lingard *et al.* (2000) and Teo (2000) have reinforced this view. Their findings suggest that negative attitudes towards waste management prevail on construction projects due to time and cost constraints, poor leadership and a lack of experience in dealing with environmental issues. In this context it would seem that the challenge to the construction industry of making itself more sustainable is more fundamental than merely creating policies and systems to monitor work-practices. Rather, the problem is cultural and lies in changing people's attitudes to issues such as waste management. Two recent studies undertaken in the Australian construction industry explored attitudinal and motivational influences on waste management behaviour during the construction phase of projects. The results of these studies are presented, in the order in which they occurred, and their results are discussed jointly below.

16.6 Case study: Colonial Stadium site, Melbourne

Goal setting and feedback are intended to motivate employees to improve their performance at work by prompting and reinforcing specific, desirable behaviours. The application of goal setting and feedback procedures has been recommended as a means of eliciting pro-environmental behaviours (Geller, 1989) and the approach has previously been successful in improving solid waste recycling behaviour among academic and administrative staff in an American university (Austin *et al.*, 1993). To date, the extent of goal setting in construction solid waste management has been restricted to broad objectives of waste reduction, expressed in percentage terms, being set at company level. Such objectives are not site specific and site personnel may perceive them to be of little relevance.

Between 1 June 1998 and 30 April 1999, an experiment was conducted during the construction of a large sports stadium in the Docklands area of Melbourne. During this experiment, waste management performance was measured to ascertain a baseline level. The volumes of two key materials, timber and concrete, wasted and recycled, were measured each fortnight. These materials were selected because they are known to make up a large proportion of construction waste (Apotheker, 1992) and were relevant to the activities occurring on the construction project at the time of the experiment (erecting formwork and placing concrete). After a 6 week baseline measurement period, representatives of work crews, sub-contractors and site management met to discuss performance and set specific performance goals for improvement in the reduction and/or recycling of the two materials at different timings. For experimental purposes, timber goals were set before concrete goals. Measurement continued and performance feedback was provided through the use of four graphical feedback charts posted at prominent positions on the site. The feedback charts included the title 'waste less timber' or 'waste less concrete' and displayed the following numerical information:

- the percentage of total waste disposed of as landfill during the current fortnight
- the percentage of total waste going to landfill during the preceding fortnight
- the percentage of bin contamination during the current month
- the percentage of bin contamination during the preceding month.

The trend, i.e. whether the percentage of total waste or bin contamination during the current measurement interval was better or worse than during the previous measurement interval, was indicated next to these figures with either a 'thumbs up' or a 'thumbs down' symbol. The percentage of the key material (either timber or concrete) that was recycled each fortnight was also shown on a line chart with 30 fortnightly increments on the *x*-axis. Recycling goals set during meetings were marked on the chart in red and actual performance was marked on the chart in black. Feedback charts were updated each fortnight.

The results of the experiment suggest that goal setting and feedback can effectively reduce the total volume of material wastage in the construction industry context. When the timber goal was set and feedback was being provided, the average waste disposed as landfill fell from a high of 30 cubic metres per fortnight to 10.7 cubic metres per fortnight. Following the removal of the timber feedback, waste disposed of as landfill increased again to an average of 19.9 cubic metres per fortnight. In the period prior to the concrete goal setting meeting, the average volume of waste disposed of as landfill was 19.8 cubic metres per fortnight; this fell to 18.7 cubic metres per fortnight when the concrete goal was set and feedback was being given. In the period following the removal of concrete feedback, the average waste disposed of as landfill rose dramatically to 32.3 cubic metres per fortnight. This finding is supported by the desktop calculations of the efficiency of material usage during the experimental period. The average efficiency score, shown in Figure 16.2, was found to be significantly higher during the periods when goal setting and feedback were in place than when these conditions were absent.

The effect of goal setting and feedback on recycling performance was, however, less promising. Figure 16.3 shows that timber recycling improved when goal setting and feedback were introduced but that this improvement was not sustained and performance in timber recycling began to decline before the timber feedback was removed. Furthermore, concrete recycling performance fell from its initially high level before the

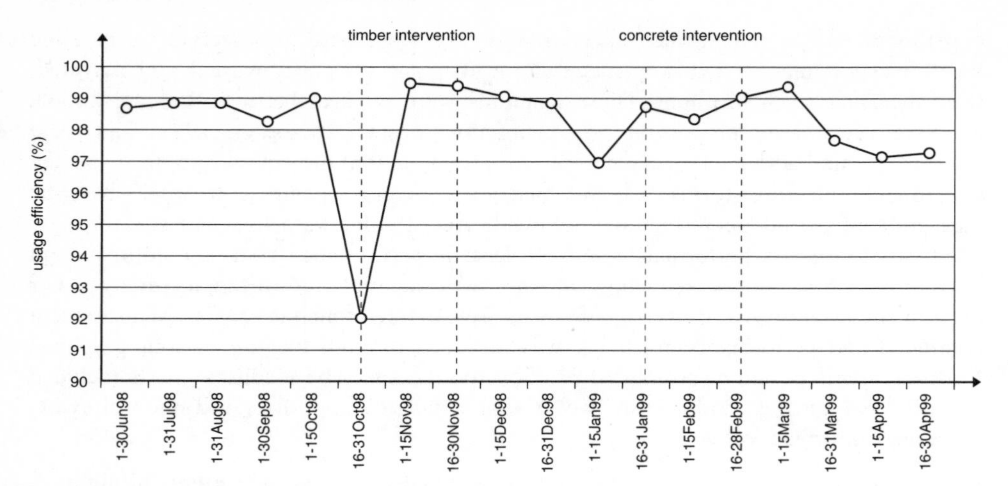

Figure 16.2 Material usage efficiency by fortnight.

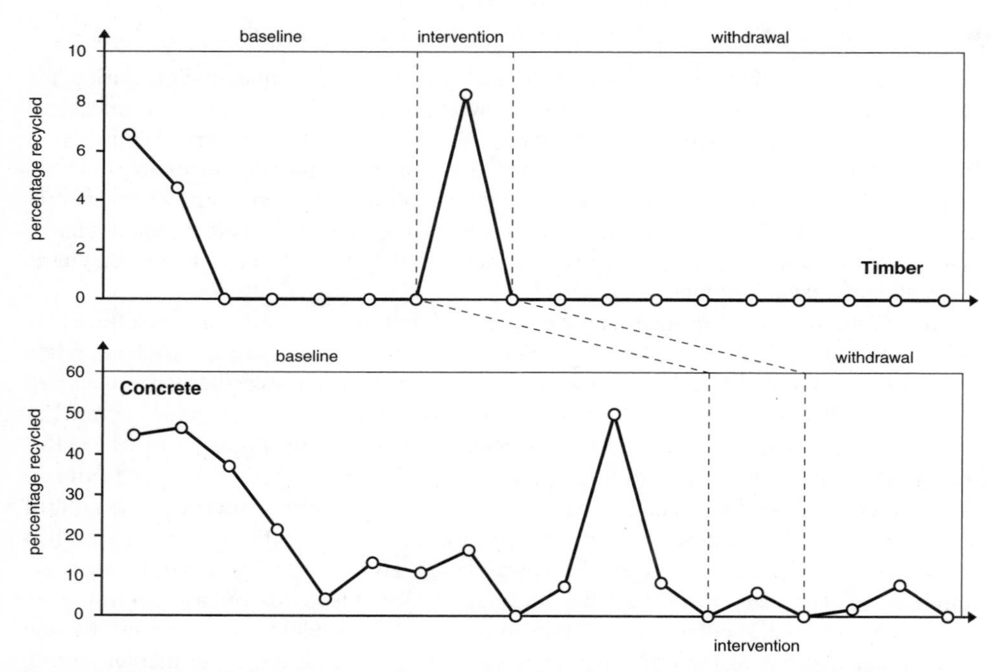

Figure 16.3 Volume of waste recycled as a percentage of total waste leaving the site (Lingard *et al.*, 2001).

concrete goals setting meeting was held and remained low during the goal setting and feedback period. It is not clear what caused the deterioration in concrete waste recycling performance during the experiment but the fact that it occurred before either of the goals were set suggests that it was the result of extraneous factors rather than the goal setting and feedback intervention.

16.7 Survey of waste management attitudes of construction site operatives

In March 2000 an exploratory study of the waste management attitudes of construction operatives was conducted on eight large commercial construction projects in central Sydney with a collective value of over Aust$2 billion (Teo, 2000). The research focused upon operatives' attitudes, and data was collected from 213 respondents using an attitudinal survey and focus group interviews. The results must be interpreted within this occupational context and similar research is needed to investigate managerial attitudes towards waste. Nevertheless, operatives have a major impact upon work-place practices, and understanding their current attitudes towards waste management is a crucial factor in any attempt to change the construction industry's environmental record for the better. The results of the survey and focus groups are summarized below.

16.7.1 Lack of responsibility

The research found a low level of voluntary support and co-operation among respondents for waste management activities on projects. In general, responsibilities for the provision of a waste management infrastructure were perceived to rest with managers and there was little sense of collective responsibility for waste management activities. The research found participants' understanding of their personal responsibility for waste management to be confused and haphazard and people's acceptance of waste reduction as part of their job responsibility varied widely.

16.7.2 Lack of managerial support

Although managers were seen as the primary source of responsibility for waste management efforts, most respondents perceived that inadequate resources were committed to this area causing frustration in terms of meeting managers' requirements. Indeed, there was a common perception that these requirements were being applied inconsistently from project to project, causing uncertainty and confusion and a certain degree of cynicism towards waste management activities.

16.7.3 Perceived costs and benefits

Cost factors emerged as a determinant of waste management's relative priority amongst project goals, waste management activities being widely perceived as economically unviable and inconvenient. Paradoxically, however, cost savings were also identified as a strong potential motivator of waste management behaviours which indicates that if attitudes towards the economic viability of waste management activities could be changed, then there would be an automatic knock-on effect into better waste management practices.

16.7.4 Inevitability

Participants indicated a strong belief that waste was an inevitable by-product of construction activity as a result of human error and project constraints and that made the issue of waste difficult to control. In general, attitudes towards waste reduction were driven by pragmatism and underpinned by a belief that the total elimination of waste is impossible on construction projects.

16.7.5 Inadequacy of training

Current levels of waste management training were perceived to be inadequate, particularly at a technical level, meaning that knowledge levels about waste reduction, reuse and recycling appear to be low. Indeed, the limited training that was provided was very general in nature and not considered detailed and technical enough to be useful in identifying waste reduction techniques in specific occupational areas. The most disturbing fact to emerge, however, was the belief among operatives that further training or knowledge is unnecessary since, in any event, they have limited capability to reduce waste levels on projects.

16.7.6 Consultation

While very few of the respondents had been consulted on how waste levels associated with their jobs could be reduced, they all felt able and willing to assist in this process. In general, the research indicates a low level of employee participation in the waste management process and an overall perception that waste management was imposed unilaterally from above.

16.7.7 Overview

The results of both of these studies suggest that there are significant attitudinal constraints working against the effective management of waste in the Australian construction industry. These are worth noting because they may also be relevant to the construction industries in other countries.

The prevailing belief that waste management is not cost effective, and performance in waste management does not receive the same recognition or reward as performance in other project objectives, such as cost or time performance, is likely to be demotivating. The goal setting and feedback techniques introduced during the Colonial Stadium experiment were based on a model of behaviour that assumes that individuals, either consciously or unconsciously, undertake a rational calculation of the costs and benefits associated with a given type of behaviour, therefore allowing behaviour to be regulated by the manipulation of rewards and/or punishments. It follows that if the benefits of behaving in a desired way are not valued, or the costs of behaving in this way are too great, goal setting and feedback will not succeed. The failure of the techniques to improve recycling performance significantly may reflect the fact that recycling incurs a cost for workers, in that the required separation of waste slows them down. In an industry that typically rewards fast production this attitude will constrain the effectiveness of any recycling programme.

The reward structure in place on construction projects and the way that this impacts upon waste management objectives needs to be considered in detail if this attitudinal constraint is to be overcome. The survey data reveal perceived cost savings as the primary motivating force for the adoption of waste reducing behaviours on projects. In this sense it is likely that monetary incentives could significantly improve the effectiveness of waste management activities in the construction industry. Anecdotal information provided by the site managers supports this view; they reported that steel waste was collected from all four quadrants of the stadium site and sold to a recycler with the money received for the steel given back to the work crews in each quadrant as 'beer money'. Site managers reported that the rate of steel recycling was consistently at 100% and even suggested that re-usable steel items were placed in recycling containers. This highlights the fact that, if monetary incentives are to be used, care must be taken not to reward recycling performance to such a degree that materials are deliberately wasted or not re-used.

In addition, any incentive scheme would need to be equitably designed because a sense of procedural fairness has been identified as an essential requisite for eliciting behaviour, such as pro-environmental behaviour, that requires individuals to act collectively to attain a common good (Blamey, 1998). This includes a belief that organizations are fulfilling their part of the bargain, everyone is 'doing their bit', and rewards are fairly distributed. The survey results reveal that there is a current perception that the direct financial benefits of waste reduction are being retained at managerial level. This perceived inequity has had the effect of perpetuating negative attitudes among construction workers by separating them from the personal benefits of their efforts. In the construction industry, where much work is sub-contracted, this may prevent the development of a sense of procedural fairness, for example, the efforts made by sub-contracted workers to reduce or recycle waste may benefit the principal contractor but this benefit may not be passed down to the sub-contracted workforce. This is likely to lead workers to resent the additional responsibility imposed upon them by management, who also impose strict cost, time and quality imperatives, and so generate a cynical attitude towards waste management.

The belief that construction waste cannot be easily reduced and the lack of adequate training or employee involvement in the waste management process are also of concern. The Colonial Stadium case study suggests that involving workers in participative goal setting and providing performance feedback can improve material usage efficiency. Such a consultative approach is likely to perform three important functions: firstly, it communicates to workers that management are interested in waste management; secondly, it generates discussion and ideas about how waste can be managed more effectively; and thirdly, it fosters a sense of ownership in the waste management process. For such an approach to be truly effective, however, it is also necessary that employees have the knowledge, skills and abilities they need to recognize opportunities to minimize waste and act accordingly, and waste management training must be improved.

16.8 Waste as a resource

The attitudinal problems identified may be partially attributed to the way waste is conceptualized. The Oxford English Reference Dictionary (OUP, 1995) defines waste as 'refuse; useless remains or by-products'. It goes on to define waste paper as 'used or valueless paper' and waste products as 'useless by-products of manufacture'. Thus laypersons' definitions of waste imply a lack of inherent value and uselessness. For waste management to be effective, waste must be regarded as having some value or use. In certain circumstances waste can be regarded as a resource and this needs to be recognized in operational definitions of waste used. For example, the Organisation for Economic Cooperation and Development (OECD) provides a guidance note on distinguishing waste from non-waste which states that 'the intended destination of a material is the decisive factor' (OECD, 1998). Using this definition, items that are discarded but destined for recovery or recycling are not waste. Furthermore, waste may cease to be waste at some point in its life; thus waste concrete that is destined to be crushed and recycled as aggregate may be a waste before it is processed, but ceases to be waste on successful completion of the recycling process. The OECD (1998, p. 12) guidelines go on to suggest that a waste ceases to be a waste when:

> a recovery, or another comparable process eliminates or sufficiently diminishes the threat posed to the environment by the original material (waste) and yields a material of sufficient beneficial use. In general, the recovery of a material (waste) will have taken place when:
>
> 1 It requires no further processing; and
> 2 The recovered material can and will be used in the same way as a material which has not been defined as waste; and
> 3 The recovered material meets all relevant health and environmental requirements.

Re-conceptualizing waste as something that can be re-used or recycled rather than something that was intended to be discarded, and recognizing that material (waste) may cease to be a waste, are important steps in the implementation of a waste management programme.

16.9 Conclusion

The construction industry has lagged behind other industries in the management of its environmental consequences. The impact of the industry is considerable and it is imperative that this impact be better managed in the future. Waste management is an important concern not only because the current usage of the earth's resources is unsustainable but also because landfill capacity is limited and the construction industry generates a large volume of the total waste disposed of in landfill sites each year.

Waste management is relatively easy to implement and does not require a full-blown EMS, making it possible for smaller enterprises, as well as major contractors, to develop waste management polices and programmes. All parties involved in the construction process can help to rectify the industry's wastefulness; an integrated effort on the part of clients, designers, suppliers, contractors, sub-contractors and waste contractors can yield significant benefits. These benefits are demonstrated by the achievements of the Australian construction industry under the WasteWise Construction programme.

There are, however, important attitudinal barriers to further improvements in waste management. These relate to the way in which costs and benefits of waste management practices are experienced, perceptions of procedural fairness and the concept of waste as possessing no value or use. These attitudes must be changed in order to further minimize waste. Evidence suggests that waste management activities can be cost-effective, and by focusing training upon the tangible economic benefits of waste reduction activities, participants in this cost-oriented industry are more likely to associate themselves with the benefits of participating in new waste management initiatives. It is important that detailed cost/benefit analyses be conducted and that training and knowledge regarding the consequences and opportunity-costs of wasteful practices are provided to the people involved. The ultimate objective should be to help people see waste as having some value.

References and bibliography

Ajzen, I. (1993) Attitude theory and the attitude–behaviour relation. In: Krebs, D. and Schmidt, D. (eds) *New Directions in Attitude Measurement* (Berlin: Walter de Gruyter), pp. 41–57.

Anink, D., Boonstra, C. and Mak, J. (1996) *Handbook of Sustainable Building: An Environmental Preference Method for Selection of Materials for Use in Construction and Refurbishment* (London: James & James (Science Publishers) Limited).

Apotheker, S. (1992) Managing construction and demolition materials. *Resource Recycling*, August, 50–61.

Austin, J., Hatfield, D.B., Grindle, A.C. and Bailey, J.S. (1993) Increasing recycling in office environments: the effects of specific informative cues. *Journal of Applied Behavior Analysis*, **26**, 247–53.

Baird, L.S, James, E.P. and Mahon, J.F. (1990) *Management – functions and responsibilities* (New York: Harper Collins Publishers Inc.).

Barrie, G. (1999) Porritt attacks industry for neglecting green agenda, *Building*, 26th March, 3–5.

Blamey, R. (1998) The activation of environmental norms: extending Schwartz's model. *Environment and Behavior*, **30**, 676–708.

CIRIA (1993) *Environmental Issues in Construction: A Review of Issues and Initiatives Relevant to the Building, Construction and Related Industries* (Vol. 2) (London: Construction Industry Research and Information Association).

Commoner, B. (1972) *The Closing Circle, Nature and Man and Technology* (Toronto: Bantam Books).

Commonwealth Department of Environment and Heritage (2000) *WasteWise Construction Program.* http://www.ea.gov.au/industry/waste/construction/wastewise/

Cosper S.D., Hallenbeck W.H. and Brenniman G.R. (1993) *Construction and Demolition Waste: Generation, Regulation, Practices, Processing and Policies* (Office of Solid Waste Management, The University of Illinois, Chicago).

Durham, M. (1999) Environmental planning. In: Best, R. and de Valence, G. (eds) *Building in Value: Pre-design Issues* (London: Arnold), pp. 95–103.

Ehrlich, P. and Ehrlich, A.M. (1986) World population crisis. *Bulletin of the Atomic Scientist*, April, 13–18.

Environment Australia (2000) *WasteWise Construction Program: Waste Reduction Guidelines*, (Canberra: Commonwealth Department of Environment and Heritage).

Environment Protection Authority v. Ampol Limited [1995] NSWLEC 16 (22 February 1995).

Faniran, O.O. and Caban, G. (1998) Minimizing waste on construction project sites. *Engineering, Construction and Architectural Management*, **5** (2), 182–8.

Federle M.O. (1993) Overview of building construction waste and the potential for materials recycling. *Building Research Journal*, **2**, 31–7.

Ferguson, J., Kermode, O.N., Nash, C.L., Sketch, W.A.J. and Huxford, R.P. (1995) *Managing and Minimising Construction Waste: A Practical Guide* (London: Institution of Civil Engineers, Thomas Telford Publications).

Geller E.S. (1989) Applied behavior analysis and social marketing: an integration for environmental preservation. *Journal of Social Issues*, **45**, 17–36.

Goldberg, S.D. and Harzog, B.B. (1996) Oil spill: management crisis or crisis management? *Journal of Contingencies and Crisis Management,* **4** (1), 1–10.

Guthrie, P. and Mallett, H. (1995) *Waste Minimisation and Recycling in Construction – A Review* (London: Construction Industry Research and Information Association).

Hassan J., Nunn, P., Tomkins, J. and Fraser I. (1996*) The European Water Environment in a Period of Transformation* (Manchester: Manchester University Press).

Hazen and Sawyer PC (1993) *Construction and Demolition Debris Reduction and Recycling: A Regional Approach* (Triangle J Council of Governments, North Carolina).

Heino, E. (1994) Recycling of construction waste. In: Kibert C.J. (ed.) *Sustainable Construction, Proceedings of First Conference of CIB TG 16* (Centre for Construction and Environment, Gainesville, Florida), pp. 565–72.

Johnson, P. (1964) Architecture is the art of wasting space. *New York Times*, 27 December, p. 9.

Kluger, J. (2000) The Big Crunch. In: Alexander, C.P. (ed.) *Earth Day 2000*, *Time*, Special Edition, April–May, pp. 44–5.

Lahner, T.E. and Brunner, P.H. (1994) Buildings as reservoirs of materials: their reuse and implications for future design. In: Lauritzen, E.K. (ed.) *Demolition and Reuse of Concrete* (London: E. & F.N. Spon).

Langston, C.A. and Ding, G.K.C. (2001) *Sustainable Practices in the Built Environment.* Second edition (Oxford: Butterworth-Heinemann).

Lemonick, M.D. (2000) How to prevent a meltdown. In: Alexander, C.P. (ed.) *Earth Day 2000*, *Time*, Special Edition, April–May, pp. 61–3.

Linden, E. (2000) Condition critical. In: Alexander, C.P. (ed) *Earth Day 2000*, *Time*, Special Edition, April–May, pp. 18–21.

Lingard, H, Graham, P. and Smithers, G. (2000) Employee perceptions of the solid waste management system operating in a large Australian contracting organisation: implications for company policy implementation. *Construction Management and Economics*, **18** (4), 383–93.

Lingard, H., Smithers, G. and Graham, P. (2001) Improving construction workers' solid waste reduction and recycling behaviour using goal setting and feedback, *Construction Management and Economics*, **19**, 809–17.

Mak, S.L. (1999) Where are construction materials headed**.** *Building Innovation and Construction Technology*, **8**, August, 34–8.

Mannermaa, M. (1995) Alternative futures perspectives on sustainability, coherence and chaos. *Journal of Contingencies and Crisis Management*, **3** (1), 27–33.

McCully, P. (1991) The case against climate aid. *The Ecologist*, **21** (6), 250–61.

McDonald, B. (1996) RECON Waste minimisation and environmental programme, *CIB TG16 Commission meetings and presentations* (Melbourne: RMIT).

McDonald, B. and Smithers, M. (1998) Implementing a waste management plan during the construction phase of a project: a case study. *Construction Management and Economics*, **16**, 71–8.

Meadows, D.H., Meadows, D.L., Randers, J. and Bahrens, W.W. (1972*) The Limits of Growth* (New York: Universe Books).

Meadows, D.H., Meadows, D.L. and Randers, J. (1992*) Beyond the Limits: Global Collapse or a Sustainable Future* (London: Earthscan Publications Ltd).

Merry W. (1990) Taking a profitable approach to recycling. *World Wastes*, July, 40–7.

Neale, B.S. (1994) Retrieving materials – the effects of EC health and safety directives. In: Lauritzen, E.K. (ed.) *Demolition and Reuse of Concrete* (London: E. & F.N. Spon).

OECD (1998) *Final Guidance Document for Distinguishing Waste from Non-Waste* (Paris: OECD).

OUP (1995) *Oxford English Reference Dictionary* (Oxford University Press).

Peng, C.L., Scorpio, D.E. and Kibert, C.J. (1997) Strategies for successful construction and demolition waste recycling operations, *Construction Management and Economics*, **15**, 49–58.

Richardson, W. (1996) Modern management's role in the demise of a sustainable society. *Journal of Contingencies and Crisis Management,* **4** (1), 20–31.

Rita, C. (1985) Building a sustainable society is possible. But it is not easy. *Natural History*, April, 85–6.

Schlauder, R.M. and Brickner, R.H. (1993) Setting up for recovery of construction and demolition waste. *Solid Waste and Power*, Jan/Feb, 28–34.

Shrivastava, P. (1992) *Bhopal: Anatomy of a Crisis* (London: Paul Chapman Publishing).

Skoyles, E.R. and Skoyles, J.R. (1987) *Waste Prevention on Site* (London: Mitchell Publishing Company Limited).

Smith, R. (1995) Improvement needs more than just compliance. *Waste Management and Environment*, December, 15.

Soibelman, L., Formoso, C.T. and Franchi, C.C. (1994) A study on the waste of materials in the building industry in Brazil. In Kibert, C.J. (ed.) *Sustainable Construction, Proceedings of First Conference of CIB TG 16* (Centre for Construction and Environment, Gainesville, Florida), pp. 565–72.

Teo, M.M.M. (2000) *Operatives Attitudes Towards Waste on Construction Projects*. Unpublished MSc Thesis, University of New South Wales, Sydney, Australia.

Thompson, D. (2000) Asphalt jungle. In: Alexander, C.P. (ed.) *Earth Day 2000. Time*, Special Edition, April–May, pp. 50–1.

UN (2000) *A Pilot Analysis of Global Eco-systems* (PAGE) (United Nations).

von Stein E.L. and Savage G.M. (1994) Current practices and applications in construction and demolition debris recycling. *Resource Recycling*, April, 85–94.

WCED (World Commission on Environment and Development) (Brundtland Commisssion) (1987) *Our Common Future* (Oxford University Press).

Wildavsky, A. (1988) *Searching for Safety* (New Brunswick: Transaction).

Witten, M. (1992) Reuse of low-end construction and demolition debris. *Resource Recycling*, April, 115–22.

17

Enterprise process monitoring using key performance indicators

Marton Marosszeky* and Khalid Karim*

Editorial comment

In 1976 a young Romanian gymnast, Nadia Comenici, scored not one but seven 'perfect' scores in competition at the Montreal Olympics. No-one had ever achieved a ten and it led many to question whether such a score could be justified: surely, no routine could ever be 'perfect'. That debate illustrated the basic problem of trying to measure something that is not naturally quantifiable. Measuring the performance of a business or enterprise in any sphere of activity can be equally difficult, particularly if comparisons are to be made, and in many situations measurement or evaluation may be quite subjective, no more than a matter of opinion.

Evaluating performance in some numerical fashion is possible, however, if suitable parameters are selected for such measurement, and benchmarks may be set through analysis of the performance of many enterprises and the better performers identified. In some situations performance measurement may be relatively straightforward, e.g., where a number of operatives (people or teams or even whole departments or organizations) are performing the same task under similar circumstances – performance measurement may be little more than counting units of output and comparing the results. In most cases, however, there are other factors that must be considered, particularly differences, large and small, in the circumstances under which production takes place. Measuring productivity in construction is a prime example as on-site circumstances may vary markedly from project to project, and the fundamental problem of each building being a unique product 'manufactured' on a unique site makes direct comparisons of on-site performance extremely difficult.

It is similarly difficult to measure the performance of other sectors of the industry, e.g., design consultants, where both the quantity and quality of the output are important but

* University of New South Wales, Australia

there are no obvious ways to measure them. In this case, the quantity of output is hard to measure given that is even difficult to make a decision as to when a design is 'complete', and how one is to measure the quality of a design: is it based on aesthetics (surely a subjective measure), functionality (hard to assess even after the building is complete and occupied, let alone beforehand), capacity to provide revenue (at least some possibility of quantitative measurement) or some combination of these (perhaps 'value for money', but how is that to be measured?).

In spite of these difficulties, methods of measuring enterprise performance have been developed which, although they may or may not provide any absolute measure of performance, do provide a basis for monitoring performance over time and for comparing the performance of one enterprise with that of others. A set of these key performance indicators and the way in which they are measured is presented in the following chapter.

The relationship of this material to value in buildings is, perhaps, not as clear as it is with that in other chapters, however, it does relate to continuous improvement in the processes of design and construction, and the measures described can be applied to construction firms as well as to design consultants. This links the process of performance monitoring to value in building in at least two ways: first, by improving the performance of individual firms that participate in the industry, and by providing a basis for clients to select between firms who are competing for their projects. The effect on value for the client is then related to the performance of the firms that perform the work – the better their performance the better the outcome for the client.

17.1 Introduction

Performance measurement systems have been one of the primary tools used by the manufacturing sector in association with the introduction of business change programs, whether based on quality management or process re-engineering concepts. Manufacturers have used performance measurement in two important ways, either before the adoption of new philosophies such as total quality management (TQM) and lean production, to help to identify weaknesses and to define current levels of performance, or following the adoption of these philosophies, to monitor the outcomes and effectiveness of various implementation strategies. Often it is used for both.

The key to the successful use of performance measurement by the manufacturing sector has been the fundamental shift from traditional management accounting-based measurement systems to new, operation-based financial and non-financial performance indicators. The use of such new formal measurement systems can help the construction industry to understand its processes and identify opportunities for improvement or to review the performance of new processes.

The literature on construction management now abounds with arguments recognizing the need for change in the productivity and quality of construction industry processes and outputs. Further, the need for industry to respond by developing or adapting new paradigms of management is also widely accepted. As such, there is no need to develop arguments to impress the need for change. There is a continuing debate, however, about the type of production and management philosophies that need to be developed, adopted, or adapted by the construction industry. Consequently, with the acceptance of the fundamental need for change as a *fait accompli*, i.e., the need for change and the subsequent need for new strategies, the aim of this chapter is to discuss one type of tool which can be helpful in the management of design and construction businesses at a strategic level. This is the tool of performance measurement and benchmarking.

In any change process, irrespective of the management approach under consideration, the question that arises is 'how is progress to be evaluated?' In fact any strategic change needs to be preceded by a phase of information gathering and analysis. Benchmarking is a concept that can be very useful in this regard. The specific advantage of benchmarking is that it informs an organization about areas of performance in comparison to objectively defined standards. These can be derived from other enterprises or other divisions of the same company. Structured comparison can confirm the existence or otherwise of problems in any part of a business.

17.2 History and overview of benchmarking

In the manufacturing sector, the adoption of new production philosophies in individual enterprises has either been preceded by benchmarking, or followed by it. While some firms introduced new performance measurement systems as a starting point for world class manufacturing, believing that the new measurements would identify the areas requiring change, others changed production methods and then developed performance measures compatible with these methods (Maskell, 1991).

Similar to re-engineering and many other management philosophies, benchmarking too has its origins in the business and manufacturing sector. Xerox Corporation in the USA is considered to be the pioneer of benchmarking. In the late 1970s, Xerox realized that it was on the verge of a crisis when Japanese companies were marketing photocopiers for less than it cost Xerox to manufacture them. However, by using benchmarking against Japanese companies, Xerox managed to improve its products and consequently its market position. Xerox initially focused on manufacturing activities and then included other areas of its business. This effort was so successful that Xerox has continued using benchmarking and is very competitive internationally (Leibfried and McNair, 1992). This example illustrates how benchmarking enabled the achievement of the kind of results expected of fundamental process change. Examples of successful use of benchmarking are shown in Table 17.1 and reinforce this view.

Table 17.1 Some reported outcomes achieved in the manufacturing sector (Marosszeky and Karim, 1997)

sales/employee/annum up 300%	customs cycle time reduced by 72%
direct/indirect staff ratio up	manufacturing waste reduced by 69%
staff changes down by 50%	team participation increased by 350%
no industrial disputes	manufacturing costs reduced by 34%
on time delivery doubled	lead time down from 6 months to 2 days
accidents down by 30%	supervisory personnel reduced by 50%
absenteeism halved	inventory reduced by 60%
lead time reduced by 83%	service readiness up from 73% to 91%
customer rejects reduced by 85%	optimum cycle down from 240 to 24 hours

Basically, in the management of businesses, benchmarking can be looked on as a decision support tool. It enables a comparison of inputs and outputs, either separately or in conjunction with each other, between entities within an organization, between organizations within a particular industry sector, or with other industries. An evaluation of these comparisons then provides an indication of whether something needs to be changed.

In the business process re-engineering literature, it is recorded that many attempts at change have failed to realize the anticipated benefits (Hammer, 1990; Lorenz, 1993). It is essential that first of all existing levels of performance be measured and problem areas identified through a formal information gathering and analysis phase. Business strategy change, if required, will follow this evaluation. Performance measurement and benchmarking focus on processes and not the organization, i.e., the process is evaluated and then decisions about the organization are made. Further, benchmarking also helps to monitor the outcomes that are achieved as a consequence of process change.

Performance measurement and benchmarking have also become popular tools in recent years, driving industry reform in areas such as product quality, environmentally responsible design and general business practices. For example, an early application was the development of CONQUAS (CONQUAS, 2001), a quality measurement instrument developed by the Singaporean Construction Industry Development Board (CIDB, now the Building and Construction Authority – BCA). This comprehensive system of physical measurements assesses quality in structure, architectural finishes and mechanical and electrical services at various stages of a construction project and leads to a composite score for a building. Government clients in Singapore give cash bonuses for outstanding quality identified using this tool. The BCA is also able to monitor quality improvement across the sector over time and hence assess the success of its policies.

In relation to environmental design, the Building Research Establishment (BRE) in the UK developed a system of performance measurement called BREEAM (BREEAM, 2001). This measure has had similar results in relation to environmentally responsible design. It defines the parameters that influence design performance and hence clients can specify levels of performance for their projects, and industry can work to develop products and design that enhance the environmental performance of buildings as a whole.

Companies and agencies in the UK have been the leaders in the use of performance measures as a tool to drive industry development at a whole-of-industry level. Arising out of the Egan Report, the Movement for Innovation (Movement for Innovation, 2000) was established and as a part of this reform strategy, thirty-eight performance measures were defined within the Construction Best Practice Programme. Of these, twelve are noted as *headline* performance measures and the others as either for *operational* or *diagnostic* purposes. The main or headline measures are in the areas of time and cost performance, customer satisfaction, defects, profitability, productivity and safety. The definitions of these are detailed on the M4I web site (Movement for Innovation, 2001a) and the results of M4I projects compared to other projects are reported in the 2001 annual report (Movement for Innovation, 2001b). The same set of performance measures have been used to drive industry reform in the housing sector through the Housing Forum (Housing Forum, 2000). More recently the movement has developed environmental performance measures in relation to CO_2 emissions and

CO_2 embodiment, water consumption, solid waste generation, biodiversity and transport (Movement for Innovation, 2001c).

In relation to enterprise performance, the Australian Centre for Construction Innovation (ACCI, 2001) has developed a set of performance measures for both design and construction enterprises, the method for developing these are described in detail in the next section of this chapter and in the subsequent section, data collected from design companies are presented in detail. This illustrates both the definition of criteria and the method of presenting performance information in an anonymous format.

17.3 Key performance indicators for construction and design enterprises

The aim of the ACCI project in developing key performance indicators (KPIs) for the construction sector was to stimulate improvement in efficiency. It was designed to identify those key areas in which comparison of performance with *best practice* would provide insight to management. The outcomes sought were:

- establishment of a basis for objective comparison in identified key areas
- implementation of performance comparisons (benchmarking)
- enterprise improvement.

KPIs were developed over a period of 2 years with the assistance of three industry working parties. To guarantee relevance, industry involvement was sought from the very beginning. The development of KPIs comprised the following five steps:

- the project was introduced through the distribution of a discussion paper and a series of industry seminars
- the next stage involved interviews with clients, consultants, contractors and specialist contractors. The purpose was to record current practice and identify future requirements of performance measurement in the industry
- a workshop reported initial findings to stakeholders and identified the priority areas of each of three groups for KPI development. Three workgroups were established to assist the further development of relevant KPIs
- these workgroups met regularly over a period of 8 months. During this period, data-collection definitions and rules for KPIs were developed. Each KPI was discussed in detail to ensure that all the relevant parameters had been identified
- once the KPIs had been developed, the firms participating in the workgroup conducted a trial survey. Feedback from this trial enabled the KPIs to be fine-tuned.

The business processes of contractors and consultants comprise numerous sub-processes, along with their inputs and outputs. Each one of these contributes to the success or failure of a business. It was agreed with the industry groups at the outset that developing an excessively large number of KPIs would be as bad as not having a performance measurement system at all. Consequently, it was decided to limit the KPIs to the most critical areas in terms of potential benefit and to areas where data could be generated readily.

17.4 Priority areas and key performance indicators for designers

Towards the end of 1998 consulting firms from New South Wales were invited to participate in the project by providing data pertaining to 1997–98. Forty-five firms registered for participation in early 1999. The ACCI signed individual confidentiality agreements with each firm to protect the privacy of its data, however in spite of this nearly half of the firms did not provide their data because of fears regarding confidentiality.

The sample primarily comprised architectural and engineering consultancies. The first results – the first KPIs – were developed on the basis of a survey of consultants in the construction industry. Obviously, the type of business – architectural or engineering – is bound to have an impact on these ratios, however, this level of analysis was not conducted because the sample size was not large enough to yield reliable indicators. The KPIs allow individual firms to compare their performance with others and to identify areas with potential for improvement. The results show that in most cases the spread between best and worst performers is quite large.

Wherever possible, the following discussion also contains international comparisons of the KPIs. These comparisons use data pertaining to firms in the USA, contained in the *1997 Survey & Report on Financial Performance in Design Firms* (PSMJ, 1997).

17.5 Profitability/efficiency

There are a number of KPIs that measure the various elements that contribute to the ultimate profitability and efficiency of an enterprise. Note that all monetary figures are in Australian dollars unless noted otherwise.

17.5.1 Annual revenue per staff member

This KPI is based on the total number of staff, including contract employees, on the last day of the reporting period. It is obtained by dividing the total annual revenue by the total number of staff. The value of the KPI ranges from $30 000 to $137 544, with an average of $81 789 (Figure 17.1).

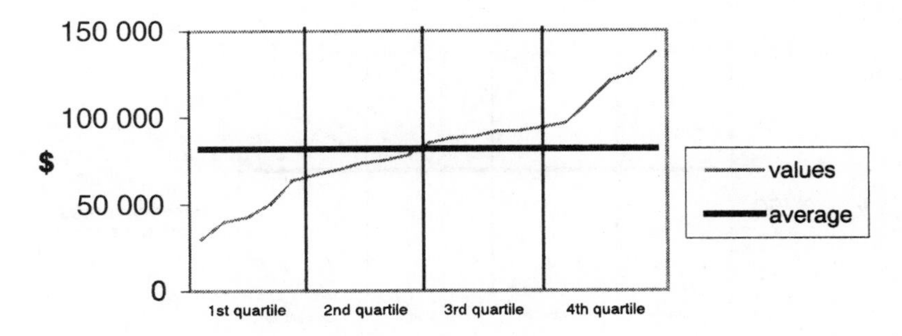

Figure 17.1 Annual revenue/total staff.

17.5.2 Annual revenue per technical staff member

This KPI is similar to the previous one, except that it is based on the number of technical staff. For traditionally structured practices the previous measure alone may be adequate, however, this KPI will be of particular value to those firms where the ratio of technical to support staff is considerably different to the norm. The annual revenue per technical staff member ranges from $52 590 to $155 178 and the mean is $100 246 (Figure 17.2).

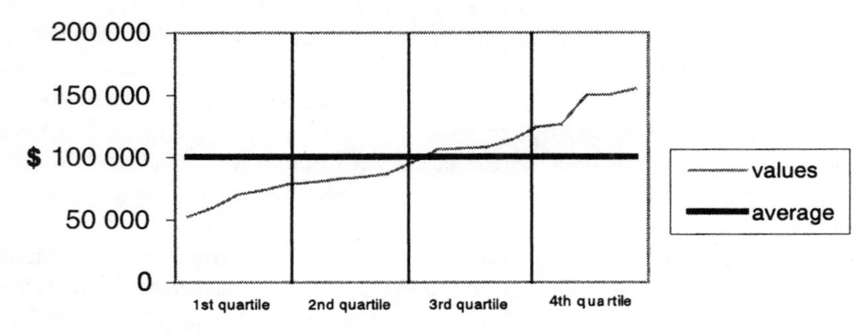

Figure 17.2 Annual revenue/technical staff.

17.5.3 Revenue per charged hour

This KPI is defined as the total annual revenue divided by the total number of hours charged to actual projects. This is an indicator that measures the integrated impact of employee efficiency. For fixed price contracts, greater efficiency should result in higher revenue.

Again the spread is quite large, with a minimum of $31.25 and a maximum of $130.67. The mean and median, $75.6 and $76, respectively, are exactly the same for all practical purposes. This is close to the charge-out rate most commonly used. These figures can, of course, vary depending on the market segment of an enterprise as well as its staff profile. Practices with a larger ratio of junior staff will have a lower than average value (Figure 17.3).

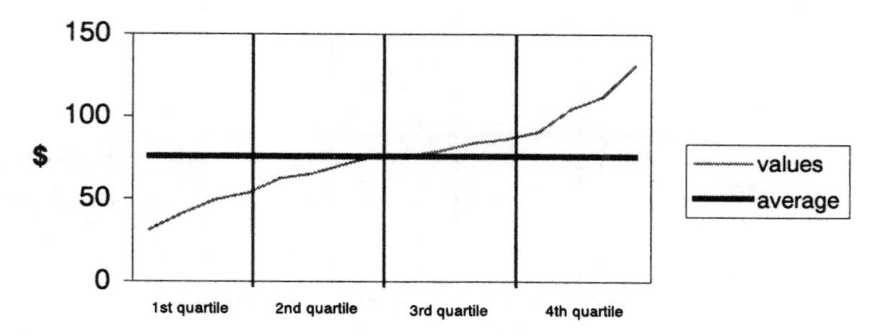

Figure 17.3 Revenue charged per hour.

17.5.4 Total payroll as percentage of annual revenue

This KPI is defined as the total payroll divided by the annual revenue, multiplied by 100. It measures the input costs that have the greatest direct bearing on profitability. Given the nature of consultancy work, the payroll cost is the most significant cost input. In the given data, this varies from 23% to 81%. With 64% of the firms above the mean, it is obvious that a smaller proportion (about one third) of firms are performing very well in this area. Discounting the tails of the distribution, the range of 40% to 60% reflects that the industry basically sells its time, and reward for value may be hard to secure (Figure 17.4).

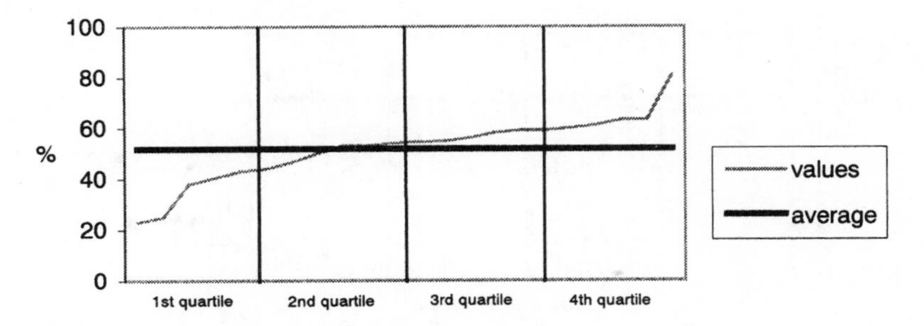

Figure 17.4 Total payroll as percentage of annual revenue.

17.5.5 Cost of winning work as percentage of annual revenue

This KPI is defined as the cost of winning work divided by the annual revenue, multiplied by 100. The cost of winning work is computed by adding the costs of marketing, speculative work and actual tendering. The total cost of winning work is also an important cost input. Its significance is further increased by anecdotal evidence suggesting that the industry is currently operating with quite low profit margins.

Again there is a large variation between best and worst performers. The values range from a minimum of 0.7% to a maximum of 13%, with a mean of 5.35%. About 41% of the firms have costs above average (Figure 17.5). Some respondents indicated that it is hard to provide accurate figures for winning work. A company's costs for this measure would be related to its ability to win repeat and referral business.

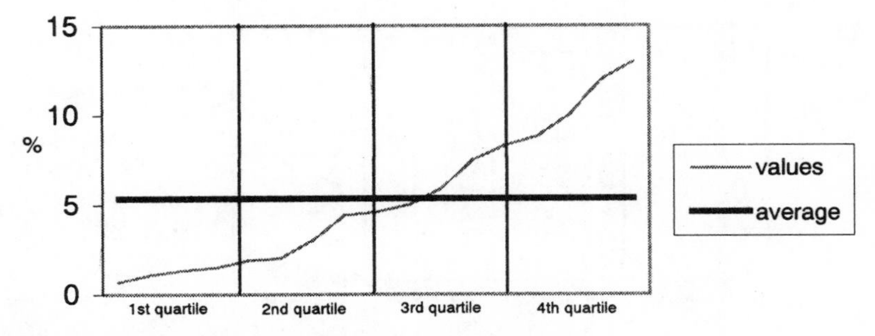

Figure 17.5 Cost of winning work as percentage of annual revenue.

17.5.6 Utilization rate

The utilization rate is the number of hours charged to projects as a percentage of the total hours worked. A higher utilization rate means more work and optimum use of staff. From the available data, utilization rates range from 35% to 92% (Figure 17.6). With the median (73%) almost the same as the mean (73.3%), the number of firms above and below average is about the same. Smaller organizations may be able to achieve utilization rates at the high end of the range. The mean utilization rate for US firms is estimated to be 68–70% (PSMJ, 1997).

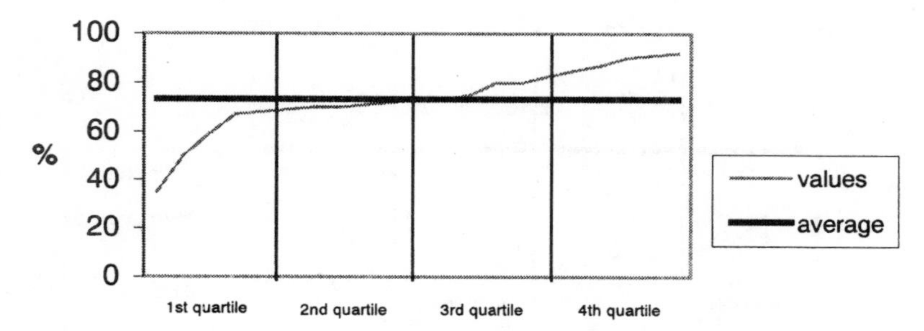

Figure 17.6 Utilization rate.

17.5.7 Contribution to overhead and profit

This KPI is obtained by deducting the direct cost of labour from the annual revenue, dividing the result by the total revenue, and multiplying by 100. The direct labour cost also includes the cost of sub-consultants and, therefore, is different from payroll costs. This

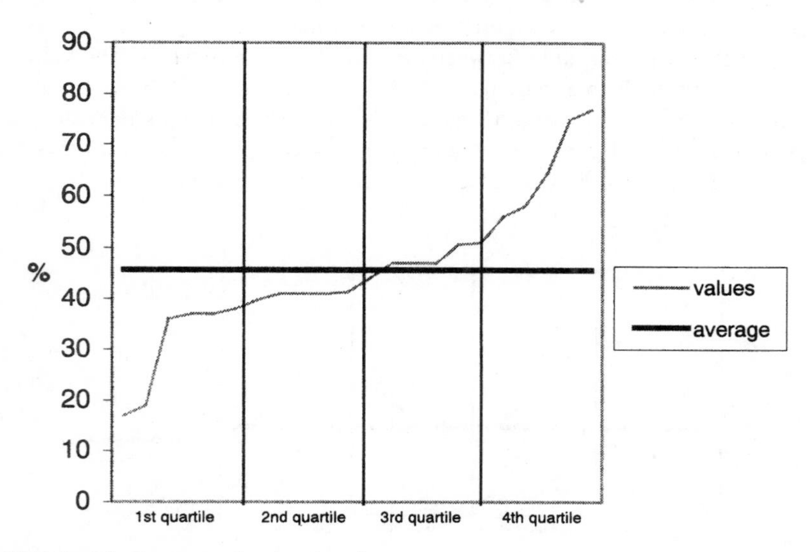

Figure 17.7 Contribution to overhead and profit.

KPI is sensitive to the efficiency of direct costs. Currently this ranges from 17% to 77%, with an average of 45.7% (Figure 17.7). This is less than the mean value of 48.7% achieved by firms in the USA (PSMJ, 1997). This measure should be looked at in association with utilization rates. High utilization rates should yield a high contribution.

17.6 Financial management

Good financial management is an outcome of good cash flow, and inventory and financial asset management.

17.6.1 Lock-up

Lock-up is made up of the inventory, i.e., work in progress plus debtors outstanding, less any fees invoiced in advance. The KPI defines lock-up in days: the lock-up divided by the average daily revenue.

Lock-up reflects the extent to which a consultant is financing the client's business. Even though a company is profitable, it can fail due to poor cash flow. The larger the lock-up, the greater the cash flow problems. From the available data, lock-up ranges between 34 and 130 days, with an average of 75.25 days (Figure 17.8). This compares well with the average of 93 days in USA (PSMJ, 1997).

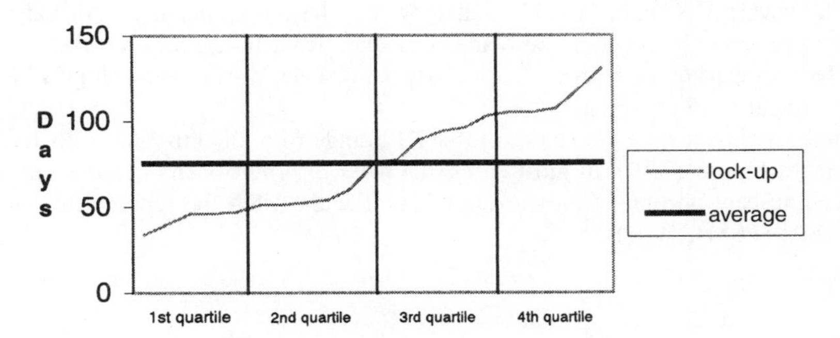

Figure 17.8 Lock-up.

17.6.2 Debt to equity ratio

This KPI relates to the long-term financial stability of an enterprise. It describes the total external long-term (repayable beyond the 12 month period) debt as a percentage of equity (equity capital, retained earnings and any loans from partners or directors).

In this case it is quite interesting to note that about 44% of the firms do not carry any debt (Figure 17.9). This position reflects the fact that current work volumes are high. Firms in this position are protected in the case of a downturn. While the average debt to equity ratio is 30.83%, about two thirds of the firms are below average, with only a few very high.

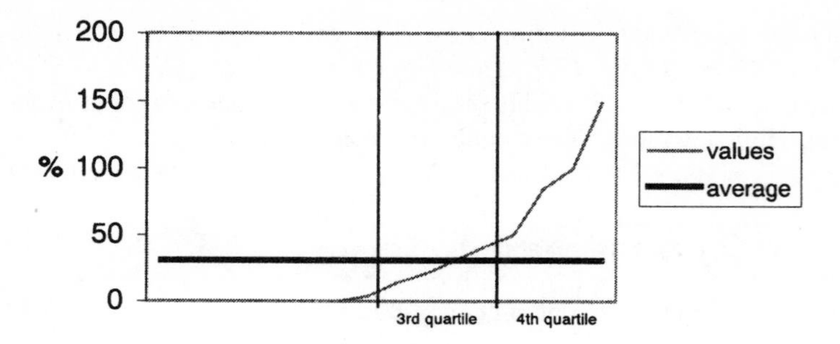

Figure 17.9 Debt to equity ratio.

17.6.3 Current ratio

This KPI is an indicator of short-term financial management and needs to be looked at very carefully. It is defined as the ratio of total current assets to the total current liabilities. *Current* is defined as falling within a 12 month time-frame. Current assets will include cash, accounts receivable, work in progress, and any other current assets. *Current liabilities* will include accounts, taxes payable, and borrowings repayable within 1 year. Generally it is accepted that if this ratio is below 1.2 a business is at risk. If the *current ratio* is greater than 1.5, it usually identifies a less than optimum utilization of financial assets. Since *work in progress* (WIP) and *debtors outstanding* are part of the current assets, however, it is quite possible that they may be a significant proportion of what appears to be a very high current ratio and consequently each firm must analyse its *current ratio* when comparing with others. Particularly, a low *current ratio* with a high WIP makes it more critical to improve cash flow.

From the available data the values of this KPI range from 0.45 to 3.66, with 1.73 being the average (Figure 17.10). In most cases, the *work in progress* and *debtors outstanding* form a significant portion of the current assets. For the USA the reported mean current ratio is 2.03 (PSMJ, 1997).

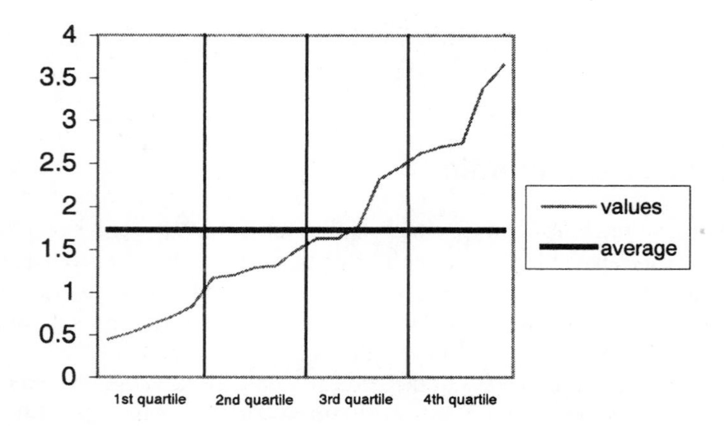

Figure 17.10 Current ratio.

17.7 Customer satisfaction

Customer satisfaction is an important contributor to business growth and sustainment. A firm with a good reputation will generate repeat business as well as new business based on that reputation. Business obtained without the process of competitive tendering is one valid measure of this and this KPI will also indicate this.

17.7.1 Ratio of negotiated jobs

Values of the KPI are obtained by calculating the dollar value of the negotiated new jobs as a percentage of the total dollar value of all new jobs. Jobs acquired through negotiation after a tendering process are not considered negotiated. Currently values range from 20–100%, with an average of 66.12%. This curve is impacted by the fact that large projects are usually put to tender (Figure 17.11).

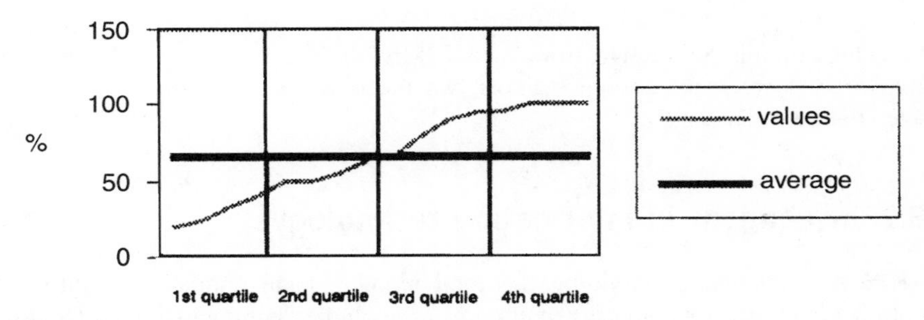

Figure 17.11 Ratio of negotiated jobs.

The data shows a moderate correlation between the number of new jobs and the cost of winning work. The greater the number of new jobs, the smaller the cost of winning work. It is anticipated that a larger data set would show a stronger correlation.

17.8 Investing in the future

The growth of Intellectual Capital (IC) is an important, perhaps the most important, element of future business growth. This, in turn, depends on the growth of Human, Structural, and Relational Capital. While the Relational Capital is captured, to a degree, by the KPI for negotiated jobs, measuring input into the remaining two categories is seen as one way of comparing the potential IC.

17.8.1 Training

Structured training is an investment in the growth of Human Capital. The hours of training per employee have been used as a KPI for this purpose. Only formal internal and external training is included and only full-time employees are counted.

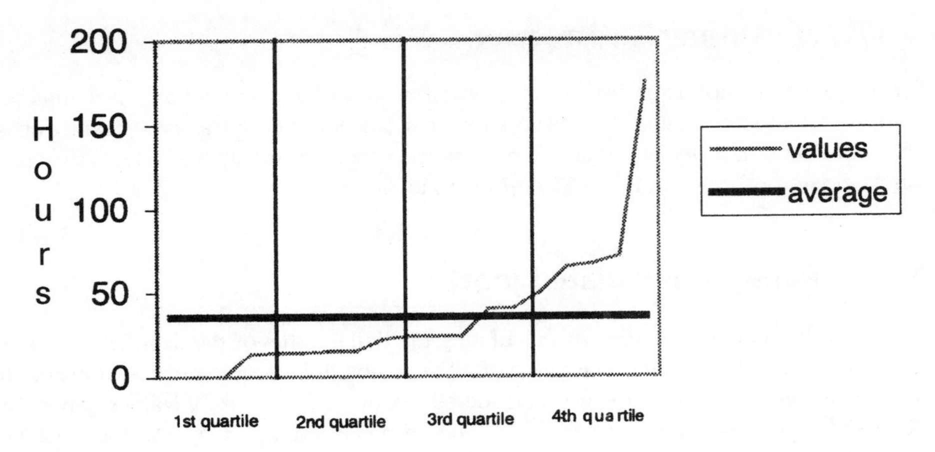

Figure 17.12 Training.

The values of this KPI range from 0 to 175, with an average of 35.5 representing approximately 1 week per annum. Almost two thirds of the reporting firms are below average (Figure 17.12).

17.8.2 Investment in information technology

This KPI is computed by dividing the total amount spent annually on information technology (IT) by the total number of employees, including contract staff, on the last day of the reporting period. The total spent on IT includes all the procurement, rental and maintenance expenses for computing hardware and software as well as the cost of employees exclusively dedicated to IT.

IT investment can act as a leading indicator as it impacts on an organization's capacity in the future. This curve might be smoother taking a 3-year average. High and low figures reflect re-investment patterns. The expenditure per employee on IT ranges from $640 to $9000, with an average of $3250 (Figure 17.13). While the highest value appears to be the result of recent upgrading there are quite a number of firms hovering around the average. About two thirds of the firms are in the range $2600–$6000. By comparison the mean expenditure per employee in USA is Aust$11 626 (not including the cost of dedicated

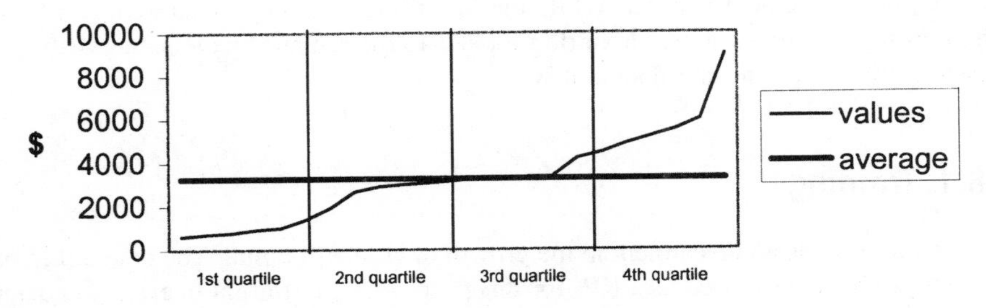

Figure 17.13 Investment in IT. IT expenditure per employee.

employees) (PSMJ, 1997). The *Yellow Pages Small Business Index* (Telstra, 1999) reports a figure of Aust$2950 per employee for the Australian business services sector including consultants.

17.8.3 Computers per employees

The number of computers per employee range from 0.5–1.25, however, the average of 0.9 seems right, since one computer per employee is the norm with 43% of firms interviewed, while 39% of the firms are below the reported average (Figure 17.14).

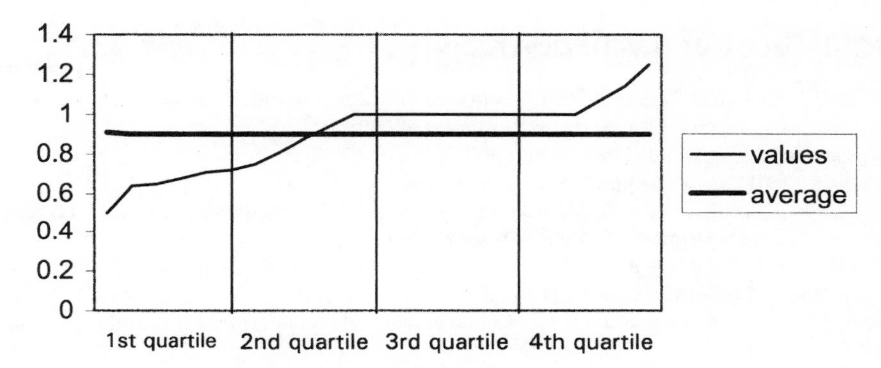

Figure 17.14 Computers per employee.

Looking over the data some interesting general observations can be made. For example, 44% of the firms involved in the study do not carry any debt – clearly an enviable position; lock-up (a function of work in progress and debtors outstanding) was found to range from 34 to 130 days. Such disparity between best and worst performers suggests that some enterprises need to intensify their efforts in prompt billing and recovery of outstanding receivables.

Another interesting outcome is that some firms obtain all their jobs through negotiation rather than competitive tendering. This may indicate that while some enterprises enjoy better relations with their customers than others, firms that negotiate all their jobs might need to broaden their client base. Experienced practitioners consider that there is a risk in having too much work on a negotiated basis as it may result in firms losing touch with market rates. Around 70% negotiated work is considered optimal.

17.9 Conclusion

It has been demonstrated that performance measurement can be used for a range of quite different objectives. Examples have been given to demonstrate its use as an instrument to drive industry reform and improvement in areas as diverse as product quality, environmental design and overall industry performance.

In addition, examples have been given of measures for targeting and assessing enterprise level performance. This technique can be used to stimulate improvement at any level. Experience has shown that objective feedback, if presented in a constructive manner, generally has the effect of encouraging improvement in both individuals and organizations. If performance data at an industry level is carefully selected it can inform industry policy development, with performance information being used to identify areas of weak performance in the sector and to monitor the effect of policy in bringing about change. Once change strategies have been implemented, ongoing data is very useful for identifying general trends, comparing future to past performance, and guiding the refinement of industry or business strategy.

References and bibliography

ACCI (2001) Australian Centre for Construction Innovation. www.fbe.unsw.edu.au/units/ACCI/

BREEAM (2001) Building Research Establishment (Watford). http://products.bre.co.uk/bream/default.htm

CONQUAS (2001) www.conquas21.bca.sg/index.htm

DPWS (1996) *The Construction Industry in New South Wales: Opportunities and Challenges* (Sydney: NSW Department of Public Works and Services).

DPWS (1997) *A Perspective of the Construction Industry in NSW in 2005* (Sydney: NSW Department of Public Works and Services).

Hammer, M. (1990) Reengineering work: don't automate, obliterate. *Harvard Business Review,* **68**, 104–12.

Hammer, M. and Champy, J. (1993) *Re-engineering the Corporation: A Manifesto for Business Revolution* (London: Nicholas Brealey Publishing).

Housing Forum (2000) www.thehousingforum.org.uk

Leibfried, K.H.J. and McNair, C.J. (1992) *Benchmarking: A Tool for Continuous Improvement* (New York: HarperCollins Publishers Inc.).

Lorenz, C. (1993) Uphill struggle to become horizontal. *Financial Times*, November 5.

Love, P.E.D., Mohamed, S. and Tucker, S.N. (1997) A conceptual approach for re-engineering the construction process. In: *Construction Process Re-engineering 1997: Concept & Applications* (Queensland: Griffith University).

Marosszeky, M. and Karim, K. (1997) Benchmarking: a tool for lean construction. In: *Proceedings of the Fifth Annual Conference of the International Group for Lean Construction* (International Group for Lean Construction).

Maskell, B.H. (1991) *Performance Measurement for World Class Manufacturing: A Model for American Companies* (Cambridge: Productivity Press Inc.).

Movement for Innovation (2000) www.m4i.org.uk

Movement for Innovation (2001a) www.m4i.org.uk/kpis/ (*m4i_kpi_report2000.pdf*)

Movement for Innovation (2001b) www.m4i.org.uk/kpis/ (*kpi_report2001.pdf*)

Movement for Innovation (2001c) www.m4i.org.uk/kpis/ (*m4i_epi_report2000.pdf*)

PSMJ (1997) *Survey & Report on Financial Performance in Design Firms*. Seventeenth edition (USA: PSMJ Resources Inc.).

Telstra Corporation (1999) *Yellow Pages Small Business Index* (Telstra Corporation).

18

Three-dimensional CAD models: integrating design and construction

Rabee M. Reffat*

Editorial comment

As the complexity of buildings has increased, particularly in the engineering services that they need in order to function effectively, so too has the fragmentation of the design process increased with engineers and other designers of various disciplines each contributing to the overall building design. Space occupied by services installations, i.e., duct risers, access floors, suspended ceilings, plant rooms, computer rooms and the like, is not rentable space and clients are always keen to maximize the amount of usable space in their buildings.

Engineers are expected to fit their systems into the smallest possible spaces, and these spaces are usually shared by a variety of services; the space under a raised floor, for example, may contain mechanical ductwork, electricity supply, data cabling, chilled water supply to air conditioning units, general water supply pipework, fans for displacement heating, ventilation and air conditioning (HVAC) systems, cabling for security systems and so on. The depth of the space under a raised floor is kept to a minimum, as more depth means greater building height, greater building height means more structure and more external wall area and, inevitably, greater cost.

The potential for these systems to get in each other's way is obvious, as the various components compete for the limited space available in plant rooms and ceiling and floor voids. It is not only services that are competing for space – the structural engineers also require some room to accommodate beams, drop slabs or column capitals, and the addition of a beam, for instance, in a ceiling space can cause huge problems for the mechanical engineer who has to run a large sheet metal duct through the same space. Unfortunately it is not uncommon for such clashes to remain undetected until the beam is already in

* University of Sydney

place, the ductwork has been fabricated and the mechanical sub-contractor comes to install it. What follows is not pretty: there are arguments about who is to blame, and who is to pay for the remedial work; ductwork has to be redesigned and new parts made, the work is delayed, tempers fray and nobody profits except, of course, the lawyers.

The difficulties arise from two situations: one is the problem of keeping all interested parties informed of changes that are made (e.g., the addition of a beam, the moving of a column) and then for anyone affected by the change to make due allowance for it in their part of the overall design; the second is the problem of the time required for someone to physically go through all the consultants' drawings looking for potential clashes. Even if time is available for such an exercise there is plenty of scope for human error and for clashes to be left unnoticed, given that anyone attempting such an exercise would have to somehow visualize, in three dimensions, all the parts of the puzzle (ducts, structure, pipework, lights, and a host of other components) that are all shown in two dimensions on different drawings (mechanical, architectural, structural, electrical and so on).

In fact these problems of co-ordination are only part of a larger problem, that of co-ordinating and integrating the whole design effort across a range of disciplines so that a holistic design solution that completely addresses and satisfies the client's brief is achieved. The value to the client, and to those who will eventually occupy the building, of such an outcome is self-evident, yet all too often the final building is not as good as it might have been because much of the design work is carried out by consultants working in virtual isolation and there is a lack of involvement by key players at crucial times in the design process. There is, however, a move now towards the use of 3-dimensional computer-aided design (3D CAD) modelling techniques to bring all participants in the design process closer together. These techniques offer the chance for all those working on a project to have easily accessible, up-to-date information about the project and the progress on the design effort at their fingertips – quite literally, as it is through their computers that such information is available.

In this chapter the author describes these modelling techniques and gives some insight into their potential utility – the potential to all but eliminate costly on-site clashes, to give greater power to the collective design effort of a large team made of people from a range of disciplines working at a variety of locations, and ultimately to provide clients with better value for their construction dollar.

18.1 Introduction

Buildings are constructed to serve the needs of their occupants. Occupants need facilities with comfortable, safe and healthy environments, utilities and technical equipment to perform their work. Building systems may be defined as a group of electro-mechanical components connected by suitable pathways for the transmission of energy, materials and/or information directed to a specific purpose. An integrated approach to building systems gives consideration to the overall objectives rather than the individual elements, components and sub-systems.

Traditionally, co-ordination of building systems is primarily that between mechanical and electrical engineers, passed down the line to contractors who co-ordinate on site. In the past two decades the integration of diverse technologies such as mechanical, electrical, bioclimatology, geophysics, optics, electronics and computer engineering has begun to play an increasingly important role in the design of building systems and the environment they control. Building systems applications may include heating, ventilation, air conditioning and cooling (HVAC), lighting, power, security, fire and life safety, building automation, audiovisual communications and computer networking of various kinds. Because of the large number of interrelated factors in these systems, there can be many solutions to the same building problem, all of which will satisfy the minimum requirements (Ahuja, 1997).

The computerization and integration of building systems technology and the information age we live in have changed the way humans perceive their habitat. Computers are mathematical instruments of enormous value in the science of building systems design. CAD systems are extremely valuable tools in drafting and modelling. Three-dimensional CAD modelling is introduced as a tool to support the co-ordination of building systems during design, construction and maintenance.

18.2 Using CAD modelling in building systems

Before the development of computers a variety of conventional media were used to represent buildings during the design process. Ancient architects used text to abstractly describe the design process (Hewitt, 1985); 2D drawings were later introduced but only expressed abstract visual thinking. Then with the widespread use of physical models in the Renaissance, the form of space and architecture was given better precision. The evolution of design tools continued with the development of perspective drawing and other attempts at representing three dimensional objects using the two dimensional medium of paper, e.g., isometric and axonometric projections.

More recently, digital technology has been developed and has matured rapidly. This growth has led to a broad understanding of the new capabilities provided by the computer that erodes the traditional boundaries of computing. It is worth employing computer technology rather than the conventional design media to bring about significant changes in the process of building systems design and maintenance. The conventional approach involves the use of drawings and models to represent the building. The type of models used in the design process can either be physical or digital – both types can be used as a means of solving complex problems that 2D drawings are unable to handle (Lin, 2001).

Three-dimensional building models are useful across the entire spectrum of architecture, engineering and construction (AEC) practices. Architects and their clients use 3D building models to observe and evaluate building designs before construction, while there is still a chance to make substantial changes at a reasonable cost. Engineers use 3D building models for energy, lighting, acoustics and fire simulations. The results of these simulations give valuable insight into building useability and safety. Professionals in the construction industry utilize 3D models to estimate costs and to plan cost-effective construction sequences. This process often leads to the early discovery of design conflicts that would otherwise result in expensive construction mistakes. Even for an existing building, it is often desirable to have a 3D model to facilitate analysis of the energy properties of the building or to predict how a potential fire might spread or to study potential changes to the building, or to identify possible uses of existing building spaces (Lewis and Sequin, 1998).

Drafting is still associated with the common perception of the application of CAD to architectural design, i.e., the concept of the 'electronic drawing board'. CAD technology has, however, progressed to a level at which it is possible to communicate design expressions representing early stage design ideas right through to detail drawings. This is quite different from CAD as an instrument for efficient production, or as a vehicle for the graphic presentation of the already designed building. Being able to use 3D CAD systems fluently is synonymous with being a good designer, rather than with being a draftsman (Szalapaj, 2001).

18.2.1 What are three-dimensional CAD models?

Three-dimensional CAD models are three-dimensional computational representations of objects drawn in the *x*, *y* and *z* axes and illustrated in isometric, perspective or axonometric views. These views are achieved simply by rotating the viewpoint of the object. Three-dimensional CAD modelling of an object in general provides the following advantages over conventional drawing:

- an object can be drawn once and then can be viewed and plotted from any angle
- a 3D CAD object holds mathematical information that can be used in engineering analysis, such as finite-element analysis and computer numerical control technology
- a 3D CAD object can be shaded, rendered and assigned various materials and finishes for visualization.

These models can be generated by the use of various types of CAD software systems such as AutoCAD, Microstation, ArchiCAD and many more.

18.2.2 The role of three-dimensional CAD models in design and construction

An understanding of the ways in which 3D CAD modelling techniques can be used to support and reflect design thinking can lead to the development of a greater integration in the building design and construction industry. Since the inception of CAD, computers appear to have played a vital role in the practice of architecture, engineering and their allied professions. This is, however, merely an illusion. It did not happen, simply because most designers in practice were not formally trained to use the computer as a productivity tool and they were therefore unfamiliar with its capabilities. In fact, there are many designers who still develop conceptual sketches for a project, then pass these sketches on to a draftsperson who creates 2D design development and construction drawings with little integration, if any, with other consultants.

The primary purpose of a 3D CAD model needs to be established at an early stage in a project. Three-dimensional CAD modelling can be used in structural, lighting, acoustic, thermal, bio-climatic and spatial analysis. Here the focus is on the use of 3D CAD modelling to support the co-ordination of building systems. There is still a common misconception that CAD systems are just drafting tools for use in the post-design stages of work rather than having a much richer role to play during designing and construction.

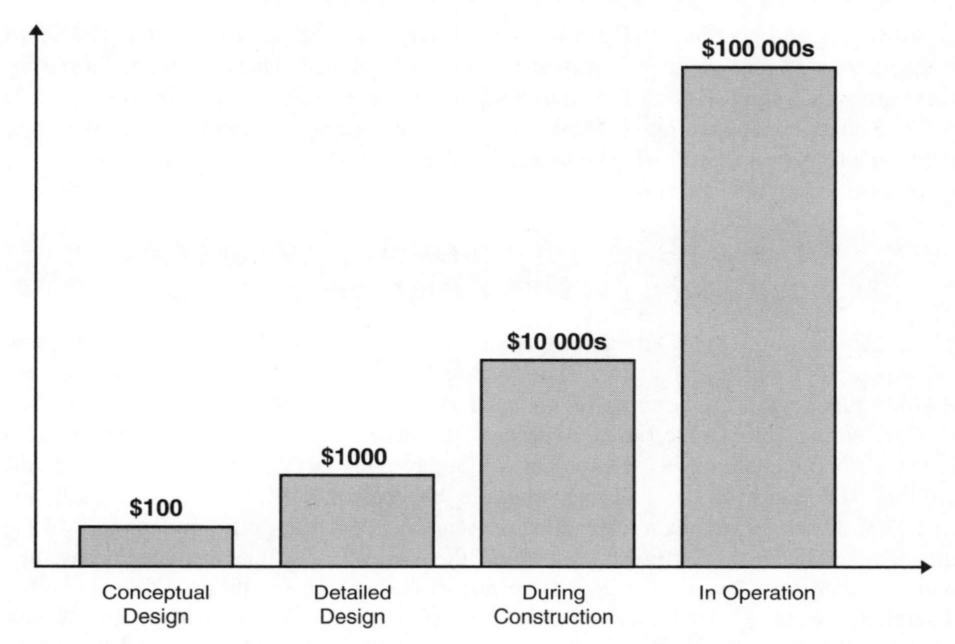

Figure 18.1 Exponential increase in repairing design and construction mistakes from conceptual design to construction operation.

Three-dimensional CAD models can help to resolve ambiguities, provide linkages to design data and present computerized visualization.

Discovering design conflicts and inconsistencies early is far less costly than repairing design and construction mistakes in buildings, as cost increases exponentially at every stage from conceptual design to construction as shown in Figure 18.1. Three-dimensional CAD modelling allows for such inconsistencies to be discovered before construction and therefore for better design and construction decisions to be made in the very early stages of the design and construction phase.

18.2.3 Interoperability and data sharing of three-dimensional CAD models

Interoperability has one important meaning: it is the capability of devices from different manufacturers to communicate and work together. Its benefits cannot be realized without establishing rules of standardization and compatibility across the different sectors of the building and manufacturing industry.

The lack of true interoperability is arguably the single largest software problem facing today's manufacturers. It can cause time-to-market delays, bottlenecks, errors, lost data, quality problems, and extensive reworking of parts. This problem also impedes the cost-effective outsourcing of design and production in global manufacturing.

There are currently a variety of software packages available designed to facilitate interoperability between the products of various manufacturers and suppliers. They assist in converting and viewing CAD formats, translating between many different CAD/CAM (computer-aided manufacturing) formats, and modifying and sharing objects. Moreover, there are Application Service Providers (ASPs) for the translation and adaptation of 3D solid models, enabling rapid sharing of 3D engineering data across design and manufacturing firms, and their clientele, regardless of their installed CAD systems.

18.3 Advantages of using three-dimensional CAD modelling in building design and construction

The advantages of 3D CAD modelling cover the whole plant-life from layout and design to construction and maintenance. The development of 3D CAD software systems has enabled planning procedures to be changed from 2D for each engineering discipline (piping, wiring, HVAC, etc.) to an integrated 3D planning procedure. Three-dimensional CAD modelling encourages users to plan work thoroughly when creating their documents. After a 3D model of the building structure has been 'built' from architectural and structural drawings, the main internal equipment is added. The draft arrangements of the different engineering disciplines determine the allocation of the available space and planning can start. Since a designer from one discipline can see the results of the other disciplines' work, a visual collision check is performed during the planning procedures. The visual collision check is currently carried out via a walkthrough inside and around the 3D CAD model. When planning is finished for the rooms involved, an automatic collision check is performed, which normally identifies the handful of collisions that need to be eliminated. Further modifications, desired by the project manager or client for example, can be accounted for (Lockau, 1996).

Three-dimensional geometric CAD models are usually in the form of wire frame and solid models. Wire frame modelling, still widely used by architects, is known as the ancestor of contemporary CAD presentation. The nature of wire frame models reveals the underlying structure of the building, thus helping architects check the project's buildability. Wire frame modelling is, therefore, used for establishing the building's skeleton in the developmental design stages of most CAD programs. Solid modelling is usually utilized for the finished model. Despite wire frame modelling's advantages, some architects prefer using solid modelling in the primary stages as well as in the final stage. The difference between wire frames and solid modelling is parallel to the difference between using paper/pencil and a 3D object like wood or foam to construct the initial design. Even if solid modelling is used from beginning to end, the first draft does not look remotely like the finished design. In the early stages, the architect is not interested in making the building look realistic, only in outlining the rationale of the design. The importance of the computer is that it acts as an interface between the physical design and the abstract ideas behind that design. In the later stages of design, realistic elements like colour, lighting and shadows are added to help communicate the building design to the client.

Three-dimensional CAD integrated systems such as *Architectural Desktop* and *Microstation Building Plant and Engineering* enable the designer to integrate mechanical and electrical design seamlessly within the building structure. This allows the exchange of accurate design information between industry professionals. With the addition of Internet capabilities, the designer can leverage this power in order to enhance co-ordination among the entire design and construction team. This will help in reducing design cycle time while offering useful tools for design and construction documentation. Furthermore, scheduling information and space planning data can be extracted and exported to external databases for further analysis to assist building owners in post-construction facilities planning and management. It enables facilities managers to track room areas, maintain assets such as furniture and equipment, track asset quantity and cost, and export this data to external databases for reporting purposes.

From a co-ordinated 3D CAD model, individual services layouts can be generated, e.g., for ductwork, mechanical pipework, public health, fire protection and electrical services, using relevant layers with text and dimensioning. These can be output for accurate tender pricing, and the contractor can use the fully co-ordinated drawings for installation purposes. The resultant model can be utilized for external and internal visualization and animation, thermal calculation, lighting design, sun studies and crowd evacuation analysis.

These are co-ordinated within the prototype and continuously tested for clashes and discrepancies. Fully co-ordinated construction drawings are produced directly from this model. This shortens the overall design co-ordination period, identifies construction problems within the computer (saving time and money), improves productivity, and reduces defects on site.

18.4 Using three-dimensional CAD modelling as a means of useful knowledge sharing

Building design is a process that has often been considered as an activity carried out only by architects and engineers. They are co-designers in proposing the end product of a building's form, materials, supporting structure and environmental control services. However, while these co-designers share the same task, that is designing a specific building, they tend to think largely about concepts with regard to their own particular interest in the evolving building solution. Their co-designing work is in the form of knowledge contribution based on their particular experience. Their method of working together will therefore be in the form of knowledge sharing in order to evolve a commonly agreed building solution. Knowledge sharing among co-designers occurs when design information representing the knowledge contribution of each individual, usually in the form of inputs and outputs, is exchanged with other members of the design team (Cornick, 1996). Examples of the knowledge shared among the co-designers include: shapes, proportions, arrangements and materials of building elements brought together in an overall building form, structural and services elements and enclosures with regard to their structural stability, and size, shape and arrangement of building services.

The development of a 3D CAD model has the essential role of supporting knowledge sharing among co-designers. One set of designers conceives the overall form and how all its parts would fit and work together, other designers then conceive how all the system parts can be engineered as an overall assembly. The degree of realism that can be created in the 3D CAD model and the ease by which 3D views can be generated are essential in knowledge sharing. This is a much richer notion than a simple demonstration of the material, finish and form of building objects.

18.5 An object oriented approach for three-dimensional CAD modelling

The representation of building objects is not limited to the graphical information that illustrates only the structure of these objects. There are various other types of non-graphical information that are no less important than the structural representation of these objects. Non-graphical information includes functional, behavioural and semantic properties. Each building object is created to perform certain functions. These functions should produce the required set of behaviours. An integrated view of both graphical and non-graphical information can be looked at as a structural, functional and behavioural scheme of object representation. *Structure* is 'what the object is', *function* is 'what it does', and *behaviour* is 'how it does it'. Furthermore, the *purpose* of creating the object or building element is 'what it is for'. The structure of an object exhibits behaviour; behaviour affects function; and function enables purpose (Rosenman and Gero, 1998). For instance, the structure of a hydraulic elevator includes a piston and oil column, its functions are to contain loads and move them vertically, its behaviours are to push loads up and hold them by compression and its purpose involves transferring people and goods from one storey to the other. This is a simplified description of the non-graphical information related to an object or a building element The non-graphical information is not limited only to the description and properties of the object but can also include the relationships of this object to other objects in the building.

Being able to exchange non-graphical information in building projects using the computer is quite useful. A general approach has been developed by employing computer techniques that were first applied in the field of artificial intelligence (AI). In this approach objects have attributes, one of which is their geometry, which can be viewed by CAD systems; databases that contain non-geometric attribute descriptions of objects use data-structures that follow agreed standards and format. Integrated software systems would assist in describing the geometry of architectural form, the geometry of the structural frame, the structural calculations of the frame, and the work sequence and time duration of construction. Using the AI approach, when the architectural form (graphical information) is changed the other three systems (graphical and non-graphical) automatically change to suit the new form. The importance of this approach lies on the capability of computer systems to recognize the graphically represented objects as real objects with attributes rather than just geometrical forms. Object-oriented databases incorporated within CAD systems allow such a facility to be utilized. Object-oriented technologies also facilitate collaborative design.

CAD systems are now becoming increasingly object based and web-based systems are moving in a similar direction. Recent trends in CAD systems development (Szalapaj, 2001) show attempts to integrate GIS (geographic information systems) with facilities management. This type of development involves making connections between graphical and non-graphical information, and object-oriented environments for supporting such integration are increasingly being used. The integration of graphical and non-graphical information is of paramount importance in achieving the best results in planning and co-ordinating building systems.

Object-oriented CAD is an important new development in the architecture and engineering professions. Traditional CAD systems were developed to mimic the processes of hand drafting and overlay graphics. A traditional CAD drawing has little more intelligence than a hand-drafted paper document. Object-oriented CAD represents the next generation of CAD applications, predicated on the concept of object-oriented design, that has been used successfully in the software industry to build much larger, more complex applications than were ever possible using older design methods. It is only recently that object-oriented design has been applied as a way to conceptualize and communicate design solutions.

The development of new CAD systems will lead architectural and engineering practices to change their working practices as the boundaries of CAD systems are redefined in the course of their applications to real projects.

18.6 Three-dimensional CAD modelling supports co-ordination

Paper-based building design and construction documentation (drawings and specifications) have been the conventional method used for co-ordination between the disciplines of the building industry. Such drawings, whether done manually or computerized, contain only unstructured graphic entities such as lines, text notations, dimensions and symbols. Through agreed conventions between various disciplines, drawings containing these graphic entities are interpreted as a coherent set of documents from which the building can be comprehended and constructed. Any inconsistencies between the various building systems, such as structural, mechanical, electrical, HVAC and security, have to be discovered and corrected. This is usually done by marking the appropriate drawings and sending them back to the appropriate consultant. It is a very challenging process that needs full understanding, imagination and visualization capacity if the co-ordination team is to discover the inconsistencies, since all drawing are projected in 2D format. The third dimension needs to be constructed and visualized in the minds of the co-ordination team members.

Creating 3D CAD models from the ground up, using architectural drawings plus structural and services drawings, makes sure that everything fits. This requires that building systems specialists should be on board, preferably at the schematic design stage, to play their roles as co-designers. Three-dimensional CAD modelling could deliver significant productivity improvements if applied early enough in a project. It is much cheaper to rub something out on a drawing than it is to sort it out on site.

Co-ordination of building systems does not mean that everything is down to the building contractors. Most construction disputes centre on time and information; when the contractor finds things do not fit, arguments over responsibilities start, the contractor does not want to accept responsibility and the whole process collapses. Building systems are the most crucial area of co-ordination and are always the biggest problem especially with complex buildings such as hospitals. If the design and construction industry commits itself to a complete 3D CAD modelling of the project, savings of at least 30% can be made by getting the co-ordination right before construction begins on site (Stokdyk, 1997). One of the cornerstones of efficient co-ordination is the removal of ambiguities in design objects and this can be easily done using 3D CAD modelling. Three-dimensional CAD models can be rendered and shaded as shown in Figure 18.2 or can be in the form of extract line drawing as shown in Figure 18.3.

Figure 18.2 3D CAD modelling helps in eliminating ambiguities in 2D drawings (Stokdyk, 1997).

18.7 Sharing a three-dimensional CAD model among various disciplines

The 3D CAD model can be organized so that each design or construction discipline has read and write access to the computer model for its own planning work, while having read-only access to the data of other design team members. This allows every discipline to carry out draft planning and to examine planning options without interfering with the work of other disciplines.

The link between architectural 3D CAD models and building systems provides the opportunity to interrogate room objects for volumes and calculations. Sharing 3D CAD models helps to transfer performance information within the integrated design process. The transfer is not limited to graphical information but could be extended to non-graphical information linked to the 3D CAD model.

Sharing the 3D CAD model and having related topical information available allows for generating cost-related components, material quantity and specification values in order to simultaneously evaluate whether the client's capital construction and facility running cost requirements are being met.

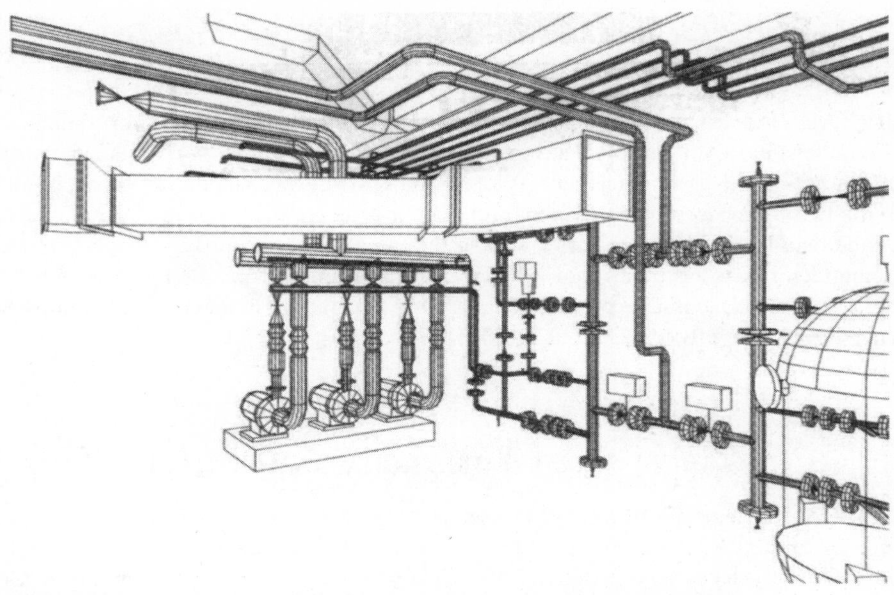

Figure 18.3 3D CAD modelling presented in an extract line drawing (LH Services, 1997).

18.8 Three-dimensional CAD modelling for building systems maintenance

The maintenance and renewal of building systems is as important as designing new systems. Reconstruction or renewal work starts, from the design point of view, with measurement and analysis of the existing situation. As-built information can be taken from the existing project documentation as long as it is available. In cases where the value of documentation has been eroded, photogrammetry with direct transfer of results into 3D CAD models can be very helpful. Photogrammetry is the technique of measuring objects (2D or 3D) from photogrammes. Commonly these will be photographs, but they may also be images stored electronically on tape or disk taken by video or digital cameras, or by radiation sensors such as scanners. Its most important feature is that the objects are measured without being touched; consequently the term remote sensing is used interchangeably with photogrammetry. Remote sensing was originally confined to working with aerial photographs and satellite images but today it includes photogrammetry, although it is still associated mostly with image interpretation.

Valve environments are often extremely congested, so a transport study is needed to prove that the clearance needed for installation and operation is available. In a real life example (Lockau, 1996), the worst case, the largest new valve left an anticipated clearance of just 10 mm. A picture was taken on the site and a 3D CAD model generated to facilitate the study of different alternatives. The advantage of generating a 3D CAD model is demonstrated in this example as various alternatives could be considered to achieve the best solution in co-ordination with the adjacent equipment.

18.9 Four-dimensional CAD modelling for effective facility management

A 4D CAD model is a 3D model with the added dimension of time. Four-dimensional CAD models allow the project team to visualize various 'what-if' scenarios and identify conflicts before the commencement of a project. This reveals scheduling inconsistencies that might otherwise remain hidden until an on-site conflict occurs. Communicating schedule details to sub-contractors is one of the main benefits of using 4D CAD models. Sub-contractors can become involved early in a project, and in 4D the models help to establish a real collaborative process with definite advantages in teambuilding and shared 'ownership' of the project (Goldstein, 2001).

18.10 Intelligent three-dimensional CAD objects

Current CAD systems are being widely used in the architectural design process, but as yet offer only limited benefits compared to their potential. The fundamental shortcoming of current CAD systems when used for the early design tasks is that a graphical CAD drawing does not represent specific technical design knowledge. It does not allow the correct and full representation, interpretation and evaluation of building systems. There has also been limited development of the use of CAD systems as integrated design environments with sufficient cross-discipline information and knowledge. This shortcoming could be overcome by the implementation of more advanced CAD systems that could capture the 'know what' (information), and 'know how' (procedures) of the design. This is what could be provided by the knowledge-based engineering (KBE) approach (Hew *et al.*, 2001).

KBE is a promising AI technique yielding the following:

- knowledge representation capability
- reasoning mechanism capability
- interface capability.

Knowledge representation capability is about controlling design knowledge through modelling. Object-oriented expression and natural language rule expression are used to describe the expert knowledge. It can also sometimes be referred to as a rule-based system. In addition, the KBE system is concerned with intelligent inferences and facilities. A reasoning mechanism capability deals with searching for and selecting application rules and with evaluating and generating logical arguments using objects. The reasoning process is often conditional involving statements in the form 'if A then B else C', where A, B, and C are propositions. The reasoning process is preceded by a repetitive sequence of message passing from one object to another. The execution of the application rules is invoked by the message received. A dynamic co-operative reasoning mechanism, as a means of communication between systems, is used to allow control of massive volumes of knowledge. KBE offers a sophisticated development environment for the developer to program the knowledge-based objects. It includes a knowledge editor, knowledge browser and a knowledge tailor. These functions offer the modeller a facility for editing and debugging the knowledge-based system (Hew *et al.,* 2001).

Intelligent 3D CAD objects have great potential and the next generation of CAD systems will be more useful for and supportive of the automated co-ordination of building systems. Intelligent 3D CAD objects are objects that can take messages and respond to them. In other words, intelligent 3D CAD objects know 'what they are' and 'what they can do' and are used as a representation unit to construct a building system. An example of an intelligent 3D CAD object is a cable tray that 'knows' its function is to distribute electric cables to various levels of the building and that can be designed in a way that allows for abrupt changes in both vertical and horizontal directions. It also knows that there are certain distances that need to be maintained between cables and the distance that the cable tray must be separated from what is below or above. These active objects may be called intelligent agents, and they are part of integrated building design systems.

These intelligent agents allow for better collaboration and co-ordination among the disciplines in the building industry and assist in discovering any inconsistencies during design and prior to construction. Each intelligent agent accommodates both graphical and non-graphical information. When an object is manipulated, corresponding changes will be made in all related objects in the building systems. If an update is not possible, at the very least, some form of alert will be sent to inform the related disciplines of a current problem that needs to be addressed. This applies equally to the discipline attempting to initiate a change and the other disciplines that may be affected by it.

18.11 Conclusion

There is obviously great potential for utilizing 3D CAD modelling to integrate the work of various disciplines in the building industry and to assist in the co-ordination of building services and systems during design and construction. Before there can be widespread adoption of this approach, however, there needs to be a change in building procurement methods and the management of the building project process.

Design and construction practitioners are gradually becoming more involved in the use and development of leading-edge CAD systems and the utilization of 3D CAD modelling and object-oriented modelling will lead to a greater efficiency in working practices between architects and other consultants, particularly engineers. This change must affect the roles of the disciplines involved and the relationships between them. The main aspect of this change will be that members of every discipline involved will have to see themselves, and be seen by others, as design partners or co-designers. If the building project design process is to be beneficially integrated then everyone needs to be of the same mind about the final product or design solution. The essential role that 3D CAD systems will play in achieving this end will be through providing the capability for simultaneously processing and, more importantly, simultaneously presenting the many different and varied information aspects of evolving building design solutions. The use of 3D CAD modelling will therefore continue to provide the visualization medium and also be a powerful front end and information-integrating mechanism by which the knowledge of each participating designer can be brought together to produce an agreed, workable and co-ordinated design solution (Cornick, 1996).

By utilizing integrated 3D CAD models, architects will go on designing building forms and materials and will be able to better demonstrate the all-round consequences of alternative proposals. Engineers will go on designing the structure and services to support architectural forms and will be in a better position to integrate their proposed solutions with those of the architect for co-ordination, appraisal and verification. Construction managers will be able to design and manage production systems for construction as an integral part of the architects' and engineers' building solution design process. Specialist trade contractors will apply their particular knowledge to produce effective, efficient and economic designs.

Integrated building design, with or without CAD, is only possible if there are forms of procurement and management that will encourage the equal involvement of all project participant designers during the building design process itself. The 3D integrated CAD modelling approach would facilitate this and help design teams arrive at building solutions that are the most satisfying for their clients.

Three-dimensional CAD models in building systems design need to be assessed not in terms of their quality of presentation but as objects that enable precise analytical and co-ordination functions. A full understanding of the ways in which 3D CAD modelling techniques can be utilized would direct the development of building designs systems from conception to completion. While the architecture of the buildings we inhabit is very much a 3D experience, design practitioners seem to be reluctant to venture themselves into 3D, often because of its perceived complexity.

Designers and engineers need to use the computers from the beginning of the design phase if they are to begin their thought processes in a 3D form. This will have a great impact on the level of co-ordination required in designing and constructing building systems. Design and engineering practices operating in a non-integrated manner will not begin to realize the true potential benefits until there is a complete change in the thinking process about technology and its place in their practices. Three-dimensional CAD modelling is an example of a computer technology that is intended to make design and construction practices more efficient through integration and it is a very valuable tool, particularly in achieving efficient, high-level co-ordination of building systems.

References and bibliography

Ahuja, N. (1997) *Integrated M/E Design: Building Systems Engineering* (New York: Chapman & Hall).

Cornick, T. (1996) *Computer-Integrated Building Design* (London: E. & F.N. Spon).

Goldstein, H. (2001) Maestros of Design and Construction Render a Virtual Masterpiece. *Construction.com,* http://new.construction.com/NewsCenter/it/features/01–20010502.jsp

Hew, K.P., Fisher, N. and Awbi, H. (2001) Towards an integrated set of design tools based on a common data format for building and services design. *Automation in Construction*, **10** (4), 459–76.

Hewitt, M. (1985) Representational forms and models of conception. *Journal of Architectural Education*, **39** (2), 2–9.

Lewis, R. and Sequin, C. (1998) Generation of 3D building models from 2D architectural plans. *Computer-Aided Design*, **30** (10), 765–79.

LH Services (1997) *Building Services Supplement*, May Issue, 10.

Lin, C.Y. (2001) A digital procedure of building construction. In: Gero, J., Chase, S. and Rosenman, M. (eds) *CAADRIA2001, Key Centre of Design Computing and Cognition*, University of Sydney, 459–68.

Lockau, J. (1996) Avoid the unexpected: Use a 3D CAD model before building or maintenance. *Nuclear Engineering International*, **41** (505), 30–2.
Rosenman, M.A. and Gero, J.S. (1998) Purpose and function in design. *Design Studies*, **19** (2), 161–86.
Stokdyk, J. (1997) The third eye. *Building Services Supplement*, May, 8–9.
Szalapaj, P. (2001) *CAD Principles for Architectural Design* (Oxford: Architectural Press).

19

Administration of building contracts

Peter Smith*

Editorial comment

Regardless of the procurement system chosen for the delivery of a construction project there will be many separate contracts that control the performance of work on site, the input from a range of consultants and, of course, the payment of those involved in the process. Added to those contracts will be a range of other transactions including those related to financing the works, having designs approved by relevant authorities, supply and testing of materials and components and so on.

On a small domestic scale project there may be no written contract, and work can proceed with just a verbal agreement between builder and client but even on the smallest project disputes can arise – the cost is higher than originally quoted, some details are not completed, additional work is needed that was not envisaged at the outset or the workmanship is not as expected. The lack of a formal written contract makes resolution of these disputes all the more difficult, and supports the view of the legendary movie producer, Sam Goldwyn, who observed, 'A verbal contract isn't worth the paper it's printed on' (Johnson, 1937).

On larger projects these problems are amplified and, in an effort to minimize their occurrence and impact, a great deal of paperwork is created, including a detailed written contract supported by extensive documentation in the form of drawings, specifications and a host of other documents. These documents are only a starting point, and subsequent records, produced as the project proceeds, including variations, quotations, tenders, site instructions and so on, must be kept up to date and stored so that they are easily accessible. In the event of a problem there must be a clear audit trail that will lead to the relevant documents, thus facilitating a speedy resolution.

Monitoring the progress of building contracts includes tasks such as keeping a register of all drawings issued, records of all site instructions and requests for information, and

* University of Technology Sydney

quantifying the progress of work on site for the purposes of preparing and assessing claims for interim payments. Naturally a great many of these tasks are now carried out with the aid of information technology and increasingly the whole process of project management is done through a centralized computer system using Internet protocols and browser software rather than with specialized proprietary software operating on dedicated networks.

Efficient contract administration is a vital part of achieving the time, cost and quality targets expected by clients as well as helping to reduce the incidence of disputes between parties, or least making the resolution of disputes faster and simpler. The value to clients extends beyond simply reducing building costs, as it also gives greater certainty to projections of cash flows and to decision-making regarding financing, allocation of resources (including staff) and business planning dependent on completion of construction work.

The value outcomes are fairly clear: reduced costs as a result of fewer disputes, minimization of interest payments as cash flows are more certain, a smoother set of transactions throughout the delivery of the project that produces a higher level of co-operation among all participants and thus a better job. This chapter looks at what makes good contract administration and just how an efficient administrator can help to ensure that projects are completed to the satisfaction of client and contractor alike.

19.1 Introduction

The administration of construction contracts is typically a complex process involving the management of a great deal of contract documentation, which invariably changes throughout the course of a project, and the flow of a vast amount of information amongst a large and diverse group of individuals and organizations.

It essentially involves the administration of progress and change on projects. In addition to a main head contract agreement between a client and a contractor, contracts that require administration on a project can include agreements with sub-contractors, suppliers, project and construction managers, designers, consultants and other participants in the process. On large projects there may be hundreds or even thousands of individual contracts, each with its own administrative peculiarities.

The administration process will vary according to the procurement method used for the project; this chapter outlines the main administrative differences that are the result of the various contract types and procurement systems. The primary focus will be on the administration of head contract agreements between client and contractor. However, the principles of these agreements can generally be applied to other contract agreements on a project.

19.2 Importance

Effective contract administration is a vital part of business success in construction. The construction industry is fraught with problems and with high levels of risk. From a contractual perspective, construction projects are characterized by the mass of contract documentation used to define their requirements. The main contract documents include the general conditions of contract, drawings and specification details (architectural, structural, services, external works and so on); other contract documents can include bills of quantities, tender conditions, tender/contract negotiations and agreements, expert reports, site and geotechnical reports, shop drawings, construction program, special conditions and amendments and addenda to the main documentation. In addition, oral agreements, implied terms and both common law and statutory legislation may form part of the contract even though not expressly included. The complexity of this documentation is exacerbated by the fact that it will invariably change throughout the course of the project. Added to this is the fact that the works will be carried out in a constantly changing environment, by a wide range and large number of individuals and organizations involved in the process, each of whom will have their own separate contract for their part of the work.

Abdelsayed (2001) undertook an analysis of the documentation requirements and the number of people and organizations involved in a variety of construction projects carried out by his company. Table 19.1 shows the average documentation requirements that he calculated for an average size project of US$10 million:

Table 19.1 Average documentation requirements for a US$10 million project (Abdelsayed, 2001)

Number of participants (companies – including all suppliers/subcontractors)	420
Number of participants (individuals)	850
Number of different types of documents generated	50
Number of pages of documents	56,000
Number of bankers boxes to hold project documents	25
Number of 4-drawer filing cabinets	6
Number of 500 mm diameter/20-year-old/15 m high trees used to generate this volume of paper	6
Number of megabytes of electronic data to hold this volume of paper (scanned)	3000

It is virtually impossible for contract documentation to reflect, fully and accurately, all the project requirements, particularly on large projects. The likelihood of errors, omissions, discrepancies and ambiguities existing in the documentation is extremely high. Not surprisingly, research has consistently shown that the vast majority of claims and disputes on construction projects stem from problems with or changes to this documentation (Choy and Sidwell, 1991).

Accordingly the majority of contractual claims on a project are legitimate with the main source of dispute relating to quantum rather than entitlement. Contractual claims are an integral part of the construction process and it makes good business sense for all parties concerned to ensure that they have claims administration procedures, processes and strategies in place to protect their financial interests in the project. From the contractor's perspective, the pursuit of claims may be thwarted by a reluctance to put their client off-side, an inability to substantiate claims, claim details being too complicated, insufficient personnel and time to prepare claims, not acting within prescribed time requirements, letting claims build up (leading to lengthy final account periods) and not mitigating the effect of a change or problem. This erodes the contractor's profit margins which, due to the highly competitive nature of the industry, are relatively low to start with when compared to most other industries. Many contracting organizations have insufficient reserves of working capital and poor claims management only adds to these capital requirements and pressures on cash flow. Large claims that take months or even years to resolve can have an extremely deleterious effect on a contractor's operations, particularly if the end result is that the contractor has the claim rejected or significantly devalued.

The contract is therefore something of a melting pot of changing documentation. One can never really be sure whether a contract is a good one until it is tested, i.e. when problems arise and the contract documents are scrutinized in detail. This is particularly so when lawyers are brought into the fray and/or when one or both parties is intent on enforcing their contractual rights rather than adopting a more conciliatory approach. It is quite remarkable how quickly the attitudes of parties to a contract can change when serious problems emerge. In recent years the industry has seen an influx of lawyers and the emergence of claims 'loophole experts' or claims consultants.

The contract documents provide the vehicle for assigning risks amongst project participants by identifying the rights and obligations, and consequently the risks, to be assumed by the parties, and so effective contract administration is necessary to protect not only a party's financial position but also their legal position.

19.3 Main forms of administration

Due to the large number of different contracts that typically exist on construction projects, administration can take many forms. The following are the main forms:

19.3.1 Head contract agreement – designated project administrator

This form of administration relates to traditional lump sum contracts between a client and a contractor whereby a third party is appointed as the Project Administrator (more commonly referred to as the Superintendent). This person or organization is not a party to the contract but is appointed as a third party to administer the contract. The project architect or engineer has traditionally carried out this role due to their intimate knowledge of the project's design. However, in recent years, many other professionals (most notably Project Managers) have assumed this role. Whilst the actual functions and powers will vary according to the terms of engagement and contract conditions, the Administrator typically has three main roles:

Agent

As agent, the Administrator is required to supervise and inspect the works and give instructions for matters covering the works generally, variations, ambiguities, errors and faults, site conditions, nominated sub-contractors and suppliers, monetary and provisional sums, omission of work, execution of work, meeting statutory requirements, workmanship and materials, opening up work for inspection, removal of work, rectification of work, material and component failure, supply of documents and generally providing further information as required. In addition, the Administrator is usually required to organize and/ or attend meetings and take care of other general administrative matters. All of these activities require accurate recording of information provided, decisions made and all correspondence sent and received. Making decisions is a key aspect of this role.

Certifier

As the project certifier, the Administrator is required to assess, value and certify contractual claims submitted by the contractor and make concomitant adjustments to the contract cost and time for completion. This typically includes claims for progress payments, variations, time extensions, delay costs and the like, contract time and cost adjustments, the enforcement of contractual provisions such as liquidated damages and the issuing of certificates for progress and final payments and Practical and Final Completion. As mentioned previously, the majority of claims stem from problems with or changes to the contract documentation. If the designer (architect or engineer) for the project is appointed as the Administrator there is a potential conflict of interest as they may well be in a position to make decisions on problems arising from their own documentation. This has led to calls from some sections of the industry for this certification role to be carried out by an independent third party such as the project manager or quantity surveyor.

Adjudicator

As adjudicator, the Administrator is required to make decisions on any disputes between the parties. In some contracts this may include referring the parties to Alternative Dispute Resolution procedures such as structured negotiation or mediation. If a party to the contract disagrees with a decision made by the Administrator, the ultimate recourse is to have the matter dealt with via arbitration or litigation.

These three administrative roles combine to make the task very onerous. It is further complicated by the fact that the Administrator is required to act impartially when making decisions despite the fact that they are normally appointed by and paid by the client.

19.3.2 Designated project administrator – multiple contracts

This role is normally filled by a person or organization under a managed (construction or project management) arrangement or design and construct arrangement. The roles and functions of the Administrator are very similar to those outlined above except that the Administrator is required to administer the many and various sub-contracts and supply and/or consultancy agreements on the project. On large projects, this may involve the administration of hundreds of sub-contract agreements, supplier/manufacturer agreements and a range of consultancy agreements.

19.3.3 Internal administration (client/contractor/sub-contractor/supplier/consultant)

This relates to administration carried out by each of the organizations involved on a project for their own internal purposes. Their administration processes will vary according to their level and type of involvement and their position in the particular procurement system chosen for the project. However, the principles and procedures of good contract administration outlined in this chapter are applicable to most, if not all, of these organizations.

19.4 Administration under different procurement systems

Administration procedures will also vary according to the procurement system used; the following is an outline of the common procurement systems.

19.4.1 'Fixed' lump sum

The administrative procedures for this system are as outlined above (in Section 19.3.1). This is the traditional and most common form of contract whereby a contractor enters into a contract with a client to carry out the construction works for a fixed lump sum price. A third party is appointed to act as agent, certifier and adjudicator for the project. The client and contractor are responsible for their own internal administration.

19.4.2 Provisional lump sum/schedule of rates

This procurement system is used when a client wants construction to proceed before the project is fully documented. A provisional lump sum contract price is agreed based on a Bill of Provisional Quantities or a Schedule of Rates for the major items of work. The administration is similar to that for a fixed lump sum contract, the distinguishing feature being the need for the completed works to be measured or re-measured to reconcile with the Provisional Quantities or Schedule of Rates.

19.4.3 Cost reimbursement (Cost-plus)

Under this procurement option, the contractor is paid the actual cost of the works plus an agreed margin. This margin might be a percentage of the total cost or a fixed fee. The administration process is dominated by the collection, submission and approval (or rejection) of all invoices and receipts for work carried out on the project. This usually requires substantial paper work and checks need to be made to ensure that all invoices and receipts relate to the actual project (and not other projects) and that they represent a fair and reasonable price for the work carried out.

19.4.4 Design and construct contracts

Design and construct contracts come in many forms. Often referred to as turnkey or fast track systems, they involve one organization entering an agreement with a client to design and construct a project for an agreed lump sum figure or some form of target or guaranteed maximum price. Novated design and construct systems give the client more control over the process by enabling them to engage their own designer to develop the design to a particular point (usually conceptual stage) and then enter into a contractual arrangement with an organization to take over the project and complete the design and construction. This organization is usually required to engage the original design team and to warrant their work. The Design, Construct and Maintain system takes the process one step further by requiring the organization to not only design and construct a project but also to maintain it for an agreed period of time after completion.

With these procurement forms the contractor effectively takes control of the project and, in a sense, becomes both the client and contractor for the project. Administration is required for both the design process (design consultants) and the construction process (sub-contractors, suppliers and manufacturers). The main distinguishing feature of this system is that the level of claims submitted to the client is normally far less than is the case with the traditional lump sum approach – this is because the design and construct organization is responsible for the development of the design documentation and cannot usually make any claims for errors or problems with this documentation.

This process is similar for BOT (Build–Operate–Transfer) and BOOT (Build–Own–Operate–Transfer) procurement systems. These systems are usually used for large projects, particularly large infrastructure projects such as tunnels and roadways, where a client has insufficient financial resources to complete a project and an agreement is reached with an organization to not only design and construct the project but also to finance it, either in part or in full. The organization receives payment for the project through remuneration from the operation of the project over a specified period. As with design and construct, the BOT/BOOT organization effectively takes on the role of client/designer/contractor and is responsible for administering the whole project.

19.4.5 Managed contracts (construction management/project management)

As outlined earlier, the administration of managed contracts is carried out by the Construction or Project Manager for the project. With this procurement form, the project is normally fast tracked with an overlap of the design and construction processes and the construction and/or project managers are paid as consultants to manage the project. Depending on the terms of engagement this may include administration of the construction process only or of the entire design and construction phases of the project. The construction/project management company may well find itself responsible for administration of not only the project as a whole but also for the administration of a large number of sub-contract, supply and consultancy agreements.

19.5 Key areas

The following outlines some of the key areas associated with the administration of construction contracts.

19.5.1 Time

Administration of time involves the notification, quantification, submission and assessment of time extension and delay cost claims and concomitant adjustments to contractual completion dates. Variations and delays are usually inevitable on projects and lead to the need for contract time for completion to be adjusted.

It may also include the application of Liquidated Damages provisions in contracts that compensate the client for late completion by the contractor. Many contracts these days also provide for bonuses to the contractor for early completion. From the contractor's perspective, the administration of time also includes monitoring the construction program and taking remedial action where necessary.

Time administration is typically a problematic area in the administration of contracts due to the difficulty of quantifying time extensions and delay costs with any precision. Good claims management in this area can make a tremendous difference to the final outcome of a building contract. The main problem is that it is difficult for the contractor to clearly quantify the effect of a delay. Most standard commercial building contracts require the contractor to submit their construction program within the first 4–8 weeks of the project. This program of forecast progress on the project may, however, bear little or no resemblance to the actual work carried out thus making it difficult to use it as the basis of time extension claims. The problems are further compounded by the fact that major delays have a ripple effect on other trades and it is not uncommon for the full effect of a delay to remain unknown until the completion of the project.

Calculation of delay costs proves even more difficult. Depending on the contract conditions, specific heads of delay cost claims can include project overheads, loss of productivity, sub-contractor delay costs, escalation of labour, plant, material and sub-contractor rates, fixed overheads, loss of profits and/or profit earning capacity, finance charges, off-site and on-site storage charges and claim preparation costs. The majority of these heads of claim cannot be calculated with any degree of accuracy and a variety of formulae are used to assist, particularly with respect to overheads.

19.5.2 Cost

Administration of cost includes the notification, quantification, submission and assessment of contractual claims for cost adjustments due to variations to the contract requirements. Cost administration also includes progressive payments to contractors during the course of the works and financial reporting. Administration by the contractor can also include the negotiation and awarding of sub-contracts.

The administration of payment on a project usually involves the submission of progress payment claims from contractors and the issue of progress payment certificates by the client or their representative culminating in the final payment certificate. Progress payment claims are normally submitted by contractors on a regular basis (usually monthly) to clients, or their representative or superintendent, for payment for work performed. These claims are assessed and progress payment certificates issued authorizing payment to the contractor. The final payment certificate is normally issued at the expiration of the defects liability period and the settlement of outstanding contract claims.

Contractor's progress payment claims typically include the following type of information: valuation of work executed, valuation of unfixed materials and goods (contract permitting), adjustments to the contract sum (variations and delay costs), retention amount (if applicable), total amount previously certified, total amount previously paid (normally same as certified) and amount claimed. Additional information or detail that may be required includes the percentage complete and value for each item of work, section, trade and sub-contract, value of nominated sub-contract work, total value of variations and evidence that nominated sub-contractors and contractor's employees' wages have been paid from previous progress payments.

Progress payment certificates include similar details, but in addition the certificate may include the amount of other adjustments to the Contract Sum, the amount of retention, the total amount previously certified and the amount certified as being due for payment to the contractor.

Cost management from the contractor's perspective involves establishing cost centre budgets and managing costs within these parameters.

Financial reporting or analysis, as part of the administrative process, involves progressive (usually monthly or quarterly) reporting on the financial status of a project incorporating the following: original contract price or budget, contingency allowances, approved variations to contract price and budget, pending claims and likely effect on price and budget, trade and cost centre budget analysis, identification of cost overruns or savings, remedial action planned to address cost overruns and total likely cost commitment (i.e., a revised budget). This is normally prepared as an internal document for the contractor, however, in managed contracts, this type of reporting is usually prepared for the client.

Financial reporting does differ for clients and contractors but the principles are the same. The client's financial reporting (often carried out by consultants) is more external in the sense that it focuses on changes to the contract sum, comparisons with original budgets and the client's total likely cost commitment. A contractor's financial reporting is more internal in the sense that they monitor their work and costs to try and ensure that they carry out the work for the price agreed and still make a reasonable profit. Generally if a contractor makes a mistake in forming their contract price they are responsible for it and cannot seek compensation from the client. It should be remembered, though, that financial reports are merely statements of monetary movements; good cost administration takes this further by incorporating cost control measures that result in effective remedial action when cost problems arise.

19.5.3 Quality

Contract administration can also include the supervision and management of the project to ensure that the finished product complies with the quality requirements as specified in the contract documentation. Checks need to be put in place to ensure that the appropriate quality is achieved.

Defective work is commonly rectified during the Defect Liability Period which is the period after Practical Completion and before the issue of the final certificate when all outstanding work must be completed and any faults, omissions, shrinkages or other defects that become apparent must be rectified. This period is commonly 3 to 6 months.

Disputes over defective work are common. Building work is subject to tolerances in terms of the quality of materials and workmanship and most disputes centre around differences of opinion as to what is or is not acceptable quality.

19.5.4 Notification

Most standard contracts place obligations on contractors to notify the client (or their representative) when any events occur that might result in a claim from the contractor. The notification period may be specified or simply described as 'within a reasonable time frame'. The purpose of early notification is to give the client as much scope as possible to minimize the adverse affects of the problem. With many contracts, failure to notify within the specified contractual time frame can void the contractor's entitlement to pursue a claim.

19.5.5 Mitigation

Under most building contracts the contractor is under an obligation (either expressly referred to in the contract or implied) to take reasonable measures to mitigate the client's potential loss in the event of problems occurring on the project. In other words, contractors cannot take advantage of the situation to maximize financial gain.

19.5.6 Supervision/Instruction/Direction

The role of Project Administrators in supervising the works and providing instructions and directions can be extremely onerous. While the actual functions carried out depend on the terms of engagement, many agreements describe a level of supervision as 'might be necessary' to ensure execution of the works 'generally' in accordance with the drawings and specification. This means that constant supervision is not necessary, but it creates a grey area between areas of work that need to be inspected and those that do not. Supervision or inspection at important stages such as prior to concrete pours and at the completion of structural framing are clear but other stages are less clear and do leave the supervisor exposed to potential liability if problems occur with sections of work that have not been supervised or inspected.

Any instruction, notice or direction should be given promptly, in writing. If oral directions are required, they should be confirmed in writing as soon as possible. Directions can take many forms, for example, under the AS 2124 (Standards Australia, 1992) contract 'direction includes agreement, approval, authorization, certificate, decision, demand, determination, explanation, instruction, notice, order, permission, rejection, request or requirement'. It should also be remembered that failure to give directions as required can result in a 'no response' being legitimately construed by the relevant party as a direction.

19.5.7 Claims management

Claims management involves the notification, quantification, submission, assessment and negotiation of contractual claims on projects. Contractual claims for cost or time adjustments may occur due to a variety of causes, of which the most common is due to variations or changes to the contract requirements as reflected in the contract documentation.

Other claims can include the following (JCC-C, 1994):

- time extension
- delay costs
- bills of quantities errors (if a bill is used)
- patent fees and royalties (not included in contract)
- deductions for uncorrected work
- latent conditions
- costs arising from site conditions
- fees paid by Proprietor
- adjustment of Prime Cost and Provisional Sums
- costs arising from employment of others on builder's default
- changes in law or statutory regulations
- extra tests
- changes in fees
- costs arising from certain instructions
- unforeseen costs due to adjoining properties
- replacement of nominated sub-contractors
- adjustment of insurance costs
- opening up work for inspection
- protection of work by separate contractors
- instructing builder to do the work of an insolvent nominated sub-contractor
- finding fossils or relics
- errors in setting out the works
- suspension of works
- change in order of works.

Effective management of the claims process is essential. Adherence to contractual requirements, particularly in terms of notification and submission of claims within required time frames, good record keeping, and effective substantiation are key aspects.

19.5.8 Negotiation and dispute resolution

The administration of contracts includes negotiation between the parties over project issues and claims and disputes are commonplace. Disputes are unfortunately an inevitable part of the construction process due to the many problems associated with the industry.

The fact that every project is different (a 'one off' prototype), the dynamic nature of activity carried out in a constantly changing environment, the great variety and volume of contract documentation, the multitude of activities and individuals/organizations involved, the difficulty in foreseeing all requirements, industrial action, changing market conditions and the traditional adversarial attitudes of parties all combine to cause high levels of dispute in the industry. Traditionally, parties with disputes that could not be resolved through negotiation sought respite through arbitration or the courts. Due to the cost and expense of taking these paths, considerable focus has been placed in recent years on the development of Alternative Dispute Resolution procedures to assist parties in resolving disputes more efficiently and cost-effectively. Traditional approaches have tended to be adversarial, assertive and aggressive. The problems created by this have led to the emergence of a new approach with greater emphasis on effective negotiation. More constructive, co-operative and collaborative approaches are being encouraged to assist the parties to work towards joint solutions. Industry changes have included the emergence of partnering and alliance procurement systems and the desire for clients to impose single point responsibility and risk on one organization through the use of design and construct type procurement systems. Alternate Dispute Resolution techniques include negotiation, mediation, conciliation, independent expert appraisal and senior executive appraisal.

Negotiation remains the most common, and preferred, method of resolving disputes. A lot of attention is now placed on the development of negotiation skills in the industry. Mediation has emerged as a very effective means of resolving major disputes. This involves a structured negotiation process whereby a neutral third party assists the parties in finding their own solution to their dispute. The mediator helps to identify the issues, develop options, identify the needs and concerns of the parties and addresses any other pertinent matters. The mediator does not make recommendations or decide on how issues should be resolved. Perhaps the greatest long-term benefit from the mediation process is that parties learn how to properly conduct negotiations thus increasing the likelihood of settling future disputes more efficiently.

The personal skills that assist in effective negotiation start with competence and expertise in the disputed area. This needs to be followed by accurate assessment carried out with due skill and care. From there, commonsense, honesty, clarity of thought and speech, confidence and 'people skills' greatly enhance the negotiation process.

19.5.9 Final accounts

The administration process usually winds down during the Defects Liability Period as minor work is completed, defects are remedied, claims are finalized and outstanding monies owed are determined. Typically a lot of negotiation between the parties takes place during this period as attempts are made to resolve outstanding issues and claims. All being well, everything is finalized by the end of this period with the issue of the Final Certificate for the final payment to the contractor.

These issues and claims are not always resolved at the completion of the defects period, and this can lead to lengthy final account periods which, in some cases, can last years. This is often the case with projects where administration has been poor, or where there have been a large number of complex claims and/or where disputes have ended up in arbitration or litigation.

19.6 Effective contract administration and claims avoidance

This section outlines recommended procedures for effective contract administration and claims avoidance. It is drawn largely from Hinds (1990; 1991a, b, c).

19.6.1 Claims avoidance

Whilst claims are inevitable on most construction projects, many measures can be taken to minimize their occurrence and, where they do occur, to enable them to be dealt with efficiently. The following outlines some recommendations to reduce the level of claims and their impact on the project.

Good project definition and contract documentation

As outlined earlier, the vast majority of contractual claims and disputes stem from problems with and/or changes to the contract documentation. Spending more time and money on the development of good documentation can go a long way towards reducing these problems. However, recent trends seem to suggest that clients are more interested in saving money during the design stage by paying reduced fees to the team of design consultants or by simply putting the responsibility of design and documentation on to the contractor via design and construct procurement strategies.

Good performance

Good performance from the whole project team will obviously reduce the level of claims on a project. The team includes the client, designers, consultants, contractors, subcontractors and suppliers.

Building relationships

Establishing good team harmony on the project will also assist; being personable and engaging appropriate people skills can thwart many potential problems.

Project organization

Site staff must be clear in their understanding of their obligations, and it is essential that good records are kept. The use of proformas to record information such as Requests for Information (RFI) and site instructions, good communication, appropriate knowledge of the contract documents, analytical record keeping and regular checks all constitute good administration practice. The use of an appropriate software system is very helpful.

Planning

Effective pre-planning should be implemented to identify possible problems and risks, and strategies implemented to deal with these problems and risks should they occur.

Potential disputes

Early action when problems develop is important. Parties should adopt an attitude of 'strengthening their position' whenever a problem emerges. This includes notification to make the other party aware of the problem as soon as possible, taking photographs and videos (dated), taking records of what activities took place (including joint records where possible) and making records of all correspondence and instructions relating to the problem. If the problem develops into a claim or dispute these records will greatly assist in the substantiation of the claim and reduce the scope for dispute.

Quantum

The predominant cause of most disputes is related to quantum, i.e. the cost or the extent of something, such as the length of a delay. The more clearly a contractor can substantiate and prove the quantum component of their claim the more streamlined the claims administration process becomes.

Quick response

Responses to all correspondence should be made as soon as is practicable. Parties should act within the contract time requirements, make necessary decisions when due and not let claims pile up in the 'too hard basket'.

Appropriate personnel

Effective administration requires the use of a sufficient number of staff with appropriate skills and experience to undertake these tasks. Due to the need for contractors to keep overheads low, there may be a propensity to minimize the employment of administration staff. This can have serious ramifications not only on the smooth running of the project but also on financial returns.

Procurement system

Many clients now adopt procurement systems that reduce their level of risk and reduce the level of claims on their project. Procurement systems such as design and construct can significantly reduce the level of claims on the project as the majority of claims stem from problems with the contract documentation and, if the design and construct organization is responsible for the preparation of this documentation, the potential for claims must be reduced. This procurement route is, however, not necessarily the panacea that is often perceived to be; this is due to the difficulties associated with developing comprehensive briefs that fully reflect the client's requirement, and ensuring that the end product represents value for money and meets the client's needs.

19.6.2 Good record keeping

Good record keeping is the foundation for the whole administration process. Records need to be kept as follows:

- complete set of contract documents
- technical reports
- samples, prototypes, inspection and test reports
- sets of drawings issued (drawing register)
- set of all construction programs
- all correspondence (at least one clean set)
- site instructions
- requests for information
- minutes of meetings (agreed and signed if possible)
- variation documents (including variation price requests, variation orders, quotes, estimates, advice)
- diaries (facts not opinions)

- daily reports (jointly signed if possible) including details of weather, contractor's work force and plant, work being undertaken, labour unrest or strikes, drawings requested and received, instructions requested and given, inspections or tests carried out and results, site visitors, discussions, delays and apparent causes, latent conditions, complaints, defective works, dangerous situations or safety problems, events that may cause future problems.

Personnel responsible for keeping these records should be clear about what is required and spot checks should be carried out to ensure that they are doing their job. Ideally the date should be recorded on all documentation and any verbal instructions or agreements should be confirmed in writing as soon as possible.

A record keeping system utilizing proformas for key information should be used. Records need to be kept in both electronic and hard-copy form and appropriate filing systems (including coding/indexing systems) put in place. On large projects this can often result in hundreds of physical files with consequent storage space problems. This space problem can become quite an issue because these records usually need to be kept for at least 6 years after completion for legal reasons. The use of micro-fiche or computer-scanned copies can assist in this respect. Security of the documents is also important as lost or destroyed documents could lead to lost claims or an inability to pursue or defend a legal action.

19.6.3 Claims risk analysis

Where dispute arises over claims involving large sums, it is prudent to undertake some form of claims risk assessment to determine appropriate courses of action. Hinds (1991c) suggests the following assessment procedure.

Establish rights and obligations

The parties should ascertain that they have not only the contractual and legal right to pursue a claim but also the technical right. Legal advice can be sought for the first two, but the latter takes on importance when the technical aspects of the claim are in question. An example might be whether site conditions encountered constitute latent conditions – a contractor will normally have the contractual/legal right to pursue a claim for extra costs due to latent conditions but expert technical advice/opinion may be necessary to clarify whether the site conditions are likely to be accepted as latent conditions. Once the contractual and technical rights have been established, it is useful to determine the quantum that would likely be awarded in the event of a successful claim.

Establish likely outcome of arbitration or litigation

Construction disputes that lead to a formal arbitration award or court judgment are rare because the vast majority of disputes are settled before any formal judgment is made or, indeed, before any commencement of proceedings (Tyrril, 1992). Matters that lead to a formal award are usually complex and, accordingly, from the outset it may be difficult to determine the likely outcome in arbitration or the courts. The more complex the case the more likely that the matter will be a drawn out affair if allowed to run its full course and the end result may well prove unsatisfactory to all concerned (particularly if costs are not awarded to the 'winning' party). Nevertheless, advice from a lawyer or arbitrator expert in the field of the dispute may prove useful.

Quality of documentation

Assessment of the quality of documentation and records held by both parties in relation to a dispute can provide a good indication of potential success. The party with the better records is more likely to be able to substantiate their position.

Economic viability of pursuing claim

Considerable time and cost is commonly associated with the negotiation and resolution of major disputes, particularly if the matter goes through the process of arbitration or litigation. Key personnel may be dragged away from other projects to assist in the preparation of claims and be involved in meetings and negotiations. Once legal counsel and expert consultants are engaged costs can skyrocket. Another consideration is whether these costs are likely to be recouped if the claim is successful.

Ongoing relationships

Protracted disputes can have a serious effect on the business relationships between the parties. This can affect the remainder of the project and the possibility of future work amongst the parties. Parties need to weigh up their short-term and long-term gains in this respect.

Attitudes of the parties

The attitudes of the parties can have a big bearing on how a dispute is handled. A party with a positive and reasonable attitude is likely to work towards a co-operative settlement whilst a party with an adversarial attitude may be intent on fighting to the end.

Consideration of the other party's situation

The resolution of disputes can be greatly assisted by gaining an awareness and appreciation of the other party's position. For example, they may be experiencing financial problems or have accountability constraints which prevent them from reaching a reasonable settlement.

Options in lieu of direct payment in full

Exploring other options in lieu of full payment in settlement of a claim can be worth consideration. This might include making concessions and settling on a lower amount to avoid the cost and time associated with a protracted dispute and to preserve business relationships. Other options might include accepting a delay in payment of a claim, requesting further substantiation of a claim (client), the offer of future work, the assurance of completion on time (contractor), joint venture offers and other commercial options.

At the end of the day, parties must also consider the effect that a protracted dispute, particularly if it is dragged through public court proceedings, may have on their reputation.

19.7 Administration trends

Given the large volume of information and data processed and stored during the administration of projects it is no surprise that computerization of this process has become commonplace. Sophisticated project administration software is now readily available and the administrative practices of most organizations are largely governed by the framework of the particular program that they use. In the past couple of years this computerization has moved to a new level by incorporating the Internet as a key administration tool.

19.7.1 Web-based collaborative project systems

The use of the Internet as the basis for information flow amongst project team members has the potential to revolutionize not only the administration process but the whole manner in which projects are procured, co-ordinated and managed.

Construction projects typically have a large number of individuals and organizations involved in the design and construction process. Traditional communication methods amongst these groups are time and labour intensive and inefficient.

> Co-ordinating the numerous parties involved to take a project from initiation through construction is often a daunting experience. Owners/developers, architects, engineers, general contractors, specialty contractors, material suppliers and government and regulatory bodies have all traditionally communicated using methods such as fax, face-to-face meetings, email, couriers and mail to exchange ideas, provide progress updates, schedule labour, deliver documents and make supply requests. (PricewaterhouseCoopers, 2001, p. 3)

This procedure is illustrated in Figure 19.1, and has an effect on the administration process.

> The complex process required to turn around a RFI (Request for Information) illustrates some of these inefficiencies: today, a RFI is hand-written by a specialty contractor, faxed to the general contractor, reviewed/rewritten and faxed to the architect. The architect may fax it to a sub-consultant (electrical, structural, mechanical) for review, who, in turn may fax it to a sub-sub-consultant (lighting, acoustical) for input. The response is formulated, documented and sent back to the sub-consultant for review, and then faxed back to the architect. Assuming no further clarification is needed, the architect faxes the RFI back to the general contractor and the owner. Once approved by the owner, the RFI is faxed back to the specialty contractor with action items. Finally, the general contractor needs to ensure that the response is received on the job site by foremen, staff, specialty contractors, suppliers, project managers and administrators, all in their respective office location. (PricewaterhouseCoopers, 2001, p. 3)

Web-based project collaboration tools and systems have emerged recently to help address this problem and improve the efficiency of communication and information flow on construction projects. Whilst these systems come in various forms, the most

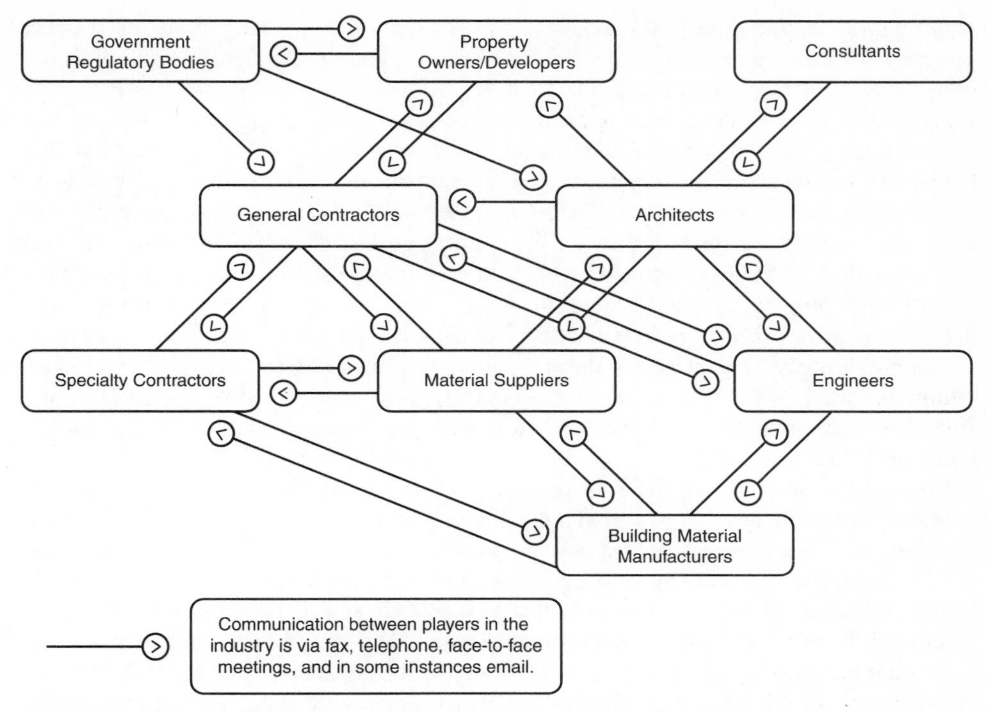

Figure 19.1 Traditional communication and work flow on construction projects (PricewaterhouseCoopers, 2001).

effective are systems based on collaborative 'extranets' which link all members of the project into an electronic network of information flow and communication. These sites host and maintain a central hub of documents that all members of the project team can access, review, amend and utilize, usually via some form of web browser software. Although many of these sites have financial backing from construction companies and other industry participants, they are essentially independent and set up to handle an almost limitless range and number of projects and clients around the globe.

Despite being in the early stages of development, web-based project systems show tremendous promise for the industry. They are often loosely described as *e-construction sites*. The number of these sites that have the capacity to manage all necessary project information has grown at a dramatic rate in the past 2 years. Warchus (2001) found that there are already well over 100 serious e-construction sites providing this type of service but contends that this is not sustainable in the medium to long term. He forecasts that this number will fall dramatically in the next year or two through mergers and business failures. Mergers are already occurring at a rapid rate and this will actually assist by reducing the number of different technology platforms and standards, leading to a more standardized playing field.

The following provides an example of one web-based system set up by Buzzsaw (2001). Buzzsaw began operations in October 1999 as a spin-off from AutoDesk (a major player in the global design and CAD software market) to explore the market potential for process automation in the construction industry. Recognizing the potential opportunities of tapping into a global industry with an average estimated turnover of US$3.9 trillion, they developed a secure online 'extranet' capable of hosting, storing and managing project documentation and facilitating electronic communication and document and data transfer amongst project team members. The growth of this web-based collaborative system that can be accessed and utilized around the globe has been quite staggering: in less than 2 years, Buzzsaw has grown to the point where it has more than 125 000 construction professionals using its services to collaborate on over 35 000 projects around the world. As a further example of the breadth of use possible with such a system, one company utilizes Buzzsaw to manage and share data on approximately 75 projects throughout the United States, Europe and Asia (Buildings.com, 2001). Buzzsaw and Autodesk have now recently merged to link Buzzsaw's administration capabilities with Autodesk's design and CAD capabilities (Buzzsaw, 2001).

Documentation and data for each project is stored in this extranet system and team members for each project are provided with security access provisions. Project team members may have access to just one project or a number of projects in which they are involved. Project documentation is posted to the extranet once and everyone in the project team has immediate access to that information and, depending on their access privileges, have the ability to review, amend and add to this documentation as well as post additional information. Theoretically the project never goes to sleep as communication and the transfer of information amongst the project team can occur 24 hours a day.

Drawings, documents, photographs, news, contact information and all other project information can be stored on the site. The main modules include Drawings, Documents, Photos and Queries. The Drawings module handles all electronic CAD drawings including all revisions. Any type of document can be loaded into the Documents module. Existing paper-based material can be scanned in whilst the system has a sophisticated filing structure to enable quick and easy searching and retrieval of information. Photographs can be downloaded and used for a range of administrative purposes including progress reviews, progress payment valuations and assessment of quality. The Queries module provides effective documentation management by providing a focus for technical queries, requests for information and the like. Additional features include project diaries, team directories, administration functionality and help via on-line reference manuals and backup telephone and e-mail support. The system also provides an effective audit trail of information flow on a project and an excellent record of what occurs on site and when. This has clear benefits for claims administration and dispute resolution.

Figure 19.2 provides an example of how web-based systems like the one developed by Buzzsaw can streamline the traditional flow of information for RFIs that was outlined earlier. The Construction Manager at the centre of the diagram is the commercial title for Buzzsaw's web-based hub which stores all the project documentation and can be accessed by all designated members of the project team. All information flow and communication occurs electronically via this central hub and records of this flow occur automatically. Project team members can access and contribute to this hub from any remote location where Internet access is available.

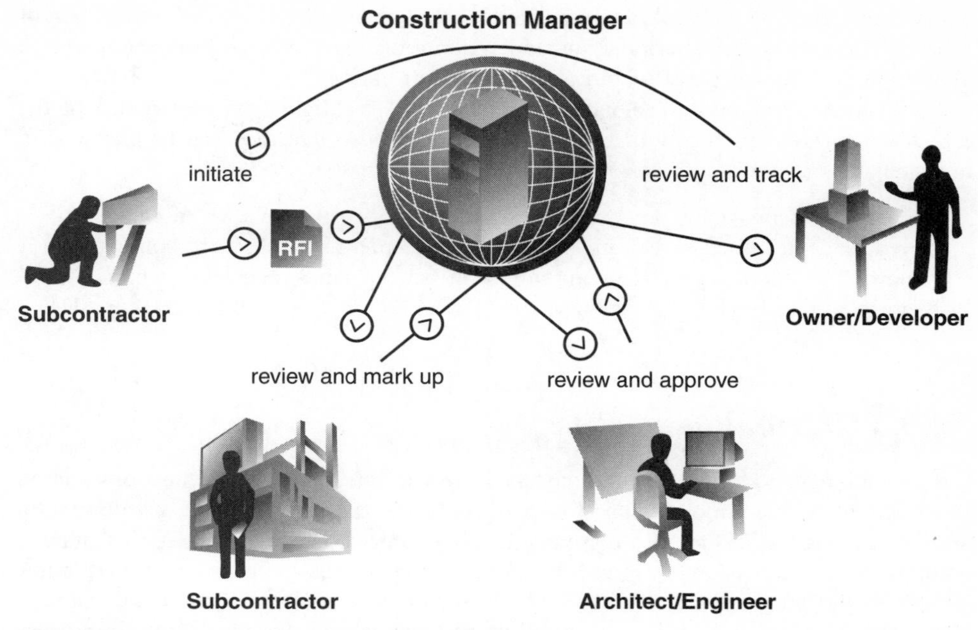

Figure 19.2 Buzzsaw RFI information flow (Buzzsaw, 2001).

19.7.2 The validity of electronic records

This use of the Internet brings into question the validity of stored electronic information not only for these web-based collaborative systems but for all forms of electronic communication such as e-mail. Legislation to address this issue is being put forward in many countries. As an example, Warchus (2001) points out that the passing of the *Electronic Communications Act 2000* in the United Kingdom and the *European Union Directive on E-Commerce* has removed the barriers to carrying out transactions and entering into contracts by electronic means.

> Not only will all contracts be validly concluded electronically without the need for signatures or formal seals (with limited exceptions), legislative confirmation that the use of digital signatures and encryption will be fully admissible in court proceedings should provide all the commercial reassurance needed for those still fond of pen and paper. (Warchus, 2001, p. 2)

Similar legislation is being introduced in Australia. In NSW, the proposed *Electronic Transactions Act 2000* gives legal validity to the transfer and storage of information by electronic means. Green (2000) points out that for the purposes of construction contracts this will mean that where contractual provisions call for or allow information to be passed from one party to the other, this can validly occur via e-mail or other electronic means. This will be so unless the contract provides to the contrary or the recipient makes it clear that they do not consent to using electronic communication.

The change will be particularly relevant to contractual provisions for extensions of time, variations, defects, termination and other provisions that require one party to give notice to another party within a specified time frame.

Given these legislative developments, Best (2001) highlights the importance of the effective management and storage of electronic records so that they can be retrieved in a systematic manner.

> Failure to do this may see companies faced with costly reviews of masses of stored data. In one US case, in 1995, the court ordered a defendant company to review 30 million pages of e-mail stored on backup tapes, a process which cost them over US$100 000. (Best, 2001, p. 11)

19.8 Conclusion

Contractual claims, problems and disputes are an integral part of the construction process. One of the major reasons for this problem is the large volume of information and data that needs to be distributed, communicated and processed. Effective administration goes a long way to alleviating many of these problems and contributes greatly to the protection of a party's financial interests and legal/contractual liability. Parties need to 'strengthen their position' at all times during the administrative process to counter possible claims and disputes. From an administration perspective, the normal project objectives of 'within time and under budget' could well be expanded to 'within time and under budget and out of court'.

References and bibliography

Abdelsayed, M. (2001) *Internal Project Analysis by Tardif, Murray & Associates Inc.* www.tardifmurray.com

Best, R. (2001) You got mail – the value of email analysed. *The Building Economist*, September, 9–13.

Buildings.com (2001) *Autodesk signs definitive agreement to purchase Buzzsaw.com.* 7th November.

Buzzsaw (2001): http://www.buzzsaw.com

Choy, W.K. and Sidwell, A.C. (1991) Sources of variations in Australian construction contracts. *The Building Economist*, December, 25–30.

Green, N. (2000) Making email notices official. *Australian Construction Law Newsletter*, December, **75**, 35–6 (Sydney: UTS).

Hinds, K. (1990) Claims presentation for maximum success. *Australian Construction Law Newsletter*, **13** (Sydney: UTS).

Hinds, K. (1991a) Principles of claims avoidance and contract administration. In: *Proceedings of Construction 91 – A New Era in Risk and Rewards,* IIR Conference, Sydney (Institute for International Research Group).

Hinds, K. (1991b) Principles of record keeping. In: *Proceedings of Construction 91 – A New Era in Risk and Rewards,* IIR Conference, Sydney (Institute for International Research Group).

Hinds, K. (1991c) Principles of risk analysis. In: *Proceedings of Construction 91 – A New Era in Risk and Rewards,* IIR Conference, Sydney (Institute for International Research Group).

JCC-C (1994) *Building Works Contract – JCC-C 1994 With Quantities* (Canberra: Joint Contracts Committee).

Johnston, A. (1937) *The Great Goldwyn* (New York: Random House).

PricewaterhouseCoopers (2001) *A Benefits Analysis of Online Collaboration Tools within the Architecture, Engineering & Construction Industry.* White Paper. http://www.pwcglobal.com

Standards Australia (1992) *Australian Standard – General Conditions of Contract AS 2124 – 1992* (Sydney: Standards Australia).

Tyrril, J. (1992) Construction industry dispute resolution – an overview. In *Dispute Resolution Workshop Notes* (Faculty of Law, University of Technology, Sydney).

Warchus, J. (2001) *E-Construction: Technology to the Rescue?* (4Projects) http://www.4projects.com/news_shadbolt.htm

20

Occupational health and safety in construction

Helen Lingard*

Editorial comment

Building is an inherently dangerous pursuit. There are many opportunities for those working on construction projects to suffer injury and even death, or to contract some form of work-related illness – workers routinely work at height, use powerful hand tools and hazardous materials and lift heavy objects, and it is only in the last century or two that the safety of workers has begun to be treated as a serious concern. It is also a relatively recent development that clients and contractors have begun to realize that the cost of avoiding accidents and illness is less than that of treating the consequences after the event, with contractors facing heavy fines, escalating insurance premiums, expensive litigation initiated by injured workers and damage to reputation all contributing to the total cost of on-site incidents.

In ancient times construction workers were seen as expendable inputs to the process, and, indeed, in some places even today individual workers are easily discarded and replaced, notably in countries where much of the on-site work is done by 'guest workers' – labourers imported from developing nations who are poorly trained and poorly equipped, and towards whom the contractors feel little loyalty or responsibility.

Governments around the world have introduced increasingly strict regulations aimed at improving safety on building sites although many countries, particularly some of those in the developing world, still lag behind countries such as Australia and the UK where worker safety is now tightly controlled. In New South Wales, for example, contractors are now required to have all electrical cables (such as those used to supply temporary power for small tools such as drills and sanders) checked and tagged every month. Those same cables must be strung above floor level so that they cannot lie in water and are out of the way of feet, barrows, pallet trucks and the like. Random inspections are not uncommon and failure to comply incurs heavy fines.

* University of Melbourne

As described in this chapter, research has shown that there are other ways, apart from such prescriptive measures, that the health and safety of construction workers can be improved and that their success is dependent on a variety of factors including the attitude of management to the problem. There is also a common perception that on-site accidents 'happen to someone else' and it requires a change in the attitude of individual workers towards safety issues before this perception will be replaced by a serious concern for improving site safety.

Work-related illness is an even more difficult problem as both cause and effect are typically latent – although a sub-contractor working on a roof may choose not to wear a safety harness, the risk is obvious and the consequences of falling are easily understood. However, when the risk is invisible (e.g., a potential health risk from outgassing from construction materials) and the consequences may not be manifest for many years (e.g., the development of tumours) it is much harder to convince workers that they need to be careful.

In this chapter the author looks at changing attitudes to occupational health and safety in the construction industry and discusses how worker safety and well-being can be further improved in the future.

20.1 Introduction

Compared to other industries, construction has a poor occupational health and safety (OHS) record. Many reasons have been cited for this, including the unique nature of the product, the constantly changing work environment and an exposure to uncertain physical conditions (King and Hudson, 1985). These are poor excuses and have hindered the adoption of innovative solutions to the construction industry's OHS problems. A recent review of the performance of the UK construction industry asserts that there is more repetition in construction work than is often supposed and that the processes of construction, repair and maintenance are repeated in their essentials from project to project (Egan, 1998). Within this context there is scope to develop a more systematized approach to OHS as it can be integrated into all project decision-making from the outset so that risks to workers' health and safety are minimized or eliminated altogether. This can be achieved through consideration of OHS in the design process through the selection of safer components, materials and construction methods, the adoption of new and safer construction technologies and increased standardization and pre-assembly. Construction work can also be better organized so as to create an environment in which safe behaviour is rewarded and unsafe behaviour is not tolerated. These changes require a re-examination of the construction process and a radical change in the industry's structure and culture. The fragmented and adversarial nature of construction must be addressed to achieve enhanced communication and co-operation between the parties to a construction project. Competitive tendering must be replaced by long-term relationships based on shared commitment to OHS, and clients must provide leadership by taking deliberate steps to ensure that OHS is properly resourced and managed on the projects they sponsor. This chapter deals with both hard (system and organizational) and soft (motivational and cultural) influences upon OHS performance, and suggests ways in which innovative solutions to the industry's OHS problems can be identified and implemented.

20.2 Occupational health and safety (OHS) legislation

In recent years, Commonwealth country governments have adopted a less interventionist approach to the regulation of occupational health and safety. Detailed prescriptive requirements of the past have been replaced with statements of the general responsibilities of employers, employees, suppliers and others. A greater reliance than ever before has been placed on industry to proactively and voluntarily manage OHS for itself. The new legislative structure allows scope for professional judgement in identifying OHS risks and selecting appropriate means of controlling these risks in order to contain them within tolerable levels.

It was envisaged that industry self-regulation of OHS would be achieved by the implementation of OHS management systems, through which organizations would identify hazards in the workplace, assess the risks posed by these hazards, then select and implement suitable measures to control these risks (Emmett and Hickling, 1995). It was expected that this process, depicted in Figure 20.1, would occur as a continuous cycle.

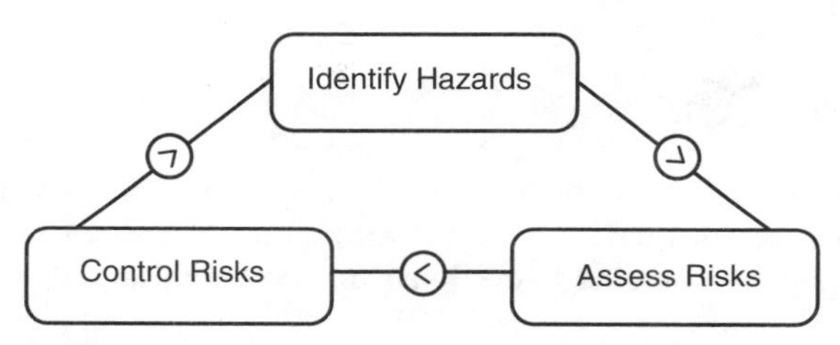

Figure 20.1 The risk management cycle.

The construction industry, however, has been slow to implement OHS management systems (Dawson *et al.*, 1988) and remains apathetic towards OHS, with investments of resources being seen as generating little tangible return. This has led some OHS law specialists to question the effectiveness of the new legislative approach (Brooks, 1988).

The structure of the construction industry may be to blame for some of the difficulties. Indeed, in Australia, the relevant construction trade union has asserted that the structure of the construction industry is ill-suited to a self-regulatory approach (Industry Commission, 1995). The construction industry in Australia is similar to that of other developed economies in that it is characterized by a small number of large firms and a multitude of small- and medium-sized enterprises (ABS, 1998). International research shows small businesses to be poorer at implementing OHS programmes (Eakins, 1992; Holmes, 1995) and to perform less well in OHS than large firms (McVittie *et al.*, 1997).

The problem is illustrated in Australia by the experience of the domestic building sector. This sector has a very poor OHS record and has shown little improvement under the new legislation (Mayhew, 1997). While large construction firms have implemented a wide range of OHS management initiatives, it seems that small- and medium-sized firms struggle to understand and meet the requirements of the new legislation.

20.3 Stages in the implementation of an occupational health and safety management system

The implementation of an OHS management system may be seen to occur in three conceptually distinct stages. Figure 20.2 illustrates these 'stages'. The occurrence of injuries and ill-health declines in 'steps' corresponding with each of these stages. The traditional approach to OHS is essentially reactive, with hazards being dealt with as they arise, and with a strong emphasis on discipline. A transitional approach is more proactive in that hazards are considered before they arise and procedures are established in an attempt to prevent occupational injuries and illnesses. At present, it is postulated that small construction firms are largely in the traditional stage in the implementation of their OHS management systems while the activities of larger construction firms fall mostly in the transitional stage. This may explain why construction accident rates have plateaued in recent years. There is a need to progress to the innovative stage of OHS management to achieve further improvement. In this stage, OHS is fully integrated into all business decision-making and every attempt is made to eliminate hazards or minimize OHS risks through the adoption of technological solutions. Attention is also paid to cultural and motivational issues and work is organized so as to encourage good OHS performance.

20.3.1 Etiology of human errors

Human error is a factor in the majority of accidents in work systems. Reason (1995) developed a scheme for examining the etiology of these human errors. Figure 20.3

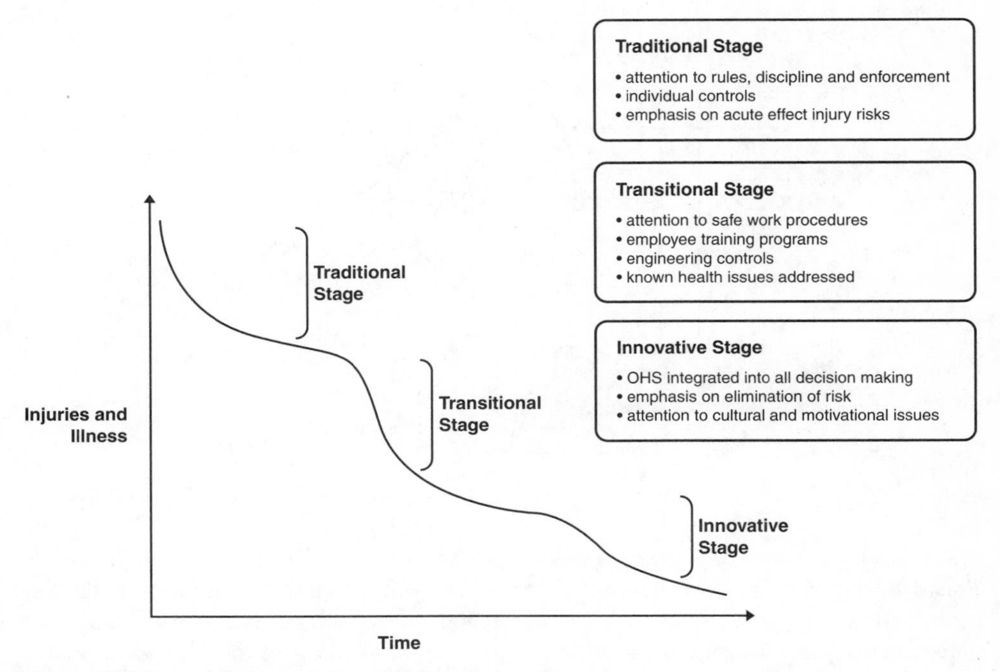

Figure 20.2 'Stages' in the evolution of OHS management responses.

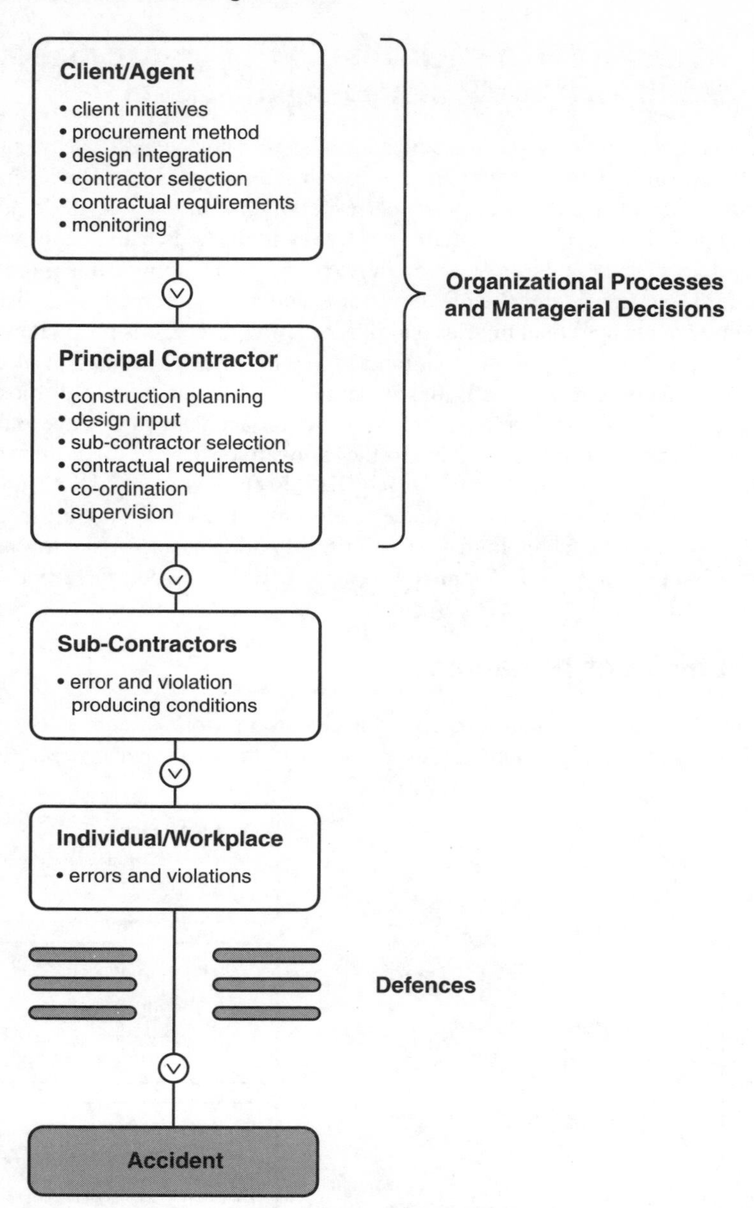

Figure 20.3 Etiology of individual errors and violations leading to accidents (adapted from Reason, 1995).

presents an adaptation of Reason's model to the environment of a construction project. It illustrates that two causal sequences for accidents exist. A failure can originate at the top of the contractual hierarchy and proceed through error-producing conditions at the various contractual levels to result in an error or violation by workers at the immediate man–equipment–material interface. A latent failure pathway also runs directly from organizational processes to deficiencies in a system's defences. In construction, both pathways may originate in a functional or organizational entity distinct and remote from the firm undertaking the construction work. Indeed, the Commission of the European Communities (1993) claims that over 60% of all fatal construction accidents can be attributed to decisions made before construction work commenced on site. Table 20.1 shows this breakdown.

Table 20.1 Source of decisions leading to fatal incidents in the construction industry (Commission of the European Communities, 1993)

Design	Organization	Site
35%	28%	37%

The existence of the pathways, identified by Reason, exposes workers at the lowest point in the contractual chain to OHS risks. Organizational and managerial processes in construction need to be carefully considered so that sources of human error may be identified, and methods of work should be designed to reduce the potential for human error so far as is possible.

20.4 Hierarchy of risk controls

A hierarchy exists to guide the selection of controls for OHS risks. An example risk control hierarchy for the risk of falling from heights is presented in Table 20.2. This hierarchy is based on the principle that control measures which target hazards at source and act on the work environment are more effective than controls that aim to change the behaviour of exposed workers (Matthews, 1993). Thus 'technological' control measures, such as the elimination of hazards, the substitution of hazardous materials or processes with less hazardous alternatives, or engineering controls, are preferable to 'individual' controls such as the introduction of safe work practices or the use of personal protective equipment. Individual controls cannot be relied upon because human beings are fallible and prone to error. Innovation in OHS should therefore focus on the identification and implementation of technological risk controls.

20.5 Organization of construction work

The selection of the preferred technological OHS risk controls can often only be achieved if OHS risks are considered prior to the actual construction of a project (Health and Safety Executive, 1995). Traditionally this was understood to be a matter for the works contractor who was engaged to undertake construction work, in accordance with design and specifications provided by an architect or design consultant, for a pre-determined tender price (Williams, 1998). This method of organizing work can, however, prohibit the identification of innovative solutions to OHS problems at the design stage.

Table 20.2 Example 'hierarchy of controls' for falls risk

Risk control category	Control measures
Eliminate the hazard	• Structures should be constructed at ground level and lifted into position by crane (e.g., prefabrication of roofs or sections of roofs).
Substitute the hazard	• Non-fragile roofing materials should be selected. • Fragile roofing material (and skylights) should be strengthened by increasing their thickness or changing their composition.
Isolate the process	• Permanent walkways, platforms and travelling gantries should be provided across fragile roofs. • Permanent edge protection (like guard rails or parapet walls) should be installed on flat roofs. • Fixed rails should be provided on maintenance walkways. • Stairways and floors should be erected early in the construction process so that safe access to heights is provided.
Engineering controls	• Railings and/or screens guarding openings in roofs should be installed before roofing work commences. • Temporary edge protection should be provided for high roofs. • Guard rails and toeboards should be installed on all open sides and ends of platforms. • Fixed covers, catch platforms and safety nets should be provided. • Safety mesh should be installed under skylights.
Safe working procedures	• Only scaffolding that conforms to standards should be used. • Employers should provide equipment appropriate to the risk like elevated work platforms, scaffolds, ladders of the right strength and height, and ensure that inappropriate or faulty equipment is not used. • Access equipment should be recorded in a register, marked clearly for identification, inspected regularly and maintained as necessary • Access and fall protection equipment such as scaffolds, safety nets, mesh, etc. should be erected and installed by trained and competent workers. • Working in high wind or rainy conditions should be avoided. • Employers should ensure regular inspections and maintenance of scaffolding and other access equipment, like ladders and aerial lifts. • Employers should ensure that scheduled and unscheduled safety inspections take place and enforce the use of safe work procedures. • Employees should be adequately supervised. New employees should be particularly closely supervised. • Employees should be provided with information about the risks involved in their work. • Employers should develop, implement and enforce a comprehensive falls safety program and provide training targeting fall hazards. • Warning signs should be provided on fragile roofs. • Ladders should be placed and anchored correctly.
Personal protective equipment (PPE)	• Employees exposed to a fall hazard, who are not provided with safe equipment (PPE) means of access, should be provided with appropriate fall arrest equipment such as parachute harnesses, lanyards, static lines, inertia reels or rope grab devices. • Fall arrest systems should be appropriately designed by a competent person. • Employees should be trained in the correct use and inspection of PPE provided to them. • Employees should be provided with suitable footwear (rubber soled), comfortable clothing and eye protection (e.g., sunglasses to reduce glare).

Source: Lingard and Holmes (2001).

Table 20.3 Suggested design solutions to common OHS problems in construction

OHS problem	Suggested design solution
Exposure to risk of falls/falling objects from working at height	• pre-fabricate building components off-site or at ground level and erect them as complete assemblies
Traffic access hazards and risk of collision between plant and machinery and/or with pedestrians or overhead power lines	• determine the location of the structure on the site so as to allow sufficient space for plant to access the site and be kept away from power lines, trenches and other hazards
Risk of falling as a result of routine cleaning/maintenance requirements	• select windows that can be cleaned from inside
Manual handling risks	• specify a maximum size of blocks to be used, e.g., all blocks must be under 20 kg
Risk of being struck by a moving vehicle during roadworks	• design roads with a wide shoulder to allow room for work crews and equipment
Insufficient means of anchorage for safety cables or lanyards	• design columns with a hole above floor level to support guard-rail cable or provide an anchor point for lanyards
Exposure to traffic risk during routine road maintenance	• upgrade initial specification to lengthen the project maintenance cycle
Exposure to the risk of health hazards associated with paints/epoxy resins	• specify water-based paints instead of organic solvent-based paints • specify products with larger molecules that give off less vapour • avoid polyurethanes or other coatings containing hazardous substances such as toluene diisocyanate

OHS risks should be assessed and control decisions made in the concept design, project planning and specification stages of a construction project (European Construction Institute, 1996). Incorporating works contractors' experience and knowledge at the design stage can improve project 'buildability' and eliminate OHS problems at source (Hinze and Gambatese, 1994). The adoption of the design and build contracting approach, in which the design and construction processes are undertaken by the same organization, has enabled closer attention to be paid to project 'buildability' issues, including OHS. Table 20.3 provides some examples of design measures which may reduce common construction OHS risks.

20.6 Buying safe construction

Choice of procurement route is only one decision for the client that is likely to influence the extent to which innovative solutions to OHS problems can be implemented. Another area, which requires further investigation, is the influence of the tendering process and contractor selection method on OHS performance. Under the competitive tendering system clients have traditionally awarded contracts to the lowest bidder (Russell *et al.*, 1992). However, in the context of intense competition and with pressure to minimize construction costs, the lowest bid may not reflect the true cost of undertaking construction work safely.

The way in which contractors price work often fails to account for OHS requirements, and it is common for the unit rate estimated for an activity to ignore safety issues (Brook, 1993). Despite the importance of providing resources for OHS, research suggests that estimators have little or no involvement in pre-construction OHS planning (Oluwoye and MacLennan, 1994) and OHS advisors are similarly excluded from the tendering process (Brown, 1996). Research suggests, however, that the inclusion of safety costs in a tender can reduce the lost time accident frequency rate from a range of 2.5–6.0 per 100 000 man hours worked to a range of 0.2–1.0 per 100 000 man hours worked (King and Hudson, 1985). To ensure that tenderers price OHS appropriately, and to facilitate fair comparisons, construction clients inviting tenders could specify the way in which prospective contractors should allocate OHS costs in their bids.

Alternative contractor selection methods can also help clients wishing to engage a safe contractor. Selective tendering, whereby contractors are subject to pre-qualification and only those that meet pre-determined performance criteria are invited to bid, allows clients to scrutinize contractors' OHS management systems and track records, and set minimum standards for the project. Regular construction clients may routinely audit contractors, maintain a list of approved contractors, and use this list to exclude contractors whose OHS performance is below standard. In some circumstances, for example, high-risk work, clients may decide to engage in negotiation with prospective contractors whose OHS performance is known to be excellent, rather than engage in competitive bidding.

There is currently great interest in new approaches to the development of business networks, such as partnering and the development of strategic alliances. Although it has been suggested that improved OHS performance is one of the benefits of partnering (Ronco and Ronco, 1996), there is no empirical evidence to support this view. It seems likely, however, that the development of long-term relationships could foster commitment to shared OHS objectives. Furthermore, participants are more likely to invest in the development of innovative solutions to OHS problems where there are long-term benefits to be gained. These outcomes will only be enjoyed, however, where construction clients take the lead in demanding high levels of OHS performance in the projects they sponsor.

20.7 Pre-planning project occupational health and safety

Pre-construction planning is essential to innovation in OHS. Once OHS risks arise on site, there are limited options for their control and these options are often individual measures at the lower end of the 'hierarchy of controls'. The need to anticipate OHS risks as early as possible has already been discussed. The processes of hazard identification and risk assessment are important to the systematic management of OHS risk. They enable decision makers to prioritize OHS risks and make rational decisions about safe methods of work and appropriate OHS risk control measures. These processes also assist decision makers in the allocation of resources to OHS risk control.

20.7.1 Identifying hazards

Standards Australia (1997) defines a hazard as 'a source or situation with a potential for harm in terms of human injury or ill-health, damage to property, damage to the environment, or a combination of these' and hazard identification as 'the process of recognizing that a hazard exists and defining its characteristics'. Hazard identification is the first step in the risk management cycle. Work should be carefully considered in the planning stage to ensure that as many hazards as possible are identified.

20.7.2 Assessing risk

Once hazards have been identified, the magnitude of the risk posed by each hazard should be evaluated. Risk may be defined as 'the combination of the frequency, or probability of occurrence, and consequence of a specified hazardous event' (Standards Australia, 1997). The process of assessing the likely impact of a risk and deciding whether a risk is tolerable is known as risk assessment. On the basis of risk assessment, risks can be prioritized according to their seriousness and decisions made about how urgently control actions are required. Quantitative methods for risk assessment exist. Ridley and Channing (1999) propose the formula:

$$risk\ rating = frequency \times MPL + probability$$

where *frequency* is the number of times a risk has been observed during safety inspections, *MPL* is the maximum possible loss rated on a scale from one (a scratch) to 50 (multiple fatality), and *probability* is rated on scale from one (once per 5 years or more) to 50 (imminent).

Another commonly used tool is a risk assessment matrix, such as the one depicted in Figure 20.4. The probability of a risk resulting in an undesirable outcome and the likely severity of the consequence of such an occurrence are represented in this two dimensional matrix. The position of a risk on the two dimensions determines whether the risk is high, medium or low and the urgency of taking control measures is based upon this position. Some risks, for example, an untidy site, may be high frequency but relatively low in consequence while others, for example, a crane striking overhead powerlines, may be

severity →

probability ↓

	catastrophic (4)	critical (3)	marginal (2)	negligible (1)
frequent (A)	high (4A)	high (3A)	high (2A)	medium (1A)
probable (B)	high (4B)	high (3B)	medium (2B)	low (1B)
occasional (C)	high (4D)	high (3C)	medium (2C)	low (1C)
remote (D)	high (4E)	medium (3D)	low (2D)	low (1D)
improbable (E)	medium (4E)	low (3E)	low (2E)	low (1E)

Figure 20.4 An example risk assessment matrix.

relatively unlikely to happen but potentially catastrophic. It is also important to recognize that the site situation will influence the position of a risk in the matrix. For example, working in an un-shored trench is potentially catastrophic but the likelihood of its occurrence depends upon soil conditions; in sand it is likely to be frequent, in clay it may be unlikely, and in rock it could be considered improbable.

Whatever method of risk assessment is used, it is important to realize that even quantified risk assessment is still essentially a subjective process, which relies on the assessors' knowledge and experience to make a sensible assessment.

20.7.3 Planning for control

Once the magnitude of OHS risks has been determined, appropriate controls can be selected. Where suitable controls cannot be found and the level of residual risk remains unacceptably high, alternative methods of working may need to be considered. Once risk control decisions have been made, these should be documented in formal plans, and responsibilities for their implementation assigned.

20.7.4 Coping with uncertainty

Although the early identification and assessment of OHS risks is important, it is not always possible to anticipate OHS risks at the commencement of a construction project, especially where design work is not complete when construction commences, as is the case with fast track or turnkey projects. The use of numerous timings and planning horizons is useful to overcome such uncertainty. The degree of detail contained in a plan should vary inversely with the planning horizon. As the planning horizon expands, the list of activities should become smaller and the specification of each activity more focused on ideas than on precise facts and numbers. Furthermore, upper management should prepare long-term plans with low levels of detail that are infrequently updated, while lower management levels should prepare detailed, short-term plans more frequently (Laufer *et al.*, 1994). This approach to OHS planning was adopted by the contractor to overcome uncertainty during the design and construction of a large infrastructure project, comprising a partially elevated freeway and bridge, in Melbourne.

20.7.5 Levels of occupational health and safety planning in the Western Link infrastructure project in Melbourne

The various levels of OHS planning carried out on the project are depicted in Figure 20.5. The initial project master plan did not include detail relating to construction methods, site layout or resource handling and thus could not deal with specific OHS issues. Design was completed in lots, and detailed construction plans were prepared for each design lot once it was complete. Construction plans covered general aspects of each element of the work. The construction plan for the bridge portion of the project, for instance, was broken down into elements such as foundations, piers, towers and deck elements. For each element the construction plan laid out engineering, labour, quality, environment and OHS issues.

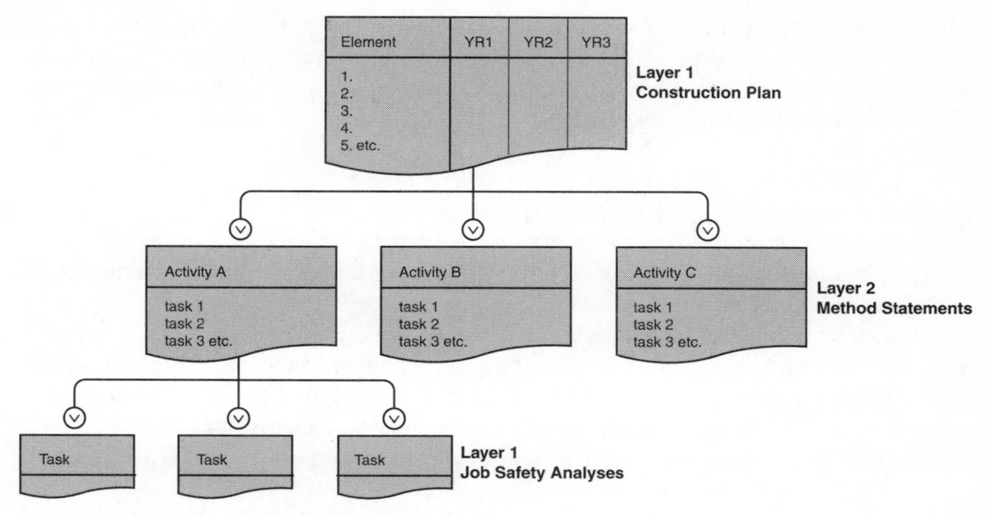

Figure 20.5 Multiple levels of OHS planning.

Construction plans were prepared at a senior level, by the project manager with overall responsibility for the work. They were long-term plans, covering the entire construction period; for example, the bridge construction plan covered a two and a half year period and was reviewed infrequently and only when required. The next level of OHS planning was in the production of method statements. Method statements were prepared by project engineers, an intermediate level of site management. They covered specific elements of the work. For example, in the bridge construction, method statements were prepared for piling, sub-structure concrete, setting up temporary works and deck pours. Method statements included highly detailed OHS procedures and were intended for use for the length of an activity. This could be a one-off activity, such as the setting up of the traveller for bridge deck construction, or a repetitive activity such as pouring deck sections. Finally, the lowest level of project OHS planning was the Job Safety Analysis (JSA). JSAs were conducted by first line supervisors with input from their work teams. They covered all new activities and were highly detailed and specific to individuals' roles and responsibilities during the work; for example, a JSA covering the setting up of a crane would identify the individual responsible for making sure the outriggers were in position.

20.8 Managing sub-contractors

The prevalence of sub-contracting is often cited as a factor contributing to the construction industry's poor OHS performance. Sub-contracting need not cause OHS problems if principal contractors exercise the same diligence in managing the activities of their sub-contractors that clients should exercise in ensuring that OHS standards are maintained on the projects they sponsor. Attention needs to be paid to the selection of safe sub-contractors and all new workers should be provided with induction training to introduce them to the risks present on site, and to any site-specific OHS rules and procedures. The activities of sub-contractors should also be co-ordinated carefully, with special attention paid to managing the interfaces between different work crews and activities. One approach to the management of sub-contractors' OHS, adopted by a large British contracting firm, is outlined in the following case study.

20.8.1 Constructing the British Library

During the construction of the British Library, the managing contractor, John Laing, demanded that their sub-contractors should outline their safety provisions and designate a site manager responsible for enforcing safety measures. Laing insisted that work was carried out in accordance with the sub-contractor's method statement; failure to do this would lead to Laing imposing 'fines'. If the sub-contractor refused to carry out requested safety activities, Laing undertook to do the work and deduct the cost from the contractor's account with the contractor's manager being removed from the site (Jones, 1988).

20.9 Adding value through occupational health and safety

Many OHS professionals argue that losses accruing from accidents in the workplace are substantial enough to provide an economic motive for employers to manage OHS costs (Bird and Germain, 1987). The costs of accidents at work can be direct costs, which can be insured against; these include employees' compensation, medical costs and property damage costs. There are also hidden costs, however, which cannot be insured against; these include reduced productivity, job schedule delays, overtime (if necessary), added administrative time, replacement and re-training costs, and damage to both workers' morale and company reputation. Research suggests that the uninsured costs of occupational safety incidents greatly outweigh the insured costs: one British study of the costs of construction site accidents found the ratio of uninsured to insured costs to be 11 to 1 (Health and Safety Executive, 1993).

It is theoretically possible to identify a level of OHS risk representing the optimum economic level of prevention and incident costs. Figure 20.6 shows the relationship between the cost of safety incidents and the cost of preventing these incidents. The accident (incident) level curve (A) is a measure of OHS risk relating to the direct and indirect costs of safety incidents. The preventive expenditure curve (P) represents money spent on OHS before the event. The cost of any level of OHS risk is the sum of A and P at that point. This will have a minimum value at a particular level of OHS risk. This risk level may be taken to be the economic optimum and is the point at which the cost benefits from improving OHS are just equal to the additional costs incurred.

The problem with such an economic analysis of OHS risk is that the economic optimum level of OHS risk may not be an acceptable level (Hopkins, 1995). This is partly due to the inequitable way in which costs are currently borne. The costs of work-related injury and disease, listed in Table 20.4, are shared between employers, workers and the community at large. The Industry Commission of the Commonwealth Government of Australia reported that employers bear around 30% of the average cost per OHS incident, workers incur 30% and the community bears the remaining 40% (Industry Commission,

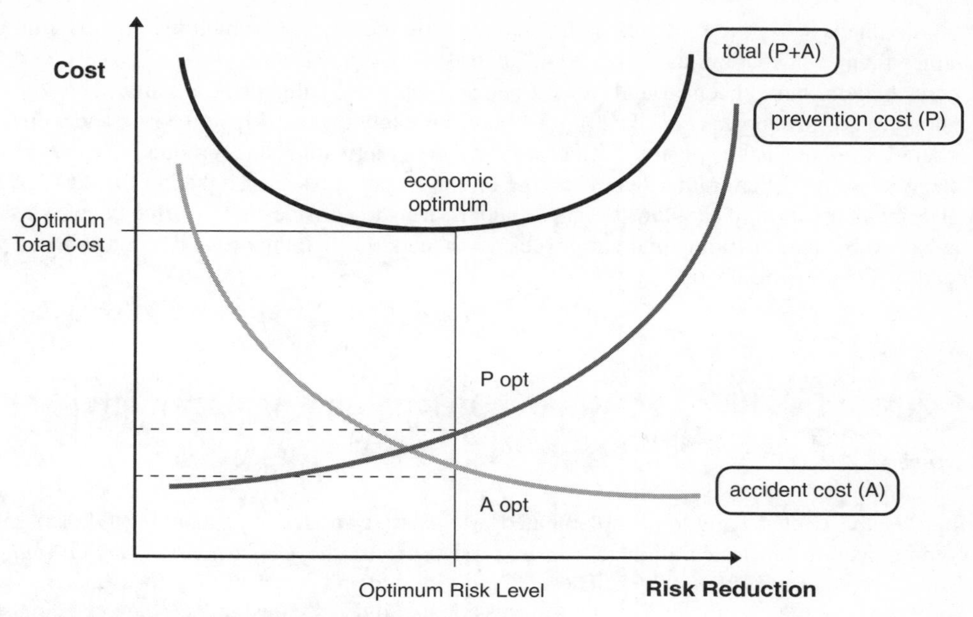

Figure 20.6 The relationship between OHS costs and risk.

1995). Similar disparities have been reported in the United Kingdom (Davies and Teasdale, 1994).

Under these circumstances a desire to reduce costs may not provide sufficient incentive for firms to develop innovative ways of minimizing OHS risk. There are, however, more positive benefits to be gained from managing OHS risk effectively. For example, Rodriguez *et al.* (1996) found a positive association between OHS performance on construction projects and overall project performance, in terms of both cost and time.

Table 20.4 Costs of occupational accidents and ill health borne by employees, employers and society

Cost bearer	Cost
Employee and family	• loss of earnings (short and long term) • hospital attendance and medical bills • pain, grief and suffering
Employer	• compensation (increased insurance premiums) • loss of output • penalties for not meeting schedule • hiring and training new staff • accident investigation costs • repair and/or environmental damage rectification • fines and/or legal costs • loss of goodwill and reputation
Society	• loss of resource deployment (labour, materials and capital) • losses resulting from major disasters e.g., fires and explosions • pain, grief and suffering

Individuals also develop global beliefs concerning the extent to which the organization values their contribution and cares about their well-being (Eisenberger *et al.*, 1986) and these beliefs have been found to influence employee diligence, commitment and innovation (Eisenberger *et al.*, 1990). It is therefore likely that working on a safe and well-ordered worksite has a positive effect on workers' motivation and productivity. These examples suggest that, rather than focusing on the negative issue of reducing the costs of OHS incidents, it may be better for construction firms to consider the positive benefits to be gained by demonstrating that employees are valued, through the provision of a safe and healthy work environment.

20.10 The UK's construction design and management (CDM) regulations experience

In 1994 the United Kingdom implemented a legislative model for enhancing OHS risk communication and integration in construction projects, the *Construction (Design and Management) Regulations 1994*. These *Regulations* identify key parties to a construction project including the client, professional advisers, designers, the principal contractor and sub-contractors or self-employed persons. Each of these parties has a defined set of statutory duties for ensuring that OHS risks are managed during the life of the project. In addition, the *Regulations* require that a planning supervisor be appointed whose role it is to co-ordinate the activities of designers, collate OHS risk information relevant to the project into a health and safety file, and inform the client as to the competence and resource allocation of designers and contractors (Martens, 1997).

Such a model may serve to overcome some of the organizational constraints currently inhibiting the selection and implementation of technological controls for OHS risks. Generally, industry participants perceive that the *CDM Regulations* have acted as a positive driving force for improved OHS in construction. There has been a perceived increase in co-operation between designers and constructors, which has led to improved buildability (Preece *et al.*, 1999) and there is a general belief that the *Regulations* will lead to less accidents and better long-term health among construction workers.

The introduction of the *Regulations* has not, however, been an unqualified success, and a number of problems associated with their implementation have been identified. In particular, compliance has generated a massive amount of additional paperwork and excessive bureaucracy (Anderson, 1998a) and the costs of compliance have proved to be higher than expected (Munro, 1996). The assessment of competence of project participants is particularly costly; for example, it was reported that, in 1996, one large contractor received 5360 pre-qualification questionnaires which cost £589 600 to complete. The same contractor received 1802 sets of tender documents which generated costs of £495 550 for the preparation of OHS responses. The total of these costs for the year was £1.085 million. These costs were incurred despite the fact that only 10% of the tender responses were successful (The Consultancy Company Ltd, 1997).

20.11 Cultural and motivational issues

A pre-requisite for achieving innovation in construction OHS is a genuine commitment to continuing improvement in OHS at all levels in the contractual hierarchy. Reason (1990) suggests that modern technology has advanced to the point where improved safety can only be achieved through attention to human error mechanisms. In construction, there is still plenty of room for further technological innovation to overcome OHS problems but in order to achieve this people must believe it is possible to improve OHS and be committed to doing so. It is therefore important to seek an understanding of the motivational and cultural issues that shape construction industry participants' beliefs about workplace risk, and their subsequent OHS behaviour.

20.11.1 Applied behaviour analysis

OHS behaviour has been improved in several industries through the use of psychologically-based techniques sometimes termed *applied behaviour analysis* (McAfee and Winn, 1989). Applied behaviour analysis involves the identification of specific behaviours representing safe and unsafe practices in an employee's job and undertaking regular and frequent measurement of employees' performance with regard to these behaviours. The behaviours are prompted and then systematically reinforced. Prompts include formal OHS goal setting, training, and fostering team competition, while reinforcers include the provision of token rewards, the provision of feedback on individual performance and the provision of feedback on group performance. Applied behaviour analysis has been implemented successfully in the difficult construction industry context (Mattila and Hyodynmaa, 1988; Duff *et al.*, 1994). An example of the application of the behavioural approach to site safety in Hong Kong is described below.

20.11.2 Applied behaviour analysis in the Hong Kong construction industry

A 9-month experiment was conducted on seven high rise public housing construction projects in Hong Kong. Four categories of safety behaviour were measured during the experiment. These were: site housekeeping, access to heights, bamboo scaffolding, and personal protective equipment (PPE). Site housekeeping was concerned with keeping the site tidy and free of rubbish. A trained observer used a safety checklist in which safe and unsafe behaviours were rated proportionally to measure behaviour on each site twice weekly. After a period of baseline measurement, participative goal setting and performance feedback were introduced to the housekeeping, access to heights and bamboo scaffolding categories at staggered intervals. PPE was maintained as a control category and thus no intervention was introduced in relation to the use of PPE.

Performance in housekeeping was found to improve on all seven sites when applied behaviour analysis was introduced. On five of these sites this improvement was highly statistically significant. On no site was there a general improvement of safety performance (e.g., improvement in all performance categories) when the housekeeping intervention was introduced. On all but one site, housekeeping performance deteriorated with the removal of the housekeeping feedback charts. On four of these sites this deterioration was statistically significant.

Safety performance in access to heights was found to improve on five of the sites with the introduction of applied behaviour analysis but the improvement was statistically significant on only two of these sites. No significant improvement in bamboo scaffolding safety performance was observed on any of the four sites at which the applied behaviour analysis was introduced to bamboo scaffolding (Lingard, 1995; Lingard and Rowlinson, 1997, 1998).

The Hong Kong results suggest that there are some limitations to the effectiveness of applied behaviour analysis in motivating construction workers to behave safely. Bradley (1989) suggests that many behavioural interventions fail because they do not take account of other aspects of the work environment. Site managers, held accountable for project schedule and cost performance, may inadvertently reward unsafe work practices on site and applied behaviour analysis is not likely to succeed in a work environment in which many conflicting and confusing motivational cues are present. Furthermore, incentive schemes, such as piece rate remuneration or bonuses, can lead to corner-cutting and unsafe work practices (Dwyer, 1991). In order for the behavioural approach to work, it is essential for construction site managers to clearly demonstrate that OHS is an important project objective, which must not be compromised in order to meet production targets. Management commitment to OHS was not assessed during the Hong Kong study but senior site managers, who initially attended the housekeeping goal setting meetings, were mostly absent from the access to heights meetings and entirely absent from bamboo scaffolding meetings, demonstrating little commitment to the applied behaviour analysis process.

One way in which management must show commitment to OHS is in the provision of a safe work environment and suitable equipment. Applied behaviour analysis acts on the behaviour of individual workers, not the work environment. It follows that, in order to change OHS behaviours, workers need to have personal control over these behaviours (Petersen, 1971). Where safe behaviour is constrained by an unsafe environment or inadequate equipment, applied behaviour analysis cannot be effective. Most of the access to heights and bamboo scaffolding items measured during the Hong Kong study would require that additional materials be provided and, most significantly, that an increased time period be allowed for the job at hand. For example, if a bamboo scaffold is to be fitted with adequate, closely boarded working platforms, guard rails and toe boards, it requires that good quality timber is available. The time required to construct a scaffold with these features is much greater than the time required to construct a bamboo scaffold without these features. The scaffold is normally erected by a specialist sub-contractor, to specifications provided by the principal contractor, at an agreed price. Where an unsafe scaffold is provided, individual workers using the scaffold to undertake external finishing work will be prevented from working safely and any attempt to modify their behaviour is likely to fail.

Applied behaviour analysis is a powerful tool and research suggests that, when the conditions are right, it can be highly effective in improving OHS performance in construction and related industries. It is not, however, a 'quick fix' solution, as some managers and safety professionals believe. It must work in concert with existing OHS systems and requires an environment in which there is clear and communicated management commitment to OHS.

20.11.3 Safety culture

Management theorists have long suggested that the performance of an organization is influenced by its culture (Likert, 1967). The Human Factors Study Group of the UK's Advisory Committee on the Safety of Nuclear Installations defines the concept of safety culture as 'the product of individual and group values, attitudes, perceptions, competencies and patterns of behaviour that determine commitment to and the style and proficiency of an organization's health and safety management' (Health and Safety Commission, 1993). In an attempt to define what safety culture means in practice, researchers have tried to identify the features of a positive safety culture (Confederation of British Industry, 1990; Lee, 1997). These features may be summarized as follows:

- genuine and committed safety leadership from senior management, visibly demonstrated by allocation of resources to safety and promoting safety personally
- giving safety a high priority through the development of a policy statement of high expectations which conveys a sense of optimism about what is possible
- long term view of OHS as part of business strategy
- realistic and achievable targets and performance standards with current information available to assess performance
- open and ready communication concerning safety issues within and between levels of the organization, with less formal and more frequent communications
- democratic, co-operative, participative and humanistic management style leading to 'ownership' of health and safety permeating all levels of the workforce
- more and better training, with safety integrated into skills training
- capacity for organizational learning, with organizations responsive to structural change and results of system audits, and incident investigations
- line management responsibility for OHS, with OHS treated as seriously as other organizational objectives
- high job satisfaction, and a perception of procedural fairness in employer/employee relations.

While the construction industry's 'macho' culture and production orientation have been identified as potential barriers to the development of a positive safety culture (Anderson, 1998b) it is not impossible, even in this difficult context, to engender a culture of safety. At a project and enterprise level, large construction firms demonstrate some, if not all, of the features of a positive safety culture. The structure of the industry does, however, militate against the achievement of an industry-wide culture of safety. One of the difficulties with the safety culture concept is that it assumes a homogeneity of workforce which does not reflect the complexity of the construction industry environment.

20.11.4 Group influences on occupational health and safety

Groups can have a direct and indirect influence on their members' behaviour (Porter *et al.*, 1983). In any organizational context, groups play an important role in determining behaviour but it is reasonable to expect that the influence of work groups will be greater where there is a high level of fragmentation. The cohesiveness of work crews, for example, has been found to influence safety performance (Hinze, 1981). Members of work crews which 'got along' were found to have better individual safety records than those who felt that there was friction within their work crew. High levels of cohesiveness do not, however, automatically lead to greater organizational effectiveness; research suggests that where employees feel a strong sense of belonging to a group, this group has a powerful influence over the behaviour of the individuals in the group. This influence may be either good or bad for an organization depending on the extent to which group members identify with, and accept the goals of, the organization as a whole (Greene, 1989). Where group values and norms are not aligned with organizational OHS objectives, cohesive work groups may hinder, rather than facilitate, effective OHS performance (Lingard and Rowlinson, 1994).

20.11.5 Attaining consensus

The attainment of consensus and trust in OHS decision-making is important if commitment to common OHS objectives is to be achieved. Active employee participation in solving safety problems and formulating safe work procedures has been identified as important to the development of a positive safety culture within an organization (Cox and Tait, 1998). In the construction industry, this communication and participation must take place both within and across organizational boundaries. All participants in the contractual hierarchy, including the client, the designer, the construction or project manager (if employed), the works contractor and sub-contractors, have a role to play. It is very important that a 'top down' approach to safety culture is not adopted. The desire to impose 'cultural control' can inhibit free and open communication and could lead to a failure to fully utilize the collective knowledge of OHS risks that the workforce has gained through hands-on experience (Back and Woolfson, 1999). Innovation in construction OHS requires a safety culture that acknowledges the existence of alternative viewpoints on OHS and does not seek to suppress or devalue these viewpoints but encourages their incorporation into project decision making.

20.11.6 Understandings of occupational health and safety risk

The uni-dimensional risk indices typically used in OHS risk management, such as annual probability of injury or death, do not reflect the way people think about risk; psychological and social processes also have a significant impact upon the way people respond to risks (Fischoff *et al.*, 1978). People understand risk as a multi-dimensional concept (Slovic, 1987). Research suggests that risks that are associated with a perceived lack of control, fatal consequences, and high catastrophic potential are considered to be more serious than risks that are unknown, unobservable and delayed in their manifestation (Slovic *et al.*, 1981). Despite the fact that more deaths are attributed to occupational illnesses than accidents, occupational health risks are not widely understood, are not readily observable, and are often delayed in their consequences. In accordance with the findings of Slovic and his colleagues, Australian research shows that construction industry participants tend to underestimate delayed effect health risks as compared to immediate effect safety risks (Holmes *et al.,* 1999). The complex way in which people understand OHS risks has implications for control strategies. Innovative solutions should be sought for both occupational health and occupational safety risks but it is probable that an effective control strategy for an occupational health risk would require a strong educational component.

Beliefs about the source or cause of an OHS risk also influence how people conceptualize risk control measures (Tesh, 1981). The attribution of causation is a logical human response to events and involves a consideration of whether the cause is understood to be internal (individual) or external (situational) to the person involved in the event (DeJoy, 1994). For example, where the cause is perceived to be internal for an OHS risk, risk control is likely to be understood in terms of individual control measures designed to act on workers' behaviour, such as procedures, training and personal protective equipment. Where the cause is perceived to be external, risk control is likely to be understood in terms of technological control measures designed to act on the work environment.

Social processes in the workplace have been found to influence perceptions of the causes of OHS risk. For example, Holmes and Gifford (1996) found that in the Australian construction industry employers believed accidents to be caused by the carelessness of their workers while the workers accepted OHS risks as an inevitable part of the job. The results of a study of understandings of risk attribution and control among small business construction participants in Australia are presented below (section 20.11.7). The results of this study suggest that both OHS risks with delayed effects and those associated with immediate effects are perceived to be attributable to individuals, and risk control is consequently understood to be the individual's own responsibility. This view is contrary to the 'hierarchy of controls' for OHS risk, which holds that risk controls that act on the work environment are preferable to those that act on the individual. Such perceptions are likely to hinder the implementation of innovative methods to eliminate or reduce OHS risk and lead to an unwarranted reliance on individual controls such as training, procedural controls and personal protective equipment. Efforts should therefore be made to change both managers' and workers' perceptions of the sources of OHS to encourage recognition of situational OHS risk factors.

20.11.7 Understandings of occupational health and safety risks and their control in the Australian construction industry

A study was undertaken to explore small business construction industry participants' understandings of OHS risk control (Lingard and Holmes, 2001). Two OHS risks were chosen as the topics for the study. One risk is associated with an immediate outcome (falls from heights) and the second (occupational skin diseases) with a delayed outcome. In-depth interviews, that were guided by a structured theme list, were conducted with fifteen participants at their workplace. The interview transcripts were coded for major themes and their content analysed. Emergent concepts for the two OHS risks were contrasted on the basis of frequency.

Participants attributed both risks to issues internal to the individual worker. For example, participants perceived that the risk of occupational skin disease was dependent upon complacency and individual susceptibility. Similarly many participants suggested that the occurrence of falls from heights is due to worker carelessness or a lack of concentration. Situational factors, such as the absence of adequate scaffolding or the specification of non-hazardous chemicals, were mentioned much less frequently, appearing to be less important sources of risk in the minds of participants. Risk controls for the two risks focused heavily on the individual controls of education, the adoption of safe work procedures and the use of personal protective equipment (Lingard and Holmes, 2001).

20.11.8 First aid training

OHS risk-taking behaviour may also be influenced by individual biases such as a belief that 'it won't happen to me'. These beliefs breed complacency about workplace OHS risks giving rise to unsafe behaviours. First aid training may be useful in helping to address this problem. Research in non-construction settings has identified an association between traditional first aid training and a lower incidence of workplace injuries (Miller and Agnew, 1973; McKenna and Hale, 1981). People trained in first aid have also expressed a greater willingness to take personal responsibility for safety and a willingness to adopt safe behaviour (McKenna and Hale, 1982). These findings suggest that first aid training may have a positive preventive effect, in addition to the traditional objective of providing laypersons with the skills to manage the consequences of incidents once they have happened.

The results of a recent experiment evaluating the effect of first aid training on the OHS behaviour of employees in small construction firms confirm this suggestion. In both individual and global performance, safety scores in three of the four measured categories were higher after participants received first aid training, indicating an improvement in safety over pre-training performance. In particular, safety scores were significantly better in 'Access to Heights', use of 'Personal Protective Equipment' and the 'Use of Tools' and site 'Housekeeping' (Lingard, 2001). Figure 20.7 shows the extent of these improvements.

The findings of this study indicate that first aid training can have a positive effect on OHS behaviour. It is likely that first aid training improves workers' motivation to avoid injury by changing the way that trainees think about risk. For example, first aid training may increase the ease with which trainees can imagine an injury occurring in their work situation or may reduce trainees' perception that they are less likely to experience an injury than fellow workers. The results of the study suggest that training larger numbers of employees in first aid could help to prevent occupational injuries. However, improvements were not observed in all performance categories and it is important that the pathway from first aid training to improved OHS behaviour be further investigated.

20.12 Conclusion

The construction industry's unenviable OHS performance contributes to its poor public image and the industry needs to increase its efforts to improve the health and safety of its workers. The current legislative framework in Commonwealth countries and members of the European Union requires the implementation of OHS management systems, yet the construction industry has been slow to implement an integrated and systematic approach to the management of workplace OHS risk. The fragmented nature of the construction

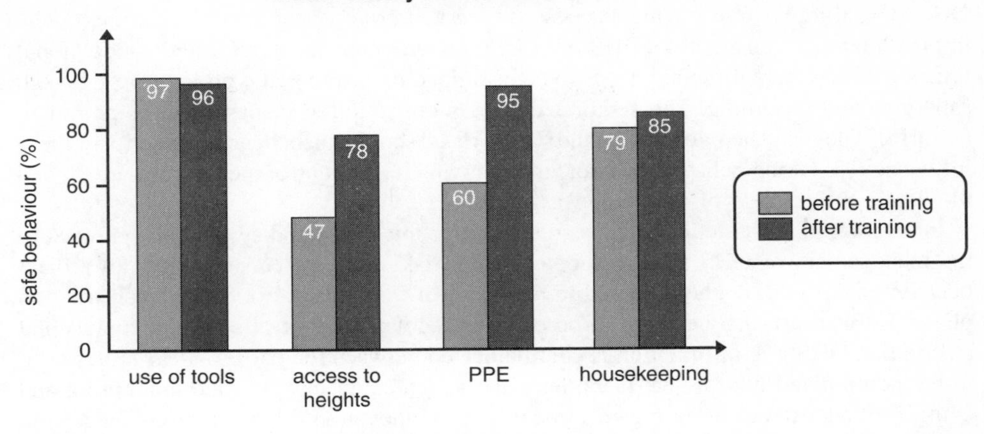

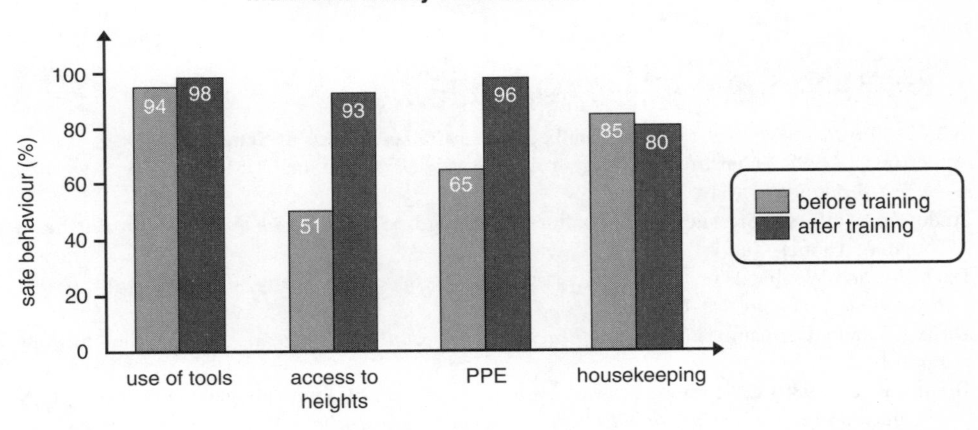

Figure 20.7 Global and individual safety performance.

industry, in which the design and construction functions are traditionally separate and which relies heavily on contracting and sub-contracting, requires a considerable effort to ensure that the risks to the health and safety of those who will undertake the work are considered. Decisions as to how best to eliminate or reduce OHS risks should be taken before the commencement of construction and, wherever possible, 'technological' solutions to OHS risks should be identified.

Clients must take the lead in ensuring that this happens on the projects they sponsor. The development of long-term relationships with responsible designers, consultants and contractors can assist in this and also provide an opportunity for strategic reduction of OHS risk, through transferring the lessons learned from one project to the next. The improvement of construction OHS also requires attention to cultural and motivational issues. The work environment needs to be examined to ensure that working safely is both satisfying and rewarding. This requires clear demonstration of management commitment to OHS. The way people understand OHS risks should also be considered, as these influence risk-taking behaviours, choices about which risk control measures to adopt, and ultimately the efficacy of these controls.

Innovation in construction OHS requires that an integrated and systematic approach to the management of OHS risk be adopted. OHS risks must be considered in all project decision-making, throughout the entire life-cycle of the project, not just the construction phase. Effort must also be made to co-ordinate the activities of project participants and ensure that OHS risk information is communicated between the parties. OHS issues need to be incorporated into project planning activities and subject to routine monitoring and control. In addition to these organizational issues, innovation relies on strong leadership in OHS and a committed and motivated project team. Workers should not be blamed for the occurrence of accidents. Instead, the pathways leading to human error should be identified and eliminated. OHS risks must no longer be accepted as 'part of the job' and psychological issues that encourage risk-taking behaviours need to be addressed.

References and bibliography

ABS (1998) *Business Register Data* (Melbourne: Australian Bureau of Statistics).

Anderson, J. (1998a) Construction safety – changes needed now to the CDM Regs, *The Safety and Health Practitioner*, May, 26–8.

Anderson, J. (1998b) Growing a safety culture. In Barnard, M. (ed.) *Health and Safety for Engineers* (London: Thomas Telford).

Back, M. and Woolfson, C. (1999) Safety culture – a concept too many? *The Safety and Health Practitioner,* January, 14–16.

Bird, F.E. and Germain, G.L. (1987) *Practical Loss Control Leadership* (Loganville: Institute Publishing).

Bradley, G.L. (1989) The forgotten role of environmental control: some thoughts on the psychology of safety. *Journal of Occupational Health and Safety – Australia and New Zealand*, **5**, 501–8.

Brook, M. (1993) *Estimating and Tendering for Construction Work* (Oxford: Butterworth-Heinemann).

Brooks, A. (1988) Rethinking occupational health and safety legislation. *Journal of Industrial Relations,* **30**, 347–62.

Brown, P.E. (1996) Total integration of the safety professional into the project management team. In: Aves Dias, L.M. and Coble, R.J. (eds) *Implementation of Safety and Health on Construction Sites (*Rotterdam: A.A. Balkema).

Commission of the European Communities (1993) *Safety and Health in the Construction Sector* (Luxembourg: Office for Official Publications of the European Communities).

Confederation of British Industry (1990). *Developing a Safety Culture* (London: CBI).

Cox, S. and Tait, R. (1998) *Safety, Reliability & Risk Management* (Oxford: Butterworth-Heinemann).

Davies, N.V. and Teasdale, P. (1994) *The Costs to the British Economy of Work Accidents and Work-Related Ill-Health* (London: HMSO).

Dawson, S., Wilman, P., Clinton, A. and Bamford M. (1988) *Safety at Work: the Limits of Self-regulation* (Cambridge: Cambridge University Press).

DeJoy, D.M. (1994) Managing safety in the workplace: an attribution theory analysis and model. *Journal of Safety Research,* **25**, 3–17.

Duff, A.R., Robertson, I.T., Phillips, R.A. and Cooper, M.D. (1994) Improving safety by the modification of behaviour. *Construction Management and Economics*, **12**, 67–78.

Dwyer, T. (1991) *Life and Death at Work: Industrial Accidents as a Case of Socially Produced Error* (New York: Plenum).

Eakins, J. (1992) Leaving it up to the workers: sociological perspectives on the management of health and safety in small workplaces. *International Journal of Health Services,* **22**, 689–704.

Egan, J. (1998) *Rethinking Construction: The Report of the Construction Task Force* (London: Department of the Environment, Transport and the Regions).

Eisenberger, R., Huntington, R., Hutchison, S. and Sowa, D. (1986) Perceived organizational support. *Journal of Applied Psychology,* **71**, 500–7.

Eisenberger, R., Fasolo, P. and Davis-LaMastro, V. (1990) Perceived organizational support and employee diligence, commitment and innovation. *Journal of Applied Psychology,* **75**, 51–9.

Emmett, E. and Hickling, C. (1995) Integrating management systems and risk management approaches. *Journal of Occupational Health and Safety – Australia and New Zealand,* **11**, 617–24.

European Construction Institute (1996) *Total Project Management of Construction Safety, Health and Environment* (London: Thomas Telford).

Fischoff, B., Slovic P., Lichtenstein, S., Read, S. and Combs, B. (1978). How safe is safe enough? A psychometric study of attitudes towards technological risks and benefits. *Policy Sciences,* **9**, 127–52.

Greene, C.N. (1989) Cohesion and productivity in work groups. *Small Group Behavior,* **20**, 70–86.

Health and Safety Commission (1993) *Third Report: Organising for Safety.* Advisory Committee on the Safety of Nuclear Installations – Human Factors Study Group (Sudbury: HSE Books).

Health and Safety Executive (1993) *The Costs of Accidents at Work* (London: HMSO).

Health and Safety Executive (1995) *Designing for Health and Safety in Construction* (London: HMSO).

Hinze, J. (1981) Human aspects of construction safety. *Journal of the Construction Division, Proceedings of the American Society of Civil Engineers,* **107**, pp. 61–72.

Hinze, J. and Gambatese, J.A. (1994) Design decisions that impact construction worker safety. In: *Proceedings of the Fifth Annual Rinker International Conference on Construction Safety and Loss Control* (Gainsville: University of Florida).

Holmes, N. (1995) *Workplace Understandings and Perceptions of Risk in OHS.* Unpublished PhD thesis (Melbourne: Monash University).

Holmes, N. and Gifford, S.M. (1996) Social meanings of risk in OHS: consequences for risk control. *Journal of Occupational Health and Safety Australia and New Zealand*, **12**, 443–50.

Holmes, N., Lingard, H., Yesilyurt, Z. and DeMunk, F. (1999) An exploratory study of meanings of risk control for long term and acute effect occupational health and safety risks in small business construction firms. *Journal of Safety Research*, **30**, 251–61.

Hopkins, A. (1995) *Making Safety Work* (St Leonards: Allen and Unwin).

Industry Commission (1995) *Work, Health and Safety, Report number 47* (Canberra: Commonwealth of Australia).

Jones, H. (1988) Safety in the Library. *New Civil Engineer*, May.

King, R.W. and Hudson, R. (1985) *Construction Hazard and Safety Handbook* (London: Butterworths).

Laufer, A., Tucker, R.L., Shapira A. and Shenhar, A.J. (1994) The multiplicity concept in construction planning. *Construction Management and Economics,* **11**, 53–65.

Lee, T. (1997) How can we monitor the safety culture and improve it where necessary? In *Proceedings of a Conference on 'Safety Culture in the Energy Industries'* (Cookham: Energy Logistics International Ltd).

Likert, R. (1967) *The Human Organization* (New York: McGraw-Hill).

Lingard, H. (1995) *Safety in Hong Kong's Construction Industry: Changing Worker Behaviour,* Unpublished PhD thesis (Hong Kong: The University of Hong Kong).

Lingard, H. (2001) The effect of first aid training on objective safety behaviour in Australian small business construction firms. *Construction Management and Economics*, **19**, 611–18.

Lingard, H. and Holmes, N. (2001) Understandings of occupational health and safety risk control in small business construction firms: barriers to implementing technological controls. *Construction Management and Economics*, **19**, 217–26.

Lingard, H. and Rowlinson, S. (1994) Construction site safety in Hong Kong. *Construction Management and Economics*, **12**, 501–10.

Lingard, H. and Rowlinson, S. (1997) Behavior-based safety management in Hong Kong's construction industry. *Journal of Safety Research,* **28**, 243–56.

Lingard, H. and Rowlinson, S. (1998) Behaviour-based safety management in Hong Kong's construction industry: the results of a field study. *Construction Management and Economics,* **16**, 481–8.

Martens, N. (1997) The Construction (Design and Management) Regulations 1994: Considering the competence of the planning supervisor. *Journal of the Institution of Occupational Safety and Health*, **1**, 41–9.

Matilla, M. and Hyodynmaa, M. (1988) Promoting job safety in building: an experiment on the behavior analysis approach. *Journal of Occupational Accidents,* **9**, 255–67.

Matthews, J. (1993) *Health and Safety at Work* (Sydney: Pluto Press).

Mayhew, C. (1997) *Barriers to the Implementation of Known Occupational Health and Safety Solutions in Small Businesses* (Canberra: Commonwealth Government of Australia).

McAfee, R.B. and Winn, A.R. (1989) The use of incentives/feedback to enhance workplace safety: a critique of the literature. *Journal of Safety Research*, **20**, 7–19.

McKenna, S.P. and Hale, A.R. (1981) The effect of emergency first aid training on the incidence of accidents in factories. *Journal of Occupational Accidents,* **3**, 101–14.

McKenna, S.P. and Hale, A.R. (1982) Changing behaviour towards danger: the effect of first aid training. *Journal of Occupational Accidents,* **4**, 47–59.

McVittie D., Banikin, H. and Brocklebank, W. (1997) The effect of firm size on injury frequency in construction. *Safety Science,* **27**, 19–23.

Miller, G. and Agnew, N. (1973) First aid training and accidents. *Occupational Psychology,* **47**, 209–18.

Munro, W.D. (1996) The implementation of the construction (design and management) regulations 1994 on UK construction sites. In: Aves Dias, L.M. and Coble, R.J. (eds) *Implementation of Safety and Health on Construction Sites* (Rotterdam: A.A. Balkema).

Oluwoye, J. and MacLennan, H. (1994) Preplanning safety in project buildability. In: *Proceedings of the Fifth Annual Rinker International Conference on Construction Safety and Loss Control* (Gainsville: University of Florida).

Petersen, D. (1971) *Techniques of Safety Management* (New York: McGraw-Hill).

Porter, L.W., Lawler, E.E. and Hackman, J.R. (1983) Ways groups influence work effectiveness. In: Steers, R.M. and Porter, L.W. (eds) *Motivation and Work Behavior* (New York: McGraw Book Co.).

Preece, C.N., Moodley, K. and Cavina, C. (1999) The role of the planning supervisor under new health and safety legislation in the United Kingdom. In: Singh, A., Hinze, J. and Coble, R.J. (eds) *Implementation of Safety and Health on Construction Sites* (Rotterdam: A.A. Balkema).

Reason, J. (1990) *Human Error* (Cambridge: Cambridge University Press).

Reason, J. (1995) A systems approach to organizational error. *Ergonomics*, **38**, 1708–21.

Ridley, J. and Channing, J. (1999) *Risk Management* (Oxford, Butterworth-Heinemann).

Rodriguez, Y.A., Jaselkis, E.J. and Russell, J.S. (1996) Relationship between project performance and accidents. In: Aves Dias, L.M. and Coble, R.J. (eds) *Implementation of Safety and Health on Construction Sites* (Rotterdam: A.A. Balkema).

Ronco, W.C. and Ronco, J.S. (1996) *Partnering Manual for Design and Construction* (McGraw-Hill).

Russell, J.S., Hancher, D.E. and Skibniewski, M.J. (1992) Contractor prequalification data for construction owners. *Construction Management and Economics,* **10**, 117–29.

Slovic, P. (1987) Perception of risk. *Science,* **236**, 280–5.

Slovic, P., Fischoff, B. and Lichtenstein, S. (1981) Facts and fears: understanding perceived risk. In: Schwing, R.C. and Walker, A.A. (eds) *Society Risk Assessment, How Safe is Safe Enough?* (New York: Plenum Press).

Standards Australia (1997) *AS/NZS 4804: 1997 Occupational health and safety management systems – General guidelines on principles, systems and supporting techniques* (Sydney: Standards Australia).

Tesh, S. (1981) Disease, causality and politics. *Journal of Health Politics, Policy and Law,* **6**, 369–90.

The Consultancy Company Ltd (1997) *Evaluation of the Construction (Design and Management) Regulations 1994* (Sudbury: HSE Books).

Williams, M.A. (1998) Designing for safety. In: Barnard, M. (ed.) *Health and Safety for Engineers* (London: Thomas Telford).

21

ISO 9000 quality management systems for construction safety

Low Sui Pheng*

Editorial comment

This chapter on quality management systems looks at the ISO 9000 standards that have been applied in the construction industry in many countries over the last few years. While the way that ISO 9000 standards are applied to the construction process and the general issues associated with quality management systems are discussed, this chapter takes a slightly different approach to quality systems in the construction industry. The chapter outlines the development of an integrated quality and safety management system. The system uses the ISO 9000 standards and the twenty clauses that make up that standard, however it shows how the standards can be developed and applied for safety management in the construction industry.

In this way, the chapter presents both an informed look at the use of quality systems in the construction industry by a recognized author in the field, but also shows how the quality systems can be applied and extended in an innovative way to other areas of importance for the industry. Although there is a chapter on safety in the book (Chapter 20), that chapter discusses the issues associated with occupational health and safety on site, and the protection of workers from industrial accidents, whereas this chapter looks at a safety management system in the context of the management techniques that are commonly associated with quality management systems.

The issue of safety in the construction industry has been a major problem for many years, and so far none of the proposed solutions to improving safety practices in the industry have been altogether effective. It is possible that the approach propounded by Low in this chapter is a good alternative and in some cases, would be extremely beneficial for both managers and workers in the industry.

* National University of Singapore

21.1 Introduction

The ISO 9000 standards have actively been promoted in the construction industry in many countries as a means of quality assurance to be provided by building contractors for the works they undertake. Some countries (for example, Australia, Hong Kong and Singapore) have gone as far as to require all contractors to be certified to ISO 9000 requirements before they qualify to bid for the larger public sector building projects. This means that Quality Management Systems (QMS) certified to ISO 9000 standards are now a ubiquitous feature in the construction industry in many countries.

Apart from quality issues, building contractors are also faced with a moral obligation to care for the well-being of workers on site. Consequently, the need for building contractors to provide a safe working environment on site cannot be overlooked. This is particularly so in the larger and more complex projects where safety becomes an even more important issue. Safety concerns life as well as property and should not be taken lightly. The relevant government agencies in the more developed countries like the United Kingdom, Australia and Singapore have legislated safe working practices which building contractors must conform to on site. The evolution of Safety Management Systems (SMS) therefore becomes the mandated means through which site safety measures can be managed more effectively.

This division between quality management and safety management, however, suggests that integrating QMS with SMS in an organization, since both management systems are common features in many construction firms, can reap immense benefits. Instead of operating two separate management systems, synergy can be achieved by integrating QMS and SMS to allow an organization to work from a common platform. In a wider context, Low (2000) observed that the development and implementation of different management systems to cover buildability principles, just-in-time productivity concepts and safety requirements within a construction firm may lead to fragmentation and confusion (Low and Chong, 1999; Low and Abeyegoonasekera, 2000; Low and Sua, 2000). To achieve synergy, Low (2000) proposed that buildability principles, just-in-time productivity concepts and safety requirements be built into existing ISO 9000 QMS, and the 20 quality system elements stipulated in ISO 9001 be used. This is similar to the existing practice of linking ISO 14000 for Environmental Management Systems with ISO 9000. The rationale for this integration arises firstly because contractors will need to deal with and address safety requirements, productivity concepts and buildability principles in their course of work. Secondly, it may not be desirable to consider all these issues separately if the organization hopes to achieve the synergy that comes through integration. Thirdly, the 20 quality system elements in ISO 9000 are generic in nature; each element may therefore be interpreted differently to meet the specific needs of an individual organization. For these reasons, it would be desirable to provide for a ‘three-in-one’ solution to achieving safety, productivity and buildability objectives through ISO 9000 QMS (Low, 2000).

This chapter focuses on how safety requirements for the construction industry may be integrated with ISO 9000 quality system elements. It examines how each of these quality system elements can be interpreted to activate safety concerns in the various ISO 9000 processes and procedures. The thrust of this chapter is shown in Figure 21.1 where an SMS is linked with a QMS to achieve an integrated Quality and Safety Management System (QSMS). The objective of this chapter is therefore to explain the criteria common

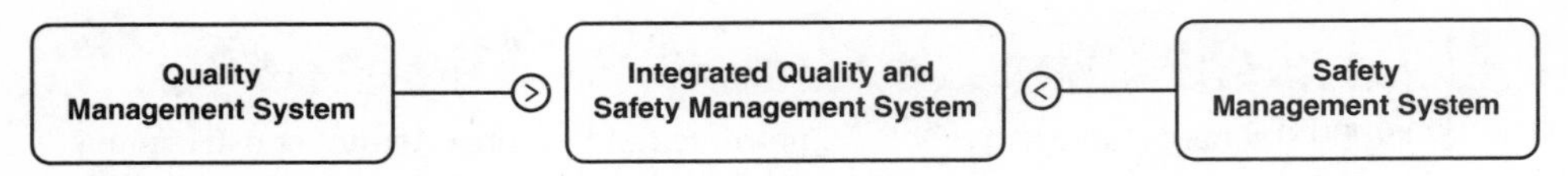

Figure 21.1 An integrated quality and safety management system.

to a QMS and an SMS and so streamline operations for the building contractor. At this juncture it should be noted that most, if not all, QMS are founded on a common template in the ISO 9000 standards regardless of the country that a building contractor works in. On the other hand, safety regulations for building operations may vary from country to country depending, in many cases, on the stage of development of the construction industry in a particular country. For this reason it is more difficult to format a QMS based on an SMS because there is no standard template to work on in different countries. It is therefore easier to integrate an SMS as a sub-set of a QMS since the ISO 9000 standards provide the global template that makes this possible. For this purpose, construction safety regulations in Singapore will be used as an example here to demonstrate this integration. The aim is essentially to achieve quality services through safety!

21.2 Construction safety regulations

In Singapore, the Factories (Building Operations and Works of Engineering Construction) (Amendment) Regulations were introduced in 1994 to include SMS and safety audit/review for the construction industry; these regulations will be referred to as the BOWEC Regulations hereafter. Subsequently on 1 July 1996, the legal requirement for worksites to employ safety officers was introduced. All these efforts lead to one conclusion – that safety must come from within an organization and from top management (Goh, 1997).

The SMS is a system of work procedures that ensures safe working practices are adhered to in order to minimize the chance of an accident occurring. It is similar in philosophy to the ISO 9000 standards. While an ISO 9000 QMS aims to achieve quality objectives through proper documentation and work procedures, the SMS, in the same fashion, aims to achieve the objective of a safe working environment through the establishment of a proper management system within which safety issues may be addressed.

The BOWEC Regulations list 13 main elements that constitute good safety management practices on construction sites. If a construction firm implemented more than half of these elements properly, it would be considered to have achieved an average SMS for its operations (Goh, 1997). The 13 elements are:

- safety policy
- safe working practice
- safety training
- group meetings
- incident investigation and analysis
- in-house rules and regulations
- safety promotion

- control of contractors
- safety inspection
- maintenance regime
- hazard analysis
- hazardous materials
- emergency preparedness.

These 13 main elements form the basic SMS guidelines in this chapter for the purpose of integration with a QMS since these are regulated in the BOWEC Regulations.

21.3 ISO 9000 quality management systems (QMS)

The International Organization for Standardization (ISO) based in Geneva, Switzerland, developed the ISO 9000 standards from BS5750 in 1987 as a way 'to encourage the international co-ordination and unification of industrial standards'. The ISO 9000 series of standards were subsequently adopted by many countries around the world. ISO 9000 is essentially a series of five standards that describe the elements for establishing and maintaining a QMS. The ISO 9000 language used is generic as it is intended to be applicable to a wide range of industries, including construction. What ISO 9000 aims to do is to spell out a comprehensive set of planned and systematic actions or procedures necessary to ensure that a product or service will satisfy given requirements for quality.

This series of standard consists of the following:

- ISO 9000 – provides guidance to the selection and use of Quality Management and Quality Assurance Standards
- ISO 9001 – sets out in details the specification for quality assurance in the Design and Development, Production, Installation and Servicing stages.
- ISO 9002 – sets out in details the specification for quality assurance in the Production, Installation and Servicing stage.
- ISO 9003 – sets out in details the specification for quality assurance in the Final Inspection and Testing stage.
- ISO 9004 – the final part of the series that explains Quality Management and Quality System Elements.

Use of the ISO 9000 standards can help to achieve the following benefits:

- everyone can now communicate on quality assurance using a common language and a common reference point internationally
- the standards contain quality assurance criteria that are sufficiently generic, comprehensive and 'implementable' so that organizations in a wide range of industries can adopt and use them
- the standards facilitate and promote third party auditing and certification. Third party certification raises purchasers' confidence as an independent body has assessed the company's quality assurance system and certified it as conforming to a recognized standard. Before this, customers will have to assess and satisfy themselves as to whether potential suppliers are quality assured (Low, 1998).

The ISO 9000 standards provide a template for the following:

- planning, implementing and maintaining quality management systems
- contractual requirements between customer and supplier
- assessments and certification
- reduced multiple customer assessment through harmonized quality standards (Low, 1998).

The ISO 9000 specifies a set of clauses that a firm seeking implementation must necessarily adhere to, however, these clauses are generic in nature. Hence, firms can fit the specified needs of their organizations into the framework of the ISO 9000. The 20 clauses specified in the ISO 9000 Standards are:

- management responsibility
- quality system
- contract review
- design control
- document and data control
- purchasing
- control of customer-supplied product
- product identification and traceability
- process control
- inspection and testing
- control of inspection, measuring and test equipment
- inspection and test status
- control of non-conforming product
- corrective and preventive action
- handling, storage, packaging, preservation and delivery
- control of quality records
- internal quality audit
- training
- servicing
- statistical techniques.

The QMS should, as far as possible, be built on a firm's existing operating systems and procedures. Where required specifically by the standard's requirements, additional procedures and documentation can be added to fill in the gaps.

The International Organization for Standardization (ISO) published the new ISO 9001: 2000 standards at the end of 2000 that emphasize continual improvement and meeting customer satisfaction rather than just compliance with standards and processes. This means that construction firms will need to be more pro-active if they are to deliver high quality work and achieve improvements progressively. This revision is, however, unlikely to affect the call for integration between quality and safety as proposed in this chapter. Instead, the new ISO 9001: 2000 standards will help to facilitate further integration with other systems relating to health, safety and the environment. At this stage, construction companies should only attempt to integrate safety requirements into their existing ISO 9000: 1994 QMS where there is already sufficient experience in implementation.

21.3.1 An integrated quality and safety management system (QSMS)

There are certain requirements in the ISO 9000 standards that can be extended for safety management in the construction industry. These form parts of the integrated QSMS and are discussed below.

Clause 4.1 Management responsibility

Management has the responsibility to ensure that a quality system exists and is effective. In quality terms, management responsibility means that management is responsible for:

- defining the policies and objectives for quality
- specifying the responsibilities and authority of all personnel whose work affects quality
- appointing a person responsible for the quality system, i.e., the Quality Manager
- ensuring the quality system is workable and effective
- demonstrating a commitment to quality.

These requirements can similarly be applied for establishing and implementing a safety policy as specified in the BOWEC Regulations.

Clause 4.2 Quality system

In quality terms, this means establishing a system that ensures work is performed and maintained to requirements. The quality system is the summation of the organizational structure, responsibilities, procedures, processes and resources for implementing quality management (Weinstein, 1997). Establishing a comprehensive quality system starts with determining the requirements of the system based on regulations and standards. After that, existing company practices should be reviewed and determined. Present and future resources need to be evaluated, including those for personnel, instrumentation, specification, acceptance standards, and quality records. After evaluating the resources a plan should be developed to implement the quality system. Control elements in the quality system plan should include program reports, action item lists, and audit reports and findings.

The quality system should be well understood and effective in meeting quality objectives. It should also provide confidence that the final output meets customer expectations. It should be able to emphasize preventive action and be documented at the same time. Another goal is that no non-conforming products should reach the customers. The same goals also apply to an SMS. While a quality system should apply to all quality-related activities, a SMS should cover all safety-related activities.

Clause 4.3 Contract review

Contract review means that contracts or orders from external customers are reviewed to ensure that the requirements are adequately defined and that the company can meet them. In terms of safety, contract review has implications for both formal contracts to supply products and services and for the implied contract with employees. With regard to safety, however, there are multiple aspects to be considered. Firstly, the company should review the contract to understand what the safety implications of meeting them are and then ensure that all safety concerns and requirements are satisfied. These safety concerns could be external (affecting the purchaser and public) or internal (affecting the employees). Secondly, there is the implied contract or agreement with the real customers of the safety management system – the employees. The expectations of the workers must be clearly defined and understood by the organization, and the organization should strive to ensure that these expectations are met. Thirdly, contract review should include law and regulations, internal requirements and voluntary standards (Weinstein, 1997).

Clause 4.4 Design control

Design control means that the product, process, and service design is planned, organized and controlled so that the resulting design meets agreed quality requirements and needs. Designing for safety is critically important because once a system is designed and installed, it is very hard to change. Superior quality and safety results are attainable only if quality and safety features are integrated in the design. Safety hazards and quality defects are most effectively and efficiently anticipated, avoided, mitigated or controlled during the initial stages of systems design or redesign.

For safety, it is important to include design control provisions to achieve the following:

- review of all customer agreements and applicable statutory and regulatory requirements at the beginning
- review of requirements to anticipate, avoid, mitigate and control hazards and assign responsibilities
- make sure that all necessary procedures are written clearly and are available where they are used.

Human factors should be considered when reviewing designs to identify which functions must be performed by man and what types of controls are needed. When correctly applied, human factor engineering improves system effectiveness, leads to fewer performance errors and accidents, minimizes rework, reduces training costs and leads to more effective use of personnel.

Clause 4.5 Document and data control

Document and data control means that controls are employed to ensure that only the valid, up-to-date documents and data are used so that quality and safety are not affected. In fact, all documents that can affect quality and safety should be suitably controlled. ISO 9000 requires the following (Weinstein, 1997):

- establishment and maintenance of data and document control procedures
- review of documents by authorized personnel
- establishment of a master list of all current documents
- establishment of a control plan for each category of document
- review of all changes to documents and data.

Clause 4.6 Purchasing

Purchasing means that controls are employed to ensure that purchased products and services that can affect quality or safety meet specified quality or safety requirements. Thus, both the purchaser and the supplier must understand what the specified requirements are. For safety, purchasing can be viewed in two separate ways: firstly, purchasing applies to materials or components that will be used in construction (the quality of the components or materials purchased must be clearly specified and verified) and, secondly, ISO 9000 requires proper purchasing procedures to be established to achieve the following:

- ensuring that purchased products meet specified requirements
- evaluation of sub-contractors
- defining the amount of control over sub-contractors
- establishing records of acceptable sub-contractors
- ensuring that purchasing documents clearly describe the products ordered, and give all relevant details.

Contractor acceptability should be based on quality and safety criteria. For a product, the quality criteria include product quality history and specifications. The same applies for safety criteria. The control exercised over sub-contractors should depend on the type of products and on the products' impact on safety. When purchasing the services of contracted personnel to work on site it is important to determine the contractor's safety track record. It is also vital that the contracted personnel are qualified and trained appropriately and that they will follow all planned quality and safety procedures.

Clause 4.7 Control of customer-supplied product

Control of customer-supplied product means managing a product owned by the customer and provided to the main contractor for use in meeting the customer's requirements. The objective is that customer-supplied products should not affect quality and safety. For safety, this is also a requirement that applies to two types of situations. Firstly, it would cover situations where workers supply their own tools or materials for use in a plant, or where prior training is required. Secondly, it covers situations where product-oriented materials are supplied by the ultimate customer and such materials might affect work safety.

Clause 4.8 Product identification and traceability

Product identification and traceability means that processes, products and services are suitably identified through all the stages of production, delivery and installation so that quality objectives may be met. Identification is the ability to distinguish one product, part, procedure or process from another so that the proper parts are used in an assembly. Identification of procedures ensures that the proper procedures are used for an activity. Traceability is the ability to track the history, application or location of an item or activity by means of recorded identification. Traceability is important when process errors are investigated. To succeed, traceability programs need to be well planned, well implemented and completely and accurately documented. For safety, identification and traceability also refer to the various types of identification required to meet regulatory safety requirements. These would include such things as material identification tags, signs, hazards and exits.

Clause 4.9 Process control

Process control means that processes are planned, executed and controlled such that the equipment, environment, personnel, documentation, and material employed always result in quality or safety requirements being met. Process control requires process identification, process planning and process maintenance. The activities that require control would depend on the types of processes involved and the inherent risks.

For all processes, controls should be in place to help ensure that all activities are carried out safely and effectively. Process control requires that any documented process hazards analysis should be reviewed and predictive and preventive maintenance procedures should be developed. In general, process hazards should be controlled by:

- avoiding the hazard entirely
- reducing the hazard by substituting less hazardous materials or activities
- providing engineering controls to prevent accidents or reduce their severity
- using personal protective equipment to shield workers
- instituting administrative controls to prevent dangerous conditions from being reached.

Clause 4.10 Inspection and testing

In ISO 9000 the clause on 'inspection and testing' means that materials, products or services received or produced, and equipment and systems used, are verified to meet requirements prior to use. Inspection and testing does not create quality, it only verifies conformance. In practice, inspection and test procedures should be designed and maintained to detect problems as early as possible. The critical stages are at the input receiving and inspection stage, and the crucial process point and output (final inspection) stage.

For safety, 'inspection and testing' applies to facilities and safety-related equipment and activities. As far as equipment is concerned, the full range of safety equipment should be included in an inspection and testing program, including personal protective equipment. With regard to activities, all safety inspection and surveillance activities should be planned for, controlled and documented. These should include activities such as normal fire and occupational safety inspection and air quality or noise level surveillance (Weinstein, 1997).

Clause 4.11 Control of inspection, measurement and test equipment

Control of inspection, measurement and test equipment means devices, techniques and standards used to verify that products or services that meet quality requirements are effectively controlled, calibrated, maintained and used. For safety, ISO 9000 requirements mean that all inspection, measurement and test equipment used in safety-related processes, or in an activity with safety implications, should be subjected to a defined control, calibration and maintenance program.

Clause 4.12 Inspection and test status

Inspection and test status means that products, processes or equipment are identified as being safe and conforming items or activities. The identification should ensure that only products, processes or equipment that have passed the required inspections are used, dispatched or installed. All safety-related materials, products or equipment should be identified to verify conformance or non-conformance with regard to tests, and inspections required either by regulations or by safety program procedures. The location and inspection status of each item should be identified, such that the status is clear to all potentially affected personnel. A common example is the placement of inspection tags on fire extinguishers and calibration stickers on pressure gauges.

Clause 4.13 Control of non-conforming product

Control of non-conforming product means that products, processes or equipment identified in the processes are acceptable. In ISO 9000, the purpose of this section is to prevent non-conforming products from reaching the customer. In terms of safety, the meaning of this section may be broadened to include all identified defects and deficiencies in activities and operations that affect safety. The safety goal ensures that work safety is not compromised by the identified defects and deficiencies.

Clause 4.14 Corrective and preventive action

Corrective and preventive action means that effective action is taken to prevent the occurrence and recurrence of non-conformities (unsafe items, processes or equipment). Corrective action is reactive, directed at the elimination of causes of actual non-conformance, while preventive action is pro-active, directed at the elimination of causes of actual non-conformance. Continuous improvement is the result of identifying and correcting the causes of non-conformances (Weinstein, 1997).

For safety management, this section refers to any safety problems or events. These could be equipment, processes, programs or human problems, test or surveillance failures or actual incidents or accidents. For example, the corrective action process could be applied to a critical pump failure, or to the finding of higher than normal toxic material concentration levels or to an accident that caused a worker to break his/her leg.

Clause 4.15 Handling, storage, packaging, preservation and delivery

Handling, storage, packaging, preservation and delivery relates to measures that identify the hazards related to specified activities. These ensure that materials, equipment and products are not so affected as to impact worker safety adversely and that there are appropriate safety procedures in place and training provided for those who conduct the activities. For safety management, the focus is on the hazards associated with materials and all of the activities used to handle, store and transport materials and products. For example, equipment should be protected from mechanical damage, water and humidity damage, temperature damage, corrosion and contamination. Materials with special storage requirements should be identified, stored and handled appropriately.

Clause 4.16 Control of quality records

Records must be established, documented and maintained to demonstrate achievement of quality or safety requirements, including regulatory requirements, and quality or safety system effectiveness. To be useful, records must be controlled, accessible and identifiable. In ISO 9000 the intent of quality records is to verify that the quality program is conducted to meet requirements. Records also verify the condition of the product and provide a history of that part, process or program.

For safety management, records are necessary for regulatory compliance, for ensuring program implementation and for providing a basis for continuous improvement through review and analysis. Effective accident investigation and analysis also require a well-documented history of activity, inspection and statistical records. Safety management records should document the following:

- accidents (injuries, production interruptions, product damage and facility damage)
- near-miss accidents
- safety and health training
- special hazards exposure
- safety meetings
- medical evaluations and histories
- safety inspections, audits and reviews
- emergency plans and drills
- hazardous chemical training, communication and monitoring
- safety, fire prevention and emergency planning procedures
- training for materials handling, certification and procedures
- noise minimization program documents.

Clause 4.17 Internal quality audits

Comprehensive audits are planned and executed to verify the existence, implementation and effectiveness of the quality or safety management system. All activities and operations should conform to requirements, and the system should meet specified goals and objectives. The aim of auditing is not only to verify compliance but also to promote continuous improvement through the identification of weaknesses and the presentation of corrective recommendations. Meeting this aim requires highly qualified audit personnel. ISO 9000 expects audits to be carried out on a schedule consistent with the importance of the activity, and personnel conducting audits should be independent of the area being audited. Once deficiencies are brought to the attention of management timely corrective actions should be taken.

The range of safety program audits includes visual inspections of conditions and activities, documentation audits for compliance and completeness, and comprehensive program audits or reviews. The audit determines whether policies and procedures are implemented as planned and whether in practice they have met the objectives set for the program. It also determines whether the objectives lead the organization to meet the program goal of effective safety and health protection. When either performance or the objectives themselves are found inadequate, revisions are made.

Clause 4.18 Training

Quality training means that the training needs of personnel whose work affects quality and safety are identified, and that training is provided so that the personnel are qualified to do their jobs. For specific tasks, qualification can be based on appropriate education, training and/or experience. Appropriate training records should be maintained and personnel performing the training should be adequately qualified. For safety programs, specific training topics will include the following:

- hazard prevention and control, including engineering controls
- safe work procedures

- administrative controls
- facility and equipment maintenance
- emergency planning and preparation
- medical programs
- workplace analysis, including hazard identification
- hazard reporting
- accident/illness investigation and injury analysis
- safety and health for employees (how to protect themselves and others)
- responsibilities of supervisors and safety officers.

Clause 4.19 Servicing

In ISO 9000, servicing means that controls are employed to ensure that servicing operations are carried out to meet requirements specified in the purchasing contract. Servicing may be routine preventive maintenance or it may be corrective maintenance (repair). In terms of safety management, servicing activities should be planned and prepared to ensure they are done safely. For safety in servicing, it is important that:

- all servicing safety requirements are established and documented, including specific procedures, personal protective equipment, road travel restrictions, emergency plans and installation instructions (including hazard notification)
- the availability and readiness of special tools or equipment is ensured
- servicing personnel are properly trained, including driver training and the use of special equipment.

Servicing is an activity performed at the site, thus servicing personnel should follow site safety requirements and should exercise care, recognizing that their knowledge of site hazards and conditions may not be complete.

Feedback is important for the planning of servicing activities. A history of problems with products or equipment should be used to help determine servicing needs in terms of personnel, supplies, training and support required. Deficiencies in performance require immediate corrective action. The performance of safe servicing activities is the result of effective planning and preparation (Weinstein, 1997).

Clause 4.20 Statistical techniques

This clause requires that measures are taken to control the selection and application of statistical techniques used in controlling processes and operations and determining the effectiveness of meeting process and operational goals. The general objective of using statistical techniques is to remove any external cause of variation in order to produce products in a consistent manner.

In terms of safety management, there are two separate objectives involved with the application of statistical techniques. Firstly, statistical techniques should be used for the proper control of all processes and production activities. Secondly, statistical techniques should be used for the management of the safety program itself. This means that all safety performance measures should be developed and statistically analysed, safety audits should be planned in a statistically sound manner and statistical techniques should be used in hazards, accident reliability and risk analyses.

21.4 Matching safety requirements with ISO 9000

The SMS is a system of work procedures that ensures safe working practices are adhered to in order to minimize the chance of an accident occurring. It is similar to ISO 9000 in that quality objectives are achieved through proper documentation and work procedures. The SMS, in the same fashion, aims to achieve the objective of a safe working environment by establishing a proper system of safety considerations.

The 13 critical elements of a SMS as set out in the BOWEC Regulations are further elaborated below for the purpose of matching them with ISO 9000 quality system elements.

21.4.1 Safety policy

The objective of a safety policy is to establish in clear and unambiguous terms the worksite management's approach and commitment to safety. These would include organizational policies and processes dealing with:

- a safe policy statement
- safety organization and safety staff
- management participation in safety
- safety performance standards
- management audits
- individual responsibility in safety
- safety objectives
- safety references and resources.

As discussed above in the ISO 9000 requirements under *management responsibility*, the requirements for safety policy can readily fit into the objective in which it is described.

21.4.2 Safe working practices

This requirement is intended to eliminate or reduce the risk of death or injury to people and damage caused to properties and assets during the execution of work. Actions should include:

- a list of works requiring safe working practices
- documentation of safe working practices
- communication of safe working practices
- implementation of safe working practices.

The relevance of safe working practices can be seen in the ISO 9000 requirements under *design control*, *document and data control* and *process control.*

21.4.3 Safety training

Safety training helps to provide knowledge, skills and attitudes that will enable personnel to perform their duties in a manner that does not present a safety hazard to themselves or others. This requirement covers the following areas:

- training needs analysis
- training programs for management staff, safety staff and workers
- training program evaluation.

ISO 9000 provides for a requirement under *training* that defines the type and scope of training, and personnel to be trained, as well as training analysis and programs. Safety training needs may be met under this requirement.

21.4.4 Group meetings

This requirement seeks to bring together personnel with specific responsibilities for health and safety so that such issues may be addressed and appropriate action can be taken where necessary. It includes the following activities:

- establishment of the safety committee including its composition, representation and frequency of meetings (the activities of the safety committee should include the owner and contractors)
- establishment of any other special committees and/or task forces
- tool box meetings where tools are checked to ensure that they meet safety standards.

This requirement may be incorporated under *process control* in ISO 9000 as part of the issues to be addressed for safety requirements. Documentation of the meetings may be kept as evidence that this function is being carried out.

21.4.5 Incident investigation and analysis

The objective of this is an investigation of ways and means by which worksite accidents may be minimized. This would include the following:

- incident investigation procedures
- scope of incident reporting and investigation
- tracking of follow-up actions
- management participation
- near-miss reporting
- analysis of accidents
- consequences
- causes
- controls.

Statistical techniques, *corrective and preventive action* as well as *control of non-conforming product* are the relevant requirements under ISO 9000 where this requirement could be addressed and met.

21.4.6 In-house rules and regulations

This requirement aims to provide all personnel with a common understanding of their obligations and responsibilities in matters of safety. It covers the following areas:

- general safety and health rules
- specific work rules
- permit-to-work system
- rules education and review program
- rules compliance checks
- use of sign postings.

Management responsibility in ISO 9000 may be utilized to address this area while the degree of compliance may be defined and checked under *process control* and *internal quality audit,* respectively.

21.4.7 Safety promotion

Safety promotion creates an awareness among personnel that encourages them to be committed to safety and health. It also emphasizes the individual's responsibilities towards that commitment. It would include the following:

- safety bulletin boards
- use of statistics and facts
- special campaigns, both internal and external
- use of awards and recognition for individuals and groups
- housekeeping promotion.

Safety promotion may be drawn into the requirement for *training* under ISO 9000 whereby the promotion of safety may be interpreted as a form of education.

21.4.8 Control of contractors

This requirement helps to ensure that all sub-contractors are aware of their obligations towards health and safety and that only sub-contractors who fulfil these obligations are employed on site. It includes the following:

- evaluation of contractors
- selection of contractors
- control of contractors
- audits of contractors' safety systems.

Contract review under ISO 9000 provides the basis for the safe and proper selection of contractors.

21.4.9 Safety inspection

The objective here is to help identify unsafe acts, hazardous operations and detrimental conditions on the worksite. This would be followed up subsequently by the following preventive and corrective actions:

- planned inspection program by management, safety staff and contractors
- tracking of follow-up actions

- critical parts inspection
- inspection of materials handling equipment.

Internal quality audit and *inspection and test status* in ISO 9000 provide the avenue to address safety inspection needs and to check that safety requirements are met accordingly.

21.4.10 Maintenance regime

This seeks to ensure that all plant, machinery and equipment used on the worksite do not present any safety hazards because of a lack of proper maintenance and repair, or unauthorized usage or operators. It takes into account the following areas:

- list of statutory equipment (owned or hired)
- preventive maintenance program
- maintenance records keeping
- identification of problems
- tracking of follow-up actions.

Maintenance regime may be matched with *servicing* in ISO 9000 for maintenance procedures and requirements to be spelt out.

21.4.11 Hazard analysis

This requirement aims to identify and evaluate potential hazards so that control measures may be implemented and the risk of these hazards occurring can be reduced to a minimum. It includes the following:

- hazard analysis program
- management of change program
- designated person.

In ISO 9000, *statistical techniques* may be utilized to a certain degree as a probability test to predict the chances of occurrence of an accident and where steps may be taken to guard against the occurrence. The technique used may in turn be translated into operable safety features under *corrective and preventive action.*

21.4.12 Hazardous materials

This looks at the receipt, movement, storage and use of all hazardous materials so that any risk connected with such materials may be managed and reduced to the lowest possible level. The areas covered in this requirement include the following:

- inventory of hazardous materials
- file for Material Safety Data sheets
- control of hazardous materials during transportation, storage, handling and use, movement control and labelling.

This requirement is concerned with the use and purchase of hazardous materials. The requirements in ISO 9000 for *purchasing, handling, storage, packaging, preservation and delivery* and *control of non-conforming product* may help to safeguard against the improper handling, disposal and treatment of hazardous materials.

21.4.13 Emergency preparedness

The objective of here is to help develop and communicate various schemes that can be effectively utilized in the event of an emergency situation occurring. It covers the following:

- emergency plan
- communication of plan
- emergency drills and exercises
- emergency power and lighting
- protective and rescue equipment
- emergency team
- mutual aid
- emergency communication.

The *corrective and preventive action* clause in ISO 9000 provides the basis for 'emergency preparedness' whereby staff and workers are trained to be prepared for crisis situations. Their structured training in this aspect may be elaborated further under the *training* clause.

21.5 Conclusion

The matching of SMS requirements (as provided for in the BOWEC Regulations) with QMS clauses (as set out in the ISO 9000 standards) suggests that it is indeed possible and beneficial to have an integrated Quality and Safety Management System within a construction firm. This match may only be peculiar to the construction industry in Singapore where the regulations apply. However, similar exercises can also be carried out taking into account the different provisions of safety legislation in other countries. For example, a similar exercise could be conducted in the United Kingdom to merge ISO 9000 requirements with the provisions of the Construction (Design & Management) Regulations 1994 (1995).

The cost to construction firms is not expected to increase substantially following the integration of an SMS with a QMS – on the contrary, with the streamlining of operations, resources should instead be utilized more synergistically with a consequential reduction in overheads. Further research work is recommended in this direction.

The resistance to change on the part of employees is likely to be another obstacle to the integration of an SMS with a QMS. Nonetheless, drastic change is unlikely to happen because QMS certified to meet ISO 9000 standards are already in place among many construction firms (Low, 1998). What is needed is therefore a shift in perception from a two-system operation to a single system on the part of not only the quality manager and the safety manager, but also all other employees who utilize the QMS and SMS. It is recommended, however, that construction firms that are developing and implementing ISO 9000 QMS for the first time should refrain from pursuing this integration from the start. They should instead concentrate on getting their fundamentals right. Integrating ISO 9000 QMS with safety requirements should only be pursued after the system has operated for some time and the learning curve is established. This recommendation also applies to construction firms with existing ISO 9000 QMS who are planning to switch over to the new ISO 9001: 2000 standards. Construction companies in Singapore are given 3 years by the relevant building authorities to switch over their existing ISO 9000 QMS to meet the new ISO 9001: 2000 standards. The approach presented in this chapter for integrating safety requirements in QMS can be adopted in the future after the switch over to the new ISO 9001: 2000 standards is completed. This methodology should help to minimize uncertainties for users where the QMS is concerned.

The study by Low and Sua (2000) confirmed that there are indeed similarities between an SMS and a QMS, making it possible to integrate these two systems and thus achieve better co-ordination and utilization of scarce resources. It is, however, necessary for construction firms to examine their SMS and QMS requirements to determine which operational aspects are more easily integrated than others. It is also necessary to appreciate that technical similarities between the SMS and QMS may not necessarily be sufficient considerations for such an integration. The non-technical or behavioural attributes of safety managers as well as quality managers are also critical in ensuring that such an integration can be successful.

From the matching exercise discussed above, it seems that there is a place for all 13 safety elements under the ISO 9000 template. The generic nature of the ISO 9000 standards is therefore beneficial here as the incorporation of safety requirements into the QMS framework is only limited by one's imagination. Many common safety requirements may be integrated into different areas and, in some cases, their functions may even overlap. How safety requirements can be addressed effectively within an existing QMS will therefore depend on how construction firms investigate and experiment until the best match and arrangement can be customized. As an extended service to industry, the QMS should therefore find its role not only within the ambit of quality management but also in providing safe working practices within the construction industry.

References and bibliography

Construction (Design & Management) Regulations 1994 (1995), S11994 No 3140 HMSO. ISBN 0 11 043845 0.

Goh, C.G. (1997) Implementation of safety management system as required under the Factories Act. *The Contractor*, **14** (1), 8–9.

Low, S.P. (1998) *ISO 9000: Practical Lessons for the Construction Industry* (Oxford: Chandos Publishing).

Low, S.P. (2000) ISO 9000 as the foundation for buildability, productivity and safety: consolidating empirical findings from the construction industry. In: Ho, S. and Chong, C.L. (eds) *Proceedings of the Fifth International Conference on ISO 9000 and TQM*, 25–7 April, Singapore, pp. 445–9.

Low, S.P. and Abeyegoonasekera, B. (2000) Integrating buildability principles into ISO 9000 quality management systems. *Architectural Science Review*, **43** (1), 45–56.

Low, S.P. and Chong, W.K. (1999) Integrating JIT into quality management systems. Construction Paper No. 104 *Construction Information Quarterly*, **1** (2), 10–21 (Chartered Institute of Building).

Low, S.P. and Sua, C.S. (2000) The maintenance of construction safety: riding on ISO 9000 quality management systems. *Journal of Quality in Maintenance Engineering*, **6** (1), 28–44.

Weinstein M.B. (1997) *Total Quality Safety Management and Auditing* (Boca Raton, US: Lewis Publishers).

PART 3

Innovation in design and construction

22

Construction innovation in globalized markets

Gerard de Valence*

Editorial comment

This chapter investigates the relationship between the procurement methods used on building and construction projects, the opportunities for innovation and innovative solutions to client requirements, and the level of competition in the industry. The economic theory of innovation suggests that returns to innovators from research and development (R&D) and positive externalities drive the process. The argument is made that traditional tendering and procurement methods used by building industry clients works against these drivers of innovation in several important respects.

The author reviews R&D in the building and construction industry, and assesses the importance of these trends for the international competitiveness of the industry in the context of alternative procurement methods and a changing regulatory environment. The analysis of the implications of these trends in deregulation and procurement for the international construction industry finds that they will determine the level of competition. The links between R&D and competitiveness are also discussed, and the author concludes that the ability of the industry to lift R&D investment will be an important determinant of competitiveness over the coming decade, and one that will see significant structural change in international building and construction industry.

Innovation is identified as one of the most powerful influences on the architecture, engineering and construction (AEC) industry, and its impact and importance are discussed in a number of chapters in this book. Here the focus is on the expansion of the industry into a global marketplace and in that context the value to clients lies in some specific areas. Globalization means greater competition, and brings, potentially at least, a broader set of skills to bear on the clients' problems. Competitors from different countries, and from different cultures and traditions, present clients with a wider choice of alternatives and push local competitors to lift their performance in all areas, including innovation and R&D.

* University of Technology Sydney

Clients therefore have the opportunity to have their projects completed more efficiently, through the utilization of the newest management systems and international best practice, and the inclusion of the best available components and materials selected from a worldwide supplier base. The overall result should be better, more valuable buildings.

22.1 Introduction

The purpose of this chapter is to look at the links between competitiveness in the construction industry and the R&D intensity and level of innovation that characterizes the industry, and relate these to the procurement systems and market structure in the industry. This discussion covers the relationship between R&D and procurement, R&D and deregulation under the World Trade Organization's Government Procurement Agreement (GPA), and the difficulty of achieving returns on R&D in an industry that is fragmented and highly dispersed. The implications of these factors for the industry are discussed.

Innovation in the construction industry has been a focus of research for over a decade. By the commonly accepted measures of innovation, such as patents, technological research papers, introduction of new products or process improvements (Freeman and Soete, 1997), the construction industry has a record of very low identifiable innovation. This is despite the industry having characteristics seen as favourable to innovation (Tatum, 1986) and the importance of innovation for competitiveness in the industry (Flanagan, 1999).

Various aspects of construction innovation have been discussed. For example, Gann (1997) on the role of government, concludes that construction requires a strong and vibrant research base partly funded by government, and that a complementary effect arises from both public and private funding. Dulaimi (1995) on attitudes, shows there is still a significant section of the industry who feel that research and development and innovation is limited to a narrow section of the industry.

Tatum (1986) identified six features of the construction industry as advantages for innovation:

- construction projects create teams presented with high levels of necessity and challenge, which promotes innovation by forcing examination of new technologies for each project
- integration of engineering, design and construction can simplify the construction process and decrease cost
- the low capital investment typical of construction firms allows high flexibility for the adoption of new technologies
- a pool of technologically experienced personnel provides depth of knowledge
- the strong emphasis on process limits barriers to imitation, because new processes can spread rapidly without patent restraints (but this may also discourage innovation)
- construction production processes do not create rigid restraints.

Four characteristics of the construction industry were later identified by Tatum (1989) as being constraints to innovation:

- the low capital intensity of the industry limits its interest in investment for automation – if a firm has adequate market share and profitability then pressure to innovate is reduced

- the institutional framework is not supportive of innovation (the number of firms, the legal incentives for technological inertia, regulatory influences, and craft organization of labour)
- building cycle volatility affects capital investment and economies of scale
- suppliers have not created technological improvements in the equipment and tools used by construction.

Tatum's analysis is descriptive of the construction industry, but lacks a model of how innovation occurs in the industry. Technological change and innovation have become central to the economic analysis of development of individual industries and economic growth in general. The state of any individual industry is seen as driven by and subject to the forces of competition, where competition is driven by changes in technological opportunities and appropriability (how firms get a return on investment in R&D and innovation) of knowledge (Dumenil and Levy, 1995). Firms typically are in a process of constant adjustment and selection, as industries adapt to entry and exit, and innovative projects or R&D succeed or fail. Nelson and Winter (1982) were the first to compare this to the biological concept of natural selection in evolution. Verspagen (1998) suggests that economic systems adapt over time but do not necessarily produce optimal outcomes, as suggested by the work of Arthur (1988) and David (1985). In the analysis of economic growth, this line of thinking led to the development of a new approach to economic growth, where investment in R&D leads to positive externalities (as many of the results of this R&D become public knowledge as new products and processes are developed) and increasing returns (Romer, 1990).

Inter-firm differences in R&D and innovation account for differing rates of increase in productivity, improvements in the quality and differentiation of products within industries. The Canther and Pyka (1998) model sees firms engaging in R&D, and what they describe as search and experimental activities, in pursuit of profits. They also see technological development as cumulative, as industries develop along specific technological trajectories, with decreasing exploitable technological opportunities. The benefits of new technology depend on how easy it is for new ideas and knowledge to be transferred to other firms in the same industry.

To date there has not been a good explanation available for the construction industry's record of innovation grounded in the broader theories of innovation. However, a model of construction innovation in a revised framework, using procurement methods and market deregulation as potential drivers of innovation, is proposed in this chapter. The argument is on two levels. Firstly, at the level of industry structure, the relationship between innovation and concentration is discussed; the issue here is whether the research intensity of construction is an outcome of industry competition rather than a requirement, as it is in high research intensity industries. Secondly, the issues of appropriability of research and innovation revenues and treatment of knowledge externalities are considered in the context of procurement methods used.

The discussion uses recent developments in two fields: from the theory of industrial organization, Sutton's (1991, 1999) theory of endogenous sunk costs and its application to fragmented, low research intensity industries is discussed. Then, using the innovation paradigm from evolutionary economics, the effects on R&D from increased international competition are highlighted. This leads to a discussion of the role of procurement methods used for building and construction projects as a determinant of the level of innovation in the industry.

22.2 Research intensity

There are major differences between industries in rates of both technological progress and productivity growth. Efforts have been made to find both theoretical and empirical explanations of the factors that drive these differences, and the variables that can be applied to inter-industry differences in technology and productivity growth. One explanation for differences among industries in technology and productivity growth is R&D intensity, or the degree of innovation that is found in the industry, and the differences between the R&D and innovative activity in one industry compared to another. Research intensity is typically measured by R&D as a percentage of sales or income.

R&D intensity in an industry is determined by two key variables. One is 'technological opportunity', which determines the productivity of R&D and the opportunities that are available for innovation. The other is the ability of innovating firms to 'appropriate' a significant share of the economic value created by innovation, or to capture the externalities created through new knowledge. Although both these variables influence R&D intensity of firms in an industry, only technological opportunity will affect the rate of technological advance in an industry in the long-run, even if both opportunity and appropriability influence R&D intensity.

Nelson and Wolff (1997, p. 207) say that 'differences across industries in their R&D intensities tend to be quite durable. This suggests that, to the extent that these are the major determining variables, differences across industries and technological opportunities and inappropriability conditions tend to be persistent.' They propose that cross industry differences in technological opportunities are due to R&D opportunities differing between industries; in turn, differences between industries in technological progress will be driven by differences in R&D intensity, appropriability and opportunity that are available. Their theory is that industries with high R&D intensity and technological opportunity must be receiving a high rate of flow of new technological opportunities to make up for those that are being exhausted.

In the industrial organization or industry economics literature, industries are usually seen in terms of a number of firms which advance along a single technological trajectory, and these firms compete in enhancing the quality of their individual versions of the same basic product (homogeneity of product). In this case, firms make decisions on how much R&D to finance, and apply that R&D to product development. This view fits some industries well; however, many industries encompass several groups of products rather than a large number of versions of a single product. The products may be close substitutes in consumption, but embody different technologies, so R&D projects that enhance products in one group may generate huge spillovers for products in other groups.

Such complex overlapping patterns of substitutability have bedevilled industrial organization analysis since Chamberlin (1932) first developed the definition of an industry as limited by the chain of substitution, where industries were defined by their product. If industries are broken into separate sub-industries in order to address this problem, the choice in R&D spending can be between any number of technologies for the development of different groups of products. The products may be close or distant substitutes for products of firms on other technological trajectories. Both of these linkages operate on the demand side. When the linkages are strong they reflect the presence of scope economies in R&D; where the linkages are weak these scope economies will be absent and there will be a low degree of substitution across sub-markets.

Applying this discussion of sub-markets to the building and construction industry raises a number of interesting issues. The first is, of course, the general lack of specialization of firms in the construction industry in terms of their product. The answer to the question 'What does the industry produce?' is varied; some believe that the industry provides services (management, co-ordination, finance), others believe the industry delivers products (buildings and structures). The former group argues that the main task of the industry is one of co-ordinating site processes while the latter are more concerned with the building itself. The building and construction industry is typically broken into the engineering, non-residential, and residential building sectors, and there are some firms that cross all of these areas, however, firms typically work in either the residential or the non-residential sectors. Many of the larger firms cover both engineering and non-residential building in their activities. Within the non-residential building sector there are 10 or 12 different sub-markets, divided into offices, retail, factories, health and so on. Some firms specialize in building particular types of buildings; in Australia, for instance, Grocon specializes in high-rise office buildings and Westfield specializes in shopping centres. However, it is more common for building contractors to apply their management skills to a range of building types and not limit themselves to specific sub-markets. In this case, for the construction industry, sub-markets are difficult to identify because firms can be highly specialized in one area, or they can be highly generalized and put up a wide range of buildings and structures.

22.3 Research and development (R&D) and market structure

The degree of monopoly power exercised by the largest firms in an industry is expressed in the concentration ratio, which is the degree to which an industry is dominated by the largest firms. Typically the concentration ratio uses the largest four firms in an industry, ranked by market share or sales as a percentage of total industry sales. The definition of the concentration ratio is the percentage of industry total sales (other measures are capacity, output, employment or value added) accounted for by the largest firms. The extent of control over prices is determined by the intensity of competition in a market, which is, in turn, determined by the number of firms and type of product.

Sutton (1999) in his development of the theory of market structure and concentration suggests that the effect of R&D spending on the technological trajectory of industries is crucial. Where the degree of substitutability across products associated with different R&D trajectories is high, concentration will necessarily be high, because if all firms have a low market share an increase in R&D spending will be profitable, and the high spending firm can capture sales from low spending rivals on its own trajectory and on others. Sutton shows that under these circumstances the number of trajectories along which firms will operate is small, since low spending firms are vulnerable to increases in R&D spending by rivals. On the other hand, if the degree of substitution across products is low, then in spite of the effectiveness of R&D spending, concentration may be low. 'This can only happen if there are many product groups, associated with different independent R&D trajectories . . . here, escalation yields poor returns, since outspending rivals can lead only to the capture of sales from products in a single, small product group' (Sutton, 1999, p. 13).

In Sutton's analysis it turns out that industries or sub-industries with a low R&D to sales ratio can have a low level of concentration, and this can continue indefinitely. The industries where there are a large number of firms in an increasing market, characterized by buyers who place different relative weights on different aspects of technical performance (product attributes), and many alternative technologies are available for those products, leads to a market where there is an indefinite number of firms, each with a small market share. For this type of industry, with low R&D spending, Sutton's theory predicts that the concentration ratio will be low and that the level of concentration will decrease as the size of the market increases. Sutton compares the characteristics of such an industry to industries where the R&D to sales ratio exceeds a high threshold value, and Sutton's theory predicts that concentration will increase as the spending on R&D increases. In a low R&D intensity industry, the market share of the largest firm will be relatively small; in a high R&D intensity industry the market share of the largest firm can be very high (Sutton, 1999).

Industry structure is important because it is the primary determinant of the level of competition, and the form that competition takes, in an industry. Related to this are the way the process of competition affects prices and profits, the ease of entry of new firms into or frequency of exit from an industry, the impact of demand shocks (i.e., the business cycle) and the effects of new technologies, such as e-business and e-commerce. For the building and construction industry, however, the methods used for tendering and procurement of projects are important determinants of the level and form of competition and distinguish the industry from many others where competition occurs through marketing campaigns, new products and so on.

22.4 Procurement and innovation

Clients are increasingly using a variety of alternative procurement methods aimed at reducing cost, achieving time schedules and milestones, shortening duration, reducing claims, and improving constructability and innovation. The overall trend is toward versions of design and build and turnkey construction because of the advantages of a project delivery system that combines designers, builders, and sometimes suppliers into a single entity (de Valence and Huon, 1999). Many surveys have established that clients perceive the design and build (D&B) approach as providing better value for money, and giving rise to less disputes than other procurement methods, and these surveys suggest that an experienced client with a clear brief can use it satisfactorily with projects of most sizes (Akintoye, 1994; Ndekugri and Turner, 1994; Songer and Molenaar, 1996).

The move away from traditional procurement systems will have significant effects on innovation; because the traditional design–bid–build method does not allow for capture of intellectual property by construction contractors in their tenders, there was a perverse disincentive to innovate. With the increased use of non-traditional procurement methods such as design and construct (D&C), D&B, build, own and operate (BOO), and build and maintain (B&M), this disincentive is removed and firms can appropriate for themselves the benefits of innovation and R&D, such as new ways of working or organizing work that improve the firm's competitiveness.

Craig (1997a) discusses innovation in D&C procurement systems, where the contractor bears single point responsibility for the complete product, like any other manufacturer.

Craig compares purchasing a car to acquiring a building, where unlike deciding which car to buy, a building is purchased through the tender process which must not only evaluate design but also production capability, time and price. Procurement of projects through the tendering process appears to limit the successful tenderer's scope to be innovative. He concludes that tendering rules or codes have been developed to maintain the integrity of the bidding process, not to encourage innovation. He also argues (Craig, 1997b) that 'alternative tenders' are potentially valuable to both clients and contractors, and to society at large. Contractors can make novel proposals to owners, and society benefits from such innovation. Tenderers can put forward more efficient and cost-effective methods of construction. However, there is not yet sufficient established custom and practice for modification of the existing 'tendering contract'.

Craig (2000) develops these issues and asks three questions on procurement and innovation. The first is: do tendering processes encourage innovation? The essential basis of the tendering code is that all tenderers are to be treated equally and fairly, that contract award criteria are established in advance and known by all parties, thus creating a transparent award process. Tendering rules produce direct price competition for a specified product. The question then becomes: can traditional tendering processes permit innovation? The answer is that a successful tenderer's scope to be innovative is very limited. There is opportunity to find novel ways of organizing work to achieve maximum profits within the tender price and opportunity exists for 'bid shopping' to drive down sub-contract prices. One tender might seek competitive advantage by offering to the owner a contract term more favourable than any from a competitor. Craig asks: what scope is there, at tender stage, to offer the client novel design (which is the bidder's intellectual property) at a saving on the original design? Bidders are not asked to put forward design suggestions, there are no criteria for evaluation of novel proposals, tenderers cannot be treated equally and fairly if one is preferred on an 'alternative' tender, which is a non-conforming tender in terms of the original invitation.

Finally, he asks: does D&B or D&C as a procurement system more easily permit innovation? The point is made above that using the tender process to competitively evaluate design, capability, time and cost is not easy. 'Competitive design is not easy to evaluate in the context of tendering. Traditionally it has been done by a two-stage process . . . a design competition . . . and production competition. Wrap this up in a single stage process and the objectivity appears to be replaced by subjectivity in picking the winner, and the apparent integrity of the bidding process is lost, unless very clear integrity criteria are established at the outset for evaluation of competing designs' (Craig, 2000, p. 33). He concludes that the traditional tendering process for building works does not encourage design innovation by tenderers, but it has always been possible for tenderers to seek competitive advantage through novel construction methods.

22.5 Regulatory environment

The construction industry fits into the category of a fragmented, low research intensity industry. As such it is responsible for few new products, and much process innovation is in response to developments in other industries, particularly materials manufacturing and the information technology and communications (ITC) industries. Sutton's theory predicts that this type of industry typically undergoes a significant increase in concentration as national markets are deregulated and opened up. This is what the GPA reached through the World Trade Organization (WTO) is doing, by opening member countries' public sector construction projects to international competition. Therefore the GPA might lead to a few large contractors dominating each national market, probably with a different mix of national and international firms in each market. The rise of these large firms might then be an important element in increasing R&D and innovation in the industry.

Governments in many countries have made changes in their policies and regulations on procurement under the regulations formulated by the World Trade Organization (Ichniowski, 1995; Mattoo, 1996; Ashenfelter *et al.,* 1997; Korman, 1997). Reforms are underway in the bidding and contractual systems used for public projects in Australia, Japan, South Korea, Europe and the US. These countries are opening their construction markets to contractors from foreign countries, often with less discriminatory and more competitive forms of tendering (Gransberg and Ellicott, 1996; Spacek, 1996; Reich 1997). Reform of the bidding and contractual system used for public projects in Japan, and government policies on further opening of the construction market to foreign countries, is discussed; as is Japan's 1995 'Action Plan', which provided for open and competitive bidding procedures to be used by public agencies for procurement of construction, design and consulting work that are valued at or above the WTO government procurement thresholds (Dunn, 1995).

Legislative changes in the US are allowing public owners the opportunity to use design–build as a project delivery option (Krizan, 1996; Loulakis and Cregger, 1996). Changes in the US federal government's procurement system, allowing federal agencies to use a limited form of design–build construction contracts, has led to the establishment of procedures for agencies to follow when they enter into a design–build project. To combat the problems inherent in traditional low-bid procurement, many states are following the example of the federal government by enacting procurement options to allow and encourage alternative procurement methods (Charles, 1996).

The experience of other industries after deregulation has been a significant increase in competition for domestic companies as new entrants, often large established international firms, come into their home markets. The larger the firm, the more likely it is to engage in R&D and innovation. Therefore, the expansion of international contractors into new markets could lead to an increase in both the size of the largest contractors and their ability to undertake R&D spending.

22.6 Competitiveness

Competitiveness drives the economic performance of countries and industries exposed to international competition. The competitiveness of nations in the global economy lies in the four broad attributes of a nation described by Porter (1985, 1990), attributes that individually, and as a system, constitute Porter's 'diamond of national advantage'. This is the playing field that each nation establishes and operates for its industries. These attributes (Porter, 1990, p. 139) are:

- factor conditions – the nation's position in factors of production, such as skilled labour or infrastructure, necessary to compete in a given industry
- demand conditions – the nature of home-market demand for the industry's product or service

- related and supporting industries – the presence or absence in the nation of supplier industries and other related industries that are internationally competitive
- firm strategy, structure, and rivalry – the conditions in the nation governing how companies are created, organized and managed, as well as the nature of domestic rivalry.

The Porter diamond (Figure 22.1) identifies the determinants of industry competitiveness (Porter, 1990, p. 140):

> These determinants create the national environment in which companies are born and learn how to compete. Each point on the diamond – and the diamond as a system – affects essential ingredients for achieving international competitive success: the availability of resources and skills necessary for competitive advantage in an industry; the information that shapes the opportunities that companies perceive and the directions in which they deploy their resources and skills; the goals of the owners, managers, and individuals in companies; and most important, the pressures on companies to invest and innovate.
>
> When a national environment permits and supports the most rapid accumulation of specialized assets and skills – sometimes simply because of greater effort and commitment – companies gain a competitive advantage. When a national environment affords better ongoing information and insight into product and process needs, companies gain a competitive advantage. Finally, when the national environment pressures companies to innovate and invest, companies both gain a competitive advantage and upgrade those advantages over time.

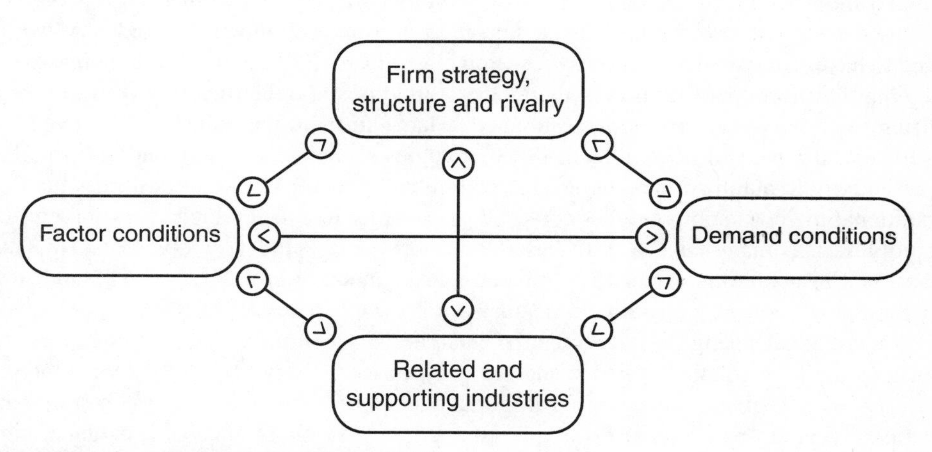

Figure 22.1 The industrial diamond (Porter, 1990).

With deregulation and the opening of national markets for public sector projects, as is the case in Australia, contractors will be subject to increasing competition. This will make the ability to enhance competitiveness a crucial factor in the long-term prospects of construction companies.

22.7 Conclusion

The importance of competitiveness is increasing as the global economy develops as both a market for construction firms and as a source of new competitors. The 'diamond' of factors that determine an industry's competitiveness are affected by government policies and attitudes. The GPA will increase openness to trade, capital and technology movements in building and construction, and this openness increases opportunities for greater economies of scale and scope and specialization of firms, two important sources of competitiveness.

The alternative methods and systems of procurement and delivery being increasingly used by clients will emphasize innovative design solutions in order to lower ownership or occupancy costs. As major projects become part of an international globalized market with deregulation of public sector procurement, domestic contractors will be competing against aggressive new entrants from other countries into their market, who will use the increasing diversity and complexity of procurement methods to win a share of the market. Also, the clients who have major projects are typically experienced, long-term investors with considerable expertise in alternative procurement methods. The traditional tendering and procurement process for building works is not amenable to innovation by tenderers. Craig (2000) suggests there are processes available for procurement of design services that allow competition on innovative design, in which design/technical matters of the tenders are evaluated separately from price and could be used for D&C procurement.

The influence of procurement methods has been a determining factor in the level of innovation in the industry, and the significance of increased use of alternative forms of procurement lies in the fact that they provide incentives for increased R&D and innovation. When combined with deregulation the effect will be to move the industry toward more R&D. In the past, the level of innovation may have been an outcome of industry structure, not a determinant; however, a more R&D-intensive industry would tend to have structure determined by research intensity. As R&D expenditure increases the potential level of concentration in the industry also rises and the interplay between these factors will determine the eventual number of large firms in the industry.

Finally, the pace of development in technologies related to building and construction (particularly in automation, software integration and general ITC) makes it possible that traditional methods and processes could be swept away in a technological revolution that rapidly makes many of the craft-based traditions in the industry obsolete. This has happened in many other industries: the automobile industry in the early 1900s and steel making at the end of the twentieth century are good examples.

Technological change is significantly changing the competitive environment in the building and construction industry and the level of technology used by the construction industry in any given country is an important determinant of the level of international competitiveness. With global economic competition the importance of creating a high value adding, high productivity growth industry increases.

The necessary conditions for the competitiveness of national industries include strong and sustained levels of productivity growth, a readiness to adopt innovation and new technology and a commitment to delivering value for clients' money. On their own, however, these conditions are not sufficient; other factors related to the attitudes of industry participants to R&D and innovation, the level of investment in R&D by the industry and the role of national governments in promoting the uptake of new technologies are also important.

References and bibliography

Akintoye, A. (1994) Design and build: a survey of construction contractors' views. *Construction Management & Economics*, **12** (2), 155–63.

Arthur, B. (1988) Competing technologies, increasing returns and 'lock-in' by historical events. *Economic Journal*, **99** (394), 116–31.

Ashenfelter, O., Ashmore, D. and Filer, R. (1997) Contract form and procurement costs: the impact of compulsory multiple contractor laws in construction. *Rand Journal of Economics*, **28** Special Quandt Issue, S5–S16.

Canther, U. and Pyka, A. (1998) Technological evolution – an analysis within the knowledge-based approach. *Structural Change and Economic Dynamics*, **9** (1), 85–107.

Chamberlin, E.H. (1932) *The Theory of Monopolistic Competition*. First edition (Cambridge, MA: Harvard University Press).

Charles, M. (1996) Congress approves new design-build law. *Civil Engineering*, **66** (3), 100.

Craig, R. (1997a) Competitive advantage through tendering innovation. Presented at the *Fourth Annual Conference of the Construction Industry Institute*, April, Melbourne.

Craig, R. (1997b) The 'tendering contract': fairness, equality and innovation. In: Davidson, C. and Mequid, T.A.A. (eds) *Procurement – A Key to Innovation*. CIB W92 Publication 203, pp. 91–100.

Craig, R. (2000) Competitive advantage through tendering innovation. *Australian Construction Law Newsletter*, **70** (March–April), pp. 32–45 (Sydney: UTS Faculty of Design, Architecture and Building).

David, P.A. (1985) Clio and the economics of QWERTY. *American Economic Review*, **75** (2), 332–7.

de Valence, G. and Huon, N. (1999) Procurement strategies. In: Best, R. and de Valence, G. (eds) *Building in Value: Pre Design Issues* (London: Arnold Publishers).

Dulaimi, M. (1995) The challenge of innovation in construction. *Building Research and Information*, **23** (2), 106–9.

Dumenil, G. and Levy, D. (1995) Structural change and prices of production. *Structural Change and Economic Dynamics*, **6** (4), 397–434.

Dunn, E. (1995) The 1994 Public Works Agreement. *Business America*, **116** (11), 15.

Flanagan, R. (1999) Winning projects and satisfying customers through innovation. In: Bowen, P. and Hindle, R. (eds) *Proceedings CIB W55/W65 Triennial Symposium, Customer Satisfaction: A Focus for Research and Practice*, September, Capetown.

Freeman, C. and Soete, L. (1997) *Economics of Industrial Innovation* (Cambridge, MA: MIT Press).

Gann, D. (1997) Should governments fund construction research? *Building Research and Information*, **25** (5), 257–66.

Gransberg, D.D. and Ellicott, M.A. (1996) Best value contracting: breaking the low-bid paradigm. *40th Annual Meeting of AACE International*. Vancouver, Canada.

Ichniowski, T. (1995) Reform advances on procurement. *Engineering News Record*, **235** (6), 7.

Korman, R. (1997) Reforms worry industry groups. *Engineering News Record*, **238** (5), 10.

Krizan, W.G. (1996) Design-build faces legal hurdle. *Engineering News Record*, **237** (7), 17.

Loulakis, M.C. and Cregger, W.L. (1996) Federal legislation will increase design-build opportunities. *Civil Engineering*, **66** (7), 35.

Mattoo, A. (1996) The government procurement agreement – implications of economic theory. *World Economy*, **19** (6), 695–720.

Ndekugri, I. and Turner, A. (1994) Building procurement by design and build approach. *Journal of Construction Engineering and Management*, ASCE, **120** (2), 243–56.

Nelson, R.R. and Winter, S.G. (1982) *An Evolutionary Theory of Economic Change* (Cambridge, MA: Harvard University Press).

Nelson, R.R. and Wolff, E.N. (1997) Factors behind cross-industry differences in technical progress. *Structural Change and Economic Dynamics,* **8** (2), 205–20.

Porter, M.E. (1985) *Competitive Advantage: Creating and Sustaining Superior Performance* (New York: Free Press).

Porter, M.E. (1990) *The Competitive Advantage of Nations* (New York: Free Press).

Reich, A. (1997) The new GATT agreement on government procurement – the pitfalls of plurilateralism and strict reciprocity. *Journal of World Trade,* **31** (2), 125–51.

Romer, P.M. (1990) Endogenous technological change. *Journal of Political Economy,* **98**, 71–102.

Songer, A.D. and Molenaar, K.R. (1996) Selecting design-build: public and private sector owner attitudes. *Journal of Management in Engineering*, **12** (6), 47–53.

Spacek, J. (1996) New law passed to allow design-build method of project delivery. *Water Engineering & Management*, **143** (7), 9.

Sutton, J. (1991) *Sunk Costs and Market Structure* (Cambridge, MA: MIT Press).

Sutton, J. (1999) *Technology and Market Structure* (Cambridge, MA: MIT Press).

Tatum, C.B. (1986) Potential mechanisms for construction innovation. *Journal of Construction Engineering and Management*, ASCE, **112** (2), 178–91.

Tatum, C.B. (1989) Organising to increase innovation in construction firms. *Journal of Construction Engineering and Management,* ASCE, **115** (4), 602–17.

Verspagen, B. (1998) Introduction. *Structural Change and Economic Dynamics*, Special Issue on the Evolutionary Analysis of Innovation, **9** (1), 1–3.

23

Management innovation – a note of caution

Martin Loosemore*

Editorial comment

Every day we hear of corporations being 're-structured', an activity often accompanied by significant 'downsizing'. New 'flatter' management structures replace apparently top-heavy existing hierarchies and those responsible believe the new arrangement, whatever it may be, will function more efficiently than the old. 'Change management', a polite euphemism intended to disguise the often brutal activity of reducing staff numbers, has almost become a profession in its own right.

All of these activities are part of the prevailing view of how organizations should be run, based on quasi-scientific principles of often doubtful merit. The aim is always to improve efficiency and thus to improve the 'bottom line' by getting more and better work from the employees of the organization. The question that must be asked, however, is: do the results achieved support the claims of those who promote these principles? Equally important must be the question: do the ends justify the means, given the great personal hardship that much of this 'economically rational' re-organization causes for many employees who find, to their dismay, that their services are no longer required?

Innovation of any sort imposes some risks, additional to those inherent in any sort of business, and one of those risks is that an innovation may not produce the benefits promised by those who promote it. Construction is a risky undertaking for all concerned, even when conventional materials and techniques are used to erect buildings of conventional design, with the possibility of industrial action, difficult latent site conditions, insolvencies, unavailability of components or materials, bad weather, interest rate variations – there are many potential problems and any attempt to innovate must add further uncertainty.

Innovation here means new ways of designing and constructing buildings. These new ways include the use of new materials and components (e.g., materials with shape

* University of New South Wales

memory), new management practices (e.g., procurement methods, dispute management, design management), new construction methods (e.g., the use of robots, 'self-erecting' buildings), new analytical techniques (e.g., energy modelling, embodied energy analysis) and more. The key driver of innovation in the industry is competition, and with the increasing globalization of business generally, including construction, competition has become fiercer than ever. Contractors and designers must now look for ever more efficient practices or other ways to gain and/or maintain a competitive edge that will enable them to survive in the global marketplace

This chapter examines some of the claims of the champions of new management, compares those claims to the results that have actually been realized and finds some serious discrepancies. The author's view is that management, and managers, should be somewhat cautious in their adoption and implementation of new methods as there is evidence to suggest that often the results will not be as good as predicted. It may be advisable for the same caution to be exercised by those who manage the design and construction of buildings as they try to push the boundaries in the hope of getting ahead or staying ahead of the competition.

The concern for those who decide to build is how innovation may increase the value of their building and how much additional risk exposure they are prepared to accept in order to gain that increase. Innovation may produce a building that will be an excellent promotional tool as it demonstrates some advance in structure or services that is widely reported, or it may by its appearance draw attention to both itself and the function or corporation that it accommodates – in some cases a building will succeed on all fronts. For example, the ING Bank headquarters in Amsterdam (Romm and Browning, 1995), through its novel appearance and environmentally sound engineering, is credited, at least in part, with the bank gaining greater market share and becoming the second largest bank in the Netherlands; while the Guggenheim Museum in Bilbao (Guggenheim Foundation, 2001), clad in titanium and resembling a huge flower, will in all likelihood, by its remarkable appearance, become as much a symbol of the city where it stands as the Opera House is of Sydney. Clearly innovation can add value but it requires some care if the potential benefits are to be realized.

23.1 Introduction

Throughout the 1980s and 1990s technological advances revolutionized buildings but the construction industry's organizational processes failed to keep pace. Crainer and Obeng (1995) reassure us that product innovation inevitably precedes process innovation. However, governments around the world have lost patience with the 'gap' which persists between product and process innovation in the construction industry and with its associated productivity problems. In the UK this led to the Latham Report (Latham, 1994) and the Egan Report (Egan, 1998), and in Australia, to the Commonwealth Government's 'Building for Growth' report (Commonwealth of Australia, 1999). These echo the same message, being critical of the industry's resistance to change, and its performance record relative to other industries. The proposed solution is the transplantation of manufacturing techniques such as lean production, partnering, benchmarking and re-engineering to the construction industry. Unlike previous critical government investigations, published during the 1960s, 1970s and 1980s, these more recent reports are being driven by a more focused political agenda and by more powerful vested interest groups, resulting in the enthusiastic embracement of their ideas and ideals (Green, 1999). For example, Marosszeky (1999, p. 343) asserted that 'for the construction industry to be truly successful, it needs to emulate this manufacturing culture, adopting similar methods and goals, as they are as relevant to construction as to any other industry'.

This enthusiasm is not surprising because most of the techniques being advocated fit very comfortably with the engineering values and scientific methods of management which have traditionally dominated the construction industry (Blockley, 1996). Almost without exception, their emphasis is upon structural and process change underpinned by neat, prescriptive and measurable solutions. Furthermore, as Watson (1994) shows, the dramatic promises of productivity and profit improvements accompanying such techniques will have been very seductive to an increasingly insecure industry, which is under considerable pressure to improve its performance in comparison to other industries. For example, Heller (1993, p. 40) reports a manager of a manufacturing company who asserted that 'there is nothing that TQM's (Total Quality Management) tools and techniques are not able to tackle' and who reported that service had improved by 30% and inventories were down by 80%. Oliver (1993) claims even more incredible benefits in Sun Life Assurance Society's implementation of TQM where 40–90% improvements in process turn-around times, 10% reductions in unit costs, 50–80% quality improvements, higher staff and customer satisfaction and increased business, were achieved. Finally, Richardson (1996) quotes Lucas Industries who claimed that by re-engineering their production processes, they reduced lead times from 105 to 32 days, doubled inventory turn-around and increased productivity by 50%.

While the newly found enthusiasm for change within construction appears to be positive, few have highlighted the substantial evidence which questions the productivity returns associated with such techniques. The only exceptions are Green (1998, 1999), who criticized lean construction and business process re-engineering, Carmichael (1999) who considered the new techniques to be recycled versions of old ideas, and Bresnen and Marshall (2000) who offered a critical view of partnering. Similar concerns have been expressed for some time in mainstream management as far back as Cleverley's (1971) *Managers and Magic*. Indeed, the controversy continues with Crainer and Obeng (1995) arguing that despite the continuing rhetoric, few modern-day management innovations have had any significant success in translating technological improvements into more profitable and reliable production and, ultimately, into better customer services. The aim of this chapter is to explore this evidence in more detail with the objective of engendering a more critical debate in this area.

23.2 The psychology of management fads

Panati (1991, p. 5) described a fad in a generic sense as a 'trivial fancy pursued for a brief time with irrational zeal'. He argued that a true test of a fad is not its durability but its desirability, and its quantity not its quality. Whereas past centuries have been punctuated by occasional fads, the late twentieth century can be charted by their continual creation and consumption. Pascale (1991) was the first to detect this trend in the field of management, considering it a natural consequence of management's professionalization which would inevitably involve developing a set of generic concepts which could underpin management everywhere. However, while Panati (1991) explained the emergence of fads in a general sense by reference to the growth of mass communications it was not until Huczynski (1993) that an explanation of this phenomenon was produced from a managerial perspective. He explained faddish behaviour in management from a demand and supply perspective, investigating the motives of managers as consumers of ideas, and of consultants as sellers of them. The following two sections discuss these findings in detail because they help to explain the construction industry's current intoxication with management fads from mainstream management.

23.2.1 The supply perspective

Huczynski showed how management consultants play a critical role in perpetuating the market for management fads. Firstly, he argued that managerial ideas, like any product, have a limited shelf-life which means that the survival of management consultants depends upon generating a constant supply of new techniques for managers to ponder. However, because most managers only read narrowly for an average of 8 hours a week, the saleability of these ideas depends upon them being packaged in a simple and convenient way. It is even better if an idea offers a unified set of principles that can be applied to a wide range of situations because most managers reduce, as far as possible, the information needed to make a decision, welcoming any relief from lengthy searches for solutions to their problems. This tendency is particularly strong if managers operate under extreme time pressures as they often do in the construction industry. Indeed, the burden of decision-making is reduced even further if a manager can follow the lead of other managers, which is why decision-makers tend to give disproportionate weighting to anecdotal information about successes and failures in other companies rather than to statistical data which involves lengthy analysis. It is no coincidence that management fads are accompanied by mythical stories of staggering success in the world's 'best practice' organizations, rather than by solid data. After all, this is what the market wants.

23.2.2 The demand perspective

The demand perspective can be considered from both individual and organizational perspectives.

At an individual level, Huczynski found several reasons why managers succumb to the temptations of the latest fads. Firstly, he argued that the adoption of a new idea is likely to be a career enhancer, increasing the visibility of the person championing it within an organization. Secondly, a new idea can be used as a managerial defence since no one can be blamed for adopting the latest management technique, even if it doesn't work. Thirdly, the idea might be seen as offering quick results to complex problems. Indeed, because many managers are drawn from technical specialities such as engineering, law and finance, and have little managerial know-how or inclination to learn their new craft, what they want are quick-fix solutions which can defend them against accusations of ignorance. Finally, Huczynski points to the increasing need to be associated with 'best practice' which is why the new ideas are often first sold to prominent companies such as ICI, IBM or BHP so that smaller organizations will follow suit, wanting to be associated with them.

As Panati (1991, p. 5) argued 'all fads, ancient and modern, depend upon the amazing phenomenon of crowd behaviour, a momentary madness of emotional contagion. Anyone taken as an individual is tolerably sensible and reasonable – as a member of a crowd, he at once becomes a blockhead.'

At the organizational level, Huczynski explained the growing obsession with management fads by pointing to problems of declining profitability and external perceptions of inefficiency and ineffectiveness. These are problems that are very familiar to the construction industry and Huczynski argued that managers faced with declining fortunes are far more willing to experiment with new ideas than those who are not. He also argued that organizations are attracted to new ideas because they reduce the boredom of work, acting as motivational devices for employees. Since the government reports cited earlier have criticized the human resource management practices of construction companies, this explanation would seem to have particular relevance to today's construction industry. These reports have also put the industry under considerable pressure to innovate and in such an environment the political impact of a new idea can be more important than its effect. This is worrying because it is quite possible that many managers in construction are not concerned with whether the techniques work or not; the most important thing it to be seen to be using them. Indeed, to get onto tender lists, construction companies are under increasing pressure to prove that they employ techniques such as partnering, value engineering and benchmarking. According to Huczynski, in such a competitive market, ignoring any potential source of new ideas could mean giving the competition an advantage and managers are under increasing pressure to prove to customers and shareholders that they are 'doing something'. Today companies must portray themselves as innovative and pioneering which means it makes no sense to adopt second-hand management techniques. Once again the danger is that the focus becomes the technique itself, irrespective of what value it can add to an organization. In this sense, much of the impetus for management fads comes from a *fear of not changing*, a new type of fear which has received relatively little attention compared to the *fear of change* that is traditionally written about in management textbooks.

23.3 The fear of not changing

The *fear of not changing* has its origins in the late 1970s and 1980s, when there was unprecedented instability in the business environment. Influential texts such as Toffler (1970), Peters and Waterman (1982), Kanter (1983), Hornstein (1986) and Peters (1989) convinced managers that their lives would never be the same and that they would be faced with the shock of accelerating and uncontrollable technological, economic and political change. Initially the panacea was to be prepared for change, however, this incremental and reactive approach was soon discredited and managers were told that success was dependent upon an organization's ability to thrive on chaos and generate change rather than wait for it to happen. Thus the path to prosperity rested upon the courageous actions of managers in challenging the status quo and in embarking upon a constant process of revolution, self-examination, renewal, innovation and radical change. Importantly Peters and Waterman (1982), in their best selling book which sold over five million copies, warned that the world's largest and most successful companies were most vulnerable to this new environment, and this message was reinforced by Pascale (1991, p. 11) who poignantly illustrated that 'nothing fails like success'. Such warnings struck fear into the heart of the business community and produced a new and lucrative market for managerial 'wonder drugs' to help managers successfully bring about organizational change. Pascale (1991) graphically illustrates that throughout the late 1980s the imagination of managers was captured by an explosion of new buzz-words such as management by objectives, decentralization, delayering, supply chain management, value chains analysis, downsizing, diversification, restructuring and quality circles.

While the message that drove the market for management fads was one of adaptation and change, there was also evidence that cast doubt on this revolution mentality. For example, Collins and Porras (1994) examined the entire life of 18 successful companies who outperformed the American Stock Market by a factor of 15 since 1926. These included 3M, American Express, Motorola, and Procter and Gamble. They found that change was never an over-riding factor in how these companies operated. Instead, the one factor that linked these companies was their zealous preservation of their core ideology which in some cases had remained unchanged for over 100 years. Importantly, these core values emphasized people rather than processes, which was very different to the emphasis of many of the management fads which developed during that time. Since then other studies have cast serious doubt upon the amazing claims attached to many management ideas and it is to these that we now turn.

23.4 The problems with business fads

Far from providing viable solutions to problems, these fads often generate their own problems.

23.4.1 Confusion

One of the problems with any business fad is the overuse, abuse and misuse that eventually surrounds it with ambiguity and confusion. For example, Oliver (1993) questions the relationships between business process re-engineering, re-engineering, business process improvement, job redesign, business transformation, incremental process improvement and core process redesign. In a construction context, McGeorge and Palmer (1997) even point to distinctions being drawn between the hyphenated and non-hyphenated versions of re-engineering. As Oliver (1993) complains, the range of offerings riding the re-engineering ticket is so vast and bewildering that it is not surprising that every company has its own understanding of what re-engineering is and that every management consultancy has its own approach to it. Indeed, in this context, it is not surprising that only 5–10% of re-engineered companies in America have seen any benefits from the process, nor that some researchers quote failure rates as high as 70% (Hammer, 1990; Lorenz, 1993).

23.4.2 Simplification

Watson (1994) recognized that managers cannot possibly find the time to engage in a detailed thinking-through of every problem they face. Consequently most managers act intuitively and without reflection, repeating actions with which they are familiar and with which they have had success. Furthermore, in an attempt to impose some order upon an increasingly chaotic world, they tend to turn to the predictability of systems, structures, rules, procedures, new initiatives, fads and techniques. While those that peddle management fads are snobbishly discounted by management scholars as 'pop stars', it must be acknowledged that they do seem to redress a fundamental problem in today's managerial world – a severe lack of time. Most disturbingly, many management fads, such as lean production, have the effect of reducing this further and so perpetuate the demands for such techniques.

Another factor which drives managers' need for simplification is the increasing complexity of the business world. A common characteristic of management fads is that they claim to reduce management to a few simple but profound truths and universal laws. As Watson (1994) argues, the process of simplification is a necessary and understandable defence against increasing complexity but it is also dangerous because it is occurs at the expense of getting people together, sharing ideas and developing solutions to shared problems. In Watson's view, management is complex only because it involves people yet, paradoxically, reliance upon convenient and easy to digest solutions excludes them. In this sense the promise of simplification is a dangerous illusion because, as Watson argues, managers need to 'see the wood as well as the trees' so that they can see the overall context that they operate within. While the truths that will emerge will not be great and profound statements that change our managerial lives, they will help managers learn more than the deceivingly simple formulas for quickly achieving career or business success.

23.4.3 Hype

Another problem with construction management fads is the almost obscene levels of evangelical hype which surround them, examples of which were provided in the previous sections. As Cleverley (1971) argued, this has caused managers to turn to management writers in the same way that the ancient Greeks turned to their gods to help rationalize the seemingly complex world they lived in and to help them cope with the potential uncertainty and dangers which it contained. This biblical analogy to worship has become surprisingly accurate in recent years with quasi-religious language being increasingly used to capture and retain the minds and hearts of modern-day managers in an increasingly competitive market for new management ideas. For example, one now becomes a 'disciple' of the Demming 'gospel', following his 14 simple steps in the same way that Christians follow Moses' 10 commandments.

To perpetuate the myths about different techniques, it is interesting to note that the literature surrounding them contain far fewer reports of failures than successes. As Hilmer and Donaldson (1996) warn, the danger with this dogma and with the grossly biased claims is that it fosters superficiality, putting the reasoning powers of otherwise sceptical, thoughtful and pragmatic managers on hold. It also produces a dangerous sense of invincibility and immunity within a manager's mind which paradoxically increases an organization's vulnerability to crises (Wildavsky, 1988). This is particularly so for the managers of small- and medium-sized enterprises which dominate the construction industry, who do not have the necessary time, knowledge or resources to grasp the underlying principles and make them a success. The consequences are rushed and ill-conceived initiatives which destroy managers' credibility and make employees cynical about future initiatives to introduce change. As Hall *et al.* (1993) warn, most management trends involve reviewing all aspects of people, technology and process in one co-ordinated approach, which costs significant amounts of time and money and has enormous implications for issues such as retraining, safety, industrial and public relations and staff morale. Indeed this was supported by Bresnen and Marshall (2000), in their critical review of partnering in construction, where they concluded that to be effective it must be carefully thought through and supported by a clear sense of strategic direction and, most importantly, senior management commitment and resources. Furthermore, it requires an understanding of the likely impact upon individual and group behaviours, motivations and interests and a full appreciation of the long-term and complex dynamics of the implementation processes. Clearly the adoption of any new management idea should not be taken lightly.

23.4.4 Ruthlessness

Arguably, the greatest problem with many management fads is their ruthless emphasis upon productivity improvement through the eradication of redundancy and waste in organizational processes with the aim of producing sinewy and lean organizations (Richardson, 1996). While Huczynski (1993) recognizes that some management fads are grounded in the human relations school of management thought, Richardson (1996, p. 20) demonstrates how 'the scientific management movement, in its many guises, is widespread, growing and extolled as the way forward'. He goes on to argue that '[s]cientific management is a "must" for modern day strategists . . . If it isn't practised then the organization cannot expect to survive.' Richardson argues that while many of these new fads are portrayed as fresh ideas, most are out of touch with contemporary management thought which has transformed the traditional view of management as a regulatory activity designed to control some objective and static scene. The contemporary view of management is that it is a social activity which involves co-ordinating purposeful individuals who are embedded in complex and constantly changing social networks, where redundancy is a necessary aspect of organizational life, particularly in coping with the ever changing business environment of today (Rogers and Kincaid, 1981; Tsoukas, 1995; Furze and Gale, 1996).

According to Sagan (1993), a key feature of all 'high reliability' organizations is redundancy and duplication. He illustrated this by referring to US aircraft carrier operations which stress the critical importance of having both technical redundancy (backup computers, antennas and the like) and personnel redundancy (spare people and overlap of responsibilities). Overlapping responsibilities may seem inefficient in modern business terms but on an aircraft carrier it can mean the difference between life and death by ensuring that potential problems which are missed by one person are detected by another. The same principles are used to manage nuclear power stations where independent outside power sources and several coolant loops are incorporated into system designs in case existing provisions fail.

Sagan also illustrates that redundancy is also a feature of our body's immune system, which is why we can survive if certain body parts are severely damaged. For example, if one kidney is removed then the other can compensate and if our spleen is removed, then our bone marrow takes over the job of producing red blood cells. It is clear that redundancy is essential to survival in a world full of potential risks, thus it seems ironic that the construction industry is attempting to create inflexible construction organizations precisely when the business environment is demanding more flexibility.

In practical terms, contemporary research indicates that managers should be attending to the spiritual needs of their organization, avoiding prescription, encouraging spontaneity and creativity, fostering and maintaining trusting interpersonal relationships and providing people with time to think and reflect (Ryan and Oestreich, 1998). The problem with many modern day management trends is that the emphasis is very much the opposite: upon processes rather than people, and upon efficiency rather than effectiveness. While many of these new trends do not necessarily mean de-layering and downsizing, in practice it very often happens and indeed, unscrupulous managers use these new buzz-words as managerial façades to keep their shareholders happy or legitimize their streamlining plans, making only a token gesture to the philosophies and processes involved and to the introduction of meaningful cultural and structural organizational change (Hall *et al.*, 1993). The human resource management implications of many management trends are enormous and in labour-intensive industries such as construction where the principal operating cost is manpower, it is very difficult to achieve any efficiency gains without some job reductions. Evidence of this is provided by Gretton (1993) who cites Xerox, one of the world's greatest advocates of re-engineering, spending US$700 million to shed 10 000 people over 3 years and the Fortune 500 industrial companies 'sweating off' 3.2 million jobs during the 1980s. This trend continued throughout the 1990s and Richardson (1996) summarizes major job losses in the UK during this period that were a direct consequence of re-engineering activities. These are depicted in Table 23.1, with associated increases in turnover that illustrate the current trend of expecting more for less.

While there is little data on the effectiveness of these companies after their downsizing activities, a study of 3500 downsized companies in Australia revealed that 60% failed to see any productivity improvements (Savery and Lucks, 2000). The findings indicate that downsizing is often undertaken without restructuring or with the new technology needed to compensate for losses and raise productivity. The result is often a severe loss of morale and initiative.

Table 23.1 The employee to sales (£000) turnover ratios of large UK companies (Richardson, 1996)

Company	1990	1994
British Telecom	244 000 employees £11 071 000 turnover	170 000 employees £13 242 000 turnover
Yorkshire Electricity	6 836 employees £1 242 500 turnover	5 764 employees £1 307 000 turnover
British Gas	80 481 employees £7 983 000 turnover	74 480 employees £10 386 000 turnover
British Airways	50 320 employees £4 838 000 turnover	42 233 employees £6 303 000 turnover

23.4.5 Pressure and stress

Despite the enormous impact that change programs have upon people, Crainer and Obeng (1995) point out that many managers find the human aspects too uncomfortable and focus instead upon system changes. Consequently, as Handy (1984) predicted, we are increasingly seeing the emergence of a transient and bitter workforce of 'corporate mercenaries' who have lost the capacity for loyalty to their employers and who coldly drift from one organization to the next with the primary aim of securing an immediate and tangible return for their efforts. Even for those who are retained within organizations, security of employment has become a thing of the past and having fewer people in the workplace means that they are required to work under intolerable pressures and perform a greater variety of tasks than previously (Gretton, 1993). For example, Hammer (1990) claims that those who thoroughly embark upon a re-engineering process will find that they can reduce their workforce by between 40 and 80% and still maintain the same level of business. However, as Ryan and Oestriech (1998) warn, work-force reduction programs send enormous messages about how people are valued in an organization and are a very effective way of instilling fear into the workplace, since the result is inevitably reduced job security, higher work loads and higher levels of work-related stress. Chaos and catastrophe theorists such as Gleick (1987) argue that such environments are the spawning grounds for dangerous tensions revolving around issues such as safety and industrial relations which can trigger unpredictable and uncontrollable crises that can rapidly destroy the viability of organizations. Indeed, one need look no further than the West Gate Bridge collapse, the Ronan Point Tower collapse and the Challenger Space Shuttle disaster to have examples of catastrophes to which elements of stress from over-working contributed (Bignell *et al,* 1977; Jarman and Kouzmin, 1990).

23.5 Conclusion

The aim of this chapter was to provide a cautionary note about innovation. It is ironic that a traditionally conservative construction industry now appears to have an insatiable appetite for the long line of management wonder drugs which have been served up by management gurus during the 1990s. This is understandable in the current economic, political and legislative environment that pervades the construction industries of most developed countries. Managers are under extreme pressure to continually improve their performance which makes it difficult for them to resist the promises of miracles offered with these new techniques. There is, however, considerable evidence that suggests that these techniques should be approached with caution since many of them give the illusion of managerial progress while widening the gap between an increasingly unstable business environment and the need for organizations to be flexible, sensitive and responsive. Paradoxically the result is an increasingly uncreative and vulnerable construction industry. Indeed the signs are already there. It was once widely thought that technological advances would transform our lives, producing a world of leisure where we would have more time to spend with our families and enjoy life to the full. As predicted, the nine-to-five day has been consigned to the past. Unfortunately it has been replaced by a longer working day rather than more leisure time. A new recruit in the construction industry can expect to work longer hours with less resources than his/her predecessors and suffer higher levels of work-related stress. The potential long-term social implications of this trend are serious and have not been considered here, however, the organizational implication is increased vulnerability, precisely when the number of threats facing organizations is increasing and becoming more acute.

References and bibliography

Bignell, V., Peters, G. and Pym, C. (1977) *Catastrophic Failures* (Milton Keynes: The Open University Press).

Blockley, D.I. (1996) Hazard engineering. In: Hood, C. and Jones, D.K.C. (eds) *Accident and Design: Contemporary Debates in Risk Management* (UCL Press).

Bresnen, M. and Marshall, N. (2000) Partnering in construction: a critical review of issues, problems and dilemmas. *Construction Management and Economics*, **18**, 229–37.

Carmichael, D. (1999) Gurus of faddish management. In: Karim, K. *et al.* (eds) *Construction Process Re-Engineering*. Proceedings of the International Conference on Construction Process Re-Engineering, July (Sydney: University of New South Wales), pp. 365–74.

Cleverley, G. (1971) *Managers and Magic* (London: Longman).

Collins, J.C. and Porras, J.I. (1994) *Built to Last: Successful Habits of Visionary Companies* (Australia: Random House).

Commonwealth of Australia (1999) *Building for Growth – An Analysis of the Australian Building and Construction Industries* (Canberra: Commonwealth of Australia).

Crainer, S. and Obeng, E. (1995) Re-engineering. In: Crainer, S. (ed.) *The Financial Times Handbook of Management* (Pitman Publishing).

Egan, J. (1998) *Rethinking Construction* (London: HMSO).

Furze, D. and Gale, C. (1996) *Interpreting Management – Exploring Change and Complexity* (London: International Thompson Business Press).

Gleick, J. (1987) *Chaos* (London: Abacus).

Green, S. (1998) The technocratic totalitarianism of construction process improvement: a critical perspective. *Engineering, Construction and Architectural Management*, **5** (4), 376–86.

Green, S.D. (1999) The missing arguments of lean construction. *Construction Management and Economics*, **17**, 133–7.

Gretton, I. (1993) Striving to succeed in a changing environment. *Professional Manager,* July, 15–17.

Guggenheim Foundation (2001) www.guggenheim-bilbao.es/ingles/home.htm

Hall, G., Rosenthal, J. and Wade, J. (1993) How to make re-engineering really work. *Harvard Business Review,* November–December, 119–31.

Hammer, M. (1990) Re-engineering work: don't automate, obliterate. *Harvard Business Review,* **68**, 104–12.

Handy, C. (1984) *The Future of Work: A Guide to a Changing Society.* (Oxford: Basil Blackwell).

Heller, R. (1993) TQM – not a panacea but a pilgrimage. *Management Today,* January, 37–40.

Hilmer, F.G. and Donaldson, L. (1996) *Management Redeemed – Debunking the Fads That Undermine Corporate Performance* (London: The Free Press).

Hornstein, H.A. (1986) *Managerial Courage* (New York: John Wiley and Sons).

Huczynski, A.A. (1993) Explaining the succession of management fads. *The International Journal of Human Resource Management,* **4** (2), 443–63.

Jarman, A. and Kouzmin, A. (1990) Decision pathways from crisis – a contingency theory simulation heuristic for the Challenger space disaster (1983–1988). In: Block, A. (ed.) *Contemporary Crisis – Law, Crime and Social Policy* (Netherlands: Kluwer Academic Press).

Kanter, E. (1983) *The Change Masters* (London: Allen and Unwin).
Latham, M. (1994) *Constructing the Team* (London: HMSO).
Lorenz, C. (1993) Uphill struggle to become horizontal. *Financial Times*, 5th November, 23–4.
Marosszeky, M. (1999) Technology and innovation. In: Best, R. and de Valence, G. (eds) *Building in Value* (London: Arnold).
McGeorge, D. and Palmer, A. (1997) *Construction Management – New Directions* (London: Blackwell Science).
Oliver, J. (1993) Shocking to the core. *Management Today,* August, 18–22.
Panati, C. (1991) *Panati's Parade of Fads, Follies and Manias – The Origins of our Most Cherished Obsessions* (New York: Harper Perennial).
Pascale, R. (1991) *Managing on the Edge* (Harmondsworth: Penguin).
Peters, T.J. (1989) *Thriving on Chaos* (London: Macmillan).
Peters, T.J. and Waterman, R.H. Jnr. (1982) *In Search of Excellence* (New York: Harper and Row).
Richardson, B. (1996) Modern management's role in the demise of a sustainable society. *Journal of Contingencies and Crisis Management*, **4** (1), 20–32.
Rogers, E.M. and Kincaid, D.L. (1981) *Communication Networks: Towards a New Paradigm for Research* (London: The Free Press).
Romm, J. and Browning, W. (1995) Energy efficient design. *The Construction Specifier,* June, 41–51.
Ryan, K.D. and Oestreich, D.K. (1998) *Driving Fear Out of the Work-Place*. Second edition (San Francisco: Jossey Bass).
Sagan, S.D. (1993) *The Limits of Safety: Organizations, Accidents and Nuclear Weapons* (Princeton University Press).
Savery, L. and Lucks, J.A. (2000) No productivity boost from downsizing. *HR News*, April, 1–2.
Toffler, A. (1970) *Future Shock – A Study of Mass Bewilderment in the Face of Accelerating Change* (London: The Bodley Head).
Tsoukas, H. (1995) *New Thinking in Organisational Behaviour* (Oxford: Butterworth-Heinemann).
Watson, T.J. (1994) *In Search of Management – Culture Chaos and Control in Managerial Work* (London: Thomson Business Press).
Wildavsky, A. (1988) *Searching for Safety* (New Brunswick: Transaction Books).

24

Four-dimensional models: facility models linked to schedules

Eric B. Collier*, Martin A. Fischer† and George Hurley‡

Editorial comment

Construction contracts almost invariably include a time for completion, and provision for the client to recover liquidated damages should the contractor fail to complete the job by the specified date. Planning construction so that projects can be completed within the allotted time is a vital part of construction management and over many years systems have been developed that help planners juggle the many and varied tasks that are part of the overall construction sequence.

Conventional systems include Gantt charts (simple bar charts), and PERT (Program Evaluation and Review Technique) and CPM (Critical Path Method). These have been further developed with the introduction of software packages that automate much of the work involved in preparing and using these techniques. While these methods are now quite sophisticated there is still considerable margin for error and to some degree their success still depends on human interpretation and understanding, particularly in dealing with logical sequencing of activities on complex projects and visualizing three dimensional (3D) objects from two dimensional (2D) documents.

Computer-based drawing programs (computer-aided design; CAD) have been in use for quite some time and are now widely used in all areas of building design and documentation. However, it is only quite recently that 3D modelling programs have begun to replace 2D drawing programs that functioned essentially as 'electronic drawing boards'. This has allowed a move towards integration of building documentation, drawings in particular, and programming or planning software. The result of the integration is '4D planning', that is, planning which combines the three dimensions of the

* Salt Lake City Airport Authority
† Stanford University
‡ DPR Construction, Inc.

buildings with the fourth dimension of time. The outcome is a very powerful set of tools that has great functionality across a number of areas of building procurement, from producing staged 3D views of the construction sequence and the completed building for the information of clients before construction ever begins, to serving as an archive of how the project was built that can be utilized by the facility managers who look after the building long after the completion of construction.

The advantages of such a system are obvious: the interdependence of various activities is more easily seen, potential clashes can be identified and steps taken to avoid them long before they become apparent on site, rework is reduced, disputes are avoided, clients can be easily acquainted with difficulties, changes, the nature and extent of temporary works and so on, staging of construction and occupation can be demonstrated visually rather than in abstract terms allowing clients to plan their activities associated with relocation, temporary location and moving in to new spaces more easily and more effectively. The value that is added is equally clear: fewer disputes, fewer variations, and more efficient construction all mean better value for money for the client as costs are reduced and sub-optimal solutions that are the result of on-site compromises are less common.

In this chapter the authors use their experience from a large project in California to explain how 4D planning works in practice and to demonstrate the benefits that it can bring to all participants in the building procurement process. It is worth noting that the project that they describe is a 'live' project and the procedures that they describe are not merely theoretical.

24.1 Introduction

The electronic display of three-dimensional (3D) objects on computer screens is becoming increasingly common for architects, engineers, and contractors. The ability to assemble and take apart such models easily and quickly makes computer modelling a valuable tool for planning the building process. Four-dimensional (4D) computer models (3D models linked to the fourth dimension of time) for planned facilities hold great potential for the improvement of construction planning.

One of the most important elements in planning construction is the development of an accurate and feasible construction schedule. Today's project management software assists in creating complex construction schedules, showing the interrelationships of activities and balancing manpower, however, visualization of activities and related geometric objects must take place in the minds of construction planners. In contrast, 3D models make the project geometry clear. The schedule and the 3D model, when linked together, animate the construction schedule with 3D objects appearing on the computer screen in their order of installation. Simulating construction with 4D models is similar to having a new 3D model of the project for each day of construction.

The following discussion of the creation and use of 3D and 4D models in construction uses a real-life example of a large and complex project, a hospital and health care centre, to illustrate how such models can be used. Their use provides opportunities for improved project representation and communication in the design and planning of facilities.

24.2 San Mateo County Health Center

The example described here is the San Mateo County Health Center which was remodelled and renovated in the late 1990s and involved 30 000 m^2 of construction, mostly comprising new buildings. Modelling the project was done by CIFE (Center for Integrated Facility Engineering) at Stanford University working with Dillingham Construction Company, Inc., the contractors responsible for the actual construction work.

The goal of this case study was to understand the applicability of 4D modelling to retrofit and new construction projects. Although conceptually the value of 4D models can be explained and understood, actual implementation, as it is for any new technology, is more difficult. Clear understanding of the capabilities and limitations of new technology requires using and testing. Members of the Dillingham management team had become familiar with 3D CAD through a small project completed previously. With the San Mateo project they were interested in modelling a large, multi-year project and in visualizing the construction sequence. This was to be accomplished by using off-the-shelf software and hardware.

24.3 Challenges of the project

The San Mateo County Health Center is a 70-year-old facility that occupies most of a city block in the city of San Mateo, California. Dillingham was selected as the construction manager for a remodelling and renovation project with a budget of US$94.5 million and a completion date in 2000. The project involved over 26 500 m^2 of new building floor area and the remodelling of over 3500 m^2. Three buildings were demolished, five new buildings were built and the main hospital building renovated.

24.3.1 Uninterrupted hospital operations

The Health Center is the only major hospital in San Mateo County. It serves hospital patients and county programs including psychiatric services and all county-sponsored food preparation for prisons, day-care facilities and the like. In consequence, hospital operations could not be interrupted during construction. Continuous access from patient and surgical rooms to all areas of the hospital had to be guaranteed until the new facilities were ready. Disturbances to the departments being moved and to other departments had to be minimized.

Given these challenges, Dillingham had to plan construction activities carefully. These plans included the location of temporary trailers on site during some phases of construction and the building of a temporary link between an old and a new building during another phase. Another major element of the plan was to build the new central hub in two halves to allow for the proper sequencing of construction without cutting hospital operations in half with construction activity. Construction activities for the various buildings overlapped to meet the necessary schedule for the use of each building.

24.3.2 Communicating the construction plan

Dillingham was fully responsible for co-ordinating all construction activities with the hospital, hence, administrators and department heads had to understand the schedule. For example, during construction, the path ambulances took to arrive at the emergency room changed a number of times. Other departments experienced similar changes during construction. All of these changes could have had ripple effects that would have affected hospital operations and so delayed the project if all parties involved did not clearly understand where they had to be and when they needed to be there.

The method used during the design phase of the project to communicate how the buildings would be built involved showing a plan view of the entire site, including all old and new buildings. The start date and the expected date of occupancy were written on each building. If a building was to be demolished, the demolition date was noted. A number of notes and arrows were added indicating what would be happening over time. In another attempt at explanation, multiple small sketches were used to show the buildings that would be built or demolished at a certain date. Neither of these approaches gave a clear idea of the complexity of the schedule nor the implications to the hospital. The first method of showing everything on one page was confusing and the second method oversimplified the process. Creating the schedule animation based on a 3D model was a much more informative method of communicating complete information about the construction phases.

24.4 Building the models

Dillingham engaged CIFE researchers to help build the 3D and 4D models for the project. The following sections detail their experience in building and using these models.

24.4.1 Three-dimensional (3D) modelling

The architect (The Ratcliff Group, Emeryville, CA) and some of their consultants supplied two-dimensional (2D) drawings of the buildings in AutoCAD's dwg format. These drawings were used as templates for 3D modelling. The first building modelled was the Central Utilities Plant, which was to be the first new building built. All the concrete foundation work including step footings, retaining walls, and piles was modelled as well as the structural steel beams and columns, the major mechanical equipment, some of the pipework, metal stud framing and exterior architectural features. The remainder of the buildings and the site itself were then modelled. The exterior architectural features of each building were modelled including individual glazing elements and window mullions, and the structure of the buildings including columns and floor slabs.

24.4.2 Four-dimensional (4D) modelling

The second phase of the project involved linking Dillingham's Primavera schedules to the 3D models using a precursor to Bentley's Schedule Simulator, which was developed based on Bechtel's CCAE (Construction Computer-Aided Engineering) software (Ivany and Cleveland, 1988; CERF, 1992). This linking is done by extracting the layer names from the AutoCAD model and linking them to the activity number from the Primavera schedules. The user can use one-to-one, one-to-many, and many-to-one linkages to link 3D CAD elements (layers) to schedule activities. The 4D modelling software then generates an animation sequence file that specifies which layers are to be turned on (for construction) or off (for demolition) at each time interval. This animation displays construction activities at the time intervals specified by the schedule and shows the 3D objects in place and under construction during each period, e.g., a day or a week (Figure 24.1). Others have also reported on the application of 4D models (Retik *et al.*, 1990; Vaugn, 1996; EPRI and Westinghouse Electric Company, 2000; Koo and Fischer, 2000).

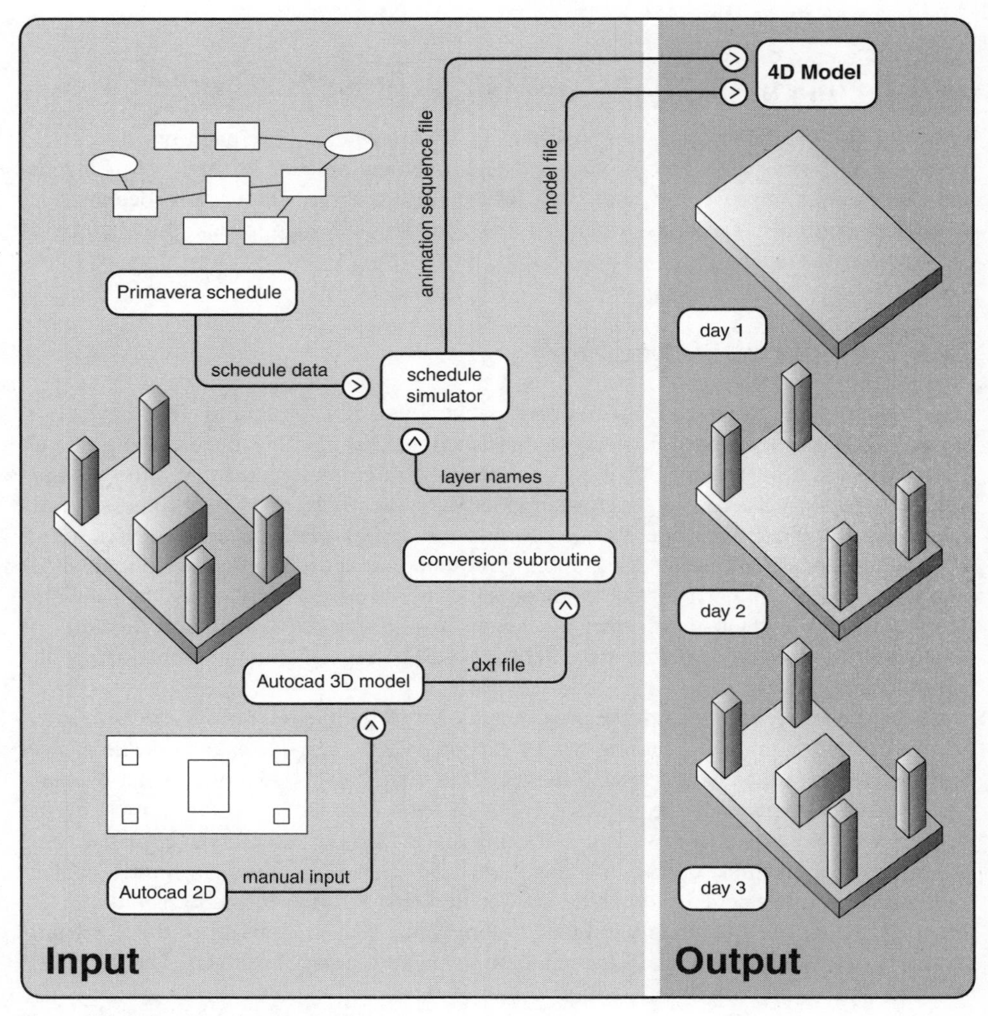

Figure 24.1 Creation of a 4D model.

The 4D animation differentiates critical path activities from non-critical activities. The animation of the schedule uses red (indicating critical path activities) and green (non-critical path activities) to highlight objects under construction at the time indicated on the screen.

The results of the modelling effort were two 4D models. The model of the Central Plant showed objects in their proper construction sequence based on an 18-month schedule. The overall site model consisted of 25 000 3D CAD objects linked to 450 activities and showed the renovation of the San Mateo County Health Center over its entire development period.

During the modelling exercise and through their participation in the planning of the project with Dillingham construction management, those involved in the modelling process observed several advantages of 3D and 4D models over conventional construction technologies and practices – these are discussed below with respect to the San Mateo project in particular and to construction projects in general.

24.5 Impact of representation of project information

Typically, project information is represented in 2D drawings, in bar chart or critical path method (CPM) diagrams, and in other text and graphical documents. Three-dimensional and 4D models improve the clarity of project information, allow early detection of geometric conflicts, serve as a unified core model and help check the quality of construction schedules.

24.5.1 Clarity of 3D geometry

Understanding the geometric relationships of architectural, structural and mechanical objects is a fundamental step in planning construction. This has been done in the past with 2D drawings and written specifications. This method of conveying graphic information is inherently inefficient as designers must produce 2D representations of their mind's eye 3D image of a building. Everyone else involved in the design, planning and construction of the structure must look at the 2D graphic representation and build their own mental 3D images, which are often incorrect or incomplete. Some people may study the drawings thoroughly and develop the correct 3D image in one day, then rebuild it mentally or remember it incorrectly another day. Sabbagh (1996) reports similar problems for the manufacturing sector.

As the modelling progressed, the project manager for Dillingham often checked on the progress of the 3D model. Coming from a meeting where certain features of the project had been discussed, he would look at the model to clarify in his own mind what the parts that had been discussed were actually going to look like. He was often struck by the difference between the hand waving and generally inadequate attempts at explanation of geometric relationships during meetings and the clarity of the 3D model. Three-dimensional models actually combine information that is often located on multiple 2D drawings; seeing the 3D model reminded Dillingham's project manager of the 2D details and the conclusions he had formed about them in his preliminary estimates. The 3D model clarified some of his questions, reminded him of others he had intended to ask, and also gave him new questions to ask.

3D models are built to the exact dimensions of the project. Modelling software allows users to select how close or how far away they want to view the objects they have drawn, i.e., the 3D model has, in a sense, an infinite number of scales. Only one model needs to be built to show the smallest details in addition to the overall view. Because the model is accurate and there is only one model for every design element, measurements can be taken directly from the model. This can translate to fewer dimensional mistakes in the field, especially as further advances are made in linking electronic measuring devices to 3D models (Anon., 1992; Edmister, 1994; Beliveau *et al.*, 1995). Local and wide area networks make it possible to share 3D and 4D models simultaneously in different locations. This means quicker transfer of geometric and schedule information than was ever possible with 2D CAD files or physical models.

24.5.2 Finding 3D conflicts before construction

Because of the difficulty of keeping in one's mind 3D information that is constructed and reconstructed from 2D information, many conflicts are not apparent until they surface during construction. Sometimes there is an omission by designers, who fall victim to the abstraction of 2D and fail to co-ordinate details through thorough 3D thinking. At other times a contractor's 3D interpretation turns out to be different from that of the designer. When an error is not detected before the installation of materials, resolving conflicts can be very costly; correcting the improper placement of concrete, for example, by jack hammering it out and pouring new concrete is an inefficient method of solving the inadequacies of poor graphic communication.

On the Central Plant, while modelling the retaining walls, it was discovered that there were two different representations for one of the retaining walls. The architect had shown the wall to be as wide as the columns while the structural engineer had shown a thinner wall with pilasters or buttresses. The architect's elevation showed a smooth surface instead of pilasters. Although a number of professionals had examined the 2D drawings, no one had discovered this discrepancy. More than six drawings on different sheets had to be consulted for the conflict to be understood and confirmed. The project had already been tendered, and there might not have been a resolution until construction had begun if the 3D model had not been built. Once the 3D model made the problem clear, the project team resolved it quickly and the hospital saved $30 000 on concrete material costs and avoided change orders. In the 3D model the wall could only be modelled one way and the conflict was obvious. This gave everyone working on the project an insight into the value of 3D modelling.

24.5.3 3D model as core information model

An important aspect of 3D models is that they can be used as an electronic geometric kernel to which non-graphic information can be added (Teicholz and Fischer, 1994); another is their ability to serve as the core model for other applications (Fischer and Froese, 1996), such as 4D modelling. The linking of electronic information to 3D models is becoming both less difficult and more powerful. Labels can be attached to objects and even the simple selection of the object can trigger established links to databases.

Eventually, all forms of information may be directly accessible from the model. This information may include material cost, availability, location, description and so on. Versions or copies of the model can be used as models for structural analysis, for lighting and acoustic analysis, for accurate quantity estimates (Clayton *et al.*, 1994), and as a shell to which even the most detailed construction elements, including each bolt and nut if desired, can be added.

24.6 Impact of four-dimensional model on communication

If a project is only to be designed but never built, or if the designer is to personally build the structure, it does not matter how incomplete the design information is or how well others understand the design intent. For most projects, however, the communication of design intent is critical to project realization. Traditionally in construction each participant uses different and distinctive media for communication; a client may only see the final model to approve the look of a building, the floor plans to approve the layout, and an unopened stack of blueprints before authorizing construction. The designer feels ownership of the original 2D documents and specifications but has little to do with the construction schedule, while the contractor takes the blueprints of the 2D documents and studies them in depth, customizing them with colour-coded markers and notes, and also creates other new documents such as the construction schedule. Once the designers finish the drawings most of the communication that goes on is non-graphical, in the form of letters or conversations that are often detached from the 2D documents.

4D models combine information from various disciplines to create a shared medium of communication. As discussed above, project information becomes more accessible to all parties since these models show more realistic views of objects and activities than abstract 2D drawings and bar charts. Anyone viewing the construction animation with 4D models, from owners to neighbourhood groups to construction workers, can use the same tool, based on the same model, to understand the project.

24.6.1 Communication with clients

The use of 4D models is especially promising for owners of construction. Clients seldom have the time or inclination to study blueprints and schedules to understand how contractors will build their facility. The 4D model, however, can not only present an overall view of how construction will take place, but can also show detailed information that can provide more complete answers to clients' questions. Animation of the schedule gives the client a better idea of the sequential nature of construction activities and a clearer view of the demands of the process. Clients often consider accelerated schedules as a construction alternative. The inexperienced might feel that the choice to accelerate or not is one of only money versus time; 4D models show the extra density of activities, equipment and workers that acceleration will cause, thus enabling clients to see that there are other factors, such as safety and quality control, that should be considered. The contractor might also have to use more lay-down area, or more points of access to the site. This can be clearly shown so that a more informed decision becomes possible.

Alternatively a sceptical client may be convinced of the feasibility of a well-planned accelerated construction plan by seeing its animation.

In almost all public, industrial, and even commercial construction 'the client' is not a single individual, but an organization, a board of directors, or an entire staff. A model that clearly shows the sequence of construction informs all members of the client team to the same degree. This helps them discuss ways of minimizing the impact of construction on operations. By understanding the construction plan more fully, they can make better plans for operations during construction. This leads to a more informed dialogue between the client and the contractor on ideas for changes in construction that lessen impacts and minimize conflicts with operations.

Figure 24.2 shows the planned construction schedule over the 61-month schedule. Month 0 shows the site with the existing buildings: the main hospital on the left, clinics building at right in the foreground, East wing between these two buildings, existing utilities plant in middle, and AIDS clinic at top right. In month 10, as work on the new central utilities plant proceeds as the critical activity, construction on the first half of the North addition begins. Trailers for temporary office space are set up next to the clinics building. The first portion of the central hub is under construction (shown in the centre of the snapshot). The hospital operations are linked through the existing East wing (shown in grey in the centre of the snapshot). In month 17, the central utilities plant and the North addition interior work are completed. The AIDS clinic is now in trailers, the old AIDS clinic has been demolished and work has begun on the new nursing wing. Month 20 sees the start of the nursing wing exterior shell construction as a critical path activity. The East wing functions are moved to the North addition. A connector is built to link the first half of the central hub to the clinics building. The first portion of the hub is finished and can now serve as a link between departments. This makes it possible to demolish the East wing (shown in the centre of the snapshot) to make room for the second half of the central hub. In month 26, the East wing has been demolished and the new clinics building and second half of the North addition are under construction in its place. The nursing wing work has moved to the interior. Month 37 brings the demolition of the old clinics building. Construction of the new diagnostics and treatment building has begun. In month 55, the remodelling of the existing hospital structure begins and the entire third floor and ancillary wings are removed. After 61 months, the new San Mateo County Health Center is completed.

At the master plan level of the San Mateo project shown in Figure 24.2, the 4D model helped clarify time–space relationships between construction and hospital operations. For instance, an early construction schedule showed that, at one stage during the project, construction activities occupied the centre of the hospital, separating operations in one part of the hospital from those in the other part. Initially, the central hub (the building with the round yard in the centre of the hospital), connecting all parts of the hospital, was supposed to be constructed as one building at one time. However, if that had happened (as was discovered with the first 4D simulation), construction would have cut hospital operations in half. It would have been necessary to put patients on gurneys (trolleys), wheel them out the door and around the block to bring them from their rooms to radiology services. This was, of course, not acceptable, and the designers and construction managers had to find a solution to maintain uninterrupted links between all hospital departments. They decided to cut the central hub in half, add a seismic joint in the middle, and build one section of the hub in the early phases of the project, as seen in Month 10. To keep hospital departments connected at

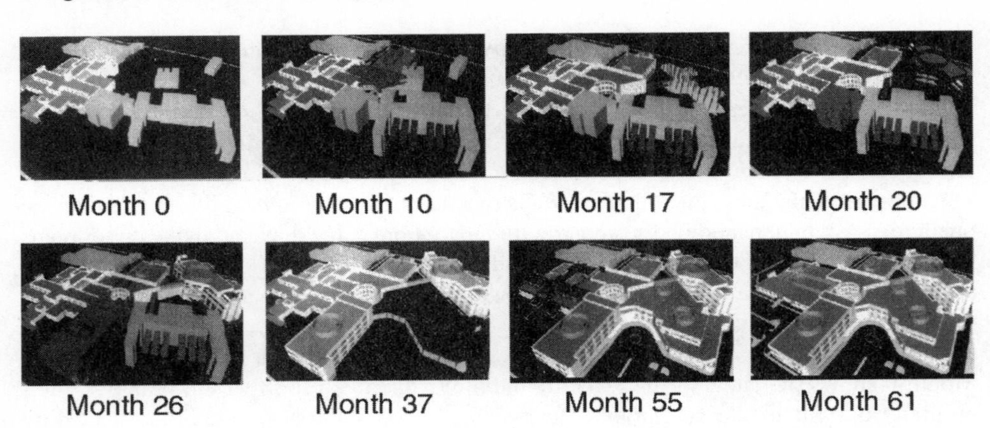

Figure 24.2 Snapshots of the 4D model.

all times, the construction manager also built a temporary link for use during months 20 to 26. The 4D model allowed the construction manager to check that his approach worked. It also enabled him to convince the hospital staff that, indeed, they would be able to get from one part of the hospital to all other parts at all times during the entire construction period without having to walk through a construction site and without going out onto the street to walk around construction areas. In discussions with the hospital and Dillingham about uses of the model it was suggested that a continuous video could be shown in a central location such as the cafeteria. This video could show a 4D model of the current construction plans and the immediate impacts on hospital operations (CIFE, 1994). This would afford a simple method of explaining construction plans in a way that all could understand. It was found that videos of the 4D model could be shown to people in many settings where it would be difficult to set up a computer display.

24.6.2 Construction planner's use of 4D

As 4D becomes more common, construction planners will use visualization of the schedule as an integrated tool in the planning process. Planners will be able to evaluate alternative schedules visually by creating different construction sequence animations. Even though the schedule and 3D model are created separately and then combined, experience has shown that once these models have been built there is relatively little additional work necessary to view any number of alternative schedules if the level of detail in the schedule or 3D model does not change too much between alternatives.

In addition to visualizing construction sequencing and constructability, 4D models can include information about spatial requirements for installation activities and the storage of materials. Every construction activity requires space for storage of objects when they are delivered, a pathway to the location of their installation, and adequate room at that location for installation (Riley, 1994). The advantage of planning with 3D and 4D objects is the same as the advantage of 3D CAD in general: representations are more complete and include volumetric and not just area requirements, and there is a greater ability to view the project as field personnel will actually experience it.

4D modelling allows viewers to check the feasibility of a construction schedule visually. It is often difficult to discover improper activity sequencing and overlaps in CPM diagrams or bar charts. On the San Mateo project the 4D model was helpful at the master plan level and at the detailed schedule level; for instance, the first time the construction simulation of the Central Plant was run the mechanical equipment appeared before the equipment pads on which the equipment was to be installed. This was an oversight that was obvious when the schedule was viewed with a 4D model, but it would have taken close study of the CPM schedule for this sequencing mistake to have been detected.

4D CAD makes possible planning that would be difficult to carry out otherwise. For example, in the installation of the cooling towers for the Central Plant, a replay of the 4D construction animation showed that all elements around the cooling tower location, including masonry walls and structural steel elements, were scheduled to be in place before the cooling towers. This alerted the installer to the fact that a crane was needed for installation of the towers; the model also showed clearly the distance they had to be lifted and moved. Such models can be used to simulate the movement and installation of items to check clearances (Williams, 1996) or to select the most appropriate lifting approach (Williams and Bennett, 1996).

24.6.3 Communication with sub-contractors

4D modelling also supports schedule co-ordination with sub-contractors. Different installation sequences can be tested and the selected sequence can be shown to the professionals performing the work. Sub-contractors can see what will be in place when they begin installation of their part of the work and where other trades will work at the same time. They can also give feedback on the plan and see clearly the impact of delays caused by them or others.

Some sub-contractors have been creating 3D shop drawings for some time (Zabelle, 1996). They can now easily animate construction and realize the same advantages as construction managers. In addition, sub-contractors can compare the actual state of construction when they are called on site with the promised state of construction in 4D on which they based their bid. Because the 4D model is the common tool, any sub-contractor will be able to see how their part fits into the larger picture.

On the Central Plant, the concrete sub-contractor was perplexed by the number of changes in elevation and the large number of step footings. He visited the construction manager's office, viewed the 4D model of the concrete installation and immediately understood the design and corresponding schedule. He said that he understood in five minutes what he had spent hours trying to figure out.

24.6.4 Communication with suppliers

Suppliers of components and materials are often left out of the loop of communications on a construction project. With 4D models suppliers can be included among the people who view the construction schedule animation. In addition to knowing when their materials need to arrive on site, they can see the intended use of their product and give feedback to the designer if a 4D model is built early enough in the project.

24.6.5 Communication with executive management

Managers face the challenge of making the most important decisions on a project while relying on information gleaned from summary reports. Often they do not have the time to study the drawings and schedules in depth or to do the mental 'piecing together' necessary for making the best decisions. With 4D models managers are able to use the same tool as all other participants to determine the desired solutions to problems. This could greatly compress the decision chain. Today, an owner who has a specific question about a product might ask the client representative, who then asks the project manager, and the question would continue from the project engineer to the sub-contractor to the supplier, with an answer returning in a few weeks or months. With the common tool of 3D and 4D CAD and the proper communication capabilities, people at either end of the chain have the opportunity to discuss the problem by viewing the model together (Glymph, 1996).

24.6.6 Communication with the public

Because construction projects affect the lives of many people not directly related to the project, 4D CAD will be especially helpful in explaining construction to those who come in contact with the project for short periods of time, such as lawyers, judges, insurance companies, approval agencies, and community groups. These professionals may have a large impact on construction projects, but currently make decisions based on very limited knowledge.

Construction events and schedules can be explained in the courtroom using 4D models. Physical models are often used in this situation because they help people unfamiliar with construction, including most jurors, to understand a project quickly (ENR, 1994). Four-dimensional modelling has the added benefit of helping the jury understand the impact of schedule delays. Hopefully, 4D modelling will help reduce the number of claims in the first place by improving the understanding among project team members.

Insurance companies are another group of potential beneficiaries; as they become more informed about the accuracy of planning using 4D modelling, they may begin to offer lower insurance rates to companies that use it, since, as was demonstrated on the San Mateo project, 4D models reduce the risk of failure, interferences and delays.

Approval agencies often require lengthy approval periods and some of this time is spent in interpreting 2D drawings. Planning departments and community groups are often concerned with how the building of a structure will impact on the community and some are actively looking for ways to incorporate 3D technology into the review process (Brenner, 1994a). Four-dimensional modelling supports the quick understanding of a project by a large number of people.

Dillingham was asked to explain the construction of the Central Plant to a neighbourhood group. This was the first building to be built on the hospital site for 40 years. At first, the group seemed quite hostile towards the upcoming construction, but the mood improved as a video of the 4D model was shown. Many people were surprised to see how quickly the outside would be completed and were able to understand that most of the 18 months of construction involved work inside the building.

24.7 Impact on management and culture

Like most new technologies, the widespread use of 4D modelling in construction planning requires many changes in processes and tools. The reality of implementing this new technology is that it is not just a streamlining of current work practices, but that its full benefits can best be realized through a change in the management and work environment of construction planning and practice.

24.7.1 Managing model production

The linking of objects and activities to create an animated construction schedule is a rather straightforward step once both the 3D model and the schedule have been created. Building the 3D model is a more time-consuming undertaking. Design managers are used to measuring design progress by the number of sheets completed. However, the production of a 3D model is not easily measured as it is an electronic model created by assembling a large number of electronic parts. Managers have much less experience in measuring progress on 3D models. However, some companies, e.g., designers of piping systems, have developed methods to measure progress by counting the number of objects modelled per unit area. They even have the computer run nightly productivity checks on the work performed that day (Celis, 1994). Many of these companies have found that they can design and document process and power systems more efficiently using 3D modelling than they could using conventional methods. These models are most commonly used to create the standard 2D documents for construction.

Many architects build 3D models of buildings but currently these are mainly used to show the completed product. Often this means that the model is built with the least number of objects possible. Detail is limited and only shown where it will be seen in predefined perspectives. The model is often not built as an integral part of design and it is difficult to determine the time necessary to develop a model with more detail.

Table 24.1 shows the productivity obtained when modelling the various parts of the San Mateo County Hospital. The numbers show significant variation in the production rates achieved for different parts of the model. Not surprisingly, the production rate was lowest on the Central Plant, since it was modelled at the highest level of detail. The model for the

Table 24.1 Hours needed to develop the San Mateo 3D model

Building	Building area (m^2)	Total modelling time (hours)	Average modelling time (hours/100 m^2)
Central utilities plant (high level of detail)	1 850	320	17.3
North addition	4 400	100	2.3
Clinics building	4 200	110	2.6
Nursing wing	10 200	200	2.0
Diagnostic and treatment center	6 000	120	2.0
Existing hospital	3 700	170	4.6
Site	n/a	180	n/a
Total	**30 350**	**1200**	**4.0**

rest of the hospital included structural and key architectural elements. It is interesting to note that modelling of the new parts proceeded about twice as fast as modelling of the existing portions of the hospital. One explanation for this difference is that the quality of the 2D drawings was superior for the new parts. While these numbers are not conclusive and more experiences like this need to be documented, they are an example of the type of managerial information needed to plan and manage the production of 3D models. A main conclusion from the San Mateo project is that the building of individual 3D objects and models is the most time consuming element of creating the 4D model: approximately two-thirds of the time spent on the project was devoted to 3D modelling. Much of this time could be eliminated if modelling firms transferred 3D models instead of 2D drawings to the construction team.

Many of those who want to develop 3D and 4D models want to model the entire project and develop a 'complete' 3D model. Even in 2D, the notion of a 'complete' set of design documents is only determined by the level of detail common in the industry. Such common levels of detail have yet to be determined for electronic models, further complicating the manager's job of predicting the time necessary to build a satisfactory 3D or 4D model. One step towards the more common use of 4D modelling is to begin by modelling small, critical parts of construction. By modelling the most important details of design, the greatest benefit is realized for the smallest effort. Additional modelling can take place as it is deemed appropriate.

The most efficient way to build the 3D model is to do it as part of the initial design and not as a post-design activity. If designers built 3D models as they designed they would benefit from having the 3D clarity of the model during design, as well as providing a basis for the contractor and other sub-contractors to build 4D models, possibly as part of pre-construction services. The additional effort required to build the electronic 3D model as opposed to simply working in 2D is becoming increasingly less significant as 3D CAD tools become more powerful even on less expensive computers. It is more efficient for designers to build the model themselves rather than have someone else decipher it from 2D; design professionals could offer this added service to clients. Once the model is created numerous schedule alternatives can be shown, objects moved to check for interferences, and walkthroughs created, all quite quickly.

24.7.2 Cultural challenges

One of the greatest barriers to the wider use of 4D computer models is the resistance to change that is typical of the building industry. Even in Silicon Valley the high-tech companies that manufacture the computers that make such modelling possible mostly still plan and build their buildings with conventional methods. On the other hand, such companies may also be the first to begin requiring 3D computer models as part of the construction delivery process (Brenner, 1994b). In general, however, there needs to be some cultural change in both the building of models and in their use before 4D models will be used more commonly.

Today, if a client requests a model of a project, they usually mean a model made of wood, paper and glue. While the computer model of the San Mateo project was being constructed, the hospital commissioned a physical model of the site. It measured 460 mm by 760 mm, cost US$100 000 and was encased in glass and displayed in the hospital lobby.

It was out of date within a few months of its completion as the hospital had made a design change and to alter the model to reflect that change would have cost an additional US$20 000. The entire electronic modelling effort, including the Central Plant, the entire site, schedule animations and video creations would have cost US$100 000 if time had been billed at US$60 an hour (including the 250 hours needed to become proficient with 3D and 4D technologies), and the change could have been done for a fraction of the estimated US$20 000. The electronic model also contains many more objects, such as all the interior walls, and can show more information, such as the colour of walls and glass, and supports the animation of the construction schedule.

The existence of the electronic model did not mean that those that could benefit from it automatically used it. Even once the model was available on site, there was some reluctance to use it. At one point, construction personnel were discussing the proper elevation for an electrical pad near the Central Plant. A point of contention was the elevation of the step footings near the concrete pad. Although the 3D model was available and would have shown the elevation of the step footing easily and clearly, the five people involved consulted the 2D drawings instead and several hours elapsed before they all agreed on the proper elevation.

The most important step towards cultural change will take place when clients begin asking for 4D models in addition to, and eventually in place of, the traditional documents. This will bring rapid change to the industry as well as even more rapid developments in technology.

24.8 Level of detail

The level of sophistication of the tools and the actual methods that are employed in building the models are important aspects of 4D modelling. The development of less expensive, more powerful computational capabilities will continue and although these improved capabilities will influence the logistics of building the model, many concerns still need to be addressed regardless of the technology. An important concern is the level of detail of the 3D and 4D model.

24.8.1 Levels of detail in 3D model

Even 2D drawings must include a determination of the level of abstraction in the drawings. Standards and conventions have developed such that two lines drawn close together with a perpendicular line at each end represent a window in a 2D floor plan (Figure 24.3). By contrast, a window may be shown in 3D as a rectangular box, but this may also represent a door. The number of lines and surfaces needed to make a box into a recognizable window can be 10 times greater than those representing the simple six faces of a 3D box.

There is a trade-off between how much detail is shown in 3D and the ease with which the model can be manipulated and viewed. Detail is demanded in 3D because abstraction is more difficult, as the case of the window and the door demonstrates. Details and connections that can be glossed over in 2D must be resolved in 3D (Coles and Reinschmidt, 1994), however, it takes valuable time to enter each detail into the computer

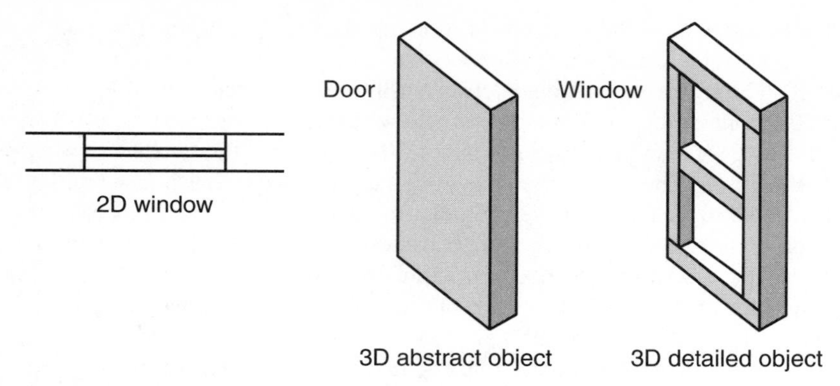

Figure 24.3 Levels of detail in CAD models.

and each object added for clarity increases the file size of the model. This means that animating objects on screen will be slower since computer speed is tied directly to the amount of information that needs to be processed for each frame.

Finding solutions to the issue of detail in the 3D model can be broken into two parts: how to generate detail, and how to view detail. One solution is to have more than one source for detail: sub-contractors could substitute detailed 3D elements for more abstract elements, just as shop drawings today show greater specificity than design documents. Material suppliers may also start supplying detailed 3D representations of their products that can be inserted easily into 3D models (Arnold and Teicholz, 1996). Perhaps most importantly, software tools are being developed that can use rules and guidelines to generate detailed 3D objects (Hodgetts, 1994). Some examples of automated applications include algorithms that take a first pass at framing a wall, including studs at 16 in (400 mm) centres, and headers (lintels) over the doorways and window openings. Other systems perform automated pipe routing given an entrance point, exit point, the geometry and location of objects in space, and the desired sizes of pipes (Howard, 1995).

It must be possible to view a model at different levels of detail. This can already be accomplished by manually turning objects off and on, but there are significant advantages if this can be done automatically. For example, software could include rules about the level of rendering detail: if the computer viewpoint is only a metre or two away from a bolt it could be displayed as a detailed geometric object, perhaps complete with threads. If the viewpoint moved back 10 m a simpler geometry could be substituted for it (Funkhouser and Sequin, 1993). Finally, if it was behind a wall or if the viewpoint was 100 m away it would not be shown at all. In the same manner a window could be displayed as a flat 2D plane from afar and as a completely detailed window from a metre away, with several levels of detail in between (Figure 24.4). This information could be accessed from databases of geometric information. Current graphic software typically does not render faces of objects that are out of sight. The further formalization of rules about viewing detail will be helpful in the long-term use of 3D and 4D models.

The San Mateo project was modelled at two levels of detail. For the Central Plant, as much detail as possible was modelled; this included step footings, steel I-beams, mechanical pads, concrete masonry units by courses, window mullions, window glazing and louvres as well as some of the mechanical pipework, showing elbows, connections

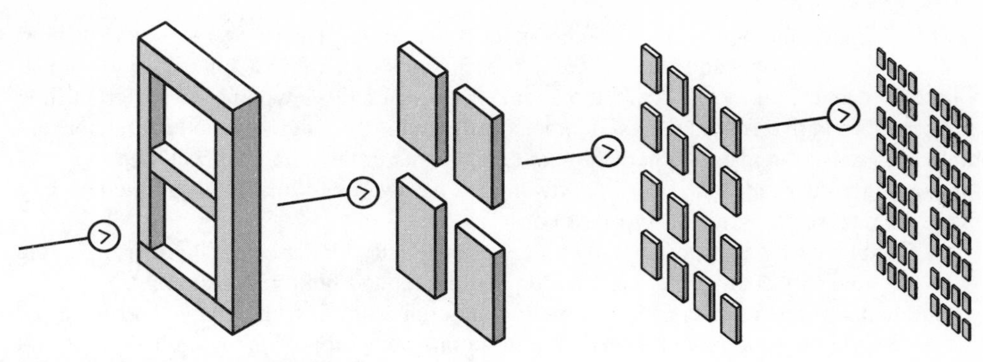

Figure 24.4 Differing levels of detail to display windows.

and pumps. This detailed modelling made the building look realistic. Less detail was shown in the remaining buildings so that the entire site could be modelled in the budgeted time. Interior and exterior walls as well as stairs were modelled, but the window mullions were only modelled as single surfaces to reduce model size.

24.8.2 Matching schedule activities and 3D elements

The value of the 4D model increases greatly as efforts are made to carefully match the schedule activities and the 3D objects. Although the 3D model for San Mateo was modelled to a detailed level, the construction schedules were still much less detailed. For example, on the Central Plant, a schedule activity was labelled simply as 'erect structural steel'; as a result, although each beam and column was modelled individually, they all appeared at once in the schedule animation. To increase the clarity of this animation, activities were added to break down the erection of the structural steel into column lines 1–3, 4–5 and 6–9. Some time was required to re-layer CAD elements and to re-link them to the new activities. Ongoing research is aimed at automating such changes in level of detail of the 3D model and schedule (Akbas and Fischer, 1999) while others focus on making 3D and 4D modelling more interactive (Krueger *et al.*, 1995; McKinney *et al.*, 1996, Fröhlich *et al.*, 1997, Schwegler *et al.*, 2000).

24.9 Conclusion

When the design of a structure is communicated with a 3D model it can be turned into a 4D model for construction planning. The construction planner can develop a sequence of construction, estimate the duration of activities and then simulate the construction sequence by having the model appear on the computer screen as it is to be built, combining the time and space representation of a project. The use of 4D models gives a graphic description of construction intent, plans become more explicit and uncertainties are reduced.

Simulating construction with 4D models is similar to having a new 3D model for each day of construction. For any given day the model can show which parts of construction will be in place and which areas will be under construction. This is as if someone were to build a physical model and take a picture of each day's construction, but with a 4D model one can repeat the process again and again, from different viewpoints, at different time increments and in different scales. In other words, architects, engineers and contractors are able to practise the installation of project components in the computer well before crews are sent into the field. Advances in software and hardware will make this an increasingly accessible technology for all types of construction.

A flexible, easily transportable 4D model dramatically increases the number of people who are able to visualize the construction and completed building accurately. Everyone from clients to the suppliers of construction materials can view the planned sequence of work. By viewing the model each can gain an understanding of the overall process and how each person's part fits in. The experience on the San Mateo project shows that this accessibility gives participants better and more complete information allowing them to work together more efficiently and create higher quality projects.

References and bibliography

Akbas, R. and Fischer, M. (1999) Examples of product model transformations in construction. *Durability of Building Materials and Components*, **8** (4). Information Technology in Construction, CIB W78 Workshop (Ottawa: NRC Research Press), pp. 2737–46.

Anon. (1992) Gehry forges new computer links. *Architecture*, **81** (8), 105–10.

Arnold, J.A. and Teicholz, P. (1996) A knowledge based information model for components in the process industry. In: Vanegas J. and Chinowsky P. (eds) *Proceedings of the Third Congress on Computing in Civil Engineering* (New York, NY: ASCE), pp. 586–92.

Beliveau, Y.J., Williams, J.M., King, J.R. and Niles, A.R. (1995). Real-time position measurement integrated with CAD: technologies and their protocols. *Journal of Construction Engineering and Management*, ASCE, **121** (4), 346–54.

Brenner, J. (1994a) Computer simulation task force. *Tracings-San Jose Chapter American Institute of Architects Newsletter*, September, 5.

Brenner, J. (1994b) Integrating CAD into the project approval process. *Tracings AIA Santa Clara Valley*, December, **5**, 7.

Celis, A. (1994) *Measuring Progress in 3D Projects*. Audiocassette (Palm Desert, CA: Convention Cassettes Unlimited).

CERF (1992) Computer aided engineering (CAE): a promising tool for improving construction site productivity. *CERF Technical Report No. 92-N6001* (Washington, DC: Civil Engineering Research Foundation).

CIFE (1994) *San Mateo County Health Center: A 3D CAD Model*. Videotape, No. 24 (Stanford, CA: CIFE).

Clayton, M.J., Kunz, J.C., Fischer, M.A. and Teicholz, P. (1994) First drawings, then semantics. In: Harfmann A.C. and Fraser M. (eds) *Reconnecting: ACADIA '94*. (The Association for Computer Aided Design in Architecture), pp. 13–26.

Coles, B.C. and Reinschmidt, K.F. (1994) Computer-integrated construction. *Civil Engineering*, **64** (6), 50–3.

Edmister, R.R. (1994) A very good day in the third millennium. *Construction Business Review*, **4** (4), 60–3.

ENR (1994) Graphics focus the issues. *Engineering News Record*, **233** (20), 23.

EPRI and Westinghouse Electric Company (2000) *Virtual Reality Construction: 4D Visualization* (Palo Alto, CA: Electric Power Research Institute; Windsor, CT: Westinghouse Electric Company,

Nuclear Systems). Available from EPRI Distribution Center, 207 Coggins Drive, P.O. Box 23205, Pleasant Hill, CA 94523.

Fischer, M.A. and Froese, T. (1996) Examples and characteristics of shared project models. *Journal of Computing in Civil Engineering*, ASCE, **10** (3), 174–82.

Fröhlich, B., Fischer, M., Agrawala, M., Beers, A. and Hanrahan, P. (1997) Collaborative production modelling and planning. *Computer Graphics and Applications*, IEEE, **17** (4), 13–15.

Funkhouser, T.A. and Sequin, C.H. (1993) Adaptive display algorithm for interactive frame rates during visualization of complex virtual environments. In: *ACM SIGGRAPH '93 Conference on Computer Graphics* (New York, NY), pp. 247–54.

Glymph, J. (1996) 3D CAD in architectural practice. Presentation at *34th Annual Spring Seminar, Northern California Construction Institute* (CA: Stanford University), May 11.

Hodgetts, C. (1994) Rubbing out the craft: architecture and fabrication in the age of information. *Architecture Californis*, **1**, 7–11.

Howard, H.C. (1995) Modelling process and form for process plant pipe routing. In: *CIB Proceedings 180, W78 Workshop on Modelling Buildings through Their Lifecycle* (CA: Stanford University), pp. 523–4.

Ivany, J. and Cleveland, A. Jr. (1988) 3-D modelling, real-time computer simulation reduce downtime. *Power Engineering*, **92** (6), 49–51.

Koo, B. and Fischer, M. (2000) Feasibility study of 4D CAD in commercial construction. *Journal of Construction Engineering and Management*, ASCE, **126** (4), 251–60.

Krueger, W., Bohn, C.A., Fröhlich, B., Schuth, H., Strauss, W. and Wesche, G. (1995) Responsive workbench: a virtual working environment. *Computer*, **28** (7), 42–8.

McKinney, K., Kim, J., Fischer, M. and Howard, C. (1996) Interactive 4D-CAD. In: Vanegas J. and Chinowsky P. (eds) *Proceedings of the Third Congress on Computing in Civil Engineering* (New York, NY: ASCE), pp. 383–9.

Retik, A., Warszawski, A. and Banai, A. (1990) Use of computer graphics as a scheduling tool. *Building and Environment*, **25** (2), 133–42.

Riley, D.R. (1994) *Modelling the Space Behavior of Construction Activities*. Ph.D. Thesis (Architectural Engineering, Pennsylvania State University).

Sabbagh, K. (1996) *Twenty-First Century Jet: The Making and Marketing of the Boeing 777* (New York, NY: Scribner).

Schwegler, B., Fischer M.A. and Liston, K.M. (2000) New information technology tools enable productivity improvements. In: *Proceedings of North American Steel Construction Conference* (Chicago, IL: AISC), 11-1–20.

Teicholz, P. and Fischer, M. (1994) Strategy for computer integrated construction technology. *Journal of Construction Engineering and Management*, ASCE, **120** (1), 117–31.

Vaugn, F. (1996) 3D & 4D CAD modelling on commercial design-build projects. In: Vanegas, J. and Chinowsky, P. (eds) *Proceedings of the Third Congress on Computing in Civil Engineering* (New York, NY: ASCE), pp. 390–6.

Williams, M. (1996) Graphical simulation for project planning: 4D planner. In: Vanegas, J. and Chinowsky, P. (eds) *Proceedings of the Third Congress on Computing in Civil Engineering* (New York, NY: ASCE), pp. 404–9.

Williams, M. and Bennett, C. (1996) ALPS: the automated lift planning system. In: Vanegas, J. and Chinowsky, P. (eds) *Proceedings of the Third Congress on Computing in Civil Engineering* (New York, NY, NY: ASCE), pp. 812–17.

Zabelle T. (1996) 3D CAD for subcontractors. Presentation at *34th Annual Spring Seminar*, Northern California Construction Institute (CA: Stanford University), May 11.

25

Future technologies: new materials and techniques

Rima Lauge-Kristensen* and Rick Best*

Editorial comment

It has been said that the pace of change is now so great that we are experiencing a technological growth similar in scale to that of the entire Industrial Revolution every 18 months. While this reflects the notion of Moore's Law, i.e., that computer chips will double in capacity (and halve in cost) every 18 months to 2 years, these rapid changes are by no means restricted to computer technology, although increased processing power is certainly a vital catalyst in many areas.

In the early 1930s Chester Gould, creator of Dick Tracy, gave his famous cartoon character a fanciful tool that helped him in his fight against crime and corruption: the two-way wrist radio. In 2001 telephone companies virtually give away WAP enabled mobile phones that make Dick Tracy's wrist radio seem like a child's toy. In his 1978 book, *Fountains of Paradise*, famed science fiction author, Arthur C. Clarke, proposed a space elevator: a cable fixed to the top of a 50 km high tower and anchored to a giant satellite in geosynchronous orbit high above the Equator. A material strong enough to make such a cable possible may already be known to science – carbon nanotubes with strength far in excess of steel and of much lighter weight can be fabricated but only at incredibly small sizes, but with further development it is clearly possible that very strong but relatively lightweight monofilament cables will be manufactured in the future.

The point of these examples is a simple one: we live in an age of extraordinary progress, where the far-fetched ideas of one decade are commonplace only a few decades later. There is no reason to suspect that the construction industry will be immune to the effects of this progress. Construction materials and processes are changing in response to many drivers: environmental concerns, globalization, increased competition, advances in information technology, computer-aided design and so on. It is interesting to see, however, that some of facets of the industry have changed very little: cottage builders still place

* University of Technology Sydney

fired clay bricks on top of one another in much the same way as the builders of ancient Mesopotamia did, and some traditional construction materials and methods are being rediscovered as people realize that the builders of antiquity had perfected many simple, effective and sustainable solutions to the fundamental problems of providing shelter and security for people and their possessions.

This chapter looks at just a few of the exciting possibilities that are being explored at present, and also some traditional methods that may provide readymade solutions in the modern age. The list is not meant to be exhaustive, and it could not hope to cover even a small fraction of the myriad products such as sealants, fasteners, connectors, plastics, protective coatings and the like that appear on world markets every day, rather it serves an introduction to some ideas that could revolutionize the way that buildings are designed and constructed, from translucent concrete to materials that have 'shape memory' and 'intelligent materials' that contain nanocomputers.

The impact of new technologies on building value can be quite profound: the use of new materials may mean that buildings are cheaper to run, are more durable, are attractive to tenants who like the 'high-tech' image that is presented, and may well mean that buildings are cheaper to build in the first instance. New processes or even the revival of traditional processes can reduce the environmental impact of completed buildings and so allow new, more stringent regulations to be met, while the industrialization of construction processes, both off-site (prefabrication) and on-site (through the use of robots), can lead to much improved efficiencies.

25.1 Introduction

Due to their complexity and number, buildings collectively have great impact upon the environment. Present day buildings are responsible for the consumption of large quantities of materials, energy and money during their construction, maintenance and use. The resource-draining features of buildings have enormous ecological, economic and, ultimately, social implications, and have become incentives for the construction industry to search for new construction materials and technologies and to industrialize the building process in the hope of improving productivity and efficiency.

The development of new construction materials and industrialized technologies is also underpinned by the desire to create 'intelligent' buildings. Smart materials and prefabricated assemblies can be used to create intelligent buildings with improved energy performance, the capability to self-monitor structural integrity, and with computerized functions and interactive systems that involve egress, security and information.

Developments in construction technology appear to be happening on two fronts. A quiet revolution has been occurring for some time in new composite material applications, for example, carbon fibre composite parts in aeroplanes and carbon fibre reinforced concrete in bridge construction, which indicates that composite materials may dominate in the building industry of the future (Enns, 2000). On the other hand, designers are also looking to indigenous and ancient construction materials and methods for inspiration. Using a mix of both new and traditional materials and techniques, the construction industry may be able to address some of the problems of scarcity of resources and energy. The twentieth century was one of extraordinary technological and social change, and this trend will continue, possibly at an even greater rate, in this century.

25.2 New materials and technologies

New materials and technologies have to be more versatile, lighter, stronger, easier to use, more durable, more energy efficient and must place less strain on natural resources than those that are already available if their use is to be justified (Scheel, 1986). Future challenges and resultant benefits of new materials may include the transmission of light, superior thermal characteristics and the ability to self monitor and self repair. The demand for low environmental impact, low toxicity and abundance may only be met by products made from renewable sources. Outstanding examples of natural materials that have superior qualities of environmental friendliness, versatility, availability and ease of use are bamboo, earth, grasses and hemp (RIC, 2000).

25.2.1 Composites

A new era of construction technology began in the 1970s with the advent of composite materials, largely spurred on by the aerospace industry. Such materials consisted of man-made fibres, particularly glass fibres, embedded in a matrix. More recently, fibres of boron, carbon and other elements have produced materials with greater stiffness and strength-to-weight ratios. Composites are mainly used in the production of lightweight but durable cladding panels for buildings, and in some cases for load-bearing walls. Although the erection of large structures almost entirely of composites is still in the future, techniques for the application of composites in civil engineering have already been developed. For example, epoxy-impregnated fibreglass sheets are used instead of steel jackets to wrap around non-ductile concrete bridge columns and rib members for seismic reinforcement or to strengthen corroded columns (Roberts, 2000). The carbon fibre materials are lighter and easier to handle, are more resistant to corrosion and require much less maintenance than steel.

Fibre-reinforced composites

Fibre-reinforced composites made with natural fibres such as hemp, ramie and kenaf are the most environmentally friendly. Isochanvre (RIC, 2001) is a composite developed in France comprising silica-coated hemp hurds, natural lime and water, forming a lightweight mixture that can be poured directly onto soil or into moulds, trowelled like cement or obtained in the form of construction boards. The product is a non-toxic replacement for cement, timber, sheetrock, plaster, insulation and acoustic tiles. Unlike cement, it is very flexible and does not crack; it is also seven times lighter than concrete. It is water-resistant, non-flammable, fungicidal, antibacterial, and rodent and termite resistant. Because of its strength and flexibility it is an ideal construction material for areas susceptible to earthquakes, tornadoes and hurricanes. The hemp fibres, bound with lime, petrify over time, becoming stronger with age.

Fibre-reinforced plastics (FRPs) are being introduced in concrete construction. The FRP can be formed into any constant cross-section, and is used to prestress or post-tension concrete structures. FRP has higher tensile strength, greater flexibility with regard to size and shape, is lighter, and has higher corrosion resistance than high-strength steel cables, but still needs improvements in some tensile properties (CII, 2000a). Another possible improvement may be the development of a vegetable-based plastic to replace the petroleum-derived component.

Smart concrete

Concrete is one of the major building materials in use today because of its mechanical strength and durability. Research into solving problems of tensile weakness and corrosion abound. 'Smart' concrete, free of these shortcomings and with some bonus attributes, may become commonplace. For instance, electrically conductive concrete could soon be used to monitor the internal condition of structures and/or to dampen vibrations or reduce earthquake damage (CII, 2000b).

Conductive concrete is made by adding carbon fibres, graphite or coke breeze to the matrix. It has been shown that deformations in the concrete affect the contact between the conductive particles and cement matrix, resulting in a change in electrical resistivity, which indicates a structural flaw. A by-product of the presence of carbon fibres is that it takes greater force for the concrete to bend and to crack (Chen and Chung, 1996). By adding enough conductive particles and fibres it is possible to increase the conductivity by several orders of magnitude thereby extending its possible applications to de-icing roads and runways, shielding computer and electrical equipment from electromagnetic interference, protecting structures against static electricity and lightning, and preventing steel in concrete structures from corroding (Construction Innovation, 1997).

Bone-shaped steel fibres have also been used to increase the strength and toughness of concrete (CII, 2000c). While the fibres control cracking by anchoring the matrix at the bulbous ends, they do not experience extreme stress due to the weak interface along their straight lengths (Figure 25.1).

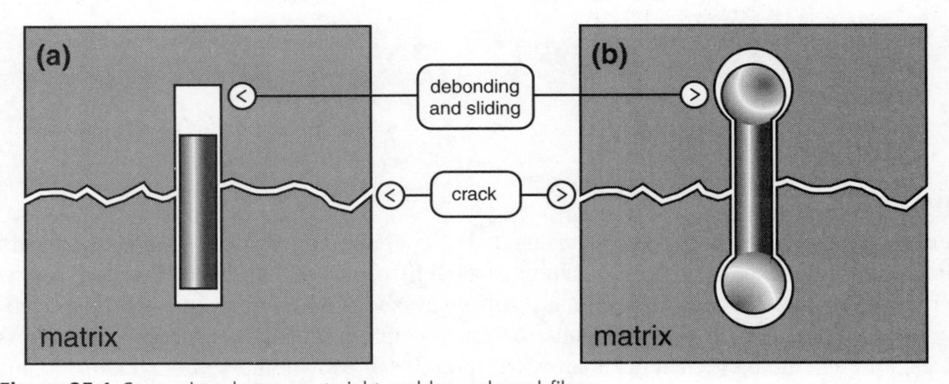

Figure 25.1 Comparison between straight and bone-shaped fibre.

Fibres are also being used to reinforce aerated concrete in the manufacture of panels for residential buildings. Fibre-reinforced cellular concrete (FRCC) is a composite of ordinary cement, a high proportion of polypropylene fibres and an expanding agent (Zollo and Hays, 1998). The panels are lightweight and insulating.

Translucent concrete

US architect, Bill Price, from the University of Houston has been developing a concrete-like material that can transmit light (Black, 2001). His idea was to transform those elements of buildings that are usually opaque, such as walls, columns and roof, into translucent elements that could still be constructed by pouring a liquid material into forms and that, once set, could support loads, insulate and be at least as durable as traditional concrete.

This revolutionary material still requires considerable development but by using crushed glass as aggregate (instead of conventional gravel), a plastic binder (in place of cement) and plastic tubes for reinforcement (rather than steel rods) Price has produced a range of translucent composites. While preliminary tests have shown that this new 'concrete' performs better in compression and flexure than conventional concrete (Shulman, 2001), there are still questions to be answered regarding other properties such as thermal performance, heat transfer, disposability and alkaline decay (Black, 2001).

One major constraint is the cost of production: the most successful mix costs around five times as much to produce as traditional concrete. The architectural possibilities, however, may well outweigh the economic problem in some large-budget projects as a material with strength and ease of moulding to shape in formwork coupled with lighting properties similar to glass would give architects a whole new way to design buildings.

Glass reinforced polymers can contain five to ten times the embodied energy of normal reinforced concrete so there is also a sustainability question to be addressed. Price is working with geologists and environmental scientists to reduce the environmental impact and he believes that the glass and plastic components of his new material may make it possible to use some recycled materials.

Describing this composite material as 'concrete' may appear misleading but the name is deliberately chosen to indicate a structural material that is poured into moulds – it just happens that the material allows the transmission of light. Price's ultimate goal is completely transparent 'concrete' but it remains to be seen if that is a realistic proposition.

25.2.2 Low density materials

Metallic foam

Foams are normally produced by causing a plastic, glass or other material to froth up with bubbles of injected gas or through self expansion by means of a chemical reaction. A new process has been developed in which a solid is mixed with microscopic, 'prefabricated' hollow spheres (called microballoons) of glass, ceramic or polymer, producing what is called an assembled (syntactic) foam (Erikson, 1999). Blown and self-expanding foams have a fairly random distribution of gas bubbles of varying sizes and shapes, while the porosity of syntactic foams is controlled by the careful selection and mixing of the preformed bubbles. Syntactic foams are isotropic, behaving the same way along every load-bearing axis. They can have cells so small (typically between 10 and 200 μm) that the material looks and behaves like a solid, but are easier to use than many other advanced materials. This technology has been applied to producing metal foams.

Although still in research stages, ceramic microsphere-reinforced metal foams suitable for high-temperature, high-pressure sandwich construction have been developed. Researchers expect that a metallic foam will generally have a density approximately half, or less, that of the parent alloy with no loss of strength. So far resin-based foams have been used in flotation objects, such as buoys, small boats and submarines, high performance aircraft, to make moulds and patterns, and to make building mock-ups. An aluminium foam that looks like solid aluminium, but is just as strong with half its weight could revolutionize construction and design.

Aerogel

Another amazing class of materials, first synthesized in the early 1930s but only recently undergoing intense R&D, are 'aerogels'. In 1997 NASA began producing this material, described as 'the lightest known solid material on Earth' (CERF, 1999). An aerogel is 'a low density, highly porous material formed by extracting the liquid from a micro structured gel' (Aspen, 2001). The parent gel may be silica, titania, alumina or one of a variety of polymers.

Aerogel has been referred to as 'solid smoke' as it is 99.5–99.8% air, with a density of only about three times that of air. It is very strong and is an extremely good insulator, being roughly 40 times as effective as the best fibreglass insulation (NASA, 1998). It can be manufactured as insulation blankets that can be cut with scissors yet it combines near weightlessness with very high R-values. One peculiarity of the material is that it is completely transparent in space but has a smoky appearance on Earth. As development progresses and production costs are reduced it is envisaged that such materials could do double duty as windows and wall insulation, allowing light in but keeping heat out during summer, and heat in during winter.

25.2.3 Substitute timber

Concern over the environmental impact of timber production and the urgency of global environmental issues has led to a new level of interest in reducing the widespread use of traditional timber products. Various approaches have been adopted, including the use of recycled timber, alternative materials, wood-based products, and fibreboards made with alternative fibres or fast growing trees (Stokke and Kuo, 2000). The AARCitecture Environmental Home, an energy-efficient home designed and built by staff and students of a university in Colorado, is made almost completely of agriculturally derived materials (RIC, 2000). The materials include structural wheat straw panels for the walls, floor and roof, soy panels with masonry units for the trombe wall, and corn, cork, jute, bamboo and other cellulose-based materials for the interior flooring, shelving, cabinetry and built-in furniture details.

Plant-based wood substitutes

Soy bean flour mixed with recycled paper products, plant-based resins (a soy flour by-product) and colorants are made into a wallboard with the appearance of granite but which is half its weight and is workable like wood. It is totally non-toxic and is non-flammable. A Spanish company has developed a wood substitute material called Maderon (RIC, 2000) derived from almond, walnut and hazelnut shells, which is itself recyclable. The shells are crushed into a powder and mixed with a resin, creating a paste that can then be moulded to make furniture items. The use of such materials is gaining in popularity due to the environmental and health advantages.

Cement-based wood substitutes

A timber substitute has been developed that is made of a cementitious material designed to replace wood lumber in construction, particularly in housing. Generically known as concrete lumber, this product is structurally strong and can be used with conventional timber tools and construction techniques (CII, 2000d). Concrete lumber is not affected by common wood problems such as defects, bowing, termites and rotting, and it has good strength (flexural, compressive and shear) properties to serve as structural members and to make pre-fabricated panels. It may become a common material used for various structural applications in the future.

25.2.4 Nanomaterials

Nanotechnology is the emerging science of manufacturing or assembling structures and even machines so small that they are measured in nanometres (nm) – there are 1000 million nanometres in a metre. Nanoparticles are usually defined as particles with dimensions of less than 100 nm; by comparison an average human hair is around 50 000 nm in diameter. The smallest nanoparticles contain only a few thousand atoms and they can have properties that are much different to those of their parent materials (MPI, 2002).

In general, the properties of nanoparticles, such as their ability to conduct electricity or to act as insulators, can be controlled by engineering their size, shape and composition. Metals with appropriate nanoparticles incorporated become stronger and harder, materials that are normally insulators become conductors, and protective coatings can become transparent (Nanotechnologies, 1999).

The use of nanoparticles could make possible non-corroding paints that cannot be scratched and metal composite car body panels that can be manufactured by pouring the material into moulds (Nanotechnologies, 1999). The capacity to manufacture materials and components at virtually an atomic level of precision would allow the production of metal components such as bearings that have none of the microscopic imperfections that are typical of conventionally produced materials. The result would be dramatic increases in strength, and much greater durability (Spence, 2001).

Buckyballs and buckytubes

In the 1980s scientists discovered that carbon could exist not only as graphite or diamond, but also in a third form, dubbed 'buckyballs' after the visionary, R. Buckminster Fuller. While little practical use has been found for these tiny carbon spheres, their discovery led to the development of 'buckytubes', single-walled carbon tubes generally around 1.1 nm in diameter (Spence, 2001). These tubes are possibly the strongest material that can be made with known matter; their tensile strength is predicted to be 100–150 times that of steel with only one quarter of the weight.

Buckytubes doped with metal are predicted to be 50–100 times more conductive than copper; epoxy composites reinforced with buckytubes can be expected to have strength to weight ratios far in excess of that of any conventional material. The challenge is to make these tubes long enough to be useful as so far developers have only managed to produce tubes around 100 000 times their own diameter; while this sounds impressive it means that their length is equal to only about twice the diameter of a human hair. Commercially produced buckytubes are being used in some microscopic applications but a great deal of development is required before ultra-thin cables comprising bundles of carbon tubes will be able to support bridges and buildings, and do the work of all the world's supercomputers at the same time (Browne, 1998).

25.2.5 Smart materials

The future will see a greater incorporation of smart materials into the building fabric; such materials will be significant components in intelligent buildings. Smart materials, which convert energy from one form to another, include shape memory alloys (SMAs), piezoelectric materials, magnetostrictive materials, polyelectrolyte gels (Ghantasaia, 1998), electrochromic and photoelectric materials.

Shape changing materials

SMAs generate a large useful force as they revert back to their original shape at a certain temperature after being strained. The wires of these alloys can be used in composite materials to apply compensating compression to reduce stress in structures due to external vibrations. SMA wires in a bridge, for instance, can counteract crack-yielding stresses, and at the same time indicate any danger of a collapse. Piezoelectric materials, which expand or contract in response to an applied voltage, have recently been used in bonded joints to resist the tension near locations that have a high concentration of strain, thus extending the fatigue life of some components. Magnetostrictive materials are similar to piezoelectric ones except that they expand in response to magnetic rather than electric fields. Ionic polymeric gels reversibly shrink or swell by changing ionic concentration (pH) or applying an electric field.

Nano-engineered composites

Smart materials are also envisaged that not only contain a variety of nano-engineered composite structures but a great many nano-scale computers – it is projected that these materials will take the form of extremely thin coatings that will be able to react to voice commands, change colour, act as computer monitors or video phones, or even act as environmental monitors, warning us of high levels of radiation or carbon dioxide (Spence, 2001).

Electrochromic materials

Smart materials that display the electrochromic effect change colour in response to changes in their ionic state. One such material, a polymer electrolyte, has been used to laminate glass panes to produce a window that can be switched from a clear and colourless low-E state that will transmit most light and heat for cold days, to a blue tinted state for hot days by applying a low electrical voltage. In the tinted state 95% of the sun's heat is blocked from penetrating the window and reaching the interior of the building (Figure 25.2).

The switchable electrochromic layer has a memory and retains the switched state without the need for continuously applied power (STA, 2000). Smart windows can lead to energy savings in excess of 30% in a wide range of latitudes.

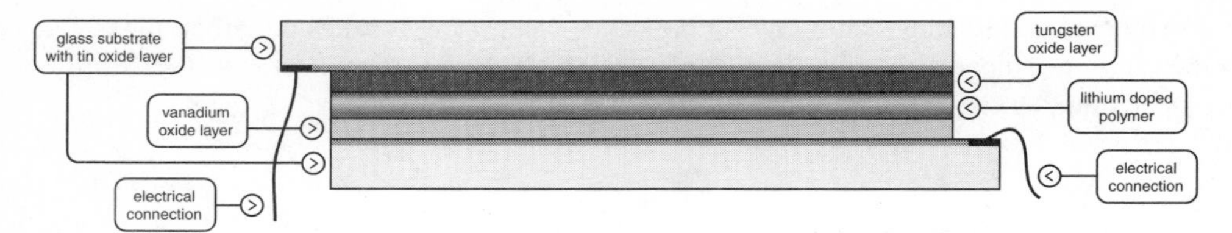

Figure 25.2 Schematic cross-section of the STA SMART Window.

Energy supply

Developments in the renewable energy field are now making it possible to construct buildings that are self-sufficient in energy. Photovoltaic materials, such as silicon or, less commonly, cadmium selenide and gallium arsenide, which convert light energy into usable electricity, are being used in solar cells that form the actual cladding of a building, including wall panels and window awnings. A solar roof incorporating photovoltaic (PV) tiles (EBN, 1999) can use the energy of the sun to meet the power requirements within a building while serving the basic architectural function of protecting the internal environment. Another photovoltaic material, a titania (TiO_2) gel, has been developed to create a coloured, transparent solar cell. The gel is sandwiched between two sheets of glass screen printed with conductive transparent electrodes. This product should excite building designers as it can be used in windows, roof lights, sun roofs and displays, as normal glass would be, with the added benefit of generating useful electrical power (STA, 2001). These cells operate over a wider range of temperatures and lighting levels and are much cheaper than the common silicon solar cell.

25.2.6 Recycled materials

Taking material out of the waste stream to re-use or to produce new construction products not only reduces landfill but also helps to alleviate natural resource depletion.

Re-use of materials

Various architects and designers have been experimenting with the use of everyday articles or materials for use in construction (RIC, 2000). For instance, recycled aluminium cans have been used as a building material at the headquarters of Solar Survival Architecture in Taos, USA, and to construct a solar-powered boat that sailed across the North Pacific in 1996. New York architect Bill Ketavalos has been erecting ‘aquatectures’ using water encased in plastic membranes since the early 1990s. Water’s many properties, such as light-transmission and good heat and sound insulation, may be utilized in novel ways in the design of low-energy buildings in the future.

Panel products

Commercially available panel products, called ‘Gridcore’, made from waste paper, cardboard and cellulose fibre, can replace timber products from endangered rainforests, or plasterboard (sheetrock) in construction. The technology, developed by the US Department of Agriculture, involves creating a slurry of fibres, which is then dehydrated and pressed into fibre mats, smooth on one side with a honeycomb grid on the other. These are glued together to form a strong yet light panel suitable for walls, roofing and floors that can be sawed, nailed and painted.

Recycled paper

Papercrete, a rediscovered building material that was developed in 1928 when timber was cheap and paper scarce, is effectively an industrial-strength papier-mâché that can be moulded into blocks or logs, used to form a sub-layer for concrete flooring, and used as a mortar or a stucco material (National Geographic News, 2001). Papercrete is made of shredded recycled newspaper, water, cement and sand, which harden to form a material that is virtually non-flammable, has high compressive strength and is an excellent insulator. The material can be dyed in the same way as cement and can be painted. Apart from its structural advantages, the use of Papercrete is an effective way of taking paper out of the waste stream and saving natural resources.

Recycled plastic and rubber

Most plastics are not biodegradable and are difficult to recycle because of contamination. Processes have now been developed which fuse and laminate or remould recycled plastics of different kinds, making them suitable for use as wallboard, furniture or noise barriers. Recycled plastic can be used as lumber for waterfront construction, in piers, wharves, pontoons and above-water applications.

There are a wide variety of different types of plastic lumber made of recycled plastic. Some plastic lumber is made entirely of high-density polyethylene (HDPE), which is used to make anything from shampoo bottles to milk jugs. Other types of plastic lumber use composites, which consist of a mixture of recycled HDPE with wood fibres, rubber, fibreglass, or other plastics (Plastic Resource, 2001). Plastic lumber composites are normally used for applications needing a stronger material or a long-lasting alternative to wood. Unlike wood, plastic lumber will not rot, crack, warp or splinter. It is denser than wood, virtually maintenance free, long-lasting (50 years or more, depending on the application), stain- and chemical-resistant, graffiti-proof, waterproof, and is UV and insect resistant. Plastic lumber also requires no painting or sealing (although some wood-composite varieties can be painted if desired) is available in almost any colour and provides a good shock-absorbing surface. An inexpensive substitute for timber, it can be recycled and does not leach chemicals.

The platforms for the bobsled/luge tracks at Lake Placid for the 2000 Winter Goodwill Games were some of the first major constructions made entirely of recycled plastic lumber (RPL) (Plastic Resource, 2000). Plastic lumber has been used by the US Army to construct vehicular bridges with a maintenance-free life expectancy of 50 years. It has been successfully trialled in the USA as railroad ties, as an alternative to traditional wood ties which need regular maintenance and replacement, and in Japan composite railroad ties made from foamed polyurethane with a continuous glass reinforcement are used to help the trains run more quietly.

Similarly car and truck tyres are being diverted from the waste stream and reconstituted into construction materials and floor coverings. These types of products provide a viable market for post-consumer resources.

Sludge bricks

Bricks made up of 70% sewage sludge are odourless and are about 20% stronger and lighter than regular bricks (RIC, 2000). They are also more porous, which means they bond to the cement very strongly. Clay and shale is added to the wet sludge and the bricks baked in an oven for several hours. Given that sludge disposal is very expensive and is often a source of odour and health problems, this is a very ecologically sound building material source.

25.2.7 Re-inventing the wheel

Many innovations rely not on technological advances in materials and software, but rather on the imaginative and innovative application of old construction materials and techniques. For instance, earth is the simplest, cheapest, most durable and most ancient building material. Earth floors and walls have excellent thermal and acoustic insulation properties, are flame-, rot- and insect-resistant, and help create stable indoor microclimates. For these reasons, its popularity is continuing to grow. Earth can be used raw or baked as bricks. Mud brick or adobe dwellings can last for centuries, and ancient cob and 'wattle and daub' type structures still exist in the UK and Europe, hidden behind modern façades. The major consideration in earth construction is protection from water and impact.

Earth construction

The term 'cob' comes from an Old English word meaning 'a lump, or rounded mass'. Cob building blocks are formed by hand by mixing earth with sand and straw. It is an inexpensive way to build as there are no forms, ramming, cement or rectilinear bricks. Mud bricks, stabilized with lime, cement or bitumen, and mixed with straw are used to build low-cost, energy-efficient, durable dwellings worldwide, as are adobe bricks made from sun-baked clay. A modern version of this ancient construction method is 'velcro adobe', first proposed in 1984 to NASA scientists for building on the moon (RIC, 2000). The structural units are sandbags filled in place with earth or sand and barbed wire is used as the 'velcro' or mortar to hold each course together. The structure is tied together with rope or metal and can be sealed (e.g., with stucco). Structurally, sand and earth have significant flexibility but the sandbags also have great compressive strength.

Rammed earth is one of the least labour-intensive of the earth building modes and requires little maintenance. Its viability, however, is site-dependent, and soils should be tested for suitability. Cement may be added to sandy soils to improve strength and stability, while clay soils are often strong enough without cement. Construction involves pouring and/or packing earth into timber forms and building up sections of wall, then removing the forms when dry. Large eaves are usually used to protect walls from damp weather. Pisé, or pneumatically impacted stabilized earth, is much like rammed earth, except that the material used may be clay and/or gravel. Pisé walls are thick and have the strength of more conventional walls.

Earth-integrated buildings can be either above or partially below ground level. For year-round comfort, earth can be piled against the walls of a conventionally constructed dwelling or, for extreme temperatures, purpose-built homes can be completely covered with earth (apart from access for people, air and light). *Leichtlehmbau* or 'light earth method' (LEM) is a newly arrived technique derived from an old European tradition of building light earthen dwellings.

'Biological' materials

Turf or grass sod roofs consist of a crop of grasses and/or herbs growing in a layer of soil over watertight, overlapping tree bark such as birch, which is near rot-proof, supported by a roof frame. Turf roofs act as effective heat storage systems and thermal and moisture barriers. Turf roofs have been tested in Germany to prove their superiority over tiles. A system of central heating has been devised to draw off excess heat from the roof and warm the interior of the house.

Gley is a biological 'plastic' membrane, naturally found in bogs and traditionally used for sealing ponds and dams that has potential for extensive use in the construction industry. The Russian process used to produce this membrane involves applying a slurry of animal waste (pig manure) over the inner base and walls of a dam in multiple, thin layers, which is then covered with vegetable organic matter such as grass, leaves, waste paper, etc. This is all then given a final layer of soil and left for several weeks to allow the anaerobic bacterial process to occur, forming a waterproof membrane.

Architectural pleaching is the ancient process of grafting growing trees of the same or similar species. The early hedgerows and living fences of Britain were made this way with low-growing or dwarf species. Up to the seventeenth century, pleaching was considered an art form in England, producing many pagodas, tree houses, and summerhouses. The technique was also used in mediaeval Europe to construct shelters and dwellings from whole grids or 'orchards' of pleached trees. The pleached branches of mature trees formed the level base of a platform that elevated the community above floodwaters while branches of outer trees formed walls and roofs.

For centuries many indigenous peoples have built the roofs and walls of their dwellings with grasses, palm fronds and reeds, particularly in areas that lack other building materials. Bundled and tied reeds can be used in a structural, load-bearing capacity, and thatched roofs are insulating, have long life-expectancy, and when densely packed, do not allow rain to penetrate. European architects of the biological building school are beginning to re-introduce these durable thatching and structural materials.

Bamboo is a grass with timber-like qualities. It is the fastest growing plant in the world and certain species can reach heights of over 30 m at rates of up to 5 cm per hour. The bamboo cable-supported Min Bridge in Szechuan is over 1000 years old. Although used by the Japanese extensively for centuries, the West is only now appreciating bamboo as an efficient, workable, versatile, cost-effective and environmentally friendly building medium. 'Bamboard' is an example of a particleboard made from bamboo that can be used in place of plywood in a wide range of applications (CERF, 1999).

Straw has been used as a building material in the form of packed straw for walls and roofing for millennia and more recently as straw bales for walls. This agricultural 'waste' has come to be regarded as a truly sustainable and energy-efficient building material. Pinned, cement-rendered and chicken wire-reinforced straw bale construction costs about half that of the nearest equivalent material (cavity brick) while having hugely superior insulation properties.

'Ecological wallpaper'

A Japanese company, Penta-Ocean Construction, has developed an absorbent wallpaper, made from traditional Japanese paper mixed with various types of diatomaceous earth. The wallpaper has proven to be very efficient at absorbing harmful chemicals, notably formaldehyde, that is given off by new materials such as particleboard. As well as reducing the potentially harmful effects of substances outgassed by materials, it is itself completely recyclable as it contains only earth and paper pulp (Penta-Ocean, 2001).

25.3 New construction technologies

While some areas of construction, such as bricklaying and basic carpentry, have changed little over hundreds and even thousands of years, other areas have undergone immense changes. These changes as well as the development of new materials, such as steel reinforced concrete and structural steel, have given designers the freedom to make buildings that bear little outward resemblance to those of the past – the massive glass-clad skyscrapers of the late twentieth century have little connection with the large masonry buildings of previous centuries. Further developments in construction technologies are being driven by the desire to reduce construction costs and to conserve the environment.

Rapid manufacturing

An emerging technology called rapid manufacturing, sometimes referred to as 3D printing, is an extension of the use of computers to design components. This fabrication technique uses the digital design files to directly control a machine that will manufacture the object. A number of techniques have been developed most of which depend on digitally 'slicing' a virtual object into thousands of very thin layers and then depositing very thin successive layers of material, usually in a powder form, fusing each layer to the whole after each deposition. In this way the object is progressively built up until complete. The main constraints on this technique at present are that relatively few materials have been developed and fabricating machines are limited in size. In spite of these limitations the US Army is now able to produce, on demand, a range of simple spare parts for equipment such as tanks, in the field, using digital files sent to the mobile fabrication plant by e-mail.

As rapid manufacturing techniques become more flexible so the range of components that can be made will increase. Development of this idea will usher in a new era, not of mass production, but of mass customization; just as publishing books on the Internet allows economical 'print runs' that consist of just a single copy of a book, so rapid manufacturing will allow production runs that produce just a single item. Customizing a design will be a simple matter of amending the digital file; there will be no need for re-tooling, no patterns, no moulds.

Pin foundations

In order to minimize significant impact on the immediate environment due to installing foundations, a technique has been developed that requires no excavation. Pin foundations, of galvanized steel, are driven into the ground using pneumatic drivers without disturbing any vegetation or surrounding soil and leaving no excavated material to be removed (EBN, 1999). Although the technique has been used for boardwalks and elevated platforms in parks, wildlife refuges and other natural areas, it is still being tested for pinning perimeter foundations for buildings.

Structural steel

Castellated beams, developed in Europe after the Second World War, have been regenerated with new automated technologies which allow them to be produced more economically (EBN, 2000). These resource-efficient, 100% recycled steel beams have the advantage of handling long spans and allowing mechanical and electrical systems to pass through the beams instead of underneath them. By cutting an I-beam lengthwise through the web in a modified sawtooth pattern and welding the two halves back together one half-step out of phase, with the narrow ends of the saw teeth joined, a hybrid of an I-beam and an open-web truss is formed that has the same weight, but that is much stronger and deeper than the original. The use of these beams can reduce building height and weight, yielding synergistic resource efficiencies.

To keep pace with the design advances in steel construction, connections have been developed that make it possible to bolt beams together without drilling or welding steel members, leaving the integrity of the steelwork intact and making the connection speedier, more simple and adjustable. These clamps secure secondary steelwork to structural members without the need for on-site power, however they are not yet suitable for connection of primary steelwork, and are not intended for permanent structural connections (CII, 2000e).

Antigravity devices

In the process of making construction faster, safer, less costly and more resource efficient, construction components are generally becoming more complicated. Imagine the consequences if a building could be freed of the need for load bearing elements – as would be the case if structures were supported and/or held together by antigravity devices. A fanciful idea perhaps but an increasing amount of research-related literature has appeared in this field. John Searle reportedly built an antigravity device in 1964, using a technology named 'The Law of the Squares', and developed it to the extent that he ran his home on a Searle Effect Generator for 30 years (Thomas, 1998). Is this something the construction industry can look forward to in the future?

25.4 Prefabrication and industrialized buildings

Historically, man's building operations were site-specific, i.e., people erected structures by assembling units of local materials. After the Second World War the need for rapid construction to produce sufficient accommodation at the lowest possible cost resulted in significant developments in prefabrication and industrialized construction technology. Today we have about 300 types of frequently used prefabricated systems (Högskolan, 1997), and new construction methods and technologies are constantly being developed to improve productivity. The industrialization of the construction process is also seen as a way to meet the demands for high quality buildings in terms of energy and resource conservation, low environmental impact, and health and comfort issues.

The change from traditional on-site methods of building to effective industrial methods based on new technology and prefabricated elements has real economic consequences, with total production costs decreasing by up to 40% (Högskolan, 1997). Prefabricated construction systems, in many cases, can be erected in 90% less time than field-applied equivalents (Picone, 1998). Prefabrication reduces the time and cost associated with scaffolding, site clean-up, on-site supervision, and capriciousness of the weather. Factory fabrication allows for better quality control and typically provides building designers with superior performance and design flexibility, as well as cost and time efficiencies.

Prefabricated assemblies and components

Prefabricated assemblies reduce on site labour and have superior performance to traditionally built assemblies. Common examples of these prefabricated assemblies include the simple reinforced, precast concrete beam, or panels of two or more materials, that form an assembly with several functions, for instance exterior and interior cladding and insulation. Glass blocks are normally labour intensive to install and often require a mason. Pre-assembled panels are easier and less expensive to fit and as they are installed like a regular glass window; glass block panels can save builders about 10% to 15% over mortared glass blocks (Miller, 1998).

New developments are also being made in high quality, safer prefabricated fixing components, such as the development of connectors and flexible anchors to reduce stresses in cladding and prefabricated concrete (Zacks, 1999). Wright (1996) describes the development of a system, comprising a glass fibre reinforced concrete (GFRC) skin cladding supported by light gauge, cold-formed steel studs, that supports the dead loads of cladding systems while neutralizing wind and seismic loads by resisting skin bending stresses (CII, 2000f).

Robotics

Innovations in the areas of computing and robotics are improving building quality and productivity. Computer-aided design in conjunction with computer-aided manufacture can improve the precision of fitting of prefabricated building components and modules on site by ensuring good dimensional accuracy (Finn, 1992). The construction industry is also taking advantage of the developments in robotics. Robots are faster and more precise than humans and are ideal for repetitive and/or dangerous tasks. Robots are being used in the factory to perform standardized repetitive tasks such as cutting, shaping, positioning and connecting members to form structural elements with great dimensional consistency. Robots can also perform finishing tasks, such as painting and trowelling, and inspect prefabricated components for flaws.

25.5 Conclusion

Innovations in the area of materials, construction technologies and prefabrication appear to fall into two categories: new technological advances and the novel application of old techniques. The growing need and demand for natural, energy-efficient and economically and ecologically sustainable structures is resulting in a trend to re-invent or apply ancient construction methods and materials in an innovative way. On the other hand, there is technological evolution in the form of smart materials, metal foams, and prefabrication and industrialization of construction methods to meet the challenges of higher productivity, increased functionality and improved safety. By the appropriate use of both new and old technologies it should be possible to create innovative solutions for the ever-increasing demands of the construction industry in the future.

References and bibliography

AEN (1998) Solar tiles. *Australian Energy News*, **8** June, 16.

Aspen (2001) *A Breakthrough in Advanced Materials*. Aspen Aerogels Inc. www.aspenaerogels.com/technology.htm

Black, S. (2001) Making light work. *RIBA Journal*, June, 60–2.

Browne. M.W. (1998) Next electronic breakthrough: power-packed carbon atoms. *New York Times*, February 17. www.nytimes.com/library/cyber/week/021798molecule.html

CERF (1999) *Celebrating Innovation for the Design and Construction Industry.* Civil Engineering Research Foundation. January. www.cerf.org

Chen, P. and Chung, D.D.L. (1996) Carbon fibre reinforced concrete as an intrinsically smart concrete for damage assessment during static and dynamic loading. *ACI Materials Journal*, **93** (4), 341–50.

CII (Construction Industry Institute) (2000a) *New Structural Material – Fibre Reinforced Plastics.* Emerging Construction Technologies. www.new-technologies.org/ECT/Civil/frp.htm

CII (Construction Industry Institute) (2000b) *Smart Concrete*. Emerging Construction Technologies. www.new-technologies.org/ECT/Civil/smartconcrete.htm

CII (Construction Industry Institute) (2000c) *Bone Shaped Short Fibre Composite*. Emerging Construction Technologies. www.new-technologies.org/ECT/Civil/boneshaped.htm

CII (Construction Industry Institute) (2000d) *Substiwood™ – Concrete Lumber.* Emerging Construction Technologies. www.new-technologies.org/ECT/Civil/swood.htm

CII (Construction Industry Institute) (2000e) *Adjustable Steelwork Connectors*. Emerging Construction Technologies. www.new-technologies.org/ECT/Civil/ibeamclamps.htm

CII (Construction Industry Institute) (2000f) *GFRC Façade Panels with Steel Stud/Flex Anchor Connection.* Emerging Construction Technologies. www.new-technologies.org/ECT/Civil/mms6.htm

Construction Innovation (1997) Conductive concrete wins popular science prize. *Construction Innovation*, **2** (3). wolf.cisti.nrc.ca/irc/newsletter/v2no3/popular_e.html)

EBN (1999) Pin foundations: no excavation required. *Environmental Building News,* **8** (10), October.

EBN (2000) Smartbeam introduced. *Environmental Building News,* **9** (3), March.

Enns, H. (2000) *Industrial Strength.* Department of Architecture, University of Manitoba. http: //cad9.cadlab.umanitoba.ca/warehouse/crit/enns.html

Erikson, R. (1999) Foams on the cutting edge. *Mechanical Engineering*, **121** (1), 58.

Finn, D.W. (1992) Towards industrialized construction, *Construction Canada,* **34** (3), 25.

Ghantasaia, M. (1998) Smart materials and micromachines. *Engineering World Magazine*, February/March. www.engaust.com.au/ew/0298smart.html

Högskolan, K.T. (1997) *The Theory for Future Building Technologies: The Theory of the Production of Architectural Space* (Stockholm: The Royal Institute of Technology). http://home.bip.net/esch/theory.html

Miller, J. (1998) Let there be light. *Builder*, **21** (11), 182.

Nanotechnologies (1999) Tiny materials, enormous innovation. www.nanoscale.com/why.asp

NASA (1998) *Aerogels Research at NASA/Marshall.* http://aerogel.msfc.nasa.gov/

NASA (2000) *Audacious & Outrageous: Space Elevators.* http://science.nasa.gov/headlines/y2000/ast07sep_1.htm

National Geographic News (2001) *'Paper' Houses May Be Trend of the Future.* http://news.nationalgeographic.com/news/2001/06/0612_paperhouses.html

Penta-Ocean (2001) *Moisture Absorbing Construction Material.* www5.mediagalaxy.co.jp/penta-ocean/english/index.html

Picone, R.J. (1998) Prefab ease. *Buildings*, **92** (10), 28.

Plastic Resource (2000) *New York Stakes Future Olympic Bid on Recycled Plastic Lumber.* www.plasticsresource.com/recycling/recycling_backgrounder/plastic_lumber.html

Plastic Resource (2001) *Plastic Lumber: Bringing New Life to Used Plastics Containers as Environmentally Friendly, Worry Free, Quality Alternatives to Wood.* www.plasticsresource.com/recycling/recycling_backgrounder/plastic_lumber.html

RIC (Rainforest Information Centre) (2000) *Good Wood Project* http://forests.org/ric/good_wood/nont_bld.htm

RIC (Rainforest Information Centre) (2001) *Good Wood Project* http://forests.org/ric/good_wood/hemp.htm

Roberts, J.E. (2000) Application of composites in California bridges. *Energy Efficiency,* Jun/Jul. www.engaust.com.au/ew/0600bridges.html

Scheel, L.M. (1986) House of tomorrow *Technocracy Section 3 Newsletter*, April, 32.

Shulman, K. (2001) X-ray architecture. *Metropolis*, April. www.metropolismag.com/html/content_0401/shulman

Spence B. (2001) Smart and Super Materials. *Nanotechnology Magazine*. http://www.nanozine.com/NANOMATS.HTM

STA (Sustainable Technologies Australia) (2000) *Smart Windows*. www.sta.com.au/sm_wind.htm

STA (Sustainable Technologies Australia) (2001) *Titania Solar Cells*. www.sta.com.au/sol_cell.htm

Stokke, D.D. and Kuo, M. (2000) *Composite Products from Juvenile Hybrid Poplars Bonded with Crosslinked Soy Adhesives*. www.public.iastate.edu/~biocom/page6.html

Thomas, J.A. Jr. (1998) *Antigravity: The Dream made Reality – The Story of John R.R. Searl* (Direct International Science Consortium).

Wright, G. (1996) Inappropriate details spawn cladding problems. *Building Design and Construction*, January.

Zacks, R. (1999) Concrete left standing. *Technology Review,* **102** (4), 30.

Zollo, R.F. and Hays, C.D. (1998) Engineering material properties of a fibre reinforced cellular concrete (FRCC), *Materials Journal*, September/October, **95** (5).

26

Construction automation and robotic technology

Gerard de Valence*

Editorial comment

A great deal of research worldwide is now devoted to the development of robots, with the ultimate aim of producing an autonomous machine with true artificial intelligence (AI). While the 'positronic brain' imagined by Isaac Asimov may never be developed, considerable advances have been made in the production of machines that can carry out many tasks either under the direct control of an operator (often from considerable distances) or following pre-programmed instructions. The robots of today are still, however, far from the cyborgs, androids and humanoid robots that science fiction writers have described for many years.

Following the manufacturing and automobile industries, where the adoption of robotic technology has made a huge impact, the construction industry is now taking advantage of developments in robotics. This chapter looks at the range and capabilities of current robotic technology in construction, which has become a large and growing field over the last two decades. It does not discuss the detail of control systems, or use of fuzzy logic or neural nets, nor does it attempt to cover specific developments in vision systems, mobility or tracking capabilities of robots – rather it looks at the range of robots that have been developed, or that are under development at present, and gives an overview of the extensive worldwide research effort that is dedicated to construction robotics.

Robots are faster and more precise than humans and are ideal for repetitive and/or dangerous tasks. Robots are being used to perform standardized repetitive tasks such as cutting, shaping, positioning and connecting members to form structural elements with great dimensional consistency. Robots can also perform finishing tasks, such as painting, trowelling, and inspecting prefabricated components for flaws.

The Japanese have been developing this technology since the early 1980s. In 1982, the WASCOR project, a collaboration between the industry's contractors and manufacturers,

* University of Technology Sydney

commenced. European and US research projects also began in the 1980s, with both universities and companies actively involved. Today there are hundreds of construction robots that have been developed and tested by these researchers.

The complexity of decision making and situational analysis necessary for a robot to be self-directing as it goes about its assigned tasks demands very high levels of processing power contained in small, lightweight units. Computer hardware is growing ever more powerful and the power can be packaged in small components, however, the ultimate future of robots may be more closely connected with the advent of computers of nanoscale dimension – computing devices built from individual molecules and available in such quantities and so cheaply that banks of these microscopic devices could be used to control the millions of individual functions necessary to allow a truly autonomous robot to function. Such projections may still be science fiction but researchers are seriously exploring such possibilities, and in the meantime the technology that exists today is being used to create machines that are able to carry out many tasks.

At the same time there has been great progress in the development of integrated automated construction systems that have produced very significant reductions in construction times and better quality buildings. Any changes in building methods or equipment that reduce time on site and improve both the quality and quantity of output must have an effect on the value for money outcome achieved by clients, as well as providing contractors with a healthy competitive advantage. As the power of computer systems increases and their cost comes down there must be great scope for further development, improvement and refinement of the sorts of robots and integrated systems discussed in the chapter.

26.1 Introduction

Despite its reputation for being technologically conservative, the construction industry is rapidly developing new robotics and automation technology that has the potential to dramatically transform the industry. Slaughter (1997) analysed 85 construction automation and robotics technologies and examined trends in the development of construction technologies and use by construction companies. She found that priorities identified for automation were close to being fulfilled, and also corresponded with opportunities for adoption. Most of these technologies attempt to reduce the complexity associated with adoption through minimal changes in existing construction tasks.

Automation and robotics were developed for both safety (e.g., nuclear and hazardous sites) and commercial reasons. Venables (1994) saw development of automation and robotics as motivated by:

- removing people from hazard by eliminating tedious and dangerous jobs
- improving safety in various ways
- providing access to difficult areas
- improving quality and consistency
- speeding up construction through extended working hours
- improving the planning of work sequences
- providing results that are independent of work sequences
- improvement of accuracy of construction
- increasing flexibility of construction processes.

The worldwide market for industrial robots increased by 15% from 1998 to 1999 (Spencer, 2001). In the United States, robot sales increased by 38% and the automotive industry, one of the largest users of industrial robots, saw a 24% worldwide increase. In the United States the automotive industry accounts for more than half of the robots in use. Ninety percent of the robots in the world work in factories and half of these help manufacture cars. In the automobile industry, the 'drivers' are increasing labour costs, robotic technology price decreases, the linking of robots with expert systems and the shortage of skilled labour. There are now some 110 000 robots working in US industrial plants, almost as many as in Japan, the two global leaders in their use. Robots and related technologies are already more accurate than the humans they replace and have improved productivity (Weimer, 2001).

Although the automotive industry was the first to install and use robots, the reasons for their use have changed over the years. Robots were initially used to increase productivity and reduce costs but the industry soon realized that it was also getting consistency and quality. Spencer (2001) identified areas where improvement and innovation in robotics benefited the automotive industry. These include seam tracking, vision systems, laser guidance and welding. Robots are also increasingly used for material handling. After using robots for more than three decades, many firms are only now beginning to take advantage of robots' potential for flexibility. Robots are also being rebuilt and used for different jobs, keeping costs down. This is a potential driver for greater use of robots in the construction industry where one of the main reasons for slow adoption is cost.

Clearly the construction industry can learn from the automotive industry. The potential of robots in construction is closely linked to improvements in computer performance over the 1990s. Alfares and Seireg (1996) discuss application of automation technology and information technology in the context of Computer Integrated Construction (CIC). Several automated construction systems and integrated information management systems have been developed in Japan as prototypes of CIC implementation.

This chapter looks at two aspects of this technology. The first is the development of robotic technology, mainly single task robots for use on building and construction sites. A number of examples of these robots are given. Secondly, there have been a number of large scale construction automation systems developed in Japan, which is the leader in this field. These are systems that create an enclosed factory environment for high rise building, and are jacked up floor by floor as construction proceeds. CIC concepts are being used, allowing control of incoming materials and components so that the construction process becomes one of automated assembly in a covered plant, which can operate in any weather and throughout day and night. Another feature is that lifts, conveyors and unmanned forklift trucks are used extensively. Gantry cranes within the 'factory' are used instead of traditional tower cranes. It was estimated that a reduction in the on-site labour force of 75% could be achieved with these automated systems (Wing, 1993).

26.2 Construction robots

Specifically designed to handle a single task, construction robots typically have a travel vehicle, a manipulator and an end-effector. The travel vehicle gives robots a horizontal mobility, the manipulator functions like a human arm, with multiple degrees of freedom (directions) of movement, and the end-effector is equivalent to a hand, performing tasks such as gripping or lifting.

Many countries are involved in construction robot research, including China, where a tunnelling robot has recently been developed (People's Daily, 2001), although the research is mainly done in the US, Japan and Europe. Also, many companies from these countries are actively participating in research projects and commercializing the technology and there are now many robots in use for specific tasks, such as demolition or earthworks.

The Japan Industrial Robot Association (JIRA) defines robots according to their degree of autonomy. There are six levels:

- manual handling devices (manually controlled)
- fixed sequence robots (which can perform only a single sequence of operations)
- variable sequence robots (where the sequence of robot movements can be easily modified by an operator)
- playback robots (which are led through the task by a human operator)
- numerically controlled robots (which can be pre-programmed with an external programme)
- intelligent robots (which can interact with the environment).

In contrast, the Robot Institute of America (RIA) defines a robot as 'a pre-programmable, multifunctional manipulator, designed to move material, parts, tools or specialist devices through variable programmed motions for the performance of a variety of tasks' (Warszawski, 1999, p. 336). Construction robots can also be classified by their functions, broken down into sub-groups based on tasks performed. The groups are structure, finish, maintenance, demolition, and others (such as level marking).

In his Introduction to *Construction Robot Systems in Japan*, Shigeyuki Obayashi said:

> During some 20 years since the introduction of the term 'construction robot', more than 550 systems for the automation, unmanned operation and robotization of construction works have been developed and tried in Japan. These systems have been reported in the journals of academic societies or other technical books. Some of them have already been improved and commercialized under different names from the original ones, while research and development efforts for some others have been discontinued due to a change in the people's sense of values.
>
> (Council for Construction Robot Research, 1999, p. 1)

The *Catalogue of Construction Robots* (IAARC, 1999) lists and describes 76 working robots and automated machines in construction. The list of companies contributing to this catalogue includes companies from Japan, France, Sweden, the UK and USA. The robots are grouped under 13 headings. *Construction Robot Systems in Japan* (Council for Construction Robot Research, 1999) reports on 164 systems operated and maintained in some form or another. In a good example of the range and diversity of Japanese construction robot research, the systems included in this report were grouped under 17 categories. Table 26.1 outlines these reports.

Demands for improved performance, increased productivity, better quality and improved site safety have led many large construction companies to invest in researching and developing autonomous and robotic machines for on-site use. In the United States, for example, the construction industry employs 5% of the workforce but accounts for 11% of

Table 26.1 Construction robot categories (Council for Construction Robot Research, 1999; IAARC 1999)

Construction Robot Systems in Japan	*Catalogue of Construction Robots*
earth work	demolition (9)
foundation work	surveying (1)
crane work	excavation and earthmoving (9)
dam construction	paving (4)
concrete work	tunnelling (12)
mountain tunnel	concrete transport and distribution (10)
shield tunnel	concrete-slab screeding and finishing (4)
marine ship/underwater work	cranes and autonomous trucks (4)
placing of reinforcement/steel-framework	welding and positioning of structural steel members (4)
finishing work of building	fire-resisting and paint spraying (5)
prefabrication of reinforcement	inspection and maintenance (8)
pavement work	integrated building construction (6)
pneumatic caisson work survey	
inspection and monitoring	
maintenance/others	
element techniques	

Note: Number in brackets is the number of robots listed in each section of the *Catalogue of Construction Robots*.
Sources: *Construction Robot Systems in Japan* (Council for Construction Robot Research, 1999) and *Catalogue of Construction Robots* (IAARC 1999).

occupational injuries and 18% of all occupational fatalities. Despite this Everett and Saito (1994) found there is considerable resistance to the use of robotics due to fear of job losses being caused by the use of robots.

26.3 European research

The range of robot development projects in Europe can be seen in Table 26.2 below. There is a diverse and widely spread research effort underway, with extensive participation by both the public sector, through European Community science funding, and private sector partners. A number of projects are wholly within company R&D departments. Kochan (2000) discusses the autonomous climbing robot ROMA; an Italian climbing and walking robot, SURFY for inspection of industrial storage tanks; and the European community project, HEROIC, which is developing a robotic tool which uses high pressure water jet cutting technology or hydro-erosion, to cut and remove concrete that is to be repaired. Examples of some of these and other projects are discussed below.

Robot research at University Carlos III of Madrid in Spain includes two systems that automate specific construction tasks. The Robot Assembly System for Construction (ROCCO) is a development of a robotized system for building construction with partners from construction machinery companies and universities in three countries. It automates tasks such as assembly of bricks, pillars and other elements in order to build block walls and external enclosures (Gambao *et al.,* 2000). The ROMA is a multi-functional mobile robot, and is autonomous, self supporting and able to carry tools and build with steel frameworks. The main applications are building and bridge inspection, nut and screw torque verification and painting (Maas and van Gassel, 2001).

Road pavers already have a high level of automation, with regulating systems for conveying and spreading asphalt, for direction of motion and paving speed and the surface

Table 26.2 European robot research

Project	Institution
High tractive power wall-climbing robot	University of Hannover, Germany
EMIR Extended multi joint robot for construction applications	Fraunhofer IFF, Germany
Mobile bricklaying robot	University of Stuttgart, Germany
TAMIR Interior-finishing works building robot	Israel Institute of Technology, Israel
Construction robot for autonomous plastering of walls and ceilings	Lulea University of Technology, Sweden
ROMA Autonomous climbing robot for inspection applications in construction	University of Carlos III of Madrid, Spain
SURFY Low weight surface climbing robot	Catania University, Italy
Automated shotcrete robot	Switzerland
Fully automated cleaning system for vaulted glass structures	Fraunhofer IFF, Germany
CIRC Computer integrated road construction	Technische Universität München, Germany
FUTUREHOME	Universities: UK ×3, Japan ×3, Germany, Spain and Sweden ×1 each; Companies: UK ×2, Japan ×3, Germany, Spain, Finland and Netherlands ×1 each
ROCCO Robot assembly system for computer integrated construction	University of Karlsruhe and Technische Universität München, Germany
SAFEMAID Semiautonomous façade maintenance device	Technische Universität München, Germany: Company Germany ×1
Remote robotic NDT research project	Universities: UK and Italy ×1 each; Companies: Italy ×2, UK and Denmark
SAPPAR Stent automatic pile positioning and recording system	Lancaster University, UK
LUCIE Lancaster University computerized intelligent excavator	Lancaster University, UK
STARLIFTER	Lancaster University, UK

accuracy of the pavement. The ESPRIT European Research Project with partners from five countries, developed the asphalt paving RoadRobot, to fully automate the road paving process. The results were presented in 1996. The computer controlled RoadRobot was the first road paver capable of navigating and steering itself (RoadRobot, 2000), with all functions necessary for road pavement construction computer controlled, from asphalt reception, conveyance and spreading through to levelling, profiling and compacting. A later European project on Computer Integrated Road Construction involves seven industrial and academic partners from five countries. Peyret *et al.* (2000) describe the experimental results of an automated compactor that uses a Global Positioning System (GPS) and is laser-guided, capabilities that are now being offered by most of the heavy equipment manufacturers in some form.

A multinational project, funded through the European Community science budget, is FutureHome. FutureHome is the development of modular construction technology for residential buildings. The CAD design environment takes into account the on-site assembly work done by a robotized crane with several sensors working within an automated control strategy. FutureHome was a 3 year, 5 million Euro project with 15 partners in six European countries and is part of the Intelligent Manufacturing Systems (IMS) program. The project commenced in December 1998 and went for 3 years. The goal of FutureHome was to apply advanced manufacturing technology to the production of houses across Europe, aiming for construction cost and time savings of at least 30% and reduction in defects on completion of 60% (FutureHome, 2000).

Shotcrete application techniques are required in tunnelling projects. Automation of the application has enabled spraying by hand or manipulator. At the Swiss Federal Institute of Technology, Zurich, research is being done to develop a fully automated process, focusing on the wet shotcrete method so the user will have an effective tool to spray concrete shells. The new robot will use three different modes: manual spraying, semi-automated and fully-automated spraying. The fully-automated mode facilitates higher performance with less danger to workers' health. The quality control is inherent in the application process in regard to layer thickness, compaction and homogeneity (Maas and van Gassel, 2001).

There has been continuing research into developing robots for the cleaning of windows. The Autowind (autonomous window cleaning robot for high buildings) program, a 2 year venture, started in 1994, was aimed at developing a prototype cable-suspended platform to be used in cleaning and maintenance tasks. The main objective of the project was to provide a cost-efficient solution for security and quality of high buildings façade maintenance. One part of the project was the development of a sensor-based platform technology for stabilization, adherence and mobility of the cleaning unit against the glass surface of the building for wireless tele-operation (Robosoft, 1996).

Another window cleaning robot project has been running since 1989. The robot itself is a four axis system, with three linear and one rotary axis, and is 3.5 m long by 1.8 m high. It is controlled from a PC with a motion controller card. The robot is able to feel and detect the edges of each window so that it can accurately locate and clean using a specially designed head. The cleaning action is self-contained and based on a conventional squeegee blade. At present, the robot cleaning system has been constructed and tested at the Building Research Establishment in the UK (ARCOW, 1997).

While on-site application of robots is still in the development stage, in the future there will be many tasks able to be carried out by robots (Figure 26.1). A prototype of a bricklaying robot is being developed in Germany (University of Stuttgart) to perform the following tasks (Pritschow *et al.*, 1995):

- remove bricks or blocks from prepared pallets
- handle different kinds and sizes of bricks and blocks
- detect and compensate for material tolerances
- calibrate the brick or block position with respect to the tool centre point
- automatically dispense the bonding material
- erect brickwork at a high level of accuracy and quality.

An automated system for surface coating removal or preparation for restoration and/or construction, BIBER's principal application is the removal of roughcast or other old

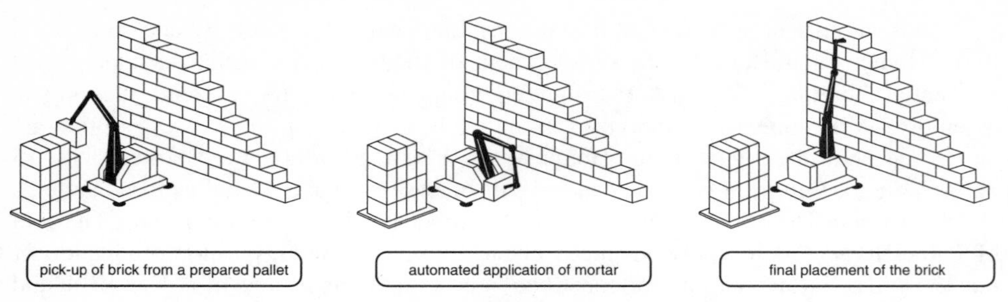

Figure 26.1 Operation sequence of automated bricklaying (CII, 2000).

coating by means of a brush. Applications are being developed for other areas, like scrubbing or brushing of large surfaces. The BIBER System consists of three components: a toolhead, a telescoping manned platform or another lifting unit, and a vacuum cleaner. The toolhead is the most innovative part of the whole system. It is the attachment and drive-housing for rapidly rotating tools, such as mills, brushes and cylinders. The system is operated by one machinist from the platform. It substitutes for as many as five to eight traditional workers, and the BIBER System can be set up by the operator and one other worker (BIBER, 2000).

26.4 United States robots

There has been an extensive development program in the US since the 1980s, and progress in US research has been rapid. When Skibniewski (1992) detailed construction robot research up to 1990 he listed robots developed in the US and Japan in two tables (pp. 348–9). US robot research included excavation, grading, concrete slipform, navigation (including GPS and laser guidance systems), interior partition and concrete masonry walls. Japanese robots listed included eight concrete work robots, three structural steel work robots, three inspection robots and three tunnelling machines.

By the end of the 1990s this had resulted in a number of commercially available robots, particularly for demolition and excavation work, and some sophisticated control and operation technology (Tucker, 1999). Lee *et al.* (1999) describe a prototype robotic excavation and pipe installation system developed by the Construction Automation and Robotics Laboratory at North Carolina State University (Table 26.3).

In another example, Stone and Pfeffer (1998) describe the sensing portion of a system used to convert an existing 30-tonne crane to computer control for automated placement of construction components. The system is designed to permit either telepresent or fully autonomous assembly of parts of buildings and industrial plants. With six degrees of freedom (directions of movement), the traditional crane cable and hook have been replaced by a cable-operated platform equipped with various manipulators. What makes this application unique is the scale of the robot: the crane's workspace is 40 m long, 23 m wide and 24 m high. This scale of operation led to a wireless control system. Stone and Pfeffer discuss the architecture of the sensor array needed to operate this large robot and the communications infrastructure needed.

Table 26.3 US robot research

Project	Institution/Company
ERMaS Experimental robotic masonry systems	North Carolina State University, USA
Robotic bridge painting system	North Carolina State University, USA
Robotic bridge maintenance system	North Carolina State University, USA
IMS Intelligent manufacturing system project	Universities: USA, Germany and Finland ×1 each; Japan ×2; Companies: USA and Switzerland ×1 each, Germany ×3, Japan ×4, Finland ×2
NGMS Next generation manufacturing system	Universities: USA ×7, Japan ×5, Germany and Ireland ×1 each; Companies: USA ×4, Japan ×10, Sweden ×3, Germany and Australia ×1 each
HIPARMS Highly productive and reconfigurable manufacturing system	Universities: USA ×1, Japan ×2, Germany and Finland ×1 each; Companies: Germany ×3, Japan ×4, Finland ×2, USA and Switzerland ×1 each
Robocrane project advanced welding system	National Institute of Science and Technology
Automated construction	University of Michigan, USA

26.5 Development of Japanese robots

In the 1980s Japanese contractors began development of construction robots and automation systems, motivated by the need to address a number of issues that were important to the future of the Japanese industry (Warszawski, 1999). In addition to improving safety and reducing labour requirements (Everett, 1994), automation and construction robots were viewed as a technology-oriented advertisement that might attract more young graduates and also allow more women to infiltrate the industry (Nielsen, 1992).

Robotic technology was introduced to Japan by the United States in 1968 (Cousineau and Miura, 1998). Originally used in the electronics and automobile industries, the success of Sony Corporation, Toyota and Nissan led to the Ministry of Trade and Industry (MITI) investigating the application of such technology in the construction industry. In one of the earliest projects, started in 1982, the WASCOR (WASeda COnstruction Robot) group was formed by Professor Hasegawa of Waseda University, Japan. There are 11 companies involved in the WASCOR group, which has completed four successive research projects to develop alternatives to various construction tasks. WASCOR IV, for instance, is an interior finishing assembly system for internal walls and ceilings (Handa *et al.*, 1996).

In 1983, the Building Materials and Construction Procedure Committee of the Architectural Institute of Japan (AIJ) internationalized this work through the International Symposium for Automation and Robotics in Construction (ISARC). This is now an annual event that attracts participants from many countries and publishes its proceedings.

Despite significant research and development effort, only a handful of Japanese construction robots have been used on building sites. Reasons for this include the complexity and non-uniformity of building structures (Nielsen, 1992; Warszawski, 1999) making improving productivity and reducing construction time and labour difficult; the tele-operated nature of the robots did not actually reduce labour because they require operators (Ueno, 1994). Other factors included the cost of production, inability to check for errors and mistakes, and the need for regular maintenance and periodical upgrading. In 1998, however, Cousineau and Miura recorded a total of 89 construction robots in use on construction sites in Japan.

26.5.1 Fireproofing robot

The fireproofing robot, initially developed by Shimizu Corporation and Kobe Steel in 1983, was the world's first construction robot (Cousineau and Miura, 1998). Fireproofing was chosen because the work is highly specialized, tedious and hazardous to health. The robot has undergone three generations of development. The first generation, the SSR-1, had a travel vehicle, manipulator arm, spray gun, oil hydraulic unit, control panel for travelling, control panel for spraying and a supply plant. The robot was battery powered and movement was guided by cables attached to the floor. The robot was pre-programmed, the playback control system allowing rapid repetition of operations along the length of the structure. Because the worker was 3 m away from the spray nozzle, exposure to chemicals was reduced. The SSR-1, however, had some serious problems: movement from one location to another was time consuming, detailed preparation work was required for each operation and changes in job scope, and workers needed to be trained in operational and maintenance procedures.

The improvements incorporated into SSR-2 included an ultrasonic sensor to allow the robot to automatically adjust its position in relation to its destination, omission of guide cables for automatic travel, a pressurized delivery system for the fireproofing mix to enhance the uniformity of spray, and an increment in the arm's length for greater coverage; in addition a special program enabling the robot to adjust its spray pattern in respect to the building structure was installed (Levy, 1990). Despite these improvements, problems remained. Movement from one location to another was still time consuming, particularly dismantling and re-assembling the robot for moving between storeys, and training was required to operate the robot. The main improvement was reduced fluctuation in the thickness of spray.

The robot's design was changed and the SSR-3 showed improved performance. Size and weight were reduced by removal of a hydraulic unit and integration of control panels, it had an off-line teaching system (OTS) which allowed manipulation by remote control, the time taken to 'teach' the robot was reduced to five minutes, the manipulator arm was improved, allowing it to trace the shape of structural components, the ability to spray while in motion allowed the SSR-3 to fireproof each floor in approximately one hour and a lifting device eliminated dismantling and re-assembling. While the robot eliminated the need for scaffolding and manual spraying, it still failed to reduce labour and created the new task of training workers (Cousineau and Miura, 1998). This is a good example of the incremental development approach used in robotic research.

26.5.2 Installation robot

A multifunctional robot for installing internal and external finishes such as glass panelling and precast concrete panels, was developed by the Fujita Corporation, a specialist contractor, a robot manufacturer and a construction equipment dealer (Miyama *et al.*, 1994). The robot has a high degree of flexibility with a working arm, end-effectors and a travelling carriage. The working arm, with six degrees of freedom, is fitted to the travelling carriage to lift prefabricated components vertically. An end-effector on the arm holds the component by pressurization and can be changed to suit the component. The carriage has a four-wheel steering system with 360 degree mobility and retractable wheels to increase stability so the robot can negotiate tight angles and obstacles on site.

This robot can be battery or mains powered. The operator controls the robot through a hand-held remote and the computer system mounted on the carriage (Miyama *et al.*, 1994). The speed of the arm is adjustable to permit installation work with a high level of accuracy. It is also equipped with two safety measures: a limiter on the working radius in respect to the handling load automatically self-terminates when the limit is exceeded, and the pressure required to hold the materials in place whilst stationary is monitored through sensors at the end-effector. A drop in level triggers an alarm, warning the operator.

On construction sites, this robot has been successful in reducing labour used and improving the quality of finishing and safety (Miyama *et al.*, 1994). The main disadvantage is the requirement for a level operating surface, however, this is not a major problem as finishing work is usually carried out after concrete has been set and levelled.

26.5.3 Concrete finishing robot

Examples of concrete floor finishing robots are 'Surf-Robo' by Takenaka Corporation and 'Flat-kun' by Shimizu Corporation. In 1984, Kajima Corporation and Tokimec pioneered the first generation of the concrete floor finishing robot known as Mark I (Levy, 1990; Cousineau and Miura, 1998). It had a travel vehicle, a three-blade steel trowel end-effector, manipulator arm, host computer, control panel and an electric power panel, with a navigation system for automatic operation. Although satisfying many requirements, the robot was heavy and required time consuming and complex preparation by the operator.

In the development of Mark II in 1986 (Levy, 1990), the components were reduced to a travel vehicle, trowel end-effector and bumper. The navigation system had a gyrocompass, measuring rollers and an 8 bit microcomputer. The Mark II robot also had an on-board diagnostics system to assist the operator in case of failure. After information such as the length and width of coverage, the lapping width, and speed of action had been input to the control panel, the robot carried out the task. An obstacle sensor allowed the robot to identify and avoid columns and other obstructions. The robot increased the productivity rate from 300 to 500 square metres per hour (Cousineau and Miura, 1998).

The third generation modified Mark II had a cable handler, increasing the number of components to four. The robot was marketed as 'Kote-King' and in 1990 Tokimec sold 30 sets at US$50 000 each (Cousineau and Miura, 1998). Tokimec later produced a simplified version of the Kote-King known as 'Robocon', the main improvements being a further reduction in weight, a quieter engine and an increased productivity rate of about 700–800 square metres per hour. Robocon was marketed at half the price of Kote-King.

26.6 Automated building systems

By the early 1990s Japanese contractors had recognized the potential for integration of construction activities through automation, and this led to the development of Automated Building Systems (ABS), based on the Flexible Manufacturing System (FMS) that was being adopted in factories (Skibniewski and Wooldridge, 1992). When seen as a manufacturing system, building is a series of repetitive tasks that can be done using automated equipment and robots. ABS emphasize this characteristic of the industry, and typically have an overhead shelter or hat-truss, an automated jacking system, an automated material handling system, and a central control station. The overhead shelter or hat-truss (usually the top floor) is constructed first and elevated to the required working height for the construction of lower floors using an automated jacking system. Once the structure is completed, the hat-truss is jacked up again for the construction of upper floors. The hat-truss provides a safe environment allowing construction work to be carried out below. Materials are moved horizontally and vertically by an automated material handling system consisting of automated lifts and conveyors.

In automated systems, major components such as columns, beams and slabs are fabricated off-site, allowing use of Just-In-Time (JIT) (Tucker, 1992; Webster, 1994), to minimize non-value adding activities in the handling process, such as those found in the traditional method of receiving, inspecting and storing of materials (Akintoye 1995). The new system allows materials to be ordered and purchased in accordance with daily requirements, with improved quality and reduced wastage.

Since the late 1980s, the Japanese 'Big Five' contractors have developed their own ABS. A total of eight systems are available, seven of which are designed for structural steel construction and one for precast concrete construction. These systems have many features in common, for example, Fujita has developed an ABS aimed at providing a high quality building at a low cost. This system first builds the uppermost floor of the building on the ground, the various automated mechanical devices necessary for structural work are installed, and the structural components of the building are constructed one floor at a time. The floor is then raised by a jack system to the next level (Tanijiri *et al.,* 1997). Although structural steel framing is the common form of construction in Japan it is not in other countries (notably Australia, the United Kingdom and China), and this means that Kajima and Obayashi will probably be better placed than the other contractors in these markets

Table 26.4 Automated Building Systems in Japan (Cousineau and Miura, 1998)

System	Company	Year	Structure
Push-Up	Takenaka Corp.	1989–1991	structural steel
SMART	System Shimizu Corp.	1991–1994	structural steel
ABCS	System Obayashi Corp.	1991–1994	structural steel
T-Up	System Taisei Corp.	1992–1994	structural steel
MCCS	Meada Corp.	1992–1994	structural steel
Akatsuki 21	Fujita Corp.	1994–1996	structural steel
AMURAD	Kajima Corp.	1995–1996	steel-reinforced concrete
Big Canopy	Obayashi Corp.	1995–1997	pre-cast concrete

because of their development of an ABS for the steel-reinforced concrete buildings that are typically found in these countries.

26.6.1 The SMART System

Developed since 1990 by Shimizu Corporation, the Shimizu Manufacturing System by Advanced Robotics Technology (SMART System) was the world's first automated building system to be used in the construction of an office building, the Juroku Bank Building in Nagoya, Japan, begun in 1991 and completed in 1994.

The SMART System is for structural steel buildings. Yamazaki and Maeda (1998) describe the SMART System as an integrated system that automates erection and welding of steel-frames, laying of concrete floor planks, installation of exterior and interior wall panels, and installation of other units. It also has an information management system to integrate design, planning and management of the project.

The construction process begins with assembly of the hat-truss at the ground floor. An operating platform for transport and assembly of structural components is built at the same time. The hat-truss is then supported and manoeuvred by a lifting system made up of four jacking towers, operated by hydraulic cylinder jacks. After the hat-truss is completed, it is jacked up to working height for the construction of the first storey. It is jacked up again once the lower floor is completed and the cycle is repeated until the building structure is finished. The floor cycle is about 9 days per floor. The hat-truss is lifted after the external panels of the lower floors are installed to form a protective shield for finishes work. The hat-truss is also enclosed with a protective sheet. During lifting, the hydraulic jacks on the towers are synchronized: the lift-up mechanism is first lifted and attached to the upper storey beams, with that in place the jacks are operated again to lift the tower and the working platform to the next working height. The whole procedure is completed when the tower base is secured to the upper storey beams. Each lift is about 4 m high with a lifting time for the 1200-tonne platform of about 1.5 hours (Maeda, 1994).

Structural components are prefabricated off-site, and on arrival at the site they are handled using the Automated Conveying System, developed for the SMART System, comprising 10 overhead travelling cranes, five trolley hoists (for horizontal conveying) and a crane for vertical lifting. Prefabricated components have bar codes for identification so, once scanned, the computer identifies the storey and position to which it is to be delivered. The component is then transported by the vertical lifting crane, transferred to the overhead cranes and moved (horizontally) to its position. Management of the construction process is by an information control system operated from a central station inside the hat-truss. The whole process is co-ordinated from the operating rooms on the ground floor and the working platform. The computer displays the movement of the component in both control rooms throughout its journey.

The SMART System has been a success for Shimizu. Labour requirements were reduced by 30%, and by up to 50% when the workers became competent in handling the systems. The workload on site management was reduced by the information control system, improving productivity. Construction waste saving was estimated at 70% (700 tonnes) of the normal load in a typical building (Maeda, 1994). During construction of the Rail City Yokohama Building in 1994 the floor cycle was reduced from 9 to 7 days (Cousineau and Miura, 1998).

26.6.2 The T-Up System

The T-Up System is one of the seven automated building systems designated for steel structures. Developed by Taisei Corporation and Mitsubishi Heavy Industries, the Totally Mechanized Construction System (T-Up System) was first used in 1992 for construction of Mitsubishi Heavy Industries Yokohama Building (Phase 1). It is a combination of a 'self-building system' and a 'project management system' (Cousineau and Miura, 1998). The three components of the system are the support, the base and a manipulator. The T-Up System can be varied in design to suit the building structure.

T-Up has a hat-truss, a production platform, two 150-tonne jib cranes, two 10-tonne overhead cranes, guide columns with 300-tonne hydraulic jacks, and a control room. The construction process begins with erection of the building core, built of either structural steel or reinforced concrete, as the support (Skibniewski and Wooldridge, 1992; Cousineau and Miura, 1998). Next the production platform (base), weighing approximately 2000 tonnes, is constructed at the ground level. Eight guide columns and hydraulic jacks are positioned and activated to elevate it to the desired floor. The jib cranes and overhead cranes are fixed to the top and underside of the platform, respectively.

The floor cycle of the T-Up System is 3 days per floor (Sakamoto and Mitsuoka, 1994). On the first day, the building core is constructed using the jib cranes. Structural steel members are conveyed into the interior of the hat while prefabricated concrete wall panels are installed along the perimeter by overhead cranes. On the second day, the steel members of the core are tightened and welded while the building structure below the hat-truss is constructed. By the third day, the welding of structural steel members is complete and the hat is ready for lifting. The climbing process is synchronized, within a difference of 5 millimetres. The system is sequentially jacked at 800 mm per stroke to a working height of about 4 m. Using the building core as a support base, it takes five strokes to complete the entire lifting phase. Once in position, the lower floor is constructed by repeating the floor cycle process. After the building structure is completed, the production platform is lowered and forms the building roof. The systems and components are then dismantled and lowered to the ground.

Post-construction evaluations carried out at the Mitsubishi Heavy Industries Building showed improved safety and an increased speed of construction. Construction time was reduced from 30 months to 24 with only 6 months required to build the steel structure (Sakamoto and Mitsuoka, 1994).

26.6.3 Roof push-up method

The roof push-up method, developed by the Takenaka Corporation, was applied to the last three floors of the 12-storey Mitsui Building in 1990 (Skibniewski and Wooldridge, 1992; Cousineau and Miura, 1998). Improvements allowed this system to be used in construction of the 14-storey Dowa Fire Insurance Company Building in 1993 (Cousineau and Miura, 1998). The method is based on constructing the top floor first and jacking it up for the construction of lower floors; the roof floor then serves as the working platform.

Using the push-up method, the roof floor and its supporting columns were lifted separately by independent hydraulic jacks. Once the construction for the desired storey was completed, the roof floor was jacked. The floor was supported by temporary supports while the supporting columns were jacked up using column jacks. At the same time, the column was replaced by a new structural column inserted from below. Safety concerns required that this process be carried out on a column-by-column basis.

Evaluation showed that extra work was required for the joining and dismantling of supporting columns each time the roof was lifted and as a result, a simplified push-up mechanism was developed for vertical lift, with a weight load of 120 tonnes per column.

Construction procedures were altered to make use of the push-up mechanisms. Supporting columns were eliminated and the roof floor loading was transferred to the structural columns of the buildings by installing a jib crane on the roof top and lifting the columns to join with the lower floors. On completion, the roof floor was jacked up and rested on the upper structural columns.

The entire climbing process was automated and controlled by computer. To further improve the working conditions, the working space was enclosed with a mesh sheet. Modification enabled the construction work to be carried out under sheltered conditions and with improved safety. The use of automation eliminated movement of heavy components by the workers as well as reducing required labour content, thus improving overall efficiency and productivity.

Evaluation of the case studies showed that further improvements would be necessary for wider application. The main limitation of an automated building system is the standardization of building design and components; any variation in plan shape will mean additional costs. Matsumoto (1998) suggests that these systems are only ideal for simple buildings of more than 20 storeys if they are to be cost effective. It is one area identified by Ueno (1994) that requires further research and development.

26.6.4 The Big Canopy System

The Obayashi Corporation developed the Automated Building Construction System (ABCS), first used in the construction of a 10-storey dormitory in 1989 (Cousineau and Miura, 1998). The ABCS is a weatherproof automated factory sitting on the building top and incorporating all the construction tasks. The factory floor is then raised when the lower floor is completed (Webster, 1994).

The ABCS consisted of a Construction Factory, an automated storage system and automated delivery system (Shiokawa and Noguchi, 1993, cited in Cousineau and Miura, 1998). It is a production platform equipped with overhead cranes, jacking system and control station. The overhead cranes lift the structural steel members to the desired floor and installation point where jointing is then carried out using welding robots. Its unique feature is that it allows the working platform to be elevated to two storeys instead of one, by alternating the jacks for the climbing process, with half used for the first level and the other half for the next level. The administration of the work process and environment is monitored using the Work Flow Control System (WFCS) which handles the co-ordination of the daily construction activities, and the Equipment Control System (ECS) which focuses on equipment use and status (Cousineau and Miura, 1998).

Unfortunately the labour requirements were high, work productivity was not improved by the use of welding robots, and the system was costly to operate. Addressing these problems, the Obayashi Corporation developed the Big Canopy Construction System, the world's first automated system applied to a precast concrete structure (Cousineau and Miura, 1998). Trialled in the construction of the 26-storey Yachiyodai Condominium in 1995, the system is distinguished by its four tower masts and a gigantic canopy at the top. The main improvement in the design was the transfer of canopy loading to the masts instead of the structural columns of the building. The masts are located outside the building on the corners and this allowed a higher degree of design flexibility (Cousineau and Miura, 1998).

The Big Canopy works by lifting prefabricated material from the ground floor to the constructing floor and conveying it to the installation point by gantry cranes fixed to the underside of the canopy. Workers using a hand-held joystick control the manoeuvring of these components; use of this system reduces the technical training required of the workers. Once the precast members are joined, the canopy is jacked up two storeys and the work cycle is repeated. The system produced an improvement in overall productivity and a 60% reduction in labour requirement (Cousineau and Miura, 1998).

Use of the Big Canopy showed a number of other positive results. Firstly, the construction period was reduced and productivity improved, with the floor cycle reduced from 14 days to 5 or 6 days per floor for the upper storeys. The system also allowed the gradual reduction of labour force from 120 at the initial stages to 50 during operation of the canopy and to just 25 during the latter stages of construction. The volume of construction waste generated was reduced, with waste disposal carried out only once every 2 days instead of the five times per day common with conventional construction methods. Working conditions were improved as workers were under cover and delays due to bad weather were avoided.

Finally, the quality of the work was improved through the use of prefabrication and quality control. Quality measurement, based on zero defects, performed on site found a 100% result in respect of precast members. Zero defects in this context means that when the precast members are joined together they are square. In 1997 Obayashi won the contract for the 33 storey, 42 000 m^2 DBS Square development in Singapore, the first time an ABS had been used outside Japan.

26.6.5 Future Automated Construction Efficient System

The Future Automated Construction Efficient System (FACES) is an all-weather, highly automated, computer-controlled system for constructing tall buildings in congested urban environments (Penta-Ocean, 2001). Developed by Japanese construction contractor Penta-Ocean, FACES was used for the first time in the construction of an office building and community centre complex in Hamacho, in Tokyo's waterfront district, in 1997. The FACES System consists of a steel frame unit that covers the length and width of the building to be constructed covered by an all-weather canopy that houses a computer control centre, crane robots and other automated construction equipment, all attached to the ceiling of the structure. As the construction of each floor is completed, the frame is raised up on pillars that become part of the superstructure, to the next floor level, where construction resumes. This sequence is repeated until the top floor is completed. The ceiling structure, the canopy, equipment and computer centre are then disassembled and removed from the structure.

26.7 Conclusion

This chapter has outlined the development and current status of construction robotic and automation technology. Despite the industry's reputation for slow technological development, the research that has developed these systems has been underway for two decades and is beginning to produce equipment used on-site or, in the case of automated building systems, replacing the conventional site.

Construction robots have demonstrated their potential to greatly improve the three building parameters of time, cost and quality. While it is still early days for these technologies, their pace of development is rapid and becoming self-reinforcing, so that as one capability is refined another becomes possible. Combined with the advances in IT and computer integration that were a feature of the 1990s, construction robots are beginning to appear on sites in specialized applications.

The large scale automated building systems developed in Japan are an order of magnitude beyond the research efforts of Europe and the US. These systems are now being deployed around the Asian market and have the potential to transform the construction of office and high-rise buildings in the same way that the production line transformed car manufacturing nearly 100 years ago. Site productivity, safety and efficiency can be dramatically improved with these systems, due to the reduction in labour content. The use of prefabricated components reduces construction cost and delivers a better quality product. The computerized control systems and overhead canopies allow co-ordination of site activities and minimizes delays. This in turn reduces claims, disputes and litigation. The reduced labour content coupled with the reduction in time significantly increases productivity. As the Japanese contractors become more familiar with use of these systems, their cost of operation (and more importantly the cost per square metre of building) will come down.

Technological development tends to find and follow its own path, which is often not obvious until the right combination of performance and price becomes available. Construction robotic and automation technology is now on a path of rapid development and increasing capability, as is robotic research generally. This will result in outcomes that will surprise many in the industry who have regarded these new technologies as largely irrelevant to the serious business of building and construction; in fact this research has shown that many of the tasks and activities associated with building and construction can be both mechanized and automated.

References and bibliography

Akintoye, A. (1995) Just-in-time application and implementation for building material management. *Construction Economics and Management*, **13** (2), 105–13 (London: E & FN Spon).

Alfares, M. and Seireg, A. (1996) An integrated system for computer-aided design and construction of reinforced concrete buildings using modular forms. *Automation in Construction*, **5** (4), 323–41.

ARCOW (1997) *Automated Robot Cleaning of Windows*. www.deltatau.com/clients/The%20ARCOW%20Project.htm

Bernold, L.E., Abraham, D.M. and Reinhart, D.B. (1990) FMS approach to construction automation. *Journal of Aerospace Engineering*, **3** (2), 108–12.

BIBER (2000) *Surface preparation system.* http://www.iaarc.org/frame/publish/biber.htm

Bock,T. and Prochiner, F.O. (1999) Automation systems in housing construction. In: *Proceedings ISARC '99* (Madrid).

CII (Construction Industry Institute) (2000) *Mobile Bricklaying Robot.* Emerging Construction Technologies. www.new-technologies.org/ECT/Other/brickrob.htm

Council for Construction Robot Research (1999) *Construction Robot Systems in Japan* (Tokyo: CCRC).

Cousineau, L. and Miura, N. (1998) *Construction Robots: The Search for New Building Technology in Japan* (ASCE Press).

Everett, J. (1994) Ergonomics, health and safety in construction: opportunities for automation and robotics. In: Chamberlain, D.A. (ed.) *Automation and Robotics in Construction XI*, Proceedings of the 11th International Symposium on Automation and Robotics in Construction (ISARC), Brighton, UK (Elsevier Science BV), pp. 19–26.

Everett, J. and Saito, H. (1994) Automation and robotics in construction: social and cultural differences between Japan and the United States. In: Chamberlain, D.A. (ed.) *Automation and Robotics in Construction XI*, Proceedings of the 11th International Symposium on Automation and Robotics in Construction (ISARC), Brighton, UK (Elsevier Science BV), pp. 223–230.

FutureHome (2000) www.iaarc.org/frame/features/projects/futurehome.htm

Gambao, E., Balaguer, C. and Gebhart, F. (2000) Robot assembly system for computer-integrated construction. *Automation in Construction*, **9** (5–6), 479–87.

Handa, M., Hasegawa, Y., Matsuda, H., Tamaki, K., Kojima, S., Matsueda, K., Takakuwa, T. and Onoda, T. (1996) Development of interior finishing unit assembly system with robot: WASCOR IV research project report. *Automation in Constructions*, **5** (1), 31–8.

Hasegawa, F. and the Shimizu Group Fs (1988) *Built by Japan: Competitive Strategies of the Japanese Construction Industry* (New York: John Wiley & Sons).

Howe, A.S. (1997) Designing for automated construction. In: *Proceedings of the Fourteenth International Symposium on Automation and Robotics in Construction* (The Robotics Institute), pp. 167–76.

IAARC (1999) *Catalogue of Construction Robots* (London: International Association for Automation and Robotics in Construction).

Kochan, A. (2000) Robots for automating construction – an abundance of research. *Industrial Robot: An International Journal*, **27** (2), 111–13.

Lee, J., Lorenc, J. and Bernold, L.E. (1999) Saving lives and money with robotic trenching and pipe installation. *Journal of Aerospace Engineering*, **12** (2), 43–9.

Levy, S.M. (1990) *Japanese Construction: An American Perspective* (Van Nostrand Reinhold).

Lorenc, S.J., Handlon, B.E. and Bernold, L.E. (2000) Development of a robotic bridge maintenance system. *Automation in Construction*, **9** (3), 251–8.

Maas, G. and van Gassel, F. (2001) *International Status Report on Aspects of FutureSite*, CIB Publication 265.

Maeda, J. (1994) Development and application of the SMART system. In: Chamberlain, D.A. (ed.) *Automation and Robotics in Construction XI*, Proceedings of the 11th International Symposium on Automation and Robotics in Construction (ISARC), Brighton, UK (Elsevier Science BV), pp. 457–64.

Matsumoto, S. (1998) R&D activities on future construction systems in Japan. In: Poppy, W. and Bock, T. (eds) *Automation and Robotics: Today's Reality in Construction,* Proceedings of the 15th International Symposium on Automation and Robotics in Construction (ISARC) (Munich: Messe Munchen International), pp. 89–93.

Miyama, A., Sawada, M. and Suzuki, I. (1994) Development of a multifunctional robot for installation of building exterior and interior finishing materials. In: Chamberlain, D.A. (ed.) *Automation and Robotics in Construction XI*, Proceedings of the 11th International Symposium on Automation and Robotics in Construction (ISARC), Brighton, UK (Elsevier Science BV), pp. 709–17.

Nielsen, R.W. (1992) Construction field operations and automated equipment. In: Knasel, T.M. (ed.) *Automation in Construction,* **1** (1) (Elsevier), pp. 35–46.

Penta-Ocean (2001) *Future Automated Construction Efficient Systems.* www5.mediagalaxy.co.jp/penta-ocean/english/index.html

People's Daily (2001) China's Robot Development Industry on Fast Track. People's Daily, 19 February. http://english.peopledaily.com.cn/200102/19/eng20010219_62758.html

Peyret, F., Jurasz, J. Carrel, A., Zekri, E. and Gorham, B. (2000) The computer integrated road construction project. *Automation in Construction*, **9** (5–6), 447–61.

Pritschow, G., Dalacker, M., Kurz, J. and Gaenssle, M. (1995) Technological aspects in the development of a mobile-bricklaying robot. In: Budny E. (ed.), *Automation and Robotics in Construction*, Proceedings of the 12th International Symposium on Automation and Robotics in Construction (ISARC) (Warsaw) pp. 281–90.

RoadRobot (2000) www.iaarc.org/frame/publish/roadrobot.htm

Robosoft (1996) *AutoWind: Autonomous Window Cleaner Robot for High Buildings.* www.robosoft.fr/SERVICE/03_WindowCR/01AutoWind/AutoWind.html

Sakamoto, S. and Mitsuoka, H. (1994) Totally mechanised construction system for high-rise buildings (T-Up System). In: Chamberlain, D.A. (ed.) *Automation and Robotics in Construction XI*, Proceedings of the 11th International Symposium on Automation and Robotics in Construction (ISARC), Brighton, UK (Elsevier Science BV), pp. 465–72.

Skibniewski, M.J. (1992) Robot implementation issues for the construction industry. In: Rahimi, M. and Karkowski, W. (eds) *Human–Robot Interaction* (London: Taylor and Francis), pp. 347–66.

Skibniewski, M.J. and Wooldridge, S.C. (1992) Robotics materials handling for automated building construction technology. In: Knasel, T.M. (ed.) *Automation in Construction*, **1** (3) (Elsevier), pp. 251–266.

Slaughter, S.E. (1997) Characteristics of existing construction automation and robotics technologies. *Automation in Construction*, **6** (2), 109–20.

Sontheimer, R. and Saal, H. (1998) Automated positioning of welding torch with the fabrication of steel structures with robots. *Journal of Constructional Steel Research*, **46** (1–3), 39.

Spencer, R. (2001) A driving force: use of robotics remains strong in auto industry. *Journal of Robotics World,* **19** (1), 18–21.

Stone, W.C. and Pfeffer, L.E. (1998) Automation infrastructure system for a robotic 30-ton bridge crane. In: Proceedings *Robotics '98 Conference* (American Society of Civil Engineers), pp. 195–201.

Tanijiri, H., Ishiguro, B., Morishima, Y. and Takasaki, N. (1997) Development of automated weather-unaffected building construction system. *Automation in Construction*, **6** (3), 215–27.

Tucker, R.L. (1992) Japanese construction industry. In: Knasel, T.M. (ed.) *Automation in Construction*, **1** (1) (Elsevier), pp. 27–34.

Tucker, R.L. (1999) Construction automation in the U.S.A. In: *Automation and Robotics in Construction XVI*, Proceedings of the 16th International Symposium on Automation and Robotics in Construction (ISARC) (Madrid).

Ueno, T. (1994) A Japanese view on the role of automation and robotics in next generation construction. In: Chamberlain, D.A. (ed.) *Automation and Robotics in Construction XI*, Proceedings of the 11th International Symposium on Automation and Robotics in Construction (ISARC), Brighton, UK (Elsevier Science BV), p. 641.

Venables, R. (1994) UK construction automation and robotics. *Journal of the Industrial Robot*, **21** (4), 3–4.

Warszawski, A. (1999) *Industrialised and Automated Building Systems: A Managerial Approach* (London: E&FN Spon).

Webster, A. (1994) *Technological Advance in Japanese Building Design and Construction* (New York: ASCE Press).

Weimer, G. (2001) Robots take over automotive plant floors. *Material Handling Management*, **56** (7), 55–6.

Wing, R. (1993) Recent Japanese progress in construction robotics. *Journal of the Industrial Robot*, **20** (4), 32–4.

Yagi, J. (1999) Automation and robotics in construction in Japan, state of the art. In: *Proceedings ISARC '99* (Madrid).

Yamazaki, Y. and Maeda, J. (1998) The SMART system: an integrated application of automation and information technology in production process. *Computers in Industry*, **35** (1), 87–99.

27

Some innovations in building services

Edward L. Harkness and Rick Best*

Editorial comment

The following extract is from the Eighth Edition of *The Australian Builders & Contractors' Price Book* (Mayes, 1914, p. 291) – it provides a marvellous snapshot of the state of engineering services in buildings less than 100 years ago.

Air–Gas Plants

For Lighting, Heating, Cooking

> The mixture called air–gas is said to contain 95 per cent of air and 5 per cent of petrol vapour; it is also said to be non-explosive. It gives a brilliant white light when used with special small incandescent burners and mantles; and a blue non-luminous flame, or vaporous heat, when used for stoves and heating.
>
> Cost of burning 5 lights of 60 c.p. burners at 1d. per hour, or a 40 c.p. light burning 25 hours cost 3d.
>
> The motive or driving power is by automatic regulated weights, which are wound up daily like a clock, the average unwinding of which is about 4 ft. 6 in. per hour, according to consumption more or less.

An air–gas plant to power 10 lights, each 60 c.p., cost around £26 in 1914, while a 200 light plant cost £153.

The striking point is that less than a century ago some buildings that were lit and heated with the latest technology were reliant on a sort of clockwork apparatus that required daily winding, and cooking, lighting and heating still depended on direct combustion of fuel at the point of use. Air–gas plants were listed as being suitable for 'homesteads, churches,

* University of Technology Sydney

laundries, factories, etc.' Today, of course, most of the things in Mayes' 1914 price book seem quaint, yet many others are remarkably similar: reinforced concrete, bricks, plumbing fittings, timber framing, roof tiles and so on.

An area that has seen enormous advances, however, is that of building services – the 1914 book contains very little related to space conditioning: a few fans, and a variety of gas and solid fuel heaters and boilers. Communications is similarly treated: a 'long distance telephone, wall pattern, half to 2 miles, two magnet generator' cost £2 10s, while the same item but '2 to 100 miles, as used by Government' was £4 10s – there is little more offered. All of this is a far cry from the sophisticated, highly integrated systems now available that control most aspects of the indoor environment, and provide us with virtually instantaneous worldwide communications through telephones, the Internet, satellite links, wireless networks and the like.

Building owners and occupants now demand the highest levels of engineering services in their buildings and building value is very closely related to the standard of services provided or contained in a building. Areas of most concern are fast and reliable vertical transportation, maintenance of the indoor environment (including indoor air quality and suitable lighting) such that productivity is maximized, communications infrastructure (e.g., data cabling, riser space) that is extensive, expandable and easily reconfigured, and building control systems (including fire and security systems) that allow the occupants to use the building to greatest effect as they pursue their normal activity, whether that is commercial, residential, institutional or whatever. Buildings that fulfil these needs best will be of greater value to their owners and be more attractive to those who may occupy them.

In the last 100 years there have been huge advances in all facets of building services, and the following chapter looks briefly at just a few of those that have occurred in the recent past, and looks ahead at some possibilities for the future.

27.1 Introduction

Innovations have been made in the systems used to provide shelter for man since the earliest times. Some innovations, whilst exploiting available technology, have proved to be environmental liabilities. Innovations of the late twentieth century include some that are considered to be energy-efficient and cost-effective as well as being environmentally sustainable and safe. As is the case with so many recent developments, many of the innovations are related to the unprecedented advances made in computing over the past 30 years. Just a few of the major developments made in recent years in building services, and some that are expected to emerge in the near future, are discussed below.

27.2 Heating, ventilation and air conditioning

Air-conditioning became necessary as the value of land in cities justified constructing more floors below a maximum permissible total building height. Not only did the reduced area of operable window limit cross ventilation, but ventilation through windows became ineffective when adjacent buildings were built close to or up to site boundaries. Light wells served to provide some ventilation in low-rise developments, but high rise buildings required ducted mechanical ventilation and, eventually, air conditioning.

It is only since the 1930s that the technology that can provide a cool indoor environment in summer with acceptable humidity levels has been available (Parlour, 1994). While there has, since then, been much development of the technology, a number of problems have emerged including sick building syndrome (SBS), ozone-damaging refrigerants, high energy use contributing to high operating costs and air pollution, and high capital cost associated with complex systems. In spite of the great improvements achieved the occupants of many modern buildings complain about the air conditioning. Some recent advances are described below.

27.2.1 Absorption chillers

Absorption chillers produce chilled water using heat instead of electricity as the main energy source; they provide an efficient solution in cases where there is surplus heat available and there is a requirement for cooling or air conditioning. Absorption chillers are, therefore, a key technology in district energy and some cogeneration systems as they utilize thermal energy (hot water or steam) that would otherwise be wasted, to produce cooling. Their other advantage is that they operate without the use of environmentally harmful refrigerants such as CFCs and HCFCs. Instead they use a refrigerant/absorbent pairing, typically water and lithium bromide, or ammonia and water. Ammonia has zero global warming impact and has no effect on the ozone layer, making it the most environmentally acceptable refrigerant. A schematic of a typical water/lithium bromide chiller is shown in Figure 27.1.

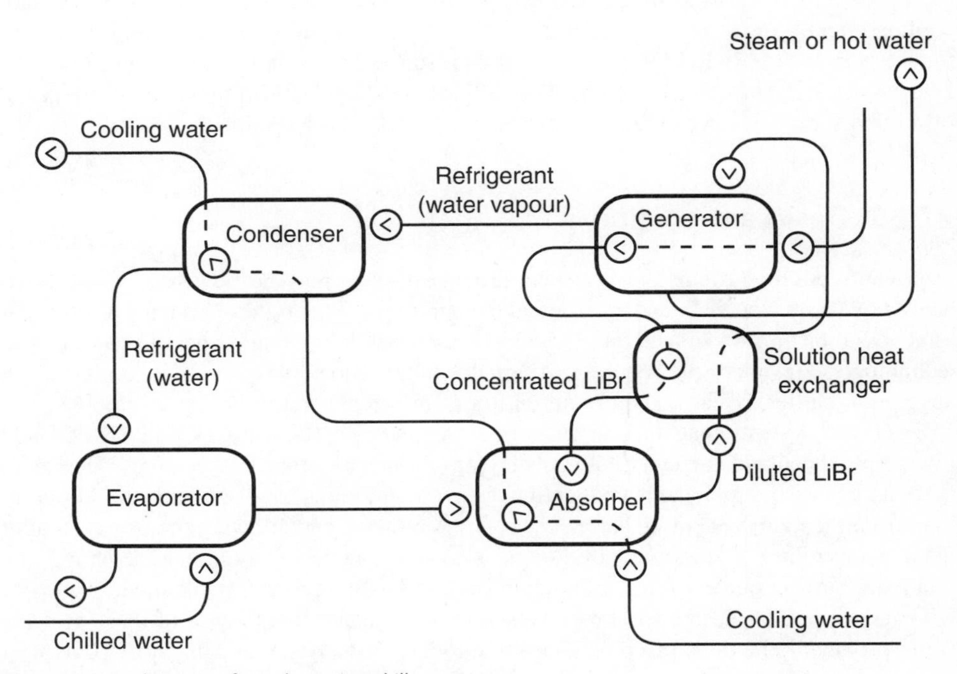

Figure 27.1 Schematic of an absorption chiller.

The absorption cycle illustrated above can be summarized as follows:

- steam or hot water is supplied to the generator boils a solution of refrigerant (water) and absorbent (lithium bromide) – refrigerant is released as a vapour and the absorbent solution is concentrated
- the refrigerant vapour is condensed using external cooling water
- liquid refrigerant passes into the evaporator where the drop in temperature cools the refrigerant – the cooler refrigerant draws heat from the return chilled water, cooling the chilled water and vapourizing the refrigerant
- refrigerant vapour is drawn from the evaporator to the absorber due to the reduction in pressure caused by the absorption of the refrigerant into the absorbent
- external cooling water removes the heat released when the refrigerant vapour becomes liquid in the absorption process
- the diluted absorbent is circulated back to the generator through the heat exchanger
- the heat exchanger transfers heat from the relatively warm concentrated solution being returned from the generator to the absorber and the diluted solution being circulated back to the generator – this reduces both the amount of heat which has to be added to the generator and the amount of heat which has to be rejected from the absorber.

Absorption chillers may be one or two stage; two-stage chillers need a higher quality thermal source as the heat derived from the first-stage generator is used to boil out additional refrigerant in a second generator. As a result the co-efficient of performance (COP) for two-stage chillers is considerably higher.

The COP is a measure of the efficiency with which the chiller transforms heat into cooling effect – it can be simply described as the 'energy out' (the energy removed from the refrigerant liquid) divided by the 'energy in' (the energy supplied to the system by the heat source, i.e., hot water or steam). Typically a single-stage absorption chiller has a COP of around 0.7, while two-stage systems may have a COP of up to 1.1.

27.2.2 Desiccant cooling

Desiccant cooling systems are also heat-driven and thus may be powered by gas, solar energy or waste heat such as that from a cogeneration plant. In these systems a desiccant removes moisture from the air, which is then cooled using conventional cooling technologies (evaporative cooling, normal compression type cooling or even an absorption chiller). Heat is then required to dry out (regenerate) the desiccant.

Desiccant systems that include evaporative coolers do not use CFCs or HCFCs and reduce peak demand for electricity as they are driven by other heat sources. Desiccant systems can also supplement conventional air conditioning by displacing the humidity load from the main system, reducing energy requirements and allowing the use of smaller plant. Controlling humidity in buildings is a key function – Figure 27.2 shows the relatively narrow range within which comfort and health are best maintained.

Under the US Desiccant Cooling Program it is estimated that the use of these systems could produce reductions of up to 24 million tonnes of CO_2 per annum by 2010 as well improving indoor air quality in buildings (NREL, 2001).

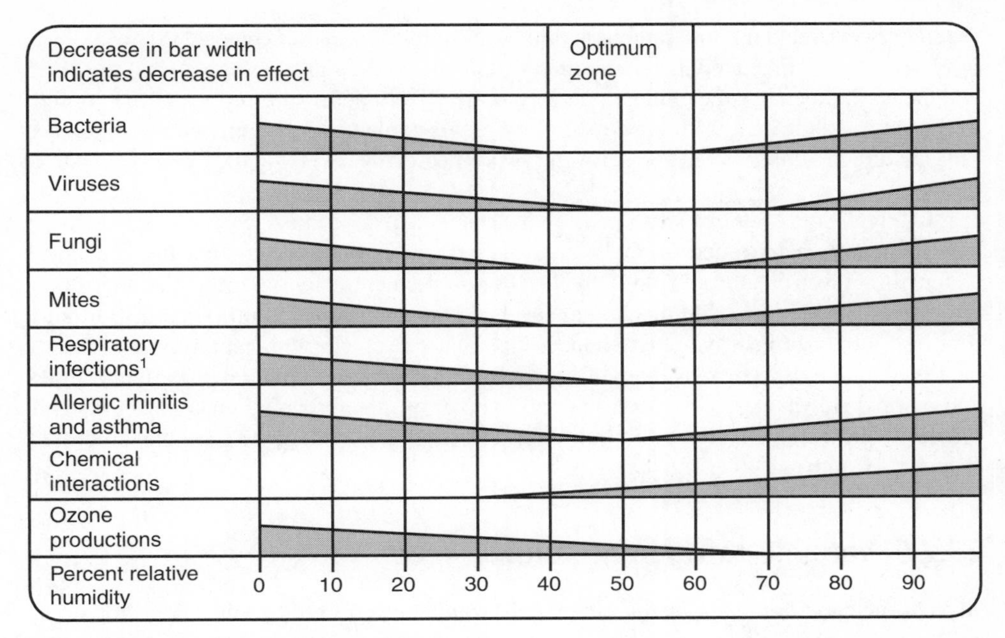

[1] Insufficient data above 50% R.H.

Figure 27.2 Effect of room relative humidity on selected human health parameters (NREL, 2001).

27.2.3 Chiller sets

In the 1960s a single, very large chiller (with an equally large stand-by chiller) met the cooling requirements of many major buildings. At the beginning of the twenty-first century it is acknowledged that most buildings require a number of chillers of different sizes, so as to be able to match the cooling requirements of the building whilst maintaining the chillers at approximately peak load, at which they operate most efficiently.

The innovation, therefore, in terms of chiller design in the last decade or two of the twentieth century was to install chiller sets rather than just one very large chiller. Such chiller sets often consist of chillers of more than one type, using a variety of fuels and technologies to meet different parts of the demand profile.

27.2.4 Geo-exchange systems

Geo-exchange systems (formerly referred to as geo-thermal systems) date back to the 1930s in North America. These systems, that use the earth as both heat source and heat sink, fell into disfavour because of the failure of plumbing systems. Plastic pipework is now available that has a projected long-term low maintenance life and geo-exchange systems are again popular.

The Australian Geological Survey Office (AGSO) building in Canberra incorporates a geo-exchange system, and air-cooled condenser systems where special function zones require environmental temperatures within tight limits. The geo-exchange system consists of plastic pipes, embedded in a compound that ensures no air gaps exist between the pipes and the earth that carry the chiller condenser water 100 m below the surface of the ground. In the summer, the heat from the condenser water is cooled by being passed through these pipes, while in winter the condenser water is pre-heated by reversing the cycle (CADDET, 1999a).

Large bodies of water can also be used as heat sources and sinks: Sydney Opera House has, for many years, relied on the waters of Sydney Harbour for cooling the condenser water from its air-conditioning system. The oceans contain many times the amount of heat required to heat all the buildings in the world and research aimed at utilizing this virtually limitless supply of heat is being pursued. One major problem with capturing the ocean's heat is related to the corrosiveness of seawater which leads to rapid deterioration of pipes and pumps. However, experiments with all manner of materials may make these systems viable in the future, at least for buildings located close enough to suitable bodies of seawater (Coony, 1996).

27.2.5 Variable refrigerant volume

Typical air-conditioning systems are based around large central chilled and hot water systems that are costly to run and maintain. Part of the reason for these high costs is that systems are often operating at partial load while most plant runs most efficiently at or near full load. As working patterns have changed in recent years, many commercial buildings are now occupied on a 24 hour/7 day basis with greatly varying occupancy rates so air-conditioning systems are often operating at well below their optimum load.

Variable refrigerant volume (VRV), or variable refrigerant flow (VRF) systems distribute heating and cooling throughout buildings via small refrigerant lines rather than water pipework or bulky air ducts delivering conditioned air from central plant that requires large fans, cooling towers and pumps. They are also easily integrated into building management systems (BMS) and so are suitable for use in modern 'intelligent' buildings.

27.2.6 Thermal storage

In response to high tariffs during times of peak load various systems have been devised to allow buildings to use off-peak power to generate heat or cooling that can then be stored for later use. Thermal energy may be stored in a variety of ways including ice and chilled water storage tanks, hot water tanks, or even beds of hot rocks. There are various reasons for storing thermal energy:

- demand for heating, cooling, and/or electricity is seldom constant over time, and the excess generation available during low demand periods can be used to produce hot water, chilled water or ice for use during high demand periods. This reduces peak demand levels, thus reducing the peak capacity required in an installation and results in a higher load factor on the units in the system. The cost of thermal storage is thus offset to some extent by the cost savings associated with lower chiller capacity and smaller cooling towers

- cogeneration plants are generally operated to meet the demands of the thermal load, which often results in excess electric generation during periods of low electricity use – by incorporating thermal energy storage, greater flexibility of plant operation is possible
- off peak production of heat or cold reduces the energy costs for heating and cooling, e.g., ice produced at night is stored for use in cooling during the day, particularly at times of peak demand
- stored energy is available in the event of plant failure thus increasing system reliability
- thermal storage facilities (hot or chilled water tanks) may double as water storage for fire protection and thus increase the cost-effectiveness of the installation
- chiller capacity necessary to meet base-load demand can be fully utilized round the clock, effectively doubling capacity in many instances.

By using ice storage the cooling demand of the building is decoupled from the operation of the chiller and so the chiller can operate at night, when the outside air temperature is lower and chillers can operate more efficiently, to meet a daytime cooling demand. This flexibility means that a smaller chiller can satisfy a larger peak cooling load as well as shifting the cooling demand to off-peak hours when electricity is cheaper.

Ice storage has advantages over chilled water storage in some situations, primarily where space is tight, as ice stores more energy in a smaller space. However, on a large scale, chilled water storage can be centralized in a large tank and costs less per unit of cooling energy than ice. For example, the McCormick Place Convention Center in Chicago has a tank of 8.5 million gallons capacity – over 32 million litres – which stores chilled water using a sodium nitrate/sodium nitrite brine cooled by ammonia (Bukowski, 1997).

27.2.7 Refrigerants

Although there are alternatives to conventional vapour compression refrigeration for cooling, many such chillers are still being installed in buildings. In recent years there has been much work done on developing alternative refrigerants to replace first CFCs and then HCFCs, both of which are believed to damage the ozone layer in the earth's atmosphere. Production of some of these refrigerants has already ceased and production of R-22, the refrigerant now in common use in air conditioning and heat pump systems, has been steadily reduced since 1996 and will be completely phased out by 2020, with equipment designed for use with R-22 ceasing production from 2010.

A number of replacements have been developed including R-410A, R-134A and some FICs (fluoroiodocarbons), which are all chlorine free and therefore do not harm the ozone layer.

As existing refrigerants are phased out building owners must upgrade their systems as, even before refrigerants such as R-22 are phased out, their cost increases dramatically as production is progressively reduced.

27.2.8 100% outside air

Generally only about 10% of the cross-sectional area of supply ducts was expected to carry the outside air component in air-conditioning systems prior to the 1980s. However, major new buildings now have the capability of drawing in up to 100% outside air. This extends the period of time that heating and cooling equipment does not need to operate, reduces energy requirements and extends the life of the equipment. The additional use of outside air also improves air quality and provides greater efficiency in flushing buildings in which toxins, such as those resulting from outgassing from furnishings, have been introduced.

27.2.9 Evaporative cooling

It has long been recognized that as water evaporates from a surface, even the skin of a human being, the temperature of that surface is reduced, as heat is required to vaporize the water. By spraying water into an air stream the temperature of the air is reduced, as heat is taken from the air as the water evaporates. This provides an economical method of cooling air in buildings that is also very healthy as it can be used entirely with outside air, thus eliminating the recirculation of air containing contaminants such as smoke and harmful bacteria. Evaporative cooling may be direct or indirect: in a direct system water is sprayed directly into the supply air and therefore may not be acceptable if it results in uncomfortable humidity levels. However, indirect evaporative cooling separates the supply air from the wetted air in a heat exchanger so that the cooled supply air cannot take up moisture as it is cooled.

Direct and indirect systems can be used in tandem and may also be coupled with a desiccant system, so that the problem of excessive humidity can be overcome as the desiccant reduces humidity, then the evaporative cooler lowers the air temperature. Hybrid systems that also include some conventional vapour compression cooling can be used in conditions that are more severe and where evaporative cooling alone cannot supply air that is sufficiently cool and dry.

27.2.10 Variable speed pumps and compressors

Electric motors that drive fans, pumps and compressors in buildings can now operate at a range of speeds allowing speed to be matched to load with resultant increases in efficiency and reductions in energy use. These motors also last longer than conventional fixed-speed motors, and are quieter in operation as well as providing improved comfort levels due to the greater levels of control that are possible. Tests have shown typical energy savings of 20–50% using these high-efficiency electric motors (Greentie, 2001).

27.3 Operable natural systems

An innovation of significance in Europe and the United Kingdom in the latter decades of the twentieth century saw empirical experiments in designing commercial buildings without full air-conditioning. Operable systems are mostly related to ventilation although occupants may also be able to control daylight levels through the use of blinds and other shading devices. Influenced by those experimental buildings in Europe and the United Kingdom, MGT Architects (Sydney) designed the Red Centre at the University of New South Wales (Figure 27.3), without air conditioning (except for computer rooms). The building has been designed to be 'operated' by the occupants. The operable systems primarily relate to

Figure 27.3 The 'Red Centre' at the University of New South Wales.

ventilation: classrooms have glass louvres, academic offices have operable casement windows and vents and there are wind turbines that generate pressure differentials and effect ventilation through a number of levels. There are space heating units in the large studio spaces, and large south facing windows provide abundant daylight in the studios. Overall the building works reasonably well although the range of temperatures experienced by occupants is somewhat larger than occurs in air-conditioned buildings.

27.3.1 Natural ventilation

The majority of modern buildings, other that small residential buildings, are still fully air-conditioned. However, the number of new buildings appearing with only partial systems, or that rely entirely on natural ventilation, is increasing. The move away from sealed, air-conditioned buildings has been prompted by a variety of factors: the desire for lower energy use, the incidence of 'sick building syndrome', recognition of the preference that people have for openable windows and the emergence of vastly more sophisticated and reliable computer modelling systems for use in building design have all been important.

Natural ventilation systems involve a range of techniques and components including wind turbines, atria, ventilation 'chimneys', the use of narrow floor plans, and operable windows. If this strategy is to be successful it is necessary for the overall building design to be driven by it with an integrated approach. Extensive computer modelling is usually essential to ensure that the building will perform as expected; buildings relying on natural ventilation typically have little margin for error and rectification work, such as the addition of air-conditioning after construction is completed, is prohibitively expensive.

For natural ventilation systems to work effectively it is usually necessary to control heat gain and loss through the building envelope. It is particularly important to control heat gain in the summer months if internal temperatures are to be kept at comfortable levels without the need for mechanical cooling. Of most concern is the heat that enters through windows; this can be controlled in a number of ways including careful attention to size and positioning of windows, particularly those facing the west and the north (in the southern hemisphere) and the use of sunshading devices both external and internal.

27.3.2 Sunscreens

In the Middle East sunscreens in the form of mashrabiyas were built of timber in front of windows. These screens provided the essential privacy required in the culture, permitted a partially obscured view out of the building, shaded against the direct component of solar radiation, and admitted diffused daylight and ventilation. Porous vases containing water effected evaporative cooling as the incoming breeze passed over the moist surface of the vases.

In most Western countries, unobstructed views are usually expected to the extent that special glass types have been developed to permit a view whilst reducing heat gain. Glazing that reduces direct solar gain, while reducing cooling loads in buildings, allows so little light into the interior that high levels of artificial lighting are often required even on clear sunny days. This has been offset to some extent in recent years by the development of glass that allows light in but blocks heat. External sunscreens of various types have been used extensively and are becoming more common as the move away from the curtain wall façades of the 1960s and 1970s continues.

Glass and the need for shading

Developments in glass, aimed at reducing heat transmission, have been directed towards reducing heat transfer by conduction (that is, resulting from the difference in air temperature on each side of the glass) and instantaneously transmitted heat gain from solar radiation. In a climate in which there is an abundance of solar radiation, the amount of heat conducted through single or double glazing or glazing with additional air spaces filled with special gases (e.g., argon) is small compared to the heat gain transmitted by radiation.

Glass treated with a low emittance film, referred to as Low-e glass, still instantaneously transmits around 20% of incident radiation. If the radiation level were, say, 600 watts per m^2, the instantaneously transmitted heat gain would be 120 watts per m^2. Added to this would be the much lesser amount of heat conducted, say 2 watts per m^2 per degree difference in the temperature between the outside and inside environments; if these were 32 degrees and 22 degrees the result would be 20 watts per m^2 of conducted heat. This is one sixth of the heat transmitted instantaneously by radiation. Clearly, if the glass could be shaded against the direct component of solar radiation, the heat gain would be significantly reduced.

Non-redundant sunscreens

Reasons for using sunscreens include making admitted daylight usable by excluding the glare of direct solar radiation, reducing the radiant heat load on occupants, reducing the

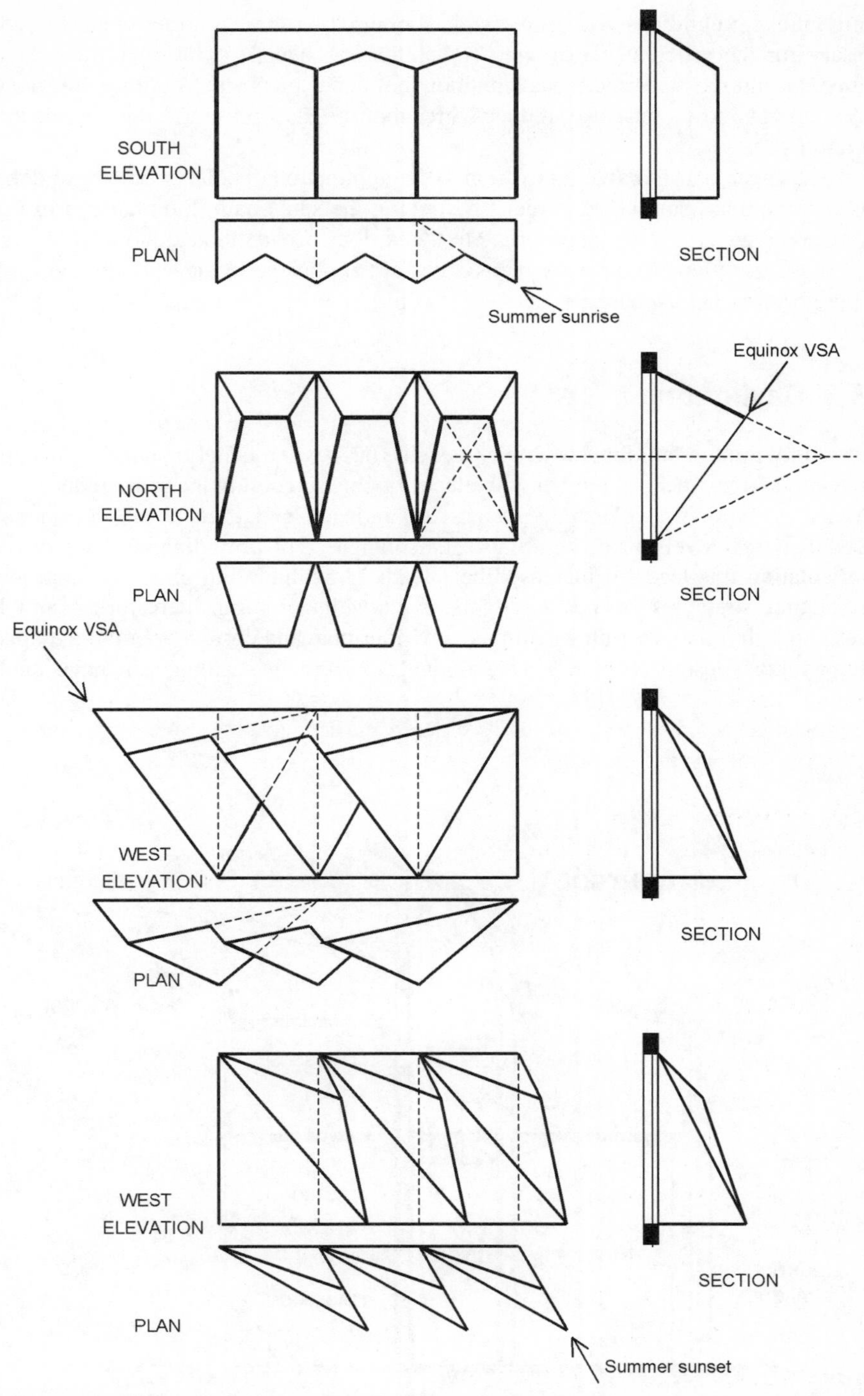

Figure 27.4 Sunscreens designed for specific orientations to exclude direct solar radiation for specified months at 32°S latitude.

cooling loads on buildings and, in particular, minimizing the variations in peak loads so necessary for the efficient performance of chillers. A benefit of using diffuse daylight admitted through the sunscreens is a reduction in reliance on electric lighting; this not only reduces energy costs but also reduces greenhouse gas emissions due to electricity generation.

There has been considerable innovation in the application of solar geometry to defining forms for minimum surface sunscreens by tracing the sun's path. Summarized in Figure 27.4 are some generic forms appropriate to the various orientations at latitude 32 degrees south. The orientations of the sunscreens noted are applicable to the southern hemisphere – in the northern hemisphere the geometry would be mirror imaged.

27.3.3 Daylighting

Using natural light in buildings wherever possible has several benefits: building occupants are happier, and less heat is produced by electric lights so cooling loads are reduced – less electricity is required, so operating costs are reduced and greenhouse emissions are reduced. Daylight levels inside buildings can be increased by providing more window area but in isolation this leads to increased heat loads in summer and increased heat loss in winter. Other strategies include skylights, light wells or atria, clerestories, and light shelves and light tubes. Buildings with narrow floor plan and correct orientation, aided by careful window design with light shelves and appropriate shading, can have daylight

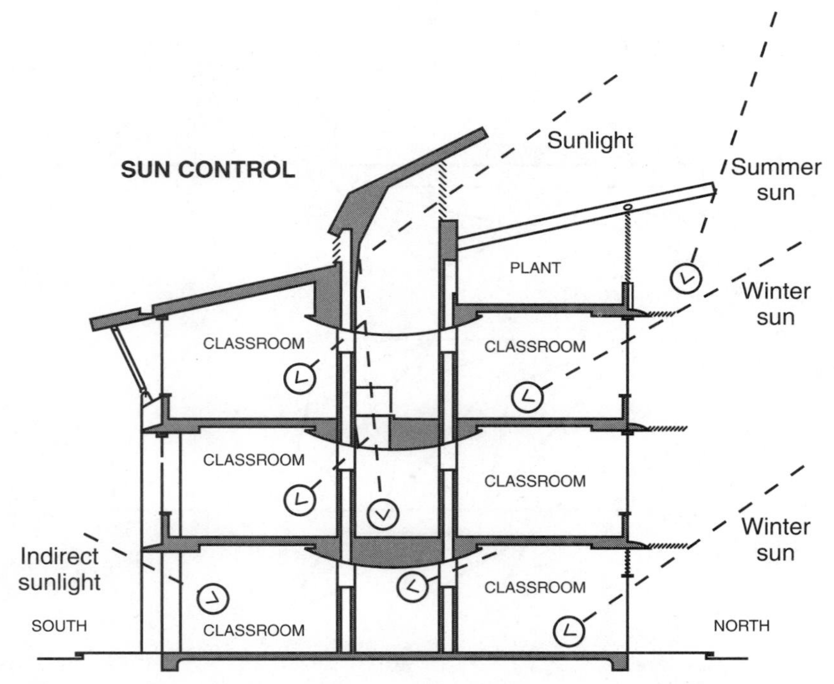

Figure 27.5 Section showing how sunshading protects the interior from direct solar radiation during summer – Languages Building, University New South Wales, Sydney.

Figure 27.6 Horizontal sunshades on the northern façade.

Figure 27.7 Daylit interior.

penetrating deep into interior spaces and need much less artificial light during daylight hours – often all that is necessary is low power task lighting at workstations to supplement the natural light that is available.

Sydney architects Jackson Teece Chesterman Willis used daylight illumination effectively in the Languages Building at the University of New South Wales in Sydney in 2000. Effective shading excludes the direct component of solar radiation and admits diffuse light from the sky vault (Figures 27.5, 27.6 and 27.7).

27.4 Energy supply

Energy efficiency in buildings has become increasingly important as awareness of the potential effects of continued emissions of CO_2 and other greenhouse gases has grown. One of the ways that emissions can be reduced is by generating electricity at the point of use using one of a variety of alternative generation systems. There have been significant advances in this area in recent years, and research is continuing in an effort to achieve greater reductions in greenhouse emissions. Areas of most interest are photovoltaics that convert solar radiation into electricity, fuel cells that convert fuel to electricity by means of chemical reaction rather than combustion, and the simultaneous production of electricity and useful heat from a single fuel source (cogeneration or combined heat and power: CHP).

27.4.1 Photovoltaics

Although the process was first discovered in the 1830s, it is only relatively recently that the generation of electricity using photovoltaics (PVs) has started to become a viable alternative to conventional power generation. Large improvements in conversion efficiency and manufacturing now see the cost of electricity produced using PVs approaching that of conventional systems. The advantages of solar power are well-known: it is noiseless, there is little or nothing to wear out, it is environmentally benign (Green, 1982). Conversion efficiencies up to 17% are now possible and solar energy is abundant – in fact, a PV generation installation 140 km × 140 km with an efficiency of just 10%, sited at an average location in the USA, would serve all the electricity needs of the entire country (Zweibel, 1990). Just a few square metres of PV on the roof of every house in Australia would eliminate the need for any new power stations.

Photovoltaic systems are modular and therefore incremental power additions are easily accommodated: as demand increases so extra panels are added to the array. This is quite different to the case of central power generation where a small increase in demand may necessitate the construction of a multi-megawatt plant that will mean there is surplus generating capacity for some time into the future. This is exacerbated by the fact that generating capacity is determined by peak loads that typically occur only during the late afternoon in summer, when air-conditioning loads are highest.

Germany has initiated a program that has already seen solar panels installed on more than 2000 family homes and similar programs are underway in Japan and the USA (UNSW, 2001). As production and conversion efficiencies increase it is expected that PVs will become a truly cost-effective option for electricity generation; their attractiveness will be further enhanced with the adoption of greenhouse abatement options, such as the introduction of carbon taxes and tradeable emissions certificates. The Commonwealth government in Australia now offers a cash rebate to householders and community groups who install grid-connected or standalone PV systems (AGO, 2001).

One major advance is the incorporation of PVs into building components such as façade panels and roof tiles.

27.4.2 Photovoltaic roof tiles

Solar cells have typically been available only in standard modules, designed to be installed on or above the building structure – usually on the roof. This is not always the most efficient position for them, particularly where tiled roofs are common, as their performance is reduced at high temperatures. Rooftops are an obvious location for PVs, as they are generally in full sun and not shaded by surrounding trees. In response to this situation a solar roof tile has been developed at the Photovoltaics Special Research Centre at the University of New South Wales in Sydney (UNSW, 2001). As the cost of PVs is still the main factor that limits the uptake of technology, the option of making the tiles perform two tasks, waterproofing the building and generating electricity, is an attractive option.

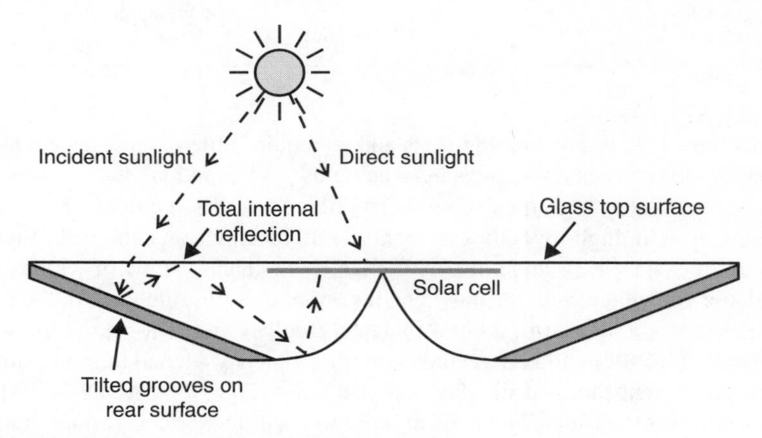

Figure 27.8 Section through PV roofing tile (UNSW, 2001).

The tile that has been developed uses a 'static concentrator' with the result that around 25% of the area of the tile is solar cell while the rest of the tile is designed to concentrate the solar energy onto the cell. Figure 27.8 illustrates the concept. The solar cells respond to light falling on both surfaces and they are interconnected down the centre of the module. The final result is a roof that doubles as a solar power plant.

27.4.3 Fuel cells

In 1839 Sir William Grove devised a way of generating electricity from hydrogen and oxygen with water as the 'waste' product. This is the basis of the modern fuel cell, originally developed by NASA to produce electricity, heat and water in space. Fuel cells have a number of advantages: they are clean and quiet, they are modular, so capacity is easily increased as load increases, they are small and can be sited almost anywhere that power is required. This allows the waste heat that they generate to be used, as well as reducing transmission losses, unsightly power reticulation lines and possibly harmful electromagnetic radiation. Apart from development of fuel cells for powering vehicles and even small devices such as video cameras, there is continuing development of fuel cells for use in buildings. Table 27.1 shows the status of development of commercial fuel cells for various applications.

Table 27.1 Status of fuel cell technology development (Fuel Cell World, 2001)

Application	Commercial plant available from	Fuel cell type
Commercial cogeneration <5 MW	1996	Phosphoric acid (PAFC)
Automotive	2002	Proton exchange membrane (PEMFP)
Commercial and residential cogeneration <500 kW	2003	
Portable/backup power	1999	
Distributed power-cogeneration	±2005	Molten carbonate and solid oxide (MCFC and SOFC)
Industrial cogeneration	±2005	
Central generation	Unknown	

Fuel cells have few, if any, moving parts and so require little maintenance. Due to their exceptional reliability they are increasingly being used in situations such as hospitals and computer centres where high quality uninterruptible power is required. When used in a cogeneration system high overall conversion efficiencies are achieved. Figure 27.9 illustrates a novel system in Japan that is fuelled by methanol, some of which is a waste product of the manufacturing plant where the system is installed (CADDET, 2001a); Figure 27.10 shows a system in a hotel in Japan, fuelled by town gas with a maximum overall efficiency of 89%, and an average overall efficiency (depending on utilization of waste heat at different times of the day and year) of over 70% (CADDET, 2001b).

It is expected that the use of fuel cells in buildings will increase as further development improves the technology and as production increases and initial costs decrease.

27.4.4 Cogeneration (CHP)

Cogeneration is the simultaneous production of electricity and useful heat from the single burning of a fuel such as natural gas. In conventional power stations up to 70% of the heat produced by fuel combustion is lost into the surrounding environment as waste heat. Cogeneration captures some of that waste heat and uses it before it is dumped. The heat may be used directly as low temperature heat for space heating, or as hot water or steam for process use; some may be used to raise steam to run a secondary electrical generator (a combined cycle). A typical family car is an example of a cogenerator: by burning fuel (petrol or diesel) shaft power is produced that drives the wheels, makes electricity to run

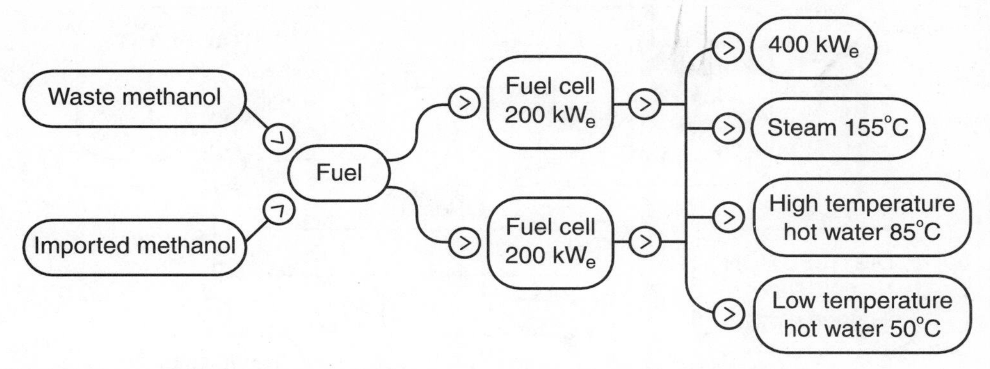

Figure 27.9 Flow diagram for fuel cell cogeneration scheme – overall efficiency 80%.

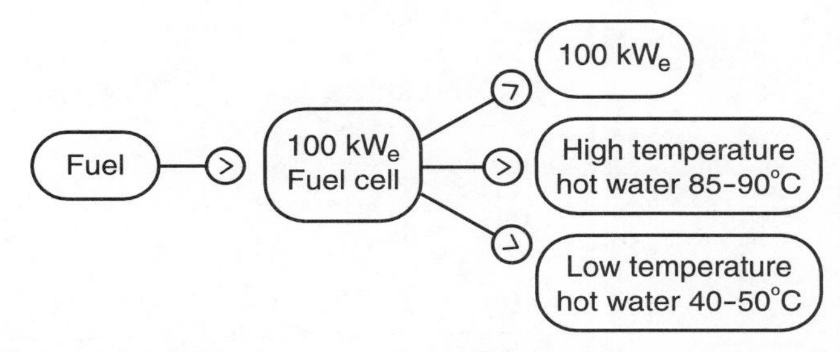

Figure 27.10 Fuel cell cogeneration system in a Japanese hotel – maximum overall efficiency 89%.

the onboard electrical system and often also drives an air-conditioner. It is, however, the capture of the waste heat from the engine's cooling system, used to heat the car's interior, that represents true cogeneration, as this is heat that is being put to good use that would otherwise be dumped into the atmosphere.

Cogeneration plant ranges from the very small (less than 150 kW) to major power suppliers of 200 MW and more that supply electricity for on-site use and export to the grid, as well as supplying steam/hot water to adjacent sites. Cogeneration units based on fuel cell technology or directly fired by natural gas are being developed that will serve individual buildings. Gas-fired microturbine systems are now available that can power a medium-sized commercial building with waste heat being available for space heating and cooling (Mitchell, 2001). Any buildings with a significant heat requirement, such as hospitals or public swimming pools have been candidates for cogeneration, and with advances in heat-driven cooling technology the potential for building-sized cogeneration is now expanding.

27.4.5 District energy

District energy (DE) is another technology that has had something of a renaissance in recent years in some countries, while it has been commonplace in others for many years. DE was originally restricted to district heating: customers were connected to a central boiler plant

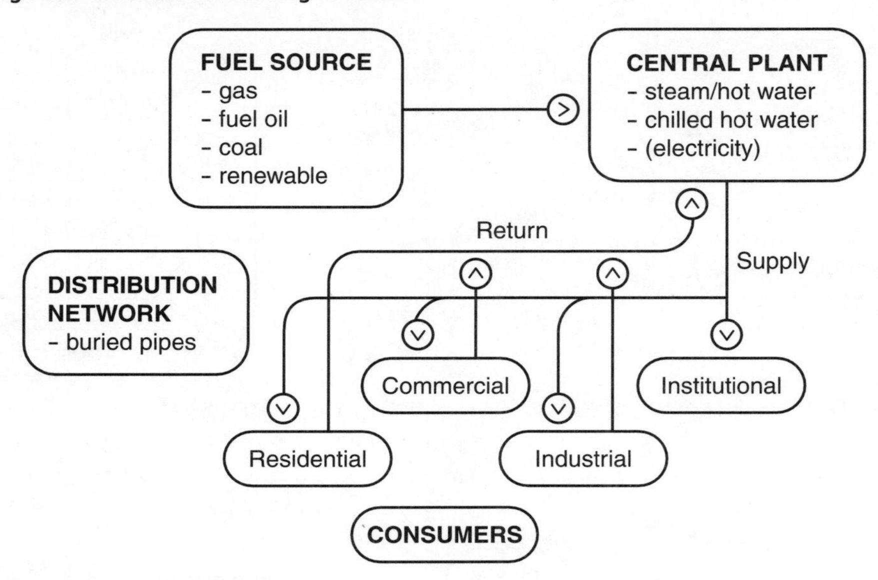

Figure 27.11 A generic district energy scheme.

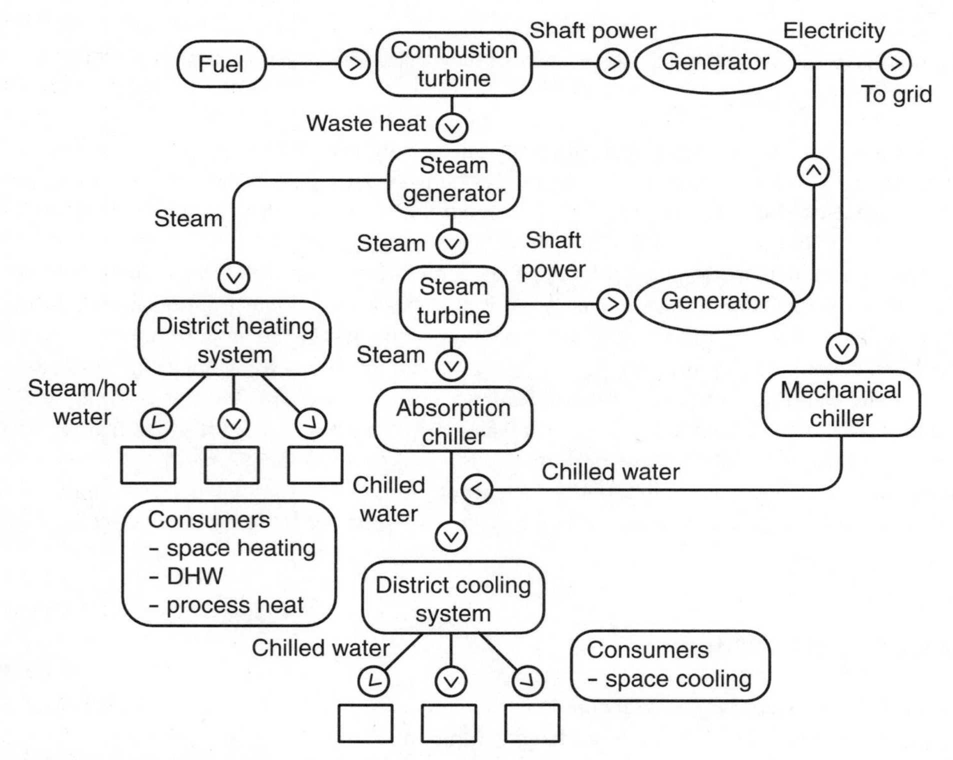

Figure 27.12 A typical trigeneration-based district energy scheme.

that supplied hot water or steam for space heating and domestic hot water (DHW). Modern DE, however, includes district cooling, based on both heat driven and conventional chiller technology, and is very often combined with electricity production (Figure 27.11). Systems where all three outputs are present, viz., heat, cold and electricity, are referred to as 'trigeneration' systems; Figure 27.12 shows a schematic for a trigeneration-based district energy system. Many configurations are possible depending on load profiles, climate, customer mix, available fuels and a host of other factors (IDEA, 2001).

These systems have definite advantages for consumers and society, as overall efficiencies of up to 90% are possible, emissions and fuel requirements are cut, consumers have greater reliability of service and much reduced maintenance costs. Buildings require smaller plant and less space to accommodate plant; health problems associated with bacteria in cooling towers are eliminated; plant is sized to meet peak loads over a wider area, so redundancy in plant is reduced; large plant is inherently more efficient than small plant, and so on (Spurr, 1996). The influences that may be expected to make DE more attractive to consumers in the future are the same as those driving fuel cell and cogeneration uptake, i.e., emissions controls, emissions trading, carbon taxes and the like.

27.5 Plumbing

The twenty-first century will be a time of substantial change in the area of plumbing (Lehr, 1990). Developments in the application of electronics with artificial intelligence providing finer levels of adjustment and diagnostics and maintenance, monitored and prescribed by microprocessors, will greatly change the way plumbing systems work. In particular there will be more use of electricity associated with plumbing. The plumbing in major buildings will probably be replaced twice in the first half of the twenty-first century, as advances in plumbing technology render existing systems obsolete.

Electronics will control pumps, sewage ejectors and other equipment, and facilitate heat recovery, make systems safer, and improve alarms and emergency remedial actions so that damage is minimized when problems occur. For instance, microprocessor-controlled mixing valves can turn water in showers and baths on and off, with water temperature selected according to the user's personal preference and with exact water temperature displayed on a control panel (NASA, 2001).

Shape memory alloys are now used to instantly control the flow of hot water in showers and sinks, thus avoiding scalding accidents. A patented valve reacts to excess temperature and turns the flow off when hazardous temperatures are sensed, restoring flow when the temperature drops below danger level (NASA, 2001). A similar system can be installed that provides emergency shut down of process control lines that handle flammable and toxic fluids and gases. This system works with pneumatically controlled valves; if excessive heat is sensed it vents the pressure at the valve, which closes the supply line. Memory metals may also be used for sprinkler systems without sprinkler heads; instead, when the pipe is subjected to heat above a certain temperature the pipe itself will perforate and act as sprinklers.

Other suggested advances include smart pipes that can monitor leaks and report back to a central computer-controlled BMS that can then respond by operating valves that will stop the flow in the leaking pipe. Another kind of smart pipe has an element that heats water as it passes through it – one major advantage of this is that it eliminates the need for both hot and cold water reticulation, as only the last metre or so of the supply pipe is used to heat the water.

A British company, Twyford Bathroom, is developing a revolutionary new toilet (Park, 2001). The freestanding pedestal that is commonplace today was actually invented by Thomas Twyford in 1883 – the new Versatile Interactive Pan represents the first major advance in toilet design since then. It will be voice-activated, its height will be adjustable and it will use little water as it will rely on vacuum to transport waste away from the bowl. In addition it will take samples of human waste and analyse them for health problems such as diabetes and colon cancer and possibly even e-mail the results to the owner's doctor. Japanese and German companies are now pursuing similar ideas, including models that check body fat and sugar levels and test for pregnancy.

27.6 Communications

Messages have been sent by man over varying distances, limited by the available technology, using hand signals, sounds, smoke signals, mirrors to send flashes reflected from the sun, and the heliograph, which has two moveable mirrors, one to catch the light from the sun and the other to reflect it in any direction. The use of light for communication was limited by the fact that it travels in a straight line. The curvature of the earth, air pollution and the inverse square law, all placed limits on the distance that light could be used for communications.

27.6.1 Optical fibres

Optical fibres have been developed that transmit light through 'light pipes' and around corners. An inner transparent fibre is used to transmit the light while the outer walls of the fibre act as a mirror. Individual fibres are thinner than a human hair, and cables made of these fibres may be laid underground or beneath the sea. Because of the speed of light, approximately 300 000 km per second, messages may be sent anywhere in the world with negligible delay and using very little energy.

Optical fibre communications networks use lasers to generate a suitable light source. Lasers can produce very tightly focused pulses of light many times a second that are sent along the fibres. These pulses of light are picked up at the other end of the cable by a light-sensitive cell that converts the pulses of light into pulses of electricity. These pulses of electricity are then decoded by a computer and the message revealed. It is possible to send the entire contents of the *Encyclopaedia Britannica* through an optical connection in less than 1 second, and by using a combination of codes, many messages can be sent along an optical fibre at the same time (Australian Academy of Science, 2001).

27.6.2 Wireless networks

Wireless computer systems are already a reality – individuals connected to their office LANs do not need to have their computers physically connected to the system. Wireless application protocol (WAP) enabled mobile telephones and palmtop devices can operate within reasonable distances of receiving nodes, allowing access to Internet services such as e-mail from anywhere within range of a suitable network. Home networks are now feasible and costs are falling to levels where they can be expected to be within the reach of average homeowners in the not-too-distant future. These networks offer the opportunity to perform functions such as turning on air conditioners and other household appliances using a WAP telephone and even includes the possibility of using the domestic refrigerator as the centre of a home network, given that it is the one appliance that is always running (LGE, 2001).

Wireless networks in offices allow great freedom for designers when configuring workplaces and makes re-organizing existing space quick and easy. It is anticipated that wireless applications will continue to grow although there are still some concerns about the ease with which unauthorized people may be able to access wireless networks by monitoring wireless signals remotely.

27.7 Security systems

The availability of advanced electronic systems has provided great opportunities for improving security in many areas but has also presented many new challenges to security. For example, it is now very easy for documents to be forged – digital technology including low-cost high-quality scanners and printers and sophisticated software that can be used by children means that identity papers, passports and the like can be manufactured at will. As fast as solutions are devised to combat such activities new methods of baffling security measures are introduced. Fear of credit card fraud is a major factor in the slower than expected uptake of e-commerce, particularly for people buying goods on-line, and elaborate measures for incorporating digital signatures into electronic documents are necessary if the security of net-based transactions is to be ensured. It is thought that the future development of very tiny but very powerful nanocomputers, or even biological computers, may mean that even the most elaborate encryption keys could be broken very quickly, thus requiring other security measures to be devised.

A variety of physiological characteristics apart from signatures can be used for identification purposes; these include face, fingerprints, footprints, voiceprints, DNA, hand geometry and the vein patterns of faces, wrists, retina and iris. These characteristics can be measured or recorded using a variety of sensors, then digitized and stored in a computer database for reference. The technique of identification by comparison with the stored record is called biometrics.

Various kinds of biometric identification are now in use and can be expected to become commonplace in the near future, as conventional means of identifying individuals become less and less reliable. Security systems are available that can recognize faces by comparing a stored image with a 'live' face and so allow or deny access in the same way that a human guard would do. The BioMouse (Ankari, 2001) is a small computer attachment that matches a user's fingerprint rather than a password to allow access to the computer. Some versions incorporate the fingerprint reader into the normal computer mouse and thus allow for automatic encryption of files, as well as controlling access, using this function.

27.8 Elevators

The introduction of elevators (or lifts), in the nineteenth century, was one of the major factors that allowed the development of the modern skyscraper, and there have been steady improvements in all aspects of elevator technology since Elisha Graves Otis demonstrated the first safety passenger lift in 1852. With the introduction in 1903 of the gearless traction electric elevator it became possible to service truly high-rise buildings.

Once again the advent of computers has made possible many improvements in elevator systems. Microprocessors have allowed for more efficient despatch logic with reduced round trip times; fuzzy logic despatching tells passengers which car will answer their call as little as one second after they press the call button, and digital closed-loop position and velocity feedback systems produce a smoother, faster ride with consistent and precise levelling. New protection systems on lift entrance doors ensure that passengers are not injured by closing doors: infrared scanners create an invisible 'net' across doorways – if a beam is broken the doors re-open instantly without touching the passenger. These systems also extend into lift lobbies so that they can detect passengers as they approach the doors. Remote elevator monitoring (REM) is part of intelligent building management systems; it uses microprocessors and internal building data communications to provide continuous diagnostic and detection functions so that malfunctions are recognized, located and reported as they occur.

27.9 Electric lighting

In 1880 Thomas Edison patented the incandescent light bulb and basically the same technology still lights most homes today. There are, however, a number of other lamp types available for both domestic and commercial use, including compact fluorescent, tungsten halogen and metal-halide.

27.9.1 Lamps

Compact fluorescent lamps have approximately five times the efficacy of tungsten incandescent lamps, use less energy and last much longer. They are available with the same socket configuration as an incandescent lamp, so they are interchangeable, however, their higher initial cost has restricted their adoption by many people.

In tungsten halogen lamps, tungsten molecules driven off filaments combine with the halogen gas. The newly formed tungsten halogen molecules diffuse towards the filament, with the tungsten being deposited back onto the filament, while the halogen is available for a further cycle. This regenerative cycle keeps the bulb clear for the entire life of the lamp. Higher gas pressures are used in the lamps, which minimizes tungsten evaporation, thereby improving the overall quality of the lamp. The colour of a tungsten halogen lamp is slightly whiter than that of a standard incandescent lamp. They have virtually replaced larger output incandescent lamps because of their higher efficacy (+15%), compact size, improved clarity and longer life.

Metal-halide and high-pressure sodium lamps have virtually replaced mercury lamps because of their high efficacy and improved colour. Metal-halide lamps produce a bright white light, produce 50% more light than equivalent wattage mercury lamps and have a much higher colour rendering ability.

At the beginning of the twenty-first century two noteworthy innovations in lamp design are the triphosphor fluorescent lamp and the electrodeless induction lamp. Triphosphor lamps offer a high output of light (12% to 15% higher than that of standard coating lamps) combined with a high colour rendering ability. They are made by combining together a thin layer of a conventional halophosphate phosphor with a thin layer of a triphosphor blend. A triphosphor blend is a combination of three narrow-band rare phosphors that reproduce the three primary colours of the spectrum, each with a very narrow bandwidth. Because of the very narrow bandwidth, the peaks are very high, thus these lamps produce a high output of light. The triphosphor lamp combines the high quality of the narrow-band rare phosphors with the economy of a conventional halophosphate coating. By replacing standard lamps with energy-efficient, reduced-wattage lamps with triphosphor coatings, the energy consumption of the fluorescent lighting load can be reduced by 10% or more and the colour rendering ability of the lamps will be significantly improved.

Current electrodelesss lamps give a light output of 1100 lumens at 23 watts (an efficacy of 48 lumens/watt) and have a life of 10 000 hours, whereas an incandescent reflector lamp gives a light output of 1000 lumens at 100 watts (an efficacy of 10 lumens/watt) and has a life of 1000 hours. A new electrodeless induction lamp will shortly be available that has an efficacy of 80 lumens/watt, a high colour rendering ability and an extremely long life of 60 000 hours.

Other technologies, however, go against the trend: it is estimated that use of halogen uplighters (which use a 250–300 watt globe) in the 1990s increased to the point where their use will soon have wiped out all the savings gained through the residential use of compact fluorescents since the 1980s (CADDET, 1999b).

27.9.2 Luminaires

A significant change in illumination design in the three decades following the 1970s introduction of the microprocessor was in the selection of luminaires and reflectances of interior furnishings. Prior to the 1970s, interior illumination design included the use of luminaires with high coefficients of utilization, broad spread polar distribution of light flux, and interior surfaces that had the characteristics of being diffusely reflecting and highly efficient in reflecting light. This approach to illumination design, however, resulted in disability glare on computers screens. Black screens that could be placed in front of computer screens were considered a remedial application only and were not popular. A different approach to illumination design was required, one that would minimize disability glare on computer screens.

Interior illumination design now requires luminaires with narrow polar distribution of light flux, and low reflectances on interior room surfaces. This has led to a low coefficient of utilization of light flux and consequently a low efficiency in the use of electrical energy with potential increases in greenhouse gas emissions. Fortunately, parallel with this necessary low co-efficient illumination design, there have been other developments that help to balance the disadvantages. These include the development of lamps and phosphors with increased efficiency, and luminaires with improved efficiency in reflecting light out of the fittings themselves (Figure 27.13). Another balancing influence was a decision to

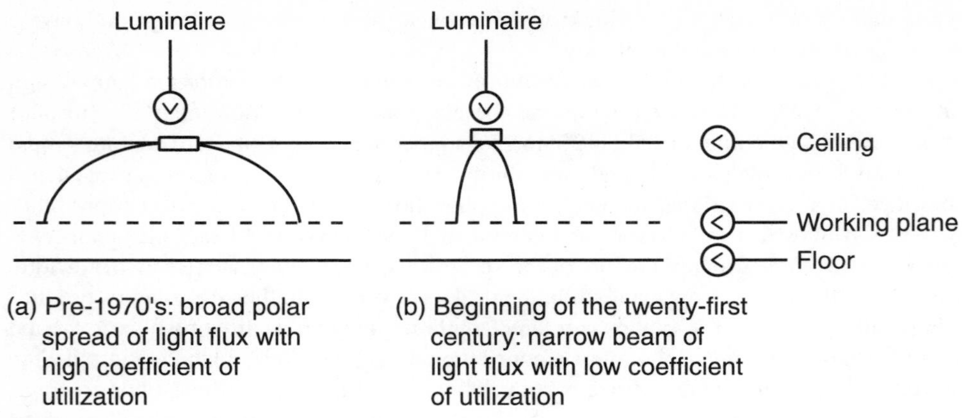

Figure 27.13 Luminaire design.

provide lower level background illuminance and additional task lighting. As a result lighting in offices now uses less electricity than it did three decades ago.

27.9.3 Lighting controls

With the greatly increased complexity of lighting systems in modern buildings has come the need for much more sophisticated lighting control systems. Lighting can now be controlled by computer software that forms part of the building management system. Typically lighting controls are fully integrated with other services such as air conditioning and security and fire systems. Sensors in the building provide data to the central control and specialized lighting arrangements can be programmed and stored for special events, different times of the day or in response to changing conditions. Graphic displays can show lighting configurations overlaid on floor plans of the building.

Several examples of modern control systems are described briefly below (Dynalite, 2000).

Orchard Cineleisure Centre, Singapore

The centre contains state-of-the-art cinemas, nightclubs, themed shopping and a virtual reality arcade. Robotic statues are programmed to come to life at various times of the day with different themes created on individual floors, and co-ordinating the lighting with the animatronics was the key requirement. Control of all lighting takes place from a central control room and requires an accurate timeclock and a failsafe communications network.

A large diversity of lighting types was required to create the animatronic and special event effects. Theatrical fittings, metal-halide fixtures, fluorescent and incandescent lamps were combined to create a wide variety of preset scenes, and the lighting sequences are all computer controlled.

The Sydney Superdome

The Superdome, part of the Sydney Olympics complex, is Australia's largest indoor sports and entertainment centre, seating up to 18 000 people. As with all Sydney Olympic venues, environmental initiatives were a high priority, particularly with a view to reducing energy consumption. With these considerations in mind, equipment was chosen for dimming/switching control and system monitoring.

Media coverage had a significant influence over the lighting design at the Superdome, as events would be captured using still photography and television. The specification called for both local and remote control capabilities. The lighting control system can be reconfigured and reprogrammed by the user, via a PC located in the media control room. About 300 metal halide floodlights and 100 tungsten halogen floodlights are controlled. Metal-halide lamps were chosen primarily for their ability to provide excellent colour characteristics.

Sydney Olympics International Broadcast Centre

During the 2000 Olympics, Sydney was host to the world's largest broadcast centre. The construction, located in a 70 000 m^2 refurbished warehouse, was unique in a number of ways. It was designed so that at the conclusion of the Games, 80% of the building could be recycled, in accordance with the commitment to staging an environmentally friendly Olympic Games.

The centre, capable of housing up to 11 000 journalists and media correspondents at peak periods, required over 2000 km of cabling for lighting, power and communications. It contains a mixture of both architectural light fittings, for day-to-day operations, and theatrical style light fittings for television interviews and broadcasts. The specification called for a control system which could effectively control both types of fittings and interface with the dimmers, using control panels as the interface for everyday lighting and 'plug-ins' for the theatrical lighting desk, enabling control of the stage lighting for major events.

27.10 Conclusion

It is less than 150 years since the introduction of the most common building services: the telephone, electric lighting and passenger lifts, and less than 100 years since Frank Lloyd Wright, in 1904, introduced a primitive form of air conditioning (that used blocks of ice to cool the air) in the Larkin Building in Buffalo. Developments in all of these services, and the introduction of others such as automatic fire systems, security systems, 'clean' rooms and so on, progressed at an astounding pace through the twentieth century. Contemporary 'intelligent' buildings now contain highly sophisticated systems that monitor and control many aspects of their operation, and at their heart lie the microprocessors that have revolutionized so many aspects of day-to-day life in developed countries around the world.

Much of the research now being done is directed towards improving the performance of the engineering services in buildings, both to provide greater levels of comfort for occupants and to reduce dependence on the pollution generating, energy hungry systems that proliferated in the second half of the last century. We are, indeed, approaching a time when many features that were previously confined to science fiction stories will be reality:

wearable computers, voice-activated building management systems, integrated communications and security systems capable of identifying individuals by recognizing their faces, maybe even building materials and components that contain many thousands of nanocomputers that can respond to commands and stimuli and change colour or shape on demand.

References and bibliography

AGO (2001) *Photovoltaic Rebate Program*. Australian Greenhouse Office. www.greenhouse.gov.au/renewable/pv/

Ankari (2001) *BioMouse – Security Beyond Passwords*. www.ankari.com/biomouse.asp

Australian Academy of Science (2001) www.science.org.au/nova/021/021key.htm

Bukowski, J. (1997) District energy provides a cool solution. *Consulting-Specifying Engineer*, September, 70–2.

CADDET (1999a) AGSO building taps into energy 'down under'. *CADDET Energy Efficiency*. http://caddet-ee.org/nl_pdf/99S_05.pdf

CADDET (1999b) Residential lighting and appliances: small amounts mean massive demand. *CADDET Energy Efficiency Newsletter*, **4**, December, 15–17 (Netherlands National Team).

CADDET (2001a) *Fuel Cell System Fed with Waste Methanol*, CADDET Technical Brochure No. 155. www.caddet-re.org/assets/no155.pdf

CADDET (2001b) Hotel cogeneration system uses highly reliable fuel cell. *CADDET Energy Efficiency Newsletter*, January. www.caddet-ee.org/techpdf/r407.pdf

Coony, J. (1996) Advances in the development of energy efficient technologies: sea water air conditioning (SWAC). In: *Proceedings of Eighty-Seventh Annual IDEA Conference*, Washington, DC (International District Energy Association).

Dynalite (2000) *Dynalite at Work*. http://dynalite-online.com/projects.asp

Fuel Cell World (2001) http://fuelcellworld.org/

Green, M.A. (1982) Solar Cells – Operating Principles, Technology, and System Applications (Englewood Cliffs, NJ: Prentice-Hall).

Greentie (2001) www.greentie.org/class/ixg08.htm

Harkness E.L. (1999) Tall building form and texture: explorations in an architectural design studio. In: Hayman, S. (ed.) *Proceedings of the 33rd Conference of the Australian and New Zealand Architectural Science Association*. University of Sydney, pp. 13–29.

Harkness E.L. and Mehta M. (1978) *Solar Radiation Control in Building* (London: Applied Science Publishers).

IDEA (2001) International District Energy Association. http://www.districtenergy.org/

Kuijpers, J.M. (1990) UNEP assessment of the Montreal protocol: refrigeration within the framework of the technology review. *Revue International du Froid*, **13**, March, 95–9.

Lehr, V.A. (1990) Plumbing in the 21st Century. *Heating/Piping/Air-conditioning*. October, 43–7.

LGE (2001) LG Electronics. www.lge.com

Mayes, C. (1914) *The Australian Builders & Contractors' Price Book*, Eighth edition (Sydney: E.W. Cole).

Mitchell, S. (2001) Gas generates hi-tech research. *Second Byte, The Australian*, July 31, 3.

NASA (2001) http://spaceflight.nasa.gov/shuttle/benefits/golf.html

Nimitz J., Glass S., and Dhooge P.M. (1998) Energy efficient, environmentally friendly refrigerants. *Twentieth National Industrial Energy Technology Conference*, Houston, April.

NREL (2001) *Advanced Desiccant Cooling & Dehumidification Program*. National Renewable Energy Laboratory. www.nrel.gov/desiccantcool/index.html

Park, M. (2001) More than an average Joe's 'John'. *Fox News*, Thursday, August 9. www.foxnews.com/story/0,2933,31677,00.htm

Parlour, R. (1994) *Building Services* (Sydney: Integral Publishing).

Perry, B. (2001) Review of the architecture of the Sydney 2000 Olympics. *The Architecture Show and Francis Greenway Society Green Buildings Conference,* Sydney, July.

Spurr, M. (1996) District energy/cogeneration systems in U.S. climate change strategy. *Climate Change Analysis Workshop*, Springfield, Virginia, June. http://www.districtenergy.org/

Stone J.L. (1993) Photovoltaics: unlimited electrical energy from the Sun. *Physics Today*, **46**, September, 22–9.

UNSW (2001) Photovoltaics Special Research Centre. www.pv.unsw.edu.au/info/roofinfo.html

Zweibel, K. (1990) *Harnessing Solar Power – The Photovoltaic Challenge* (New York: Plenum).

28

The crystal ball

Gerard de Valence* and Rick Best*

28.1 Introduction

CECIL GRAHAM: What is a cynic?
LORD DARLINGTON: A man who knows the price of everything and the value of nothing.

(Oscar Wilde, *Lady Windermere's Fan*, Act 3
St James Theatre, 20 February 1892)

Lord Darlington's reply illustrates, quite succinctly, a basic principle which is the foundation of much of the discussion in this book, i.e., that the value of something is a function of much more than its price (or cost) alone. In this book, the second in the 'Building in Value' series, the theme has once again been that of maximizing value in buildings, and in it the authors have examined a range of issues and activities related to the design and construction phases of the process of building procurement.

In some chapters the link between the specific topic and the value of the finished building is clear and obvious, and often is to do primarily with controlling cost, e.g., a result of good cost planning or contract administration. Conversely, in some cases, the link is not so easily identified, in chapters such as those dealing with the relationship between innovation and globalization, or site safety, for example. The thread is there, however, and an effort has been made to show how value is affected by each of the topics addressed by the individual authors.

In the final chapter of the first 'Building in Value' book, which dealt with issues of importance that relate to the pre-design phase of procurement, seven recurring themes were identified. Some or all of these themes were central to the discussions in the various chapters and formed a series of continuous threads that ran throughout the book and tied the various discussions together.

* University of Technology Sydney

The salient points that were identified were (Best and de Valence, 1999):

- greater integration of project teams throughout the procurement process
- a move towards a single point of responsibility and the adoption of alternative delivery systems
- a broadening of the concept of value for money, highlighted by greater use of life-cycle planning
- the use of systematic project analysis as a basis for decision-making
- increased complexity of buildings and greater use of passive engineering principles
- increased awareness of the effect of indoor environment on the health and productivity of building users
- increasing emphasis on reducing the environmental impacts of buildings.

These points are certainly relevant to the discussion here, but in this book there is a somewhat different supporting framework. While it has been less easy this time to identify similar common themes, there are some factors that overarch the specific concerns addressed by the individual authors. In closing it is appropriate to look at them and consider how they are driving change in the procurement process and how those changes are reflected in the value of the buildings that are produced.

28.2 The management of design and construction

Two important points should be noted here: the first is that design and construction, which were largely divorced from each other over the past century or so, are going through something of a reconciliation; the second is that, although building projects have some peculiarities that distinguish them from other projects, the management of the processes of building design and construction is now being treated in much the same way as the management of any other sort of project or business venture. It is worth elaborating a little on these two points before moving on.

It was only relatively recently, i.e., during the last 150 to 200 years, that the functions of building design and construction were separated, both functionally and temporally, to the point where the building design was completed before the builder became involved in the process at all, and much more recently that the architect, acting as superintendent of the works, was largely replaced by specialist construction and/or project managers. There has been, to some extent, a re-marriage of design and construction with the establishment of integrated design teams, which often include the contractor early in the process, and changes in procurement systems that have seen design and construct (D&C), or design and build (D&B), become a favoured method of project delivery. The roles of architect and contractor have changed substantially and new disciplines have emerged in the areas of construction and project management, often with a single organization providing expertise in design, construction and overall project management and co-ordination.

The view that the management of building design, procurement and construction is just a specialized area of the generic discipline of project management is closely connected to the introduction of contemporary management practices such as lean production, just-in-time, process re-engineering, benchmarking, supply chain management and human resources management to the AEC industry. The application of this sort of management theory to construction is blurring the boundaries between the management of construction projects and the management of any other sort of project. The peculiarities of construction projects that set them apart have been mentioned often: the 'one-off' nature of the product, the complexity of the product, geographically dispersed production, the number of unrelated firms, many of them small, that form a loose association ('the project team') that is dissolved once the job is complete, and so on.

With these general points in mind, it is then appropriate to consider the main activators of change in specific areas of the design and construction process.

28.3 Change drivers

There are arguably four main factors at work that have recently driven change, and are continuing to drive change, in the industry:

- information and communications technology (ICT)
- client needs and demands for better performance
- ecologically sustainable development (ESD)
- innovation in both products and processes.

There is considerable overlap between these factors and many interdependencies between them; innovation, it could be argued, is common to the other three drivers. However, it is presented separately here as it can be said that innovation drives itself, as the quest for the new is often seen not as a means to an end, but as an end in itself. In spite of the overlaps between them, the identification of the four strands does provide a useful framework for the following discussion in which the strands and the connections between them are explored in more detail.

28.3.1 Information and communications technology

In the last 20 to 30 years computers have changed forever the way that people do many things. In many facets of the AEC (architecture, engineering and construction) industry, however, computers have only been used to streamline the performance of processes and activities that had previously been done by hand, or with the aid of some sort of mechanical or electrical device. Slide rules and comptometers, for example, were replaced first by electronic calculators, then by personal computers; typewriters were replaced by word processors and so on. The actual tasks that were carried out, however, remained largely unchanged. This sort of substitution of computers for pen and paper, or other less sophisticated tools, while leaving the work processes and tasks unchanged, is recognized as the typical first stage of ICT adoption.

Quantity surveyors, for example, were quick to adopt computerized methods for the preparation of bills of quantities, a labour intensive process involving a number of separate stages culminating in a massive typing operation. Early quantity surveying software removed a good deal of the drudgery associated with manual calculation of quantities, and laborious typing and re-typing of the final documents, but it did not develop much beyond that for a long period and the underlying processes of measurement, usually done by manual taking off from printed drawings, remained the same. Thus, while the computer eased the burden of much of the routine work, the real value-adding processes were still done in the traditional way.

More recently there has been a move towards fundamental change in the way that many tasks are done, with work processes being re-designed in response to the availability of cheap and powerful technology rather than the traditional tasks being done using a computer rather than the conventional aids/tools. The use of CAD is a prime example: designers are now beginning to design in 3D using modelling software instead of just using a computer to produce 2D drawings from architects' sketches (the 'electronic drawing board' approach). It is widely believed that the next stage will be the automatic extraction of quantities from the 3D model and ultimately the automatic generation of comprehensive cost control documents – not traditional bills of quantities perhaps, but new forms of cost plan/estimate/bill or whatever that are more compatible with new methods of project delivery that utilize shared 3D project models.

Such changes in approach are part of the trend towards new ways of doing business and managing projects. They are also a part of the globalization of the design process, with advanced communication and information exchange made possible by the rapid expansion of the Internet. Online plan rooms, that allow real-time modification of drawings by geographically dispersed teams, are one of the key developments made possible by the new technology.

Computers have also made advances in other areas possible, such as 4D modelling, cost modelling and energy modelling, and enabled substantial change in the operation of building services. As the technology develops and performance improves, many tasks that were considered beyond the scope of ICT will be affected, and some of these technologies (like mobile phones and e-mail) have become widespread very quickly. There are many other possibilities: instant messaging to replace phone and e-mail tag, peer-to-peer networking driving collaborative work, mobile computing using third-generation cellular networks, microprocessors embedded in all manner of components and even in the very fabric of buildings. Nanocomputers, many times more powerful than current chips, may, in the future, even be built into basic building materials and so monitor and control many aspects of building performance and operation.

28.3.2 Client needs and demands

Client expectations have had an understandably strong influence on changes in the industry, as it is clients who drive the whole process. In this respect the industry is no different to any other: clients are customers and will look for products, services and providers that best satisfy their wants and needs, and will often pay a premium to get what they want. In recent years clients have started to become far more definite about what they want from the industry, for example, productivity improvement or use of lean production techniques. The most important of these are:

- the desire to have a single entity manage the whole procurement process so that the client deals with a single point of responsibility, typically a project manager (whether an individual or a firm)
- the push for shorter completion times – 'time is money' is the adage and attempts to compress project durations have been driven by that idea, and continue to propel the search for faster delivery methods

- the development of innovative contractual and financing arrangements has occurred as clients have gained access to global markets, and look for ways to spread risk associated with very large and complex projects, particularly in relation to work in countries other than their own
- the demand for greater building 'intelligence' in its various forms: flexibility, responsiveness, communications infrastructure and so on
- the demand for more environmentally benign buildings.

It is not only architects that must respond to these demands – all sectors of the industry have to respond appropriately if they are to remain in the game. The important point here is that clients are not merely looking for the lowest price – value for money is their goal, and they are well aware how value is achieved, and that may mean that they will spend more if they are confident that greater value will accrue to them in the long run.

It is this sort of thinking that is making clients demand new guarantees from their project teams, covering such things as responsible waste management, maintenance of site safety standards and analysis of the comparative environmental impacts of substitutable materials. Satisfaction of these client requirements will depend on the responses of all the designers, as well as the project managers, contractors, sub-contractors, and even the legal personnel who draw up the contracts and the financiers who structure the complex project finance arrangements.

28.3.3 Ecologically sustainable development

Although this has been suggested above as a client requirement, and it certainly is receiving increasing attention from clients, it actually has a much broader influence on the industry. Pressure from governments and the public is increasing and forcing changes in many aspects of life in general and the AEC and property industries are no exception. From the earliest stages of procurement, satisfaction of environmental controls is fundamental to the progress of any building project and these controls are becoming increasingly strict.

In response to demands for buildings that are less environmentally damaging, designers are pursuing a number of approaches:

- reducing operational energy through greater use of passive engineering principles (natural ventilation, mixed mode air conditioning, increased daylighting), more efficient plant and equipment (pumps, fans, chillers, lights and electrical appliances/ equipment such as photocopiers), and improved controls (occupant sensors, building management systems, advanced switching systems)
- reducing embodied energy in buildings by recycling materials and components, through adaptive re-use of existing buildings, and selection of less energy-intensive materials
- making greater use of renewable energy sources, such as photovoltaics, and more efficient energy conversion systems such as fuel cells, often as part of a cogeneration system.

The development of computerized modelling tools has been instrumental in assisting designers to achieve these aims. Energy modelling, for example, demonstrates the sort of interdependencies that are characteristic of this discussion: the general goal of ESD is set by public and government pressure, the specific goals (e.g., lower operating costs) are set by clients, designers propose innovative solutions, and these are tested and evaluated using modelling tools that are made possible by developments in IT.

28.3.4 Innovation

The search for new materials, systems and methods is partly a response to the other drivers, but it is also something that is pursued for itself. The quest for something new, 'to boldly go where no man has gone before', is a human trait that has pushed mankind to explore and innovate throughout history and it continues to be a significant factor, driving innovation in the AEC industry.

Innovation in this context includes advances in both the abstract and the 'concrete' – new management theories are developed in parallel with new materials and techniques and at the same time as those that already exist are being refined. Again there are dependencies: the microprocessor is at the heart of much innovation (e.g., in services, robotics, intelligent buildings), ESD goals provide the incentive in many areas (e.g., in services, materials, energy modelling), while clients demanding improved performance give the impetus to developments in management (e.g., 4D modelling, performance measurement, supply chain management).

Competition, which has been the primary force behind change in the industry, now has an extra dimension as globalization, riding on the back of the IT revolution, gives companies the opportunity to compete for work in a global market but also confronts them with the problem of increasing competition from overseas players.

28.4 Conclusion

To date the impact on the building and construction industry, of many of the new ideas on management and production that came to light toward the end of the twentieth century, has been limited. Nevertheless, their effect will be significant in coming years. The goals of many in the industry, and their clients, are greater efficiency, improved industry relationships, better quality products, improved time and quality outcomes and more effective process re-engineering. The firms that succeed in this transformation will be the ones to survive and thrive in the intense competition that a globalized industry faces. The conditions for competitiveness include productivity growth, an openness to innovation and new technology, and a commitment to process re-engineering, as well as the capacity to deliver high quality projects, on time and on cost. However, other factors that relate to the attitudes of industry participants and their ability to work in co-operation with each other will also determine success. The role taken by major clients in their use and choice of procurement and delivery methods will be another significant factor.

All this presents major challenges to managers in construction companies. Their primary concern has traditionally been with time and cost measures of performance and the demands of the construction process itself. Now there are emerging areas that go well beyond the building site that are going to be important aspects of their work. Management of design, including energy analysis, and the ability to improve the constructability of designs is already necessary as procurement methods move toward single-source collaborative forms. Logistics and supply chain management are becoming crucial to meeting, or beating, project schedules and time milestones. Quality, safety, finance, waste and recycling are other emerging areas. Some clients are also looking at these new areas and applying a wider range of criteria when selecting the contractors, designers and suppliers for their projects, whether individually or together in consortia or strategic alliances.

All this makes understanding both the why and the how of these areas important for managers in the industry. Where most construction management books are focused on the management of site work and construction companies, this book has taken a wider view and addressed a number of topics that we see becoming as important as those addressed in books in the mainstream.

Reference

Best, R. and de Valence, G. (1999) *Building in Value: Pre-Design Issues* (London: Arnold).

Index